Biotechnology Annual Review

Volume 1

Biotechnology Annual Review

Volume 1

Editor:

M. Raafat El-Gewely
Department of Biotechnology, University of Tromsø, Tromsø, Norway

1995

ELSEVIER
Amsterdam – Lausanne – New York – Oxford – Shannon – Tokyo

No responsibility is assumed by the Publisher for any injury and/or damage to persons or property as a matter of products liability, negligence or otherwise, or from use or operation of any methods, products, instructions or ideas contained in the material herein. Because of rapid advances in the medical sciences, the Publisher recommends that independent verification of diagnoses and drug dosages should be made.

Special regulations for readers in the USA – This publication has been registered with the Copyright Clearance Center Inc. (CCC), 222 Rosewood Drive, Danvers, MA 01923, USA. Information can be obtained from the CCC about conditions under which photocopies of parts of this publication may be made in the USA. All other copyright questions, including photocopying outside the USA should be referred to the copyright owner, Elsevier Science B.V., unless otherwise specified.

Biotechnology Annual Review 1

ISBN 0 444 81890 1

This book is printed on acid-free paper.

Published by:
Elsevier Science B.V.
P.O. Box 211
1000 AE Amsterdam
The Netherlands

Transferred to digital printing 2006

In order to ensure rapid publication this volume was prepared using a method of electronic text processing known as Optical Character Recognition (OCR). Scientific accuracy and consistency of style were handled by the author. Time did not allow for the usual extensive editing process of the Publisher.

Printed and bound by CPI Antony Rowe, Eastbourne

Scope

The *Biotechnology Annual Review* aims to cover the various developments in biotechnology in the form of comprehensive, illustrated and well referenced reviews. Chapters are written by experts in the particular fields of biotechnology. With the expansion in the field of biotechnology, coupled with the vast increase in the number of new journals reporting new results in the field, the need for a publication that is continuously providing reviews is urgent. Such a publication will help students as well as teachers, researchers as well as administrators, to stay knowledgeable with all the relevant issues in biotechnology. Naturally, all aspects of biotechnology cannot be reviewed extensively in each issue every year, but each volume will have a number of reviews covering different aspects of biotechnology. Reviewed topics will include biotechnology applications in medicine, agriculture, marine biology, industry, bioremedation and the environment. Fundamental problems dealing with enhancing the technical knowledge encountering biotechnology utilization regardless of the field of application will be particularly emphasized. Examples of such vital topics are promoters, vectors, media, induction, genetic stabilization during heterologous gene expression and relevant new techniques. Essential information dealing with the utilization of data banks such as protein and nucleic acid data banks will be reviewed. Homology studies as related to biotechnology, as well as issues dealing with the characterization of motifs and motif data bases will be also dealt with. New developments in protein engineering, optimization of protein function and protein design will be addressed. These problems dealing with protein functionality are important not only for the production of active recombinant proteins and enzymes, but also for the purpose of drug development and design based on screening using such proteins, whether by employing in vivo or in vitro assays. Drug screening and discovery using proteins of cloned and expressed genes of interest, is one of the major biotechnology activities in recent years. Newly discovered open reading frames or proteins identified by 2-D gel electrophoresis will be updated whenever the opportunity arises. Additional problems dealing with policy and regulation of biotechnology as well as the problems of development in the developing countries as related to biotechnology will be included in the various issues.

Biotechnology Annual Review

Volume 1

Editor

Dr. M.R. El-Gewely
Department of Biotechnology
Institute of Medical Biology
University of Tromsø
MH-Bygget
N-9037 Tromsø
Norway

Associate Editors

Dr. R.H. Doi
Section of Biochemisty and Biophysics
University of California, Davis
Davis, CA 95616-8535
USA

Dr. F. Felici
IRBM
(Istituto di Bologia Molecolare P.
  Angeletti)
via Pontina Km 30.600
I-00040 Pomezia
Rome
Italy

Dr. K. Inouye
Department of Food Science and
  Technology
Faculty of Agriculture
Kyoto University
Sakyo-ku
Kyoto 606-01
Japan

Dr. G. Krupp
Institut für Allgemeine Mikrobiologie
Christian-Albrechts-Universität
Am Botanischen Garten 9

D-24118 Kiel
Germany

Dr. E. Olson
Department of Biotechnology
Warner-Lambert
2800 Plymouth Road
Ann Arbor, MI 48105
USA

Dr. S.B. Petersen
SINTEF UNIMED
MR-Center
N-7034 Trondheim
Norway

Dr. J. Preiss
Michigan State University
Department of Biochemistry
Biochemistry Building
East Lansing, MI 48824-1319
USA

Dr. H. Prydz
University of Oslo
Biotechnology Centre of Oslo
P.O. Box 1125, Blindern
N-0371 Oslo 3
Norway

Dr. G.K. Rosendal
The Fridtjof Nansen Institute
P.O. Box 326, 1324 Lysaker
Norway

Dr. M. Tepfer
Laboratoire de Biologie Cellulaire
INRA - Centre de Versailles
F-78026 Versailles Cedex
France

Contributors

Dr. H.W. Anthonsen
MR-Center
SINTEF UNIMED
N-7034 Trondheim
Norway

Dr. A. Baptista
MR-Center
SINTEF UNIMED
N-7034 Trondheim
Norway

R.A. Van Bogelen
Department of Biotechnology
Parke-Davis Pharmaceutical Research
Division of Warner-Lambert
2800 Plymouth Road
Ann Arbor
MI 48105
USA

Dr. T.M.S. Chang
Artificial Cells and Organs Research
    Centre
Faculty of Medicine
McGill University
Montreal Quebec
Canada H3G 1Y6

Dr. R. Conrad
Department of Chemistry
Indiana University
Bloomington
IN 47405
USA

Dr. R.H. Doi
Section of Molecular and Cellular
    Biology
University of California
Davis, CA 95616
USA

Dr F. Drabløs
MR-Center
SINTEF UNIMED
N-7034 Trondheim
Norway

Dr. M.R. El-Gewely
Department of Biotechnology
Institute of Medical Biology
University of Tromsø
MH-Bygget
N-9037 Tromsø
Norway

Dr. A.D. Ellington
Department of Chemisty
Indiana University
Bloomington
IN 47405
USA

Dr. F. Felici
IRBM
(Istituto di Biologia Molecolare P.
    Angletti)
via Pontina Km. 30.600
I-00040 Pomezia
Rome
Italy

Dr. M.A. Goldstein
Section of Plant Biology
University of California
Davis
CA 95616
USA

Dr. W.-D. Hardt
Institut für Biochemie
Abteilung Prof. V.A. Erdmann
Freie Universität Berlin
Thielallee 63
D-14195 Berlin
Germany

Dr. R.K. Hartmann
Institut für Biochemie
Abteilung Prof. V.A. Erdmann
Freie Universität Berlin
Thidallee 63
D-14195 Berlin
Germany

Dr. S. Hashida
Department of Biochemistry
Medical College of Miyazaki
Kiyotake
Miyazaki 889-16
Japan

Dr. K. Hashinaka
Department of Biochemistry
Medical College of Miyazaki
Kiyotake
Miyazaki 889-16
Japan

Dr. S. Henikoff
Howard Hughes Medical Institute
Fred Hutchinson Cancer Research
   Center
Seattle, WA 98104
USA

Dr. N. Inoue
Central Laboratories for Key
   Technology
Kirin Brewery Co. Ltd.
1-13-5 Fukuura Kanazawa-ku
Yokohama
Kanagawa 236
Japan

Dr. E. Ishikawa
Department of Biochemistry
Medical College of Miyazaki
Kiyotake
Miyazaki 889-16
Japan

Dr. G. Krupp
Institut für Allgemeine Mikobiologie
Christian-Albrechts-Universität
Am Botanischen Garten 9
D-24118 Kiel
Germany

Dr. F. Larsen
Biotechnology Centre of Oslo
P.O. Box 1125 Blindern
N-0310 Oslo
Norway

Dr. J. Lundeberg
Department of Immunology
Institute for Cancer Research
The Norwegian Radium Hospital
Montebello
N-0310 Oslo
Norway

Dr. A. Luzzago
IRBM
(Istituto di Biologia Molecolare P.
   Angeletti)
via Pontina Km. 30.600
I-00040 Pomezia
Rome
Italy

Dr. P. Martel
MR-Center
SINTEF UNIMED
N-7034 Trondheim
Norway

Dr. P. Monaci
IRBM
(Istituto di Biologia Molecolare P.
   Angeletti)
via Pontina Km. 30.600
I-00040 Pomezia
Rome
Italy

Dr. A. Nicosia
IRBM
(Istituto di Biologia Molecolare P.
  Angeletti)
via Pontina Km. 30.600
I-00040 Pomezia
Rome
Italy

Dr. H. Ohashi
Pharmaceutical Division
Central Laboratories for Key
  Technology
Kirin Brewery Co. Ltd.
1-13-5 Fukuura Kanazawa-ku
Yokohama, Kanagawa 236
Japan

Dr. E. Olson
Department of Biotechnology
Parke-Davis Pharmaceutical Research
Division of Warner-Lambert
2800 Plymouth Road
Ann Arbor, MI 48105
USA

Dr. S.B. Petersen
MR-Center
SINTEF UNIMED
N-1034 Trondheim
Norway

G.K. Rosendal
The Fridtjof Nansen Institute
P.O. Box 326
1324 Lysaker
Norway

Dr. M. Sebastião
Laboratorio de Engenharia Bioquimica
Instituto Superior Tecnico
1000 Lisboa
Portugal

Dr. M. Sollazzo
IRBM
(Istituto di Biologia Molecolare P.
  Angeletti)
via Pontina Zm. 30.600
I-00040 Pomezia
Rome
Italy

Dr. T. Suzuki
Pharmaceuticals Division
Central Laboratories for Key
  Technology
Kirin Brewery Co. Ltd.
1-13-5 Fukuura Kanazawa-ku
Yokohama, Kanagawa 236
Japan

Dr. M. Takeuchi
Central Laboratories for Key
  Technology
Kirin Brewery Co. Ltd.
1-13-5 Fukuura Kanazawa-ku
Yokohama, Kanagawa 236
Japan

Dr. C. Traboni
IRBM
(Istituto di Biologia Molecolare P.
  Angeletti)
via Pontina Km. 30.600
I-00040 Pomezia
Rome
Italy

Dr. L. Vaz
Limetree Road 19
Liverpool, NY
USA

Contents

Introduction

M. Raafat El-Gewely

Department of Biotechnology, Institute of Medical Biology, University of Tromsø, Tromsø, Norway

Biotechnology and the utilization of biologically based technologies has a long rooted history in human activities. Ancient cultures as documented 4,000 years ago by ancient Egyptians for example, knew some secrets of fermentation and leaven bread-making, but without identifying the exact microorganisms involved. Today, a lot is known about *Saccharomyces* and its utilization not only in fermentation, but also as a host for heterologous gene expression and as a model organism for eukaryotic molecular biology. The first sequence of a complete nuclear chromosome of any eukaryote was that of *Saccharomyces cerevisiae* [1–4]. Nevertheless, most of the yeast genes are not identified, in spite of massive efforts to sequence all of the individual chromosomes or segments (European Yeast Sequencing Program, for example). The issue of what proteins the discovered open reading frames (ORFs)/genes are coding for or how are they regulated will remain a mystery, at least for a while. Naturally all of this information on yeast as well as on other organisms will increase our understanding of biological systems in addition to being assets for further progress in biotechnology.

Currently, at least 25% of all copper produced worldwide is bioprocessed, but the Romans presumably were the first to use bacterial action for biomining in Rio Tinto Copper mine in Spain 2,000 years ago [5]. Other minerals as precious as gold or as cheap as phosphate are being bioprocessed. There is a lot to be learned in this field of application which promises to increase efficiency without the use of toxic and dangerous chemicals. Hopefully, biomining could deliver the required minerals with environmentally sound procedures at a feasible cost that permits its use even with lower grade ores.

Our knowledge about human genes as a result of the human genome project [6,7] or by rapid cDNA sequencing, expressed sequence tags (EST) [8–10], will eventually increase, thus providing a wealth of information about our genetic makeup and open the potential for providing better diagnostic tools, therapy and a powerful basis for drug discovery.

The potential for improving plants, animals, marine organisms using modern biotechnology might be the only hope to improve food resources globally.

In the field of biotechnology, in spite of its current momentum and expansion and deeply rooted history, modern applications are still young relative to the potential and expectations.

Address for correspondence: Department of Biotechnology, Institute of Medical Biology, University of Tromsø, 9037 Tromsø, Norway. Tel.: +47 776 44654; Fax: +47 776 45350.

The *Biotechnology Annual Review* aims to cover the various developments in biotechnology in the form of comprehensive, illustrated and well referenced reviews. Chapters are written by experts in the particular fields of biotechnology. With the expansion in the field of biotechnology, coupled with the vast increase in the number of new journals reporting new results in the field, the need for a publication that is continuously providing reviews is urgent. Such a publication will help students as well as teachers, researchers as well as administrators, to stay knowledgeable with all the relevant issues in biotechnology. Naturally, all aspects of biotechnology cannot be reviewed extensively in each issue every year, but each volume will have a number of reviews covering different aspects of biotechnology. Reviewed topics will include biotechnology applications in medicine, agriculture, marine biology, industry, bioremedation and the environment. Fundamental problems dealing with enhancing the technical knowledge encountering biotechnology utilization regardless of the field of application will be particularly emphasized. Examples of such vital topics are promoters, vectors, media, induction, genetic stabilization during heterologous gene expression and relevant new techniques. Essential information dealing with the utilization of data banks such as protein and nucleic acid data banks will be reviewed. Homology studies as related to biotechnology, as well as issues dealing with the characterization of motifs and motif data bases will be also dealt with. New developments in protein engineering, optimization of protein function and protein design will be addressed. These problems dealing with protein functionality are important not only for the production of active recombinant proteins and enzymes, but also for the purpose of drug development and design based on screening using such proteins, whether by employing in vivo or in vitro assays. Drug screening and discovery using proteins of cloned and expressed genes of interest, is one of the major biotechnology activities in recent years. Newly discovered open reading frames or proteins identified by 2-D gel electrophoresis will be updated whenever the opportunity arises. Additional problems dealing with policy and regulation of biotechnology as well as the problems of development in the developing countries as related to biotechnology will be included in the various issues. Suggested review contributions can also be sent to the editors.

In the first issue of this series the following topics are covered:

1. Biotechnology domain (M.R. El-Gewely)
2. Application of 2-D protein gels in biotechnology (R. VanBoglen and E. Olson)
3. Prokaryotic promoters in biotechnology (M.A. Goldstein and R.H. Doi)
4. Comparative methods for identifying functional domains in protein sequences (S. Henikoff)
5. Peptide and protein display on the surface of filamentous bacteriophage (F. Felici, A. Luzzago, P. Monaci, A. Nicosia, M. Sollazzo, C. Traboni
6. Aptamers as potential nucleic acid pharmaceuticals (A.D. Ellington and R. Conrad)
7. Towards a new concept of gene inactivation: specific RNA cleavage by endogenous ribonuclease P (R.K. Hartmann, G. Krupp and W.-D. Hardt)

8. Artificial cells with emphasis on bioencapsulation in biotechnology (T.M.S. Chang)
9. The production of human erythropoietin (N. Inoue, M. Takeuchi, H. Ohashi and T. Suzukui)
10. Lipases and esterases, a review of their sequences, structure and evolution (H.W. Anthonsen. A. Baptista, F. Drabløs, P. Martel, S.B. Petersen, M. Sebastião and L. Vaz)
11. Solid phase technology (J. Lundeberg and F. Larsen)
12. Ultrasensitive enzyme immunoassay (S. Hashida, K. Hashinaka and E. Ishikawa)
13. The politics of patent legislation in biotechnology: an international view (G.K. Rosendal)

References

1. Wicksteed BL, Collins I, Dershowitz A, Stateva LI, Green RP, Oliver SG, Brown AJ, Newlon CS. A physical comparison of chromosome III in six strains of *Saccharomyces cerevisiae*. Yeast 1994;10:39–57.
2. King GJ. Stability, structure and complexity of yeast chromosome III. Nucleic Acids Res 1993;21:4239–4245.
3. Bork P, Ouzounis C, Sander C, Scharf M, Schneider R, Sonnhammer E. Comprehensive sequence analysis of the 182 predicted open reading frames of yeast chromosome III. Protein Sci 1992;1:1677–1690.
4. Oliver SG et al. The complete DNA sequence of yeast chromosome III. Nature 1992;357:38–46.
5. Moffat AS. Microbial mining boosts the environment, bottom line. Science 1994;264:778–779.
6. Guyer MS, Collins FS. The human genome project and the future of medicine. Am J Dis Child 1993;147:1145–1152.
7. Collins F, Galas D. A new five-year plan for the U.S. human genome project. Science 1993;262:43–46.
8. Adams MD, Kerlavage AR, Kelley JM, Gocayne JD, Fields C, Fraser CM, Venter JC. A model for high-throughput automated DNA sequencing and analysis core facilities. Nature 1994;368:474–475.
9. Adams MD, Soares MB, Kerlavage AR, Fields C, Venter JC. Rapid cDNA sequencing (expressed sequence tags) from a directionally cloned human infant brain cDNA library. Nat Genet 1993;4:373–380.
10. Venter JC. Identification of new human receptor and transporter genes by high throughput cDNA (EST) sequencing. J Pharm Pharmacol Suppl 1:355–360.

Biotechnology Annual Review
M.R. El-Gewely, editor

Biotechnology domain

M. Raafat El-Gewely

Department of Biotechnology, Institute of Medical Biology, University of Tromsø, Tromsø, Norway

Abstract. Biotechnology and the use of biologically based agents for the betterment of mankind is an active field which is founded on the interaction between many basic sciences. This is achieved in coordination with engineering and technology for scaling up purposes. The application of modern recombinant DNA technology gave momentum and new horizons to the field of biotechnology both in the academic setting and in industry. The applications of biotechnology are being used in many fields including agriculture, medicine, industry, marine science and the environment. The final products of biotechnological applications are diverse. In the medical applications of biotechnology, for example, the field has been evolving in such a way that the final product could be a small molecule (e.g. drug/antibiotic) that can be developed based on genetic information by drug design or drug screening using a cloned and expressed target protein.

What is biotechnology?

There are several definitions for biotechnology in the literature. This mainly reflects that the field is changing too dynamically for a definition to stay conclusive. Webster's dictionary for example defines biotechnology as "the use of data and techniques of engineering and technology for the study and solution of problems concerning living organisms". The US Congress, Office of Technology Assessment (OTA) defines biotechnology as "Any technique that uses organisms or parts of organisms to make or modify products to improve plants or animals, or to develop microorganisms for specific use" [1], while the Organization for Economic Cooperation and Development (OECD) specifies that "biotechnology is the application of scientific and engineering principles to the processing of materials by biological agents to provide goods and services" [2]. These definitions give a global description of the field without really revealing too much about its nature.

The word "biotechnology" itself was coined in German, biotechnologie, by Kar Ereky, the Hungarian engineer, in 1917, to describe all work by which products are produced from raw materials with the help of living organisms [3]. This description comes at a time of activities to use microorganisms and aseptic fermentation for much needed products such as glycerol, butanol and acetone [3].

However, the use of microorganisms to produce or to modify a product did not start at the beginning of this century. The ancient Egyptians, for example, were documenting

Address for correspondence: M. Raafat El-Gewely, Department of Biotechnology, Institute of Medical Biology, University of Tromsø, 9037 Tromsø, Norway. Tel.: +47 776 44654; Fax: +47 776 45350.

6

Fig. 1. This unique beer making scene shows the various phases of the ancient Egyptian beer making process as depicted on the entrance of the Mastaba chapel of Hetepherachet, a great official of the Egyptian Old Kingdom. Beer was made from baked bread. The bread was soaked in water and mixed with sweeteners like honey. The fermented mass was filtered and stored in large red pottery jars. Hetepherachet, the owner of the tomb lived under several pharaohs of the 5th dynasty (ca. 2400 BC). Hetepherachet was a judge and a priest of the goddess of Truth, Maat. The tomb from which this chapel scene derives stood at Saqqara, where the cities of the dead of Egypt's ancient capital Memphis are situated. The chapel is at the National Museum of Antiquities at Leiden, The Netherlands, since 1904.

the process of making bread, and the stages of the brewing process over 4,000 years ago (Fig. 1). They did not identify the exact microorganisms involved, or have all the conditions of these processes controlled, but it cannot be disputed that such "final" products could not be obtained without the use of living organisms. Robert Bud [3], in his elegant work, "The Uses of Life, A History of Biotechnology" emphasizes the last hundred years of development in the field, but argues that it would be wrong to grant equal antiquity to the concept of biotechnology. However, biotechnology, similar to many aspects of human activities, including the sciences, also has historical roots. A few decades ago, prior to the utilization of recombinant DNA in biotechnology, the field of biotechnology was quite different from the current expectations of biotechnology. We cannot by analogy, deny the status of a "biotechnology" company to all those companies that did not employ modern genetic engineering tools in their production methods. Human activities are always subject to continuous development, but we should try not to forget the roots, history, or the key mechanisms underlying progress.

Biotechnology, in its broad sense, has been utilized for centuries through its application in brewing, wine-making, bread-making, food preservation and waste treatment. Agricultural practices themselves, even before the rediscovery of Mendel's Laws of Genetics in 1900, reflect human activities in using the magic of harnessing solar energy through biological catalysis to produce valuable products directly from plants such as carbohydrates, oils, proteins, fibers and thousands of chemicals that included medicinal chemicals (Fig. 2).

The surface of the earth receives an immense amount of solar energy every day. The amount of solar energy received in 1 week equals all the estimated proven reserves of oils, natural gas, coal, and uranium (8×10^{11} tons coal equivalent) [4]. Recent estimates of the total annual amount of biomass (plant matter produced by photosynthesis) indicate about 2×10^{11} tons of organic matter, reflecting an average coefficient of utilization of the incident photosynthetically active radiation by the entire flora of the earth of only 0.27% [5]. Although the energy utilized by photosynthesis relative to the potential of utilizing the photosynthetically active radiation, plants are considered to be the biggest factory for the above compounds. The utilization of this energy by plants, algae or photosynthetic bacteria of higher efficiency theoretically could allow them to act as solar energy collectors at least in some geographical areas where water could

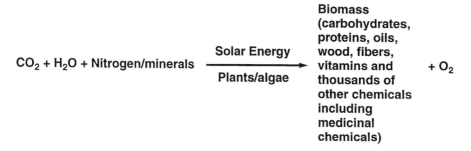

Fig. 2. The primary biological reaction in plants and algae.

8

also be available. Thus, harnessing solar energy by biological means could be a key to solar energy utilization, and it could be an area where developed and developing countries could cooperate for their mutual benefit.

It is of importance to understand that ancient civilizations did not just utilize "life forms"/microorganisms and plants, but they used also other biological materials that they considered to be of help. Such a use again was without fully understanding the exact mechanism of what they are doing or the exact product or benefit. Here I have to state the example of "Hirudin" which is now cloned and expressed in yeast cells, currently on the scale of several kilograms a year. Its market introduction as a recombinant therapeutic protein is planned for 1996 [6]. This protein was found to be a potent anticoagulant without any detectable side effects and allergic reactions [7]. Research indicated that recombinant hirudin appears to be effective as an antithrombotic [8,9] and following balloon angioplasty [10]. However, its natural use was from the ugliest and creepy organisms, *Hirudo medicinalis,* known as the leech (Fig. 3). The earliest documented record of leeches being used for medicinal/remedial purposes appears in a painting in an Egyptian tomb of around 1500 BC [11]. Leeching is also documented in medical encyclopedia from India compiled between 500 BC and AD 200 [12]. The history of the leech and leeching is exquisitely reviewed [7,11]. It is interesting to note that the company "Biopharm" was established to raise leeches for the production of chemicals and reagents from their saliva for pharmaceutical use [13]. Even microsurgery has found a use for leeching or leech products in preventing blood clotting around tissues/parts to be reattached [3]. Eglin C also produced by the leech, is a potent protease inhibitor (Fig. 4). It has been cloned and produced heterologously by *Escherichia coli* as a therapeutic protein against emphysema, septic shock, Crohn's disease, arthritis and arthrosis [14].

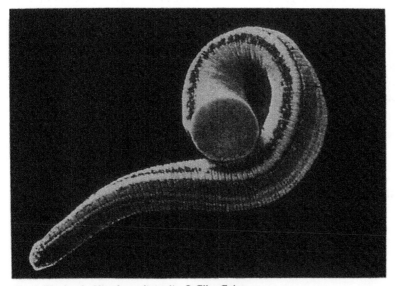

Fig. 3. The leech, *Hirudo medicinalis.* © Ciba-Geigy.

Fig. 4. Structural model of the enzyme eglin C, the lower and smaller molecule, in a complex form. © Ciba-Geigy.

This aspect of the utilization of biological life forms to produce products is still in harmony with the above definitions of biotechnology and does not really clash with the concept of biotechnology. As engineering techniques developed the whole process became much more efficient.

Early civilizations, in contrast, utilized some form of biotechnology based on unmodified organisms. This is conceptually similar to the current production of important products from unmodified organisms.

It is not the purpose of this chapter however to review the history of biotechnology. The history of biotechnology, particularly of the last 100 years, has been recently and elegantly reviewed [3,15].

If biotechnology in its utilization of microorganisms and aseptic fermentation as a form of viable and economical industry has existed for nearly a century, the word biotechnology itself became a common word only in the last two decades. This is mainly due to the use of genetic engineering in biotechnology in the mid-1970s. Modern genetic engineering and molecular biology techniques revolutionized the whole field of biotechnology. It is equally important to indicate that these techniques also revolutionized all aspects of academic research in biologically related fields, enabling scientists to do and verify what could not have been done before, at least not with the same speed. These techniques made it possible to research how genes and cellular mechanisms work by physically isolating genes, resolving their nucleotide sequence, finding regulatory elements, and verifying the protein or RNA product, in addition to providing a wealth of information about the molecular similarities or differences between genes of the same organism or between different species. These similarities not only helped in

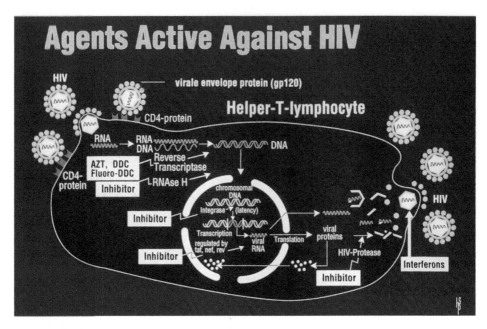

Fig. 5. Anti-HIV agents and sites of action. © F. Hoffman-La Roche Inc., Basel [15].

establishing the evolutionary lineage between related or unrelated species on a molecular basis, but, more importantly, could provide additional insight into how proteins work and how the primary sequence of a protein is correlated to its function, since the active sites of proteins are usually least subject to change.

Similarly, the application of genetic engineering techniques in the field of biotechnology made it possible to study and use important proteins that are naturally produced in normal cells but are not, for example, produced in amounts permitting investigations to elaborate on the understanding of their precise role in the cell. Several of the proteins in the pharmaceutical market now are therapeutic proteins that are produced through the application of genetic engineering. Other therapeutic proteins that are produced by biotechnology now include proteins that were utilized prior to the utilization of genetic engineering and that were produced for example from animal sources (e.g. insulin) or from the human cadaver (e.g. human growth hormone).

Recently, with further developments in the field of biotechnology, proteins that are produced by genetic engineering by themselves are not the final product, but rather are the tool in drug design and screening. As an example of this new strategy, recombinant transcription factors are the basis for the development of new pharmaceuticals [16]. Several of the antiviral strategies to develop drugs for the treatment of the HIV virus are based on the screening of inhibitors of the function of key viral specific proteins [17] (Fig. 5).

Therefore, gene technology and genetic engineering should be at the core of modern biotechnology. These technologies are continuously opening new potentials and applications. At the same time new advancements in basic research have a direct impact on the field of biotechnology. No single biological system so far has been fully character-ized in terms of our total knowledge of its genetic information or even its expressed proteins under normal or different physiological conditions. The impact of this knowl-edge to further enhance the potential of biotechnology cannot be underestimated. The application of 2-D gels to identify proteins of *E. coli* and other organisms is discussed in this volume [18].

Modern definition of biotechnology

Biotechnology is an interdisciplinary science dealing with optimization and utiliza-tion of biological catalysis and genetic information for the development of useful products or systems.

The economic impact of biotechnology cannot be ignored as a vital progressing industry. In the USA alone, in 1991, sales of biotechnology products approached $4 billion and are expected to reach $50 billion in about 10 years [19].

The interdisciplinary nature of biotechnology

Clearly, many basic fields such as genetics, molecular biology, biochemistry, cell biology, microbiology, immunology, chemistry, pharmacy, pharmacology and chemical engineering are providing basic information and depth to the field of bio-technology. Several types of products/output of biotechnology could result from such a positive interaction (Fig. 6).

Why gene technology?

Genes control biological processes in all living organisms in coordination with envi-ronmental or external stimuli. Therefore, it is only logical that the genetic approach is the most direct approach to harness and optimize bio-catalysis, or to design a product based on the function/structure of a biomolecule that can be produced by genetic engineering.

Several companies were formed that were entirely dedicated biotechnology compa-nies (DBC) in addition to the established companies that started to use biotechnology and genetic engineering methods.

Although the applications of biotechnology are expanding rather rapidly to cover a great deal of human activities, some aspects of catalysis could be accomplished chemi-cally rather than biologically. The use of biological catalysts of course could be carried out at much lower temperatures and under less extreme conditions with virtually no pollution and other health hazardous compounds that are dangerous both to workers and society. Biotec hnology itself is used to solve the problems of industrial waste or chemical pollution. Chemistry remains a very important aspect of modern biotechnol-ogy; however, several aspects of modern biotechnology could not be achieved by chemistry alone.

12

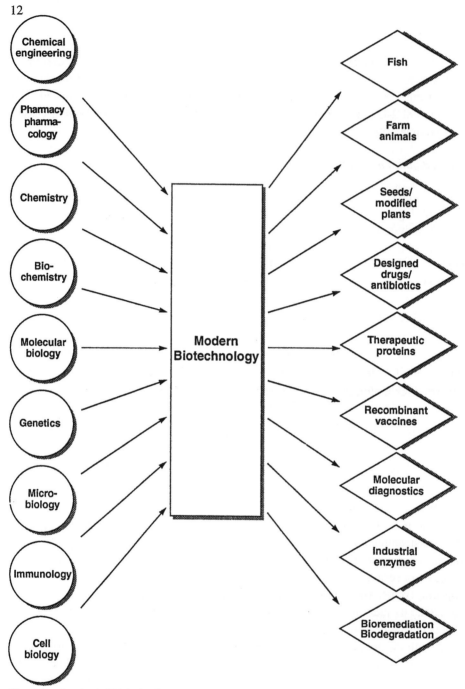

Fig. 6. Input–output of biotechnology.

Advantages of genetic engineering and modern biotechnology in comparison to classical methods of biotechnology

1. *Amplification of rare, but useful proteins.* Gene technology can help in produc-

ing proteins that exist only as a few copies per cell in large enough amounts for use as therapeutic proteins or as a target for drug design. For example, interferons inhibit the multiplication of viruses and tumor cells, thereby improving the host's immune defenses. Interferons are produced in very small amounts in human blood. With the methods of genetic engineering, it is possible to produce interferons as a therapeutic in sufficient amount to cover any special needs (see Figs. 31 and 36d) [20].

2. *Relative speed and efficiency.* By using recombinant DNA techniques, which are relatively simple, therapeutic proteins can be "cloned" and by using appropriate expression vectors, a large amount of the protein can be produced in a short time. Before gene technology, the Swedish company Kabi Vitrum (now part of Pharmacia a/s), as the major world producer of human growth hormone (HGH), used to isolate HGH from cadaver pituitaries. This hormone is used in the treatment of children with growth impairment. In September, 1978 Kabi made an agreement with Genentech (USA) to produce HGH by cloning in *E. coli* based on a 28-month project. The strain with an appropriate vector was handed over to Kabi within 7 months [2]. Subsequently, by fermentation and protein purification, enough HGH protein can be isolated which is equivalent to that obtained from several hundred thousand cadavers. Naturally, not all examples are so straightforward, added to the probability of a reduced functionality of some of the recombinant proteins due to folding and aggregation problems (see Fig. 7). The recombinant proteins have to be tested for verification and often their folding and thus their biological activities have to be restored in vitro if so required.

3. *Space saving.* A large amount of the desired product can be produced from a small space (using a limited number of fermenters), relative to the space needed to isolate the same product using traditional methods. As an example in biomedical biotechnology, the production of human insulin (by fermenters) is now taking place in Eli Lilly in a smaller space, relative to the space needed to store animal pancreases (from pigs or cows) and to purify such animal insulin, for human use in diabetes management. Another example is the production for animal feed of "Pruteen", a single cell protein (SCP) from bacteria, produced by ICI using discarded oils. Theoretically, one million tons can be produced on a land area of 100 ha, compared with the two million hectares required to grow soya beans to produce an equivalent amount of protein [21]. Ironically, market realities often shift the economic feasibility rather fast. Because of the oil crisis and middle East war in 1973, the increase in oil prices made it difficult for the British Petroleum company to produce any SCP from its proposed plant in Sardinia with a capacity of 100,000 tons using petrocarbons [15].

4. *Lower risk of contamination by unknown pathological agents.* It is common knowledge that treating humans with products isolated from human cadavers (hormones, etc.) carries a high risk of transferring viruses or other pathogenic agents. Several cases of Creutzfeldt-Jacob disease (CJD) were reported among patients who received cadaver-derived growth hormone [22,23]. The use of

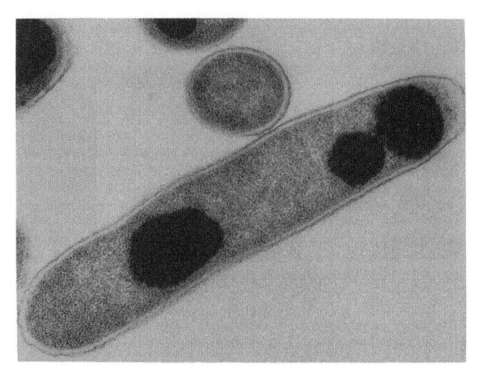

Fig. 7. Electron micrograph of interferon-producing *Escherichia coli*. The interferon-producing *E. coli* cells are packed full with stored recombinant interferon which appears black under the electron microscope. © F. Hoffman-La Roche Inc., Basel [20].

blood products for patients requiring such products could carry such a high risk and occasionally undesirable situations do occur. By using gene technology, not only could the presence of genetic material of any pathological agent be detected by molecular diagnosis and consequently not used, but also several human blood components could be produced heterologously by genetic engineering instead of by bleeding other humans for the purpose, in addition to taking such a risk. The same goes for the production of vaccines by genetic engineering; instead of using some sort of a "weakened" virus/bacteria, a specially engineered surface protein could be produced and used as a vaccine in order to avoid contamination with virus particles or other pathogenic agents. Several recombinant vaccines for human and veterinary use are now on the market.

5.   *Clean industry.* Biotechnology is considered a cleaner form of industry and it can be carried on in harmony with the environment obeying all the regulations that are put forward to control such an industry. Biotechnology methods can be used to clean the waste and pollutants of other industries.

6.   *Biocontrol versus the massive use of chemicals as pesticides.* Classical methods for managing insects or weeds in agriculture have resulted in the use of hazardous chemicals that polluted the environment in many parts of the world

and still continue to do so. On the other hand, several natural products could be subjected to biotechnology methods to produce effective compounds with a small safer "window" of affected targeted organisms.

7. *Cost effective.* Due to the use of modern methods, space saving and fast adjustability to market needs, a significant cost saving could potentially be realized using genetic engineering methods to produce a product and make this product always subject to improvement (radical product development) for new demands, which cannot be achieved by the classical methods. Using the example of eglin C, which is naturally produced from leeches, as mentioned earlier (Figs. 3 and 4), it takes about 62,000 leeches to produce 1 g at a cost of about $300,000 taking about 6 months to finish. Genetic engineering was the answer especially since the leech itself is a protected species [14].

8. *Developing totally new products or new functions for an old product.* In addition to facilitating the process of drug screening and discovery by using recombinant proteins as a target (in vitro, or in vivo), genetic manipulation can help design new proteins as is evident, for example, in the creation of new peptide or protein libraries (combinatorial libraries) that can be screened for new functions.

From genes to products

Several types of biotechnology-related products are on the market and their production and utilization are continuously subject to further development. Some examples are listed below.

Genes (DNA) and oligonucleotides

In several applications genes or parts of genes are the final potential product for commercialization. In these cases diagnostic probes [24], primer oligonucleotides for polymerase chain reactions (PCR) or antisense oligonucleotides [25,26]. The potential for using nucleic acids antisense [27] or antishape (decoy) [28] in medicine are reviewed in this volume.

Amino acids

Bioproduction of amino acids is quite extensive and many amino acids are used in medicine, food and feed additives, cosmetics, sweetener synthesis and flavor enhancers [29]. Few bacterial species are used for the production of amino acids [30–34].

Peptides

Bioactive peptides can be synthesized in vitro or in vivo (by expressing the corre-

16

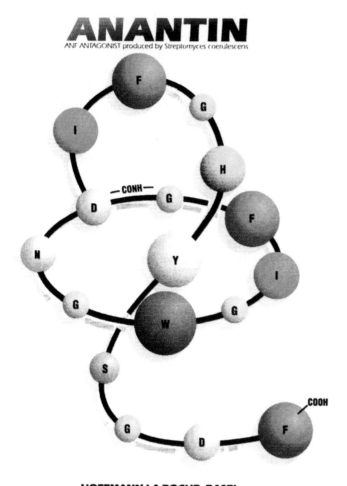

ANANTIN
ANF-ANTAGONIST produced by Streptomyces coerulescens

— CONH —

COOH

HOFFMANN-LA ROCHE, BASEL

Fig. 8. Anantin, a peptide produced by *Streptomyces coerulescens.* © F. Hoffman-La Roche Inc., Basel [36].

sponding sequence heterologously) based on prior information of their structure and bioactivity, or they can be screened out of a pool for their actions as epitope, binding ability, as an inhibitor, enhancer, or as an antibiotic [35]. The peptide, anantin, consisting of 17 natural amino acids forming a peptidic ring system, was found to be an antagonist of the atrial natriuretic factor ANF [36]. Anantin was isolated from a strain of *Streptomyces coerulescens* (Fig. 8).

Antibiotics, vitamins and other biochemicals

Many of these biochemicals including several fragrances and flavor molecules are produced microbiologically. Consequently, their genes could be manipulated for productivity or further modification of some of these compounds [34,37–41].

Therapeutic proteins

The number of therapeutic proteins on the market are on the increase in addition to many more that are under clinical trial prior to their medical use [42]. See also below.

Recombinant vaccines

Genetic approaches to design vaccines have been employed for human health as well as animal health [43–45]. See also below.

Industrial enzymes

Large-scale production of enzymes has occurred for many years. However, the new applications of modern genetic engineering methods are making it possible to modify proteins in terms of their suitability for industrial applications and making their heterologous expression possible under totally controlled conditions with potential reduction of the cost. The use of enzymes in industry is expanding [46] with continuous searching for new sources of enzymes and applications [47–49]. The feasibility of making plants as bioreactors for the production of enzymes was recently evaluated [50]. With applications of genetic engineering methods new modified enzymes for industry can be engineered and selected for. As an example, enzymes suitable for a non-aqueous environment were recently engineered [51]. There is a great demand for other types of enzymes needed for genetic engineering and molecular biology research and several of them are being produced heterologously.

Engineered antibodies

Genetic engineering tools were used to express novel antibodies with different modifications such as humanized antibodies (containing the C region from humans and a murine V domain, for example), chimeric antibodies, antibodies with catalytic groups and bifunctional antibodies. Combinatorial libraries produced in bacteriophages may present an alternative to animal immunization as a source of antigen binding specificities [52,53]. Modified antibodies could be produced heterologously in bacterial or mammalian cells.

Recombinant proteins to enhance animal productivity

Several recombinant hormones have been developed for potential animal applications [54–56]. A safety evaluation of animals treated with recombinant products, for example recombinant growth hormone, was carried out [57].

Genetically modified organisms

Transgenic plants [58], animals [59] and genetically engineered microorganisms are

18

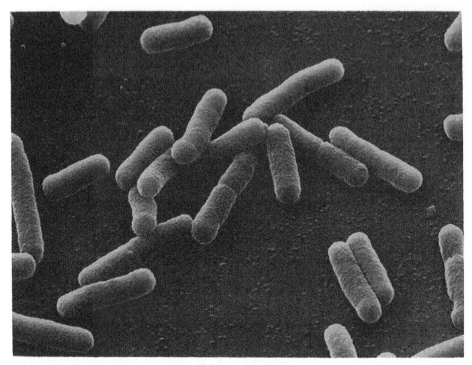

Fig. 9. Escherichia coli bacterium commonly used for heterologous gene expression. © F. Hoffman-La Roche Inc., Basel [298].

being developed both for higher productivity or to produce special recombinant products. In recent years, genetically modified animal models for different human diseases were developed in order to facilitate treatment or drug design [60–64]. In addition to the production of recombinant proteins, vaccines, antibodies, enzymes, and single cell proteins, modified and non-modified microorganisms are used in bioremediation and biodegradation [65–67]. The release of any genetically modified

Fig. 10. Bacillus subtilis, first identified in hay, is very widespread in nature. It has been used in industry, for instance in producing proteases for washing powders and for manufacturing antibiotics. *Bacillus subtilis* forms the antibiotic bacitracin, first described in 1945. Bacitracin is a metabolite which is produced in the bacterial cells when lack of nutrients causes them to form particularly resistant spores. The antibiotic effect of the protein bacitracin is based on inhibition of cell wall synthesis in other bacteria. *Bacillus thuringiensis*, which is related to *Bacillus subtilis*, forms a protein during sporulation which is toxic for many insect larvae but safe for man. This protein is already being used successfully as a biological insecticide. In genetic engineering, *Bacillus subtilis* and *Escherichia coli* are the most important bacterial strains used for culturing recombinant proteins. © F. Hoffman-La Roche Inc., Basel [298].

Fig. 11. Baker's yeast, *Saccharomyces cerevisiae*, is commonly used for the heterologous production of human and mammalian proteins. Baker's yeast is a unicellular fungus (under common lab conditions), but it has the latent capacity to undergo a dimorphic transition to psuedohyphal growth [299]. Baker's yeast is used to manufacture recombinant proteins, one important example being the hepatitis B vaccine. © F. Hoffman-La Roche Inc., Basel [298].

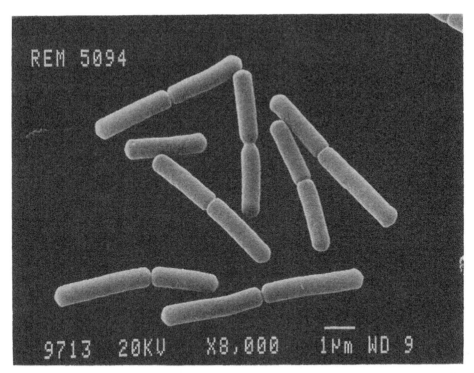

Fig. 10.

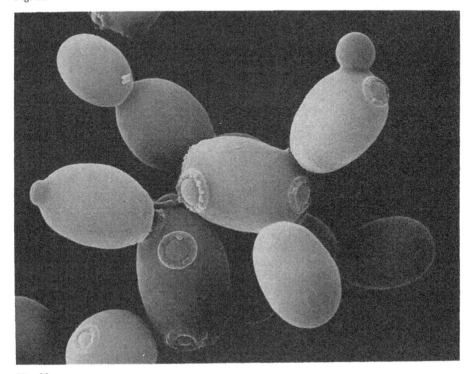

Fig. 11.

20

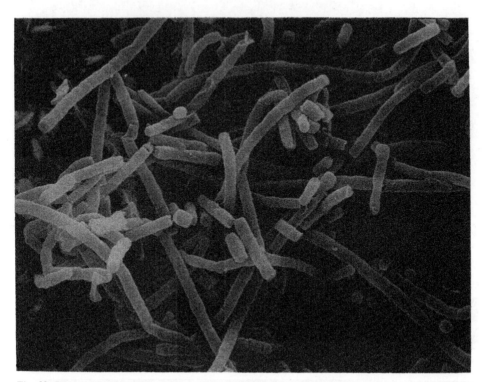

Fig. 12. *Streptomyces* produce a wide variety of biologically active substances, in particular antibiotics; 90% of the 4,000 known, naturally occurring antibiotics are produced by *Streptomyces* species, and they form a rich reservoir for scientists seeking new active substances. The strain of *Streptomyces toxytricine* shown here was found in an analysis of soil samples from Majorca; it produces a particularly interesting active substance: lipstatin. Lipstatin and its derivative, tetrahydrolipstatin (THL), inhibit lipases, enzymes which are secreted by the pancreas. In the human intestine, lipases cleave the fats which are ingested with food. If lipstatin blocks the action of these enzymes, fats are no longer cleaved, but are excreted without being digested. A secondary effect is that cholesterol is no longer absorbed in the now fatty milieu of the intestine but is also excreted. Clinical scientists at Roche are currently investigating the ability of lipstatin to lower cholesterol levels. © F. Hoffman-La Roche Inc., Basel [298].

organism is under strict regulation. Several microorganisms are used in biotechnology for the bioproduction of their own authentic products, perhaps after genetic improvement. Alternatively, they are used as hosts for heterologous gene expression of proteins of other origins. Some of these organisms are *Escherichia coli* (Fig. 9), *Bacillus subtilis* (Fig. 10), *Saccharomyces cerevisiae* (Fig. 11), and different species of *Streptomyces* (Fig. 12). Prokaryotic promoters and their use in biotechnology for heterologous gene expression are reviewed in this volume [68].

Ribozymes

Ribozymes can be engineered to interact with a specific sequence in mRNA, thus opening a whole new dimension to negatively regulate gene expression and, similar to antisense, could provide a tool for gene therapy. Their use in medicine is based on "killing the messenger" [69]. Some recent applications include the reversal of drug sensitivity in multidrug-resistant tumor cells by ribozymes [70], and suppression of neoplastic phenotype in vivo [71]. Gene inactivation by ribozymes is discussed further in this volume [27].

How genes can be manipulated

Breeding programs

Genes have been manipulated for many years in plants and animals by breeding programs that included selection, crosses, mutations for numerous economical traits that are heritable or have a high heritability. Modern plant and animal breeding benefited from molecular biology in using molecular markers for their breeding schemes.

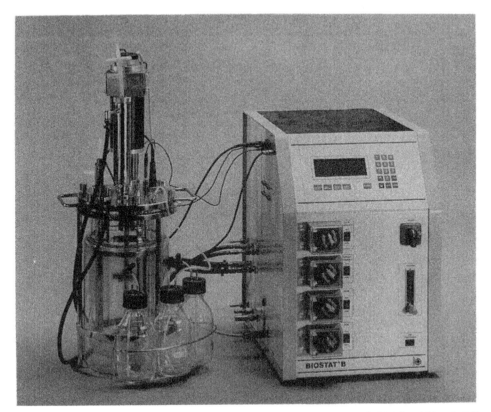

Fig. 13. An example of a bench top fermenter, the Biostat B. It is available with 2 l, 5 l or 10 l, optional, working volume vessels. Courtesy of B. Braun Biotech International, Melsungen.

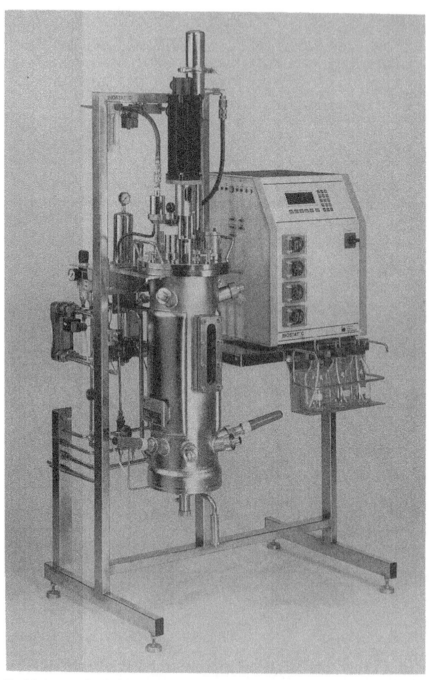

Fig. 14. An example of a laboratory scale fermenter, the Biostat C. It has interchangeable vessels (10 l, 15 l, or 20 l working volumes). Courtesy of B. Braun Biotech International, Melsungen.

Fig. 15. Pilot scale fermenter as for example the Biostat UD. The Biostat UD is available in two versions, one with interchangeable 20 l, 30 l and 50 l working volume vessels, the other with a 100 l working volume vessel. Courtesy of B. Braun Biotech International, Melsungen.

24

Fig. 16. Two 1000 l pilot fermenters with auxiliary equipment at the Biological Research Center. © F. Hoffman-La Roche Inc., Basel [300].

Recombinant DNA techniques and genetic engineering techniques

Since the early 1970s, these techniques have seen revolutionary advances permitting the isolation, manipulation and amplification of genes. These techniques have been helping basic research in addition to their utilization in biotechnology. The techniques of fermentation and down stream processing are basic to any production of recombinant proteins. Figures 13–16 show examples of fermenters with different capacities.

Examples of biotechnology applications in agriculture

Plant biotechnology

Plants are the primary source of food and fibers. Several plants are also used for their medicinal chemicals and other natural products such as rubber, petroleum substitutes, anti-microbial agents and sweeteners [72]. Until 1982 genetic improvements of agriculturally important crops relied exclusively either on sexual recombination followed by selection or to a lesser degree on mutagenesis [73]. The techniques for transferring foreign genes into plant cells and regenerating the transformed cells into fertile transgenic plants were established in 1983 [58]. Drug resistance markers such

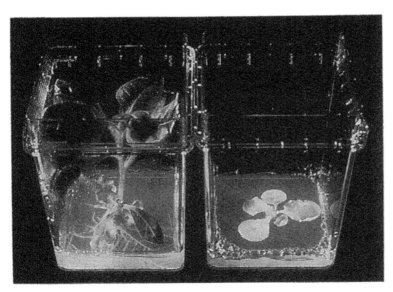

Fig. 17. Expression of the prokaryotic kanamycin antibiotic resistance marker in tobacco plants. In this cytotoxin-containing nutrient medium, only plants into which a kanamycin-inactivating foreign gene has been inserted can survive (right); untreated plants die (left). © Ciba-Geigy.

as kanamycin, are often used during the development of plant expression vectors to monitor the introduction of desired genes (Fig. 17).

The use of molecular markers in plant breeding
Although plant breeding techniques are still being used, breeding programs often utilize molecular markers that correlate to economically important traits. Hence these techniques enhance the effectiveness of the breeding program with several molecular markers employed [74,75]. PCR amplification from single seeds using only a small portion of the embryo allows the screening of large seed populations in a non-destructive manner and facilitating DNA marker-assisted breeding was developed [76].

Old and difficult questions dealing with the genetic and molecular basis of heterosis as well as the molecular markers associated with heterosis are being addressed. These results are leading the way for further improvement and predictability in utilizing heterosis in crop plants [77,78].

Non-molecular techniques are also used in modern plant breeding such as tissue culture, somaclonal variation and protoplast fusion [79–81].

Gene transformation, and genetic engineering
Plant genetic engineering has progressed to the point where most plant crops can be manipulated to improve nutritional composition, flavor and storage ability [82]. The introduction of genetic transformation methods in plants signals a new era in plant genetic manipulation. Initially, these methods utilized the plant pathogenic element

T-DNA of *Agrobacterium tumefaciens* for some dicotyledonous plants [83]. Naked DNA was used also for the transformation of plant cells after removing cell walls [79,84], or by bombarding plant tissues with microscopic metal particles coated with the desired DNA construction [85,86].

Examples of transgenic plants
Different transgenic plants are developed in research laboratories, but several steps are involved before the marketing of any genetically modified plants for any new trait. These steps range from gene cloning and expression to product development, field testing, breeding into multiple choice varieties, product characterization and regulatory review, public acceptance, and then marketing [87].

Disease and insect resistance. An increasing number of transgenic plants are now available. One of the early success stories of plant protection by genetic engineering was the over-expression of a viral coat protein gene to confer resistance in transgenic plants against the same virus [58]. However, the prospects of broadening the resistance to include other viruses are being addressed [88]. Resistance to some races of bacterial pathogens, e.g. *Pseudomonas syringae* in tomato, was conferred by introduction of a protein kinase gene (Pto) to the susceptible tomato variety [89]. The Pto gene product (serine-threonine kinase) is involved in a signal transduction pathway for plant defense.

Resistance of plants to a narrow window of insects was also engineered using an in-secticidal protein derived from *Bacillus thuringiensis* (B.T.) [90] (see Figs. 18–21).

In transgenic plants that were engineered for their resistance or productivity, the issue of diversity has to be addressed and tests should be made to evaluate if crosses with leading varieties are sufficient to ensure diversity in the species.

Direct food improvement of genetically engineered plants. Although plant genetic engineering techniques can be used to improve almost any economically important crop, the field still is under-utilized relative, for example, to applications in the biomedical field. However, with the great pressure to improve the quality and quantity of the food supply to meet the needs of the ever increasing global population, this situation will have to change. Farmers have to produce more food for the next 40 years than was produced since the beginning of agriculture [91].

Manipulation of the pathways/genes of the interconversion between sugar and starch are tolerated in plants [82,92]. Starch can be manipulated to be increased in sink/storage organs and conversely sugars can be increased in source tissue [82,93].

Plant lipids vary in their composition and content of saturated and unsaturated fatty acids. Using genetic engineering methods, attempts are made to reduce saturated fatty acids, such as linolenic acid, and/or increase the unsaturated fatty acids such as palmitic and stearic acids [94,95].

Although many crop plants produce proteins, several of these proteins have unbal-anced amino acid content and thus these proteins could be less then ideal for human nutrition. Moreover, some plants have substances which inhibit the proper digestion of protein [96].

Fig. 18. Transgenic cotton for B.T. The large boll of cotton on the left is the product of a transgenic plant. The cotton plant resistance was improved by incorporated B.T. protein to control certain caterpillars, such as the cotton bollworm, tobacco budworm and pink bollworm. Plants not weakened by insects are healthier and produce more cotton. The boll on the right was grown in the same field but comes from an unprotected plant. Courtesy of Monsanto Company.

Fig. 19. Yield of transgenic cotton for B.T. The potential increase in yield becomes evident when the leaves are stripped from a plant damaged by insects (right) and a transgenic plant that has built-in protection (left). Courtesy of Monsanto Company.

28

Fig. 20. Transgenic potatoes for B.T. Researchers at Monsanto have produced a transgenic potato plant that repels the Colorado potato beetle. Using genetic modification, a gene from a common soil bacterium, B.T., was introduced into the potato plant. The gene prompts the plant to produce a protein that controls the beetles. Efficacy is demonstrated in this Idaho field trial — the row of genetically improved plants in the center shows no insect damage, the unprotected rows on either side show heavy insect damage. Courtesy of Monsanto Company.

Fig. 21. Potatoes from plants that have been genetically improved with the B.T. gene. Courtesy of Monsanto Company.

Fig. 22. Gene technology makes it possible to breed more resistant grain varieties which can also withstand extreme climatic conditions. Grain varieties of this type will be especially important for Third World countries. © Ciba-Geigy.

Alternative strategies to enhance protein quality have been postulated [82]:

1. expressing a desirable heterologous storage protein;
2. increasing the expression of a desirable homologous protein;
3. suppressing the expression of a non-nutritional protein;
4. improving the quality of an endogenous protein by adding a synthetic region of DNA coding for peptides rich in desirable amino acids.

Other transgenic plants
Transgenic plants were also developed for tolerance to cold, freezing and other extreme climatic conditions (Fig. 22) as well as expressing heterologous proteins such as, for example, the naturally sweet proteins thaumatin and monellin [82].

Control of ripening and storage ability
Experiments were undertaken in tomatoes using antisense technology to alter the production of the plant hormone ethylene. The use of antisense ACC synthase inhibits ethylene production and thus severely inhibits fruit ripening. Ripening can then be manipulated by added ethylene [97].

Medicine

In the industrialized countries, genetic engineering and biotechnology are more developed in the field of medicine than in any other field of biotechnology. The medical applications of biotechnology include those outlined below.

Molecular diagnostics

Precise diagnoses of hereditary diseases, cancer, bacterial and viral infections are required prior to any possible treatment. Molecular diagnosis employs modern molecular biology techniques and consequently has several applications in medicine. Several molecular diagnostic techniques are standard in the clinical laboratory in identifying deleterious human mutations; however, the field is still open for new applications or automation [98–101]. In the field of infectious diseases, molecular diagnostics allow earlier detection of infection, avoiding the need for culturing infectious agents for the purpose of diagnosis. Also, molecular diagnosis can be employed in the absence of immunologic detection of the agent or if conditions for culturing the agent have not yet been developed [102]. The development of long PCR where a large DNA fragment of about 35 kb can be amplified will enhance the detection of genetic mutations as well as identify the genotype of an infectious virus [103].

Production of recombinant therapeutic proteins

Recombinant proteins are those proteins that are produced by genetic engineering. Currently the production of therapeutic proteins that represent the equivalent of the natural human protein is the predominant application of biotechnology in medicine. This trend is subject to change. However, the impact of the production of therapeutic proteins, growth factors and therapeutic monoclonal antibodies in the medical field will certainly be greater in this decade than, for example, the impact of antibodies in the 1950s. It is estimated that the human body has about 100,000 different proteins to carry on the different catalytic activities. Therefore, the potential for using proteins for therapy in cases of lower or impaired activity of any of the needed proteins is theoretically great. This potential however, is limited by the following factors:
1. mode of delivery avoiding the gastrointestinal system as the protein will be broken down if administered orally;
2. overcoming antigenicity that could result from using proteins other than human or even human proteins that are aggregated due to problems that could be associated with heterologous gene expression.

In the biotechnology industry, utilizing microorganisms for the production of therapeutic proteins requires specified and certified standards of manipulation and waste treatment (Figs. 23 and 24). Before any large-scale production, examination of proteins expressed in cells is done, for example, by gel electrophoresis (Fig. 25). Cells are harvested in special centrifuges for the industrial scale operations (Fig. 26). Recombinant proteins are often engineered to be secreted in the media, but if the proteins are programmed to be expressed and retained in the heterologous host cells more steps are needed for the purification of proteins from harvested cells. Figure 27 shows large amounts of frozen cells with genetically engineered α-2a interferon. Subsequently, the cells are thawed and the recombinant protein is purified (Figs. 28–31). In the production of the newly approved recombinant factor VIII (Kognate™), monoclonal antibodies are included in the purification steps (Fig. 32). Monoclonal antibodies were also employed in the purification of recombinant proteins as for example α-2a interferon [20]. The recombinant antihemophilic factor (factor VIII) was produced using mammalian cell lines engineered to express the protein. Mammalian cell lines are often used in biotechnology to express human proteins (Fig. 33).

Several therapeutic proteins that are developed by the different genetic engineering companies were recently reviewed [42] (see Table 1). Several more therapeutic recombinant proteins are at different stages of development or clinical testing. The use of recombinant cytokines (tumor necrosis factor, interferon, granulocyte colony-stimulating factor) in upregulating in vivo host defenses in normal or immunocompromised humans and animals was shown to be active and safe in humans [104]. The treatment of hemophilia A and B by recombinant factors VIII and IX was shown to be effective [105]. Recombinant erythropoietin, as a growth factor for red blood cells that help in certain anemia cases due to chronic renal failure, will likely become an important tool in efforts to achieve the elusive goal of bloodless cardiac surgery [106].

Application of recombinant DNA techniques made it possible to isolate genes and consequently proteins that were merely hypothetical just a few years ago. Recently, for

Fig. 23. Treatment with sulfuric acid kills all the bacteria, but the interferon alfa-2a they contain is made more accessible for subsequent processing. Concentrated sulfuric acid completely destroys the genetic material of the bacteria. Thus no bacteria can possibly survive, nor can any intact genetic information ever possibly by transmitted to other microorganisms. Treatment with sulfuric acid was required under the guidelines issued in the early 1980s, before people were entirely convinced of the safety of using genetic engineering techniques to produce interferon. Internationally recognized guidelines have since loosened this requirement in light of experimental findings. At the Biotechnology Plant, however, for technical reasons, the bacteria continue to be killed prior to further processing. © F. Hoffman-La Roche Inc., Basel [20].

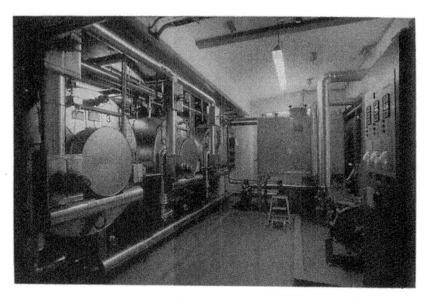

Fig. 24. Waste treatment in biotechnology in a production plant. All waste water that has been in contact with bacteria is collected in large tanks in the basement of the Biotechnology Plant and heat sterilized. Any spillage resulting from technical failures is treated in the same way. © F. Hoffman-La Roche Inc., Basel [20].

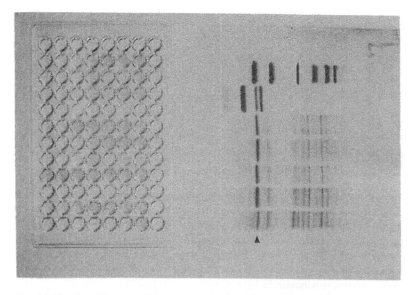

Fig. 25. The identification of human proteins by gel chromatographic separation (above) and by the color test (below). © Ciba-Geigy.

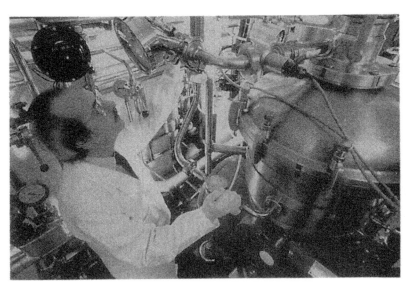

Fig. 26. Harvesting cells by centrifugation. The centrifuge effluent is piped out to the automatic waste water treatment system for heat sterilization (see Fig. 24). © F. Hoffman-La Roche Inc., Basel [20].

Fig. 27. Inactivated, centrifugally separated and deep-frozen cell mass with genetically engineered interferon after completion of a fermentation process in the Roche Biotechnology Pilot Plant. © F. Hoffman-La Roche Inc., Basel [20].

Fig. 28. The extraction of α-2a interferon from bacterial cells (see Fig. 27) in saline solution for an extended period using special agitation vessels. © F. Hoffman-La Roche Inc., Basel [20].

Fig. 29. Removing dead cells and cell debris from soluble proteins including α-2a interferon. © F. Hoffman-La Roche Inc., Basel [20].

example, a new hormone, thrombopoietin, which was postulated 30 years ago as a vital blood-cell growth factor [107] has been isolated and its role has been confirmed.

The role of thrombopoietin is believed to be more direct in stimulating the maturation of megakaryocytes, the blood cells from whose cytoplasm mature platelets are formed (Fig. 34). Several other known hematopoietic growth factors and cytokines stimulate megakaryocytopoiesis and platelet production, but with no clear exclusive effect as blood growth factor [108].

This newly discovered hormone offers great hope in the treatment of cancer patients by stimulating the production of their own platelets after radiation or chemotherapy.

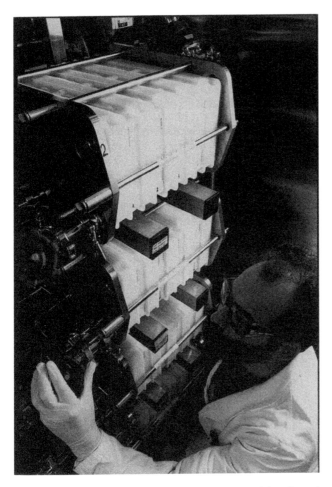

Fig. 30. Concentration of the solution containing α-2 interferon by ultrafiltration (total surface area used is 20 m^2). Following the different filtrations to remove water and salts, different chromatographic steps are used to purify the protein which is ultimately purified using monoclonal antibodies. © F. Hoffman-La Roche Inc., Basel [20].

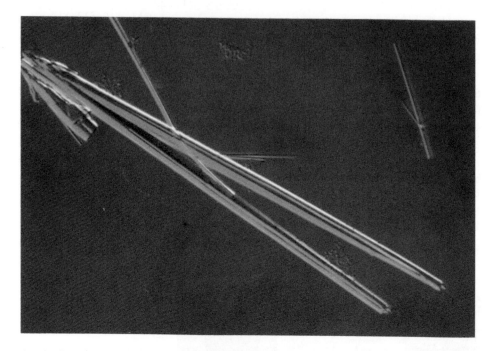

Fig. 31. Rod-shaped crystals of human alfa-2a interferon, genetically engineered in microorganisms, under polarized light. Large protein molecules like the interferons only form such crystals when they are in a highly purified state. © F. Hoffman-La Roche Inc., Basel.

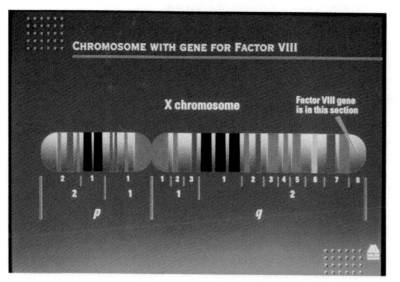

Fig. 32a. See following page for figure legend.

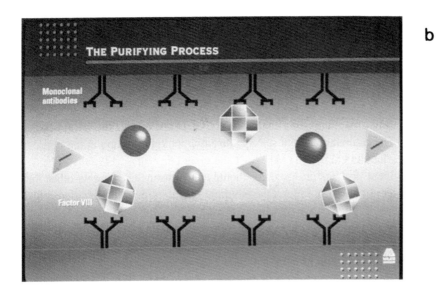

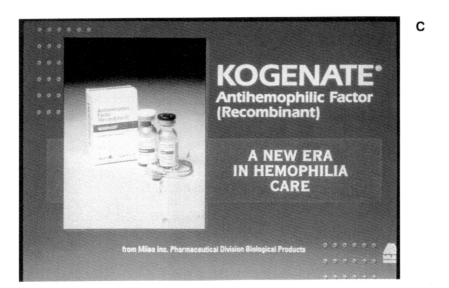

Fig. 32. Human factor VIII, from a genetic map to a product on the market. (a) Gene location of factor VIII on human chromosome X. Factor VIII gene was cloned and expressed in a mammalian cell line that was engineered to secrete factor VIII in the medium. The secreted protein is harvested for further purification. (b) The purifying process. The harvested factor VIII undergoes a rigorous purification process of anion exchange and monoclonal immunoaffinity chromatography. The material is further concentrated and most of the culture medium proteins are removed. In the very unlikely event of unde-tected virus contamination of the cell culture, the capability of removing and inactivating viruses is built into the purification process. The final product is highly purified and ready to be used. (c) The final marketable product. Kogenate. Antihemophilic factor (Recombinant). Courtesy of Miles Inc., a Bayer Company.

Radiation and chemotherapy destroy platelets and marrow cells putting cancer patients undergoing such treatment at risk of death from hemorrhage unless repeated blood transfusions are made.

Two genetic engineering companies in cooperation with a scientific institution were behind the cloning, expression and verification of its role [109–111]. Wendling et al. [112] using a recombinant receptor (c-Mpl) demonstrated that the megakaryocyte colony stimulating activity as well as platelet-elevating activity can be removed from thrombocytopenic plasma suggesting that both activities are due to the ligand that is thrombopoeitin. The estimated annual sales of thrombopoeitin could be in the range of 1 billion dollars after clinical trials [113].

Other recombinant proteins are used for therapy for reasons and targets other than that of the equivalent natural product. One example is recombinant human deoxyribonuclease I (rhDNase), the efficacy of which is being tested in cystic fibrosis (CF) patients by inhalation to improve clearance of purulent sputum from human airways [114].

Once a potential therapeutic protein is cloned and expressed, several phases of preclinical and clinical development stages have to be conducted before the recombinant protein is ready to be launched on the market [115] (see Fig. 35). Government agencies such as for example the FDA (Food and Drug Administration) have to approve the use of such a therapeutic protein. Figure 36 shows examples of recombinant therapeutic proteins already on the market.

Fig. 33. Production of recombinant protein by mammalian cells. Glaxo GIMB molecular microbiology scientist sampling a fermenter during a recombinant mammalian cell culture process. The recombinant protein resulting from such a culture is used for biochemical and biological studies and eventually for drug screening. Courtesy of Glaxo, Greenford.

example, a new hormone, thrombopoietin, which was postulated 30 years ago as a vital blood-cell growth factor [107] has been isolated and its role has been confirmed.

The role of thrombopoietin is believed to be more direct in stimulating the maturation of megakaryocytes, the blood cells from whose cytoplasm mature platelets are formed (Fig. 34). Several other known hematopoietic growth factors and cytokines stimulate megakaryocytopoiesis and platelet production, but with no clear exclusive effect as blood growth factor [108].

This newly discovered hormone offers great hope in the treatment of cancer patients by stimulating the production of their own platelets after radiation or chemotherapy.

Fig. 30. Concentration of the solution containing α-2 interferon by ultrafiltration (total surface area used is 20 m^2). Following the different filtrations to remove water and salts, different chromatographic steps are used to purify the protein which is ultimately purified using monoclonal antibodies. © F. Hoffman-La Roche Inc., Basel [20].

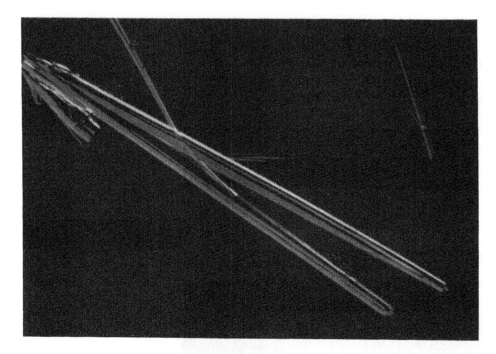

Fig. 31. Rod-shaped crystals of human alfa-2a interferon, genetically engineered in microorganisms, under polarized light. Large protein molecules like the interferons only form such crystals when they are in a highly purified state. © F. Hoffman-La Roche Inc., Basel.

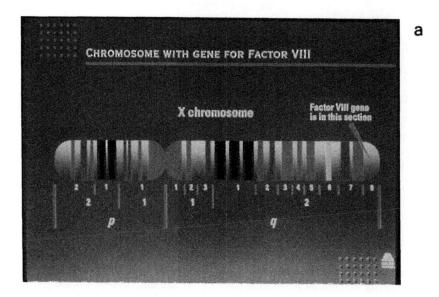

Fig. 32a. See following page for figure legend.

Table 1. Examples of recombinant proteins and vaccines on the market

Protein	Commercial name	Company	Use
Alteplase (t-PA)	Activase	Genentech	Acute myocardial infarction pulmonary embolism
Epoetin-α	Epogen	Amgen	Anemia of chronic renal failure
Epoetin-α	Procrit	Ortho Biotec	Anemia of chronic renal failure and AIDS
Factor VIII	Kognate	Miles/a Bayer Co.	Hemophilia
G-CSF	Neupogen	Amgen	Chemotherapy-induced neutropenia
GM-CSF	Leukine	Immunex	Autologus bone marrow transplant
GM-CSF	Prokine	Hoechst-Roussel	Autologous bone marrow transplant
Hepatitis B vaccine	Energix-B	SmithKline Beecham	Hepatitis B prevention
Hepatitis B vaccine	Recombivax	Merck	Hepatitis B prevention
Human growth hormone	Humatrope	Eli Lilly	hGH deficiency in children
Human growth hormone	Protropin	Genentech	hGH deficiency in children
Human insulin	Humulin	Eli Lilly	Diabetes
Interferon-α-2a	Roferon-A	Hoffmann-La Roche	Hairy cell leukemia and AIDS-related Kaposi's sarcoma
Interferon-α-2b	Intron A	Schering-Plough	Genital warts, AIDS-related Kaposi's sarcoma, non-A, non-B hepatitis
Interferon-α-n3	Alferon N	Interferon Sciences	Genital warts
Interferon-γ-1b	Actimmune	Genentech	Chronic granulomatous disease
Muromonab-CD-3	Orthoclone OKT3	Ortho Biotech	Acute kidney transplant rejection

Recombinant vaccines

Several genes coding for proteins that are expressed or constitute parts of infectious agent surface proteins (viral, bacterial or fungal) are cloned and expressed heterologously. These proteins can be used, modified or unmodified, as immunogens in vaccine preparations to protect humans from such infectious diseases. Recombinant vaccines offer the advantage of being free of the pathogenic agents as well as being subject to further genetic manipulation to increase their antigenicity and activity. Vaccines against the devastating HIV virus still face some difficulties. So far these vaccines fail to neutralize primary HIV isolates in spite of the fact that they do illicit strong immune response and neutralize laboratory strains [116]. However, two recombinant vaccines against hepatitis B became available in 1987 [117]. Two different companies are expressing the hepatitis B surface antigen in yeast (HBSAg) [118–120a]. Serum derived hepatitis B vaccine that was introduced in 1982 is no longer available in the US (committee on infectious diseases) [120b] and the UK [121].

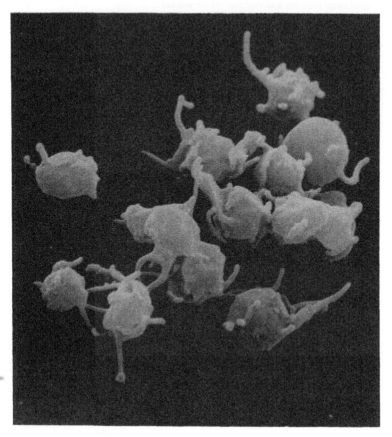

Fig. 34. Blood platelets. Blood platelets are not actually cells, but cell fragments which have been shed by megakaryocytes, the giant cells of the bone marrow. Blood platelets are of vital importance in blood coagulation and wound healing. They secrete a number of growth promoting substances including platelet-derived growth factor (PDGF). Actually, the PDGF itself is now being produced by genetic engineering and being clinically tested as a wound healing agent. © F. Hoffman-La Roche Inc., Basel [298].

Yeast-derived recombinant hepatitis B vaccines have now been given to millions of infants and children throughout the world. It appears that these recombinant vaccines have an excellent safety record and protection for children who complete a three dose vaccination [122]. Another recombinant hepatitis B expressed in Chinese hamster ovary cells (CHO) is now being tested [123].

Immunotoxins as modified therapeutic proteins

Immunotoxins are a new class of cytotoxin agents composed of bacterial or plant toxins coupled to monoclonal antibodies or growth factors as ligands. These immunotoxins can specifically target and kill cells that display the specific antigens or

Development of Biotechnical Pharmaceuticals

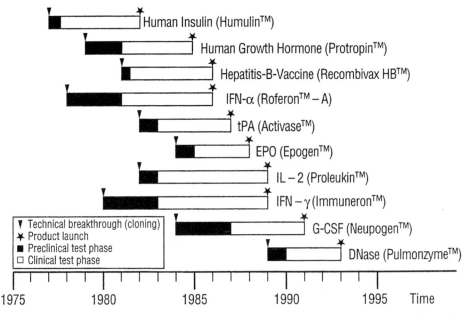

Fig. 35. The different stages of development of therapeutic proteins, from clones to marketable product. One commercially available product for each of the recombinant proteins has been selected to illustrate the time required for development. These recombinant proteins were developed by different biotechnology companies. © F. Hoffman-La Roche Inc., Basel [300].

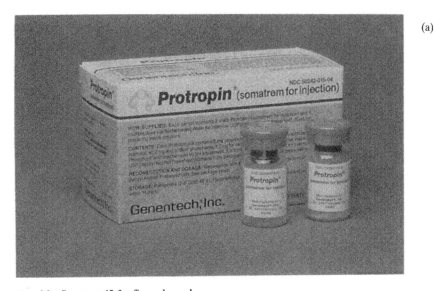

(a)

Fig. 36a. See page 43 for figure legend.

growth factor receptor on their surface. The immunotoxins can be manipulated by genetic engineering to enhance their activity or to increase their tolerance in the body [124,125].

In spite of the fact that these types of immunotoxins are in the early stages of development and clinical trials to treat different types of human tumors, they demonstrate encouraging results [126]. Problems related to the production of antibodies against immunotoxins have to be overcome. It was suggested that chemically coupling the immunotoxins to polyethyleneglycol might render them less immunogenic, as in the case of adenosine deaminase [127].

(b)

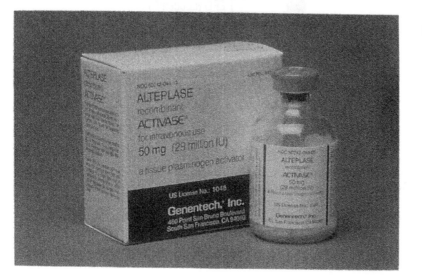

(c)

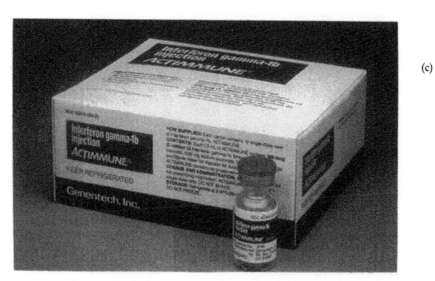

Fig. 36b,c. See page 43 for figure legend.

(d)

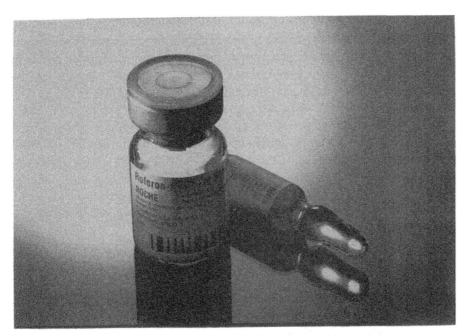

(e)

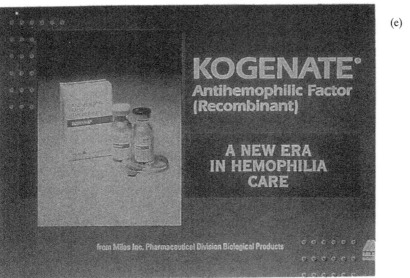

Fig. 36. Example of recombinant human product already in the market for therapy. (a–c) Courtesy of Genentech Inc., South San Francisco. (d) © F. Hoffman-La Roche Inc., Basel [20]. (e) Courtesy of Miles Inc., a Bayer Company.

Birth control and biotechnology

The need to develop a safe reliable method of contraceptive for family planing could not be greater, especially in parts of the world where the currently used methods are not acceptable or are hard to implement while the birth rate and population growth far exceed the already deteriorating economic ability to support a further increase in population in many countries. Several approaches are being developed towards this goal either as a male or female contraceptive vaccine. Talwar et al. [128] reported their experiments using non-recombinant and a recombinant vaccine against human chorionic gonadotrophin (hcG) which is essential in implantation [129,130].

Their phase II clinical trial, when they used chemically coupled tetanus toxin (tt) or diphtheria toxin (dt), demonstrated that such a vaccine is effective and reversible. Other experiments using the hcG subunit carboxylic terminal portion (hcGpCTP) showed, however, that both the efficiency and safety are still in question [131]. Other male contraceptive vaccines are also being developed. For example, a recombinant vaccine against human acrosomal protein Sp-10 is currently being tested on the basis of its tissue specificity as well as functional assays, indicating that anti-Sp-10 antisera inhibit sperm-egg interactions [132].

Recombinant proteins as a target for drug discovery

In the last few years much of the research and development in biomedical biotechnology has been focused not necessarily on the production of therapeutic proteins per se, but by using several recombinant proteins as targets for drug discovery. In this approach the basic problems of cloning, heterologous gene expression, solubility and activity of the heterologously produced proteins have to be solved as in the case of the production of therapeutic proteins. However, in this new application the purified active protein is not the final product but it can then be used as a target for drug discovery or drug screening. Several companies, for example, are cloning and expressing important proteins that are essential for the full activity of the human immunodeficiency virus (HIV), the etiological agent for the acquired immune deficiency syndrome (AIDS), in order to use such a protein, now available in sufficient amounts, to screen for inhibitors specific for the catalytic activity of such a vital protein for the virus (see Fig. 5 for example). Consequently this could lead to the development of drugs that could be used to treat AIDS patients. Inhibitors for HIV protease are currently being evaluated for the treatment of AIDS patients [133–137]. Similarly, recombinant HIV reverse transcriptase was used as a target for drug development to medicate AIDS patients [17,138,139]. Figure 37 shows a receptor binding assay for drug development studies. Recently, there was a concerted effort to use transcription factors in new strategies to aid in the search for human therapeutics, thus expanding the frontiers of pharmaceutical development [16].

Other human recombinant proteins were also used for drug screening and discovery, such as, for example, stromelysin and thrombin. Stromelysin is a metalloproteinase of connective tissue that degrades many non-collagenous components of the extracellular matrix and may play a role in the activation of latent procollagenase. It is regulated by several cytokines, growth factors and protooncogenes [140,141]. A human stromelysin catalytic domain (SCD) has been heterologously expressed in *E. coli*, purified to homogeneity, and the recombinant SCD was used for inhibitor screening. This screen-

Fig. 37. Receptor binding assay. A senior molecular biologist performing a receptor binding assay on 'T' cells in culture medium, GIMB, Geneva, Switzerland. Courtesy of Glaxo, Greenford.

ing resulted in the identification of several competitive inhibitors for SCD that are tryptophan derivatives [142]. Similarly a new class of inhibitors of thrombin (serine protease involved in thrombosis and hemostasis) was isolated using a similar strategy [143].

Although much has been written in recent years about rational drug design, no drug has been designed de novo, i.e. without using a natural substrate or inhibitor or by finding a lead as a result of screening [144]. As a result of this, drug discovery is currently dominated by drug screening for new biologically active molecules. In addition to the thousand of chemicals that have been accumulated synthetically, or are found in extracts of plants, animals, microorganisms and marine organisms, new techniques have been developing rather rapidly for the creation of new potentially biologically active molecules that can be utilized in screening schemes. Phage display libraries, representing millions of different peptides on the surface of filamentous phages, have been constructed and utilized in screening strategies for several ligands such as receptors or antibodies [145–151]. Display peptide and oligonucleotide libraries made on a solid matrix were also constructed [152–154]. Antibody display libraries that represent a repertoire of specific antibodies were developed to make immunochemical reagents that can be used in Western blots, staining of cells and epitope mapping [155]. Combinatorial revolutionary chemistry is also being developed [156]. This new chemistry promises to increase the number of chemicals to be screened, and yet they can be easily tagged for identification, to almost unlimited numbers [157–150].

Gene therapy

Gene therapy is defined as the delivery of a functional gene for expression in somatic tissues with the intent to cure a disease [160]. A decade ago, gene therapy remained a concept and only a potential that might lead to the treatment of hereditary diseases, cancer and viral diseases. Currently, gene therapy is expected to become a major clinical practice in the future [161]. Although it is estimated that humans have about 100,000 different genes, only over 4,000 distinct genetic disorders are listed in the 1991 edition of "Mendelian Inheritance in Man" by McKusick [162]. However, the molecular nature of the majority of these listed diseases is unknown [163]. On the positive side, the techniques for gene therapy have improved tremendously in recent years. Consequently, gene therapy appears to be gaining momentum as a powerful tool for the therapy of genetic diseases and cancer in the near future and it will remain a subject of scientific and clinical research.

Somatic gene therapy, in contrast to germline therapy, is the subject of the current research in gene therapy for human diseases. Somatic gene therapy involves the introduction of novel genetic material into somatic cells to express therapeutic gene products [164]. Since many mutations in humans are caused by nonsense mutations, gene therapy by suppressor tRNAs has been suggested [165]. Germline gene therapy has been demonstrated in transgenic mice, for example, using a full-length and truncated form of the dystrophin cDNA [161]. Some of the genetic diseases that are candidates for somatic gene therapy are listed in Table 2.

Modes of gene therapy
Basically, there are two general methods for the administration of genetic gene therapy, ex vivo and in vivo [164].

Ex vivo gene therapy. In this strategy, tissues such as lymphocytes, hepatocytes, tumor cells, fibroblasts or bone marrow cells are removed from the patient by surgical biopsy and the required genes are introduced into the cells after proper cell isolation and growth in culture. The genetically manipulated cells are then re-implanted in the body of the patient for gene expression or over-expression for the required gene product.

Target cells for gene therapy should be readily removable from the patient for in vitro manipulation, they should have the capacity to divide and grow indefinitely and cells containing the target gene should stably continue to express the introduced gene permanently. This approach eliminates the risk associated with rejection and the use of immunosuppressors. At present, and in spite of reported successes [200], this approach continues to be a difficult field of surgical research [164]. Moreover, the amount of the therapeutic gene product that can be expressed by ex vivo methods is limited by the number of cells that can be transplanted into the body [222].

Table 2. Examples that are potential candidates for gene therapy

Gene/protein	Disease	References
Mendelian genes		
Adenosine deaminase	Severe combined immunodeficiency	166, 167, 168
α1-Antitrypsin	Emphysema	169, 170, 171, 172
Arginosuccinate synthetase	Citrullinemia	173
CD-18	Leukocyte adhesion deficiency	174, 175
Cystic fibrosis transmembrane regulator	Cystic fibrosis	176, 177, 178, 179
Dystrophin	Duchenne muscular dystrophy	180, 181, 182, 229
Factor IX	Hemophilia B	183, 184, 185
Factor VIII	Hemophilia A	183, 186
α-L-Fucosidase	Fucosidosis	187
Glucocerebrosidase	Gaucher's disease	188, 189, 190
β-Glucuronidase	Mycopolysaccharidosis type VII	191, 192, 193
β-Globin	Thalassemia	194, 195, 196, 197
β-Globin	Sickle cell anemia	194, 195, 196
α-L-Iduronidase	Mycopolysaccharidosis type I	198
Low-density lipoprotein receptor	Familial hypercholesterolemia	199, 200, 201
Ornithine transcarbamylase	Hyperammonemia	202
Phenylalanine	Phenylketonuria hydroxylase	203, 204, 205
Purine nucleoside phosphorylase	Severe combined immunodeficiency	206, 207
Sphingomyelinase	Niemann-Pick disease	208, 209
Other		
Cancer		210, 211, 212, 213, 214, 215
Viral diseases		216, 217
Cardiovascular diseases		218, 219, 220
Aging		221

In vivo gene therapy. In this method, genetically engineered DNA using plasmid or viral vectors is administered directly to a specific organ or tissue in the patient's body. Currently, this strategy is preferred for its relative simplicity and non-invasive nature. In this approach DNA is essentially used as medicine, the same way therapeutic proteins or organic compounds are used.

In vivo gene therapy also includes the administration of genetically engineered donor cells that are transduced with the desired gene. Donor cells used should have matched HLA antigen. In vivo as well as ex vivo gene therapy using skin is being tested [223].

Gene transfer system
Several methods are used to deliver the DNA both in the ex vivo and vivo gene therapy strategies.

Non-viral transfer methods. These methods ensure the absence of viral gene sequences that are cotransfected along with the target gene.

Chemical and physical methods. Methods that are used for the transformation of DNA into animal or human cell cultures as well as tissue cultures, are also used for transformation using the ex vivo method. DNA constructions are used to transform cells that are removed from the patient prior to their in vitro culture and reimplantation into the patient's body. The standard calcium phosphate method has been used in the context of gene therapy [224,225]. However, electroporation [226] and bead transfection [227] methods appear to be more effective. These chemical and physical methods could be used to transfect cells with viral vectors.

Liposome-mediated gene transfer. Currently the transfer of DNA constructions via liposomes is used widely in clinical trials because of its relative simplicity and lower immunogenicity [163,220,228,230].

Receptor-mediated gene transfer. Receptor-mediated gene transfer is achieved by the conjugation of supercoiled DNA to a ligand, for example polylysine. The protein used as a ligand is recognized by the corresponding receptor on the surface of the target cells [231–234].

Direct injection of the DNA into target tissues. The injection of DNA directly is ultimately the method of choice for gene transfer. However, if delivery methods do not target the appropriate tissues, then appropriate use of tissue-specific control elements will be required for selective gene expression [235]. Using experimental animals the volume of the injected material appears to be critical for increased expression of reported genes [236].

Viral transfer methods. Several viral based vectors are used to transform animal cells with high efficiency, such as herpes simplex, retroviral and adenoviral based vectors [163]. Vectors based on the adeno-associated virus (AAV) which is non-pathogenic to humans, are also being tested for their potential in gene therapy [237,238]. Viral methods have the potential to be used both for the ex vivo and in vivo gene therapy experiments.

Gene therapy and ethics

Similar to many new advances in medicine, ethical limitations have to be considered in the context of benefit and risk assessment. Internationally, numerous policy statements on human genetic intervention have been published supporting the moral legitimacy of somatic-cell gene therapy for the cure of disease [239]. On the other hand, considerable debate about the ethics of human germline gene modifications are still making it difficult for this approach to be a medical objective in the near future. It has been suggested that the discussion about human germline intervention should continue so that such an approach can be carefully compared with alternative strategies for preventing genetic diseases [240]. Ethical issues related to gene therapy have been debated as political issues to analyze their different parameters [241].

Molecular pharming

Several therapeutic proteins are targeted to be expressed in the milk of farm animals. This alternative to microbial or animal cell based bioreactors started after the finding that transgenic lactating mice could secrete active plasminogen activator (t-PA) and other human proteins into their milk [242–244].

Production of therapeutic proteins from lactating farm animals (molecular pharming) has become an attractive possibility in recent years since the produced proteins are usually active to similar post-translational modifications and glycosylation and proper folding, avoiding expensive downstream processing costs [245].

The yield of the recombinant protein produced in milk is quite high (up to 35 g/l) [41]. Experiments to target the recombinant protein to be secreted in milk were made using gene sequences that have tissue-specific hormonally induced expression in the mammary gland such as that of bovine beta-lactoglobulin, beta casein or whey acidic protein [246–250]. Standard transgenic techniques involve the microinjection of the reconstructed DNA in artificially inseminated eggs followed by the implementation procedure. Screening for transgenic animals that incorporate the intact gene are done by standard PCR or blotting techniques. Animals that produce the desired proteins during lactation are selected.

Marine biotechnology

Marine biotechnology is the branch of biotechnology that is based on the utilization of marine organisms. Several marine based natural products are entering clinical trials while others are helping in medicinal chemistry and cell biology research [251].

Oceans cover about 70% of the surface of the earth, but have 99.9% of the water. Consequently there is a much richer diversity of living organisms and species in oceans than of terrestrial origin. This richness and diversity of life offers a great potential for providing a food resource, pharmaceutical, biochemical and biotechnology applications. Ideally, research and any potential utilization can be managed in harmony with the proper protection of the marine environment.

Generally, in spite of the enormous potential of marine organisms [252–254], their utilization and cultivation is far underrepresented. Actually, many of the marine organisms themselves have not been fully identified.

A few laboratories now are using sophisticated equipment to isolate microalgae and bacteria from the depth of the oceans leading the way to their characterization and evaluation for any potential biotechnological application [255,256]. Oceans also offer extreme environmental conditions that would allow us to understand molecular adaptation. Oceans cover not only the major part of the earth's surface, they also reach depths exceeding that of the height of Mt. Everest. The adaptation of organisms and macromolecules under such high pressure prompted studies on protein behavior [257] and the identification of pressure response proteins (PRP), in a similar way to heat shock proteins [258].

Biochemicals from marine organisms

Numerous food and pharmaceutical industries are historically derived from terrestrial organisms. Now over 115 drugs from soil microbial sources are in clinical use to treat a wide variety of infectious diseases, cancer and other diseases [259]. Because of the growing demand for new therapeutic agents, sea microorganisms could increase our potential to help cope with such a demand. It is clear that the new combinatorial chemistry [150,157,159,231] will have a significant role in filling the gap in drug discovery, but biochemicals from natural sources will always have a role to play because of their complexity and their availability in nature.

The bioactive molecules from marine bacteria and algae include anti-inflammatory agents [256,259,260], antifungal agents [261], compounds with antitumor activity [262], and with antibiotic [263], anti HIV virus [264] and anti-influenza virus properties [265,266].

Enzymes from the sea

A wide range of enzymes were isolated from marine organisms or from industrial fish waste. Recently, for example, a lysozyme was isolated from scallop waste [267], a new agarase from a marine *Vibrio* [268], a type II restriction modification enzyme hjaI from the marine bacteria *Hyphomonas jannaschiana* [269] and a thermostable DNA polymerase from the hyperthermophilic marine archaebacteria *Purococcus* [270]. Polysaccharide degrading enzymes were isolated from marine organisms including chitinase [271] and pullulanase [272].

Fish vaccine

As the intensive farming of fish and shellfish is becoming an economic reality in several parts of the world, there is a need to develop vaccines against bacteria and viruses that affect these organisms. This is in analogy to the development of classical vaccines in the poultry industry after it was intensified. Now, some poultry vaccine are also recombinant [273].

Efforts are being made to develop recombinant vaccines for fish against bacteria [274–276] and against viral infections [277–280]. Figures 38a and b show the scale-up of fermentation to produce bacterial antigens to be used for the vaccination of fish in fish farms. Development is being made in improving vaccines by modern genetic engineering methods.

Transgenic fish

In order to transfer or to over-express economically important traits in fish, several experiments were made using model fish such as zebra fish [281] and medaka [282, 283] as well as commercially important fish such as, for example, salmon [284], trout [285], carp [286] and tilapia [287–289].

The most frequent transgenic fish were made to over-express growth hormone genes [286,290,291]; however, transgenic Atlantic salmon were made to express antifreeze protein and thus potentiate the establishment of freeze-resistant salmon that would help in fish farms in the cold climate [284].

Fig. 38. Production of APOVAX fish vaccine. (a) Bacterial propagation for the isolation of APOVAX fish vaccines. Clinical isolates of bacteria from diseased fish in Norwegian fish farms. Bacterial stock cultures are first cultivated to a desired level of growth in a pre-cultivator before full-scale production takes place. The bacteria are propagated in a defined atmosphere with automatic control of pH, temperature and air supply. The actual vaccine components are mixed in a holding tank to form the final product. (b) Fish vaccine filling line. After all the vaccine components are mixed in the holding tank the vaccine is siphoned aseptically into an automatic filling-line, packed and finally stored at 4–8°C (39–46°F). Courtesy of Apothekernes Laboratorium A.S., Tromsø.

Methods of gene transfer in transgenic fish

Two main methods are currently used in the construction of transgenic animals.

Microinjection
Direct injection of fertilized fish eggs has been used to produce transgenic fish. Usually several million DNA copies are injected into the nuclei of fertilized eggs or oocytes. Although this method or its variation is the most commonly used, the rate of success is low [286,287,292].

Electroporation
Electroporation-mediated DNA transfer into fish eggs has been recently improved [281,293]. Several attempts were also made to transfer DNA into fish sperm by electroporation [288]. Improved electroporation methods would undoubtedly replace microinjection since it is relatively easier to use and it can be used to generate much larger numbers of transgenic fish.

Recommendations have been made to use all fish based sequences for the construction of transgenic fish for the optimization of gene expression and for safety measures [290,294].

Bioremediation

Bioremediation is the process of using living organisms, whether natural to the environment or added to the polluted site, in order for them to feed on spilled hydrocarbons to hasten their decomposition and remove the threatening danger to the environment.

One of the biggest oil spill disaster in recent years and what followed can give examples of how to be prepared for such a danger. On 24 March 1989, the tanker Exxon Valdez ran aground on Bligh Reef, Alaska, releasing nearly 41 million liters of Alaskan North Slope crude oil and causing massive oiling to 2,000 km of shores within Prince William Sound at the Gulf of Alaska [295]. This resulted in a tremendous pressure to clean up the mess as effectively as possible. This disaster provided a forum for application and testing both for the Environmental Protection Agency (EPA) and Exxon. An interesting finding was that in the Exxon bioremediation they relied exclusively on native microorganisms found to be abundant on the shoreline but adding fertilizers containing nitrogen and phosphorous [295,296]. Laboratory experiments demonstrated clearly that the added nutrients in combination with microorganisms were the key to the clean up. To asses biodegradation statistically, Bragg and his colleagues added traces of the non-biodegradable hopane so that they could use it as an index for the more labile hydrocarbons in the oil. Figures 39 and 40 show the spraying of fertilizer (Inipol) on the oiled shoreline and the results clearly demonstrate the effect of using fertilizers.

The extent of bioremediation correlated with the concentration of nitrogen present or

Fig. 39. Workers used airless sprayers to apply Inipol liquid fertilizer on oiled shorelines. Courtesy of Exxon Production Research Company, Houston.

Fig. 40. A dramatic reduction in oil remaining on the cobbled shoreline was observed in a rectangular site treated with Inipol liquid fertilizer during a 1989 bioremediation test conducted by the US EPA at Snug Harbor, Prince William Sound, Alaska. Courtesy of Exxon Production Research Company, Houston.

added oxygenation and type of shore soil might also be critical in the success of this type of bioremediation [297]. Naturally these findings do not stop efforts to engineer organisms for the task, especially when local biodegrading microorganisms are not present at the site, but it also gives the obvious message that life does not depend on carbon alone. However, a careful evaluation of the total loss of toxicity of the degraded hydrocarbons has to be made, whether mixed natural cultures or genetically engineered organisms are used.

Acknowledgements

Many people, to whom I am very grateful, have helped in this project. I would like to express my special thanks to Dr. E. K. Weibel of Hoffmann-La Roche Ltd for his support and for providing valuable material and information as well as sponsoring the cost of color reproduction of some of the photos. I also wish to thank Dr. Willy Zimmermann for his kind support of this project early on and for providing valuable information and material. Dr. G. Kishore of Monsanto kindly provided some material. Anne Aune of Bayer sent us valuable material and photos dealing with one of their newly released therapeutic proteins. Vera Lund of Apothekernes Laboratorium sent and sponsored the reproduction of some color photos. Dr. H. Kiatel of B. Braun sent some photos dealing with fermentation equipment and sponsored color reproductions. Dr. James Bragg of Exxon sent valuable material about the unique Exxon experiment of bioremediation following the unfortunate Exxon Valdez oil spill. Dr. P. Connolly of Glaxo also sent us important material as did Genentech.

Dr. H.O. Schneider, the Curator of the Egyptian Department at the National Museum of Antiquities in Leiden, sent us useful information and a photo representing the early fermentation by the ancient Egyptians.

Tove Eriksen Heen very patiently helped in typing the manuscript.

I would like to thank the members and the students of the Department of Biotechnology, University of Tromsø for many discussions.

This work would not have been possible without the support and the scholastic atmosphere at the University of Tromsø and the support and patience of Sara El-Gewely.

References

1. US Congress, Office of Technology Assessment, Commercial Biotechnology: an International Analysis, OTA-BA-218. Washington, DC: US Government Printing Office, January 1984;3.
2. Bull AT, Holt G, Lilly MD. Biotechnology, International Trends and Perspectives. Paris: Organization for Economic Co-Operation and Development, 1982.
3. Bud R. The Uses of Life: a History of Biotechnology. Cambridge, UK: Cambridge University Press, 1993.
4. Hall DO, Coombs J, Higgins IJ. Energy and biotechnology. In: Higgins IJ, Best DJ, Johns J (eds) Biotechnology Principles and Applications. Oxford, UK: Blackwells Science, 1985;24–72.

5. Hall DO, Rao KK. Photosynthesis, 5th edn. Cambridge, UK: Cambridge University Press, 1994.
6. Hull A. Biotechnology and Genetic Engineering. Basel, Switzerland: Ciba Pharma Communication, 1993.
7. Markwardt F. Past, present and future of Hirudin. Haemostasis 1991;21:11–26.
8. Topol EJ, Fuster V, Harrington RA, Califf RM, Kleiman NS, Kereiakes DJ, Cohen M, Chapekis A, Gold HK, Tannenbaum MA et al. Recombinant hirudin for unstable angina pectoris. A multicenter, randomized angiographic trial. Circulation 1994;89:1557–1566.
9. Bacher P, Walenga JM, Iqbal O, Bajusz S, Bredding K, Fareed J. The antithrombotic and anticoagulant effects of a synthetic tripeptide and recombinant hirudin in various animal models. Thromb Res 1993;71:251–263.
10. van-den-Bos AA, Deckers JW, Heyndrickx GR, Laarman GJ, Suryapranata H, Zijlstra F, Close P, Rijnierse JJ, Buller HR, Serruys PW. Safety and efficacy of recombinant hirudin (CGP 39 393) versus heparin in patients with stable angina undergoing coronary angioplasty. Circulation 1993;88:2058–2066.
11. Fields WS. The history of leeching and Hirudin. Haemostasis 1991;21:3–10.
12. Glasscheib HS. The March of Medicine. New York: GP Putnam's Sons, 1994;153–166.
13. Kennedy JM. Bloodsucking leeches discovering a new niche as a microsurgery aid. Los Angeles Times, June 29, 1987.
14. Ciba-Geigy. Biotechnology. Pharma Policy and Economics. Basel, Switzerland: Public Relations Ciba-Geigy, 1988.
15. Bud R. 100 Years of Biotechnology. Events and publications that molded the industry's place in society. BioTechnology 1993:S14-S15.
16. Peterson MG, Tupy JL. Transcription factors: a new frontier in pharmaceutical development? Biochem Pharmacol 1994;47:127–128.
17. Taylor PB, Culp JS, Debouck C, Johnson RK, Patil AD, Woolf DJ, Brooks I, Hertzberg RP. Kinetic and mutational analysis of human immunodeficiency virus type 1 reverse transcriptase inhibition by inophyllums, a novel class of non-nucleoside inhibitors. J Biol Chem 1994;269:6325–6331.
18. van Bogelen RA, Olson E. Application of 2-D protein gels in Biotechnology. In: El-Gewely MR (ed) Biotechnology Annual Review, vol 1. Amsterdam: Elsevier, 1995;69–103.
19. The FY 1993 US Biotechnology Research Initiative. Biotechnology for the 21st Century. A Report by the Federal Coordinating Council For Science, Engineering and Technology, 1993;3.
20. Unternahrer-Rosta S, Rohrer T, Ryser S. The Roche Biotechnology Plant in Basel. How the Anticancer Drug Interferon alfa-2a is Made. Basel, Switzerland: Editiones Roche, F. Hofmann-La Roche Ltd., 1992.
21. Huxley A. Green inheritance. The World Wildlife Fund Book of Plants. Garden City, New York: Anchor Press/ Doubleday, 1985.
22. Hope J. The biology and molecular biology of scrapie-like diseases. Arch Virol 1993;(Suppl 7):201–214.
23. Rappaport EB. Iatrogenic Creutzfeldt-Jakob disease. Neurology 1987;37:1520–1522.
24. Rona RJ, Beech R. The process of evaluation of a new technology: genetic services and the introduction of DNA probes. J Public Health Med 1993;15:185–191.
25. Bayever E, Iversen P. Oligonucleotides in the treatment of leukemia. Hematol Oncol 1994;12:9–14.
26. Hiddemann W, Griesinger F. Preclinical aspects and therapeutic perspectives of acute and chronic leukemias. Curr Opin Oncol 1993;5:13–25.
27. Hartmann RK, Krupp G, Hardt W-D. Towards a new concept of gene inactivation: specific RNA cleavage by endogenous ribonuclease P. In: El-Gewely MR (ed) Biotechnology Annual Review, vol 1. Amsterdam: Elsevier, 1995;215–265.
28. Ellington AD. Aptamers as potential nucleic acid pharmaceuticals. In: El-Gewely MR (ed) Biotechnology Annual Review, vol 1. Amsterdam: Elsevier, 1995;185–214.

56

29. Glick BR, Pasternak JJ. Molecular Biotechnology. Principles and applications of recombinant DNA. Washington, DC: ASM Press, 1994.
30. Patek M, Krumbach K, Eggeling L, Sahm H. Leucine synthesis in Corynebacterium glutamicum: enzyme activities, structure of leuA, and effect of leuA inactivation on lysine synthesis. Appl Environ Microbiol 1994;60:133–140.
31. Bruckner H, Becker D, Lupke M. Chirality of amino acids of microorganisms used in food biotechnology. Chirality 1993;5:385–392.
32. Herry DM, Dunican LK. Cloning of the trp gene cluster from a tryptophan-hyperproducing strain of *Corynebacterium glutamicum*: identification of a mutation in the trp leader sequence. Appl Environ Microbiol 1993;59:791–799.
33. Cordes C, Mockel B, Eggeling L, Sahm H. Cloning, organization and functional analysis of ilvA, ilvB and ilvC genes from *Corynebacterium glutamicum*. Gene 1992;112:113–116.
34. Neidleman SL. Industrial chemicals: fermentation and immobilized cells. In: Moses V, Cape RE (eds) Biotechnology, The Science and the Business. Harwood Academic Publishers, 1991;297–310.
35. Houghton RA, Appel JR, Blondell SE, Cuervo JH, Dooley CT, Pinilla C. The use of synthetic peptide combinatorial libraries for the identification of bioactive peptides. Biotechniques 1992;13:412–421.
36. Weber W, Fischli W, Hochuli E, Kupfer E, Weibel EK. Anantin - a peptide antagonist of the atrial natriuretic factor (ANF). I. Producing organism, fermentation, isolation and biological activity. J Antibiot 1991;44:164–171.
37. Glick BR, Pasternak JJ. Molecular Biotechnology. Principles and Applications of Recombinant DNA. Washington, DC: ASM Press, 1994.
38. Law J, Fitzgerald GF, Uniacke-Lowe T, Daly C, Fox PF. The contribution of lactococcal starter proteinases to proteolysis in cheddar cheese. J Dairy Soc 1993;76:2455–2467.
39. Hutchinson CR. Drug synthesis by genetically engineered microorganisms. BioTechnology 1994;12:375–380.
40. Franco CM, Coutinho LE. Detection of novel secondary metabolites. Crit. Rev. BioTechnology 1991;11:193–276.
41. Bailey JE. Toward a science of metabolic engineering. Science 1991;252:1668–1675.
42. Rubin S. Biotechnology and the pharmaceutical industry. Cancer Invest 1993;11:451–457.
43. Connell N, Stover CK, Jacobs Jr WR. Old microbes with new faces: molecular biology and the design of new vaccines. Curr Opin Immunol 1992;4:442–448.
44. Morgan RA, Anderson WF. Human gene therapy. Annu Rev Biochem 1993;62:191–217.
45. Carn VM. Control of capripoxvirus infections. Vaccine 1993;11:1275–1279.
46. Cowan DA. Industrial Enzymes. In: Moses V, Cape RE (eds) Biotechnology, The Science and the Business. Harwood Academic Publishers, 1991;311–340.
47. Yang H, Cao SG, Ma L, Ding ZT, Liu SD, Cheng YH. A new kind of immobilized lipase in organic solvent and its structure model. Biochem Biophys Res Commun 1994;200:83–88.
48. Østgaard K, Wangen BF, Knutsen SH, Aasen IM. Large-scale production and purification of K-carrageenase from Pseudomonas carrageenovora for applications in seaweed biotechnology. Enzyme Microb Technol 1993;15:326–333.
49. Berka RM, Dunn-Coleman N, Ward M. Industrial enzymes form Aspergillus species. BioTechnology 1992;23:155–202.
50. Austin S, Bingham ET, Koegel RG, Mathews DE, Shahan MN, Straub RJ, Burgess RR. An overview of a feasibility study for the production of industrial enzymes in transgenic alfalfa. Ann NY Acad Sci 1994;721:234–244.
51. Arnold FH. Engineering proteins for nonnatural environments. FASEB J 1993;7:744–749.
52. Wright A, Shin SU, Morrison SL. Genetically engineered antibodies: progress and prospects. Crit Rev Immunol 1992;12:125–168.
53. Sandhu JS. Protein engineering of antibodies. Crit Rev Biotechnol 1992;12:437–462.

54. Eppard PJ, Bentle LA, Violand BN, Ganguli S, Hintz RL, Kung Jr L, Krivi GG, Lanza GM. Comparison of the galactopoietic response to pituitary-derived and recombinant-derived variants of bovine growth hormone. J Endocrinol 1992;132:47–56.

55. Grings EE, deAvila DM, Eggert RG, Reeves JJ. Conception rate, growth, and lactation of dairy heifers treated with recombinant somatotropin. J Dairy Sci 1990;73:73–77.

56. Kato Y, Shimokawa N, Kato T, Hirai T, Yoshihama K, Kawai H, Hattori M, Ezashi T, Shimogori Y, Wakabayashi K. Porcine growth hormone: molecular cloning of cDNA and expression in bacterial and mammalian cells. Biochim Biophys Acta 1990;1048:290–293.

57. Juskevich JC, Guyer CG. Bovine growth hormone: human food safety evaluation. Science 1990;249:875–884.

58. Chua N-H. Plant biotechnology. Editorial overview. Curr Opin Biotechnology 1994;5:115–116.

59. Cioffi JA, Wagner TE. Application of biotechnology and transgenic animals toward the study of growth hormone. Am J Clin Nutr 1993;58(Suppl 2):296S-288S.

60. Hall PA, Lemoine NR. Models of pancreatic cancer. Cancer Surv 1993;16:135–155.

61. Doetschman T, Shull M, Kier A, Coffin JD. Embryonic stem cell model systems for vascular morphogenesis and cardiac disorders. Hypertension 1993;22:618–629.

62. Dorin JR, Dickinson P, Alton EW, Smith SN, Geddes DM, Stevenson BJ, Kimber WL, Fleming S, Clarke AR, Hooper ML et al. Cystic fibrosis in the mouse by targeted insertional mutagenesis. Nature 1992;359:211–215.

63. Snouwaert JN, Brigman KK, Latour AM, Malouf NN, Boucher RC, Smithies O, Koller BH. An animal model for cystic fibrosis made by gene targeting. Science 1992;257:1083–1088.

64. Tybulewicz VL, Tremblay ML, LaMarca ME, Willemsen R, Stubblefield BK, Winfield S, Zablocka B, Sidransky E, Martin BM, Huang SP et al. Animal model of Gaucher's disease from targeted disruption of the mouse gluco-cerebrosidase gene. Nature 1992;357:407–410.

65. Brunke M, Deckwer WD, Frischmuth A, Horn JM, Lunsdorf H, Rhode M, Rohricht M, Timmis KN, Weppen P. Microbial retention of mercury from waste streams in a laboratory column containing merA gene bacteria. FEMS Microbiol Rev 1993;11:145–152.

66. Kuhad RC, Singh A. Lignocellulose biotechnology: Current and future prospects. Crit Rev Biotechnol 1993;13:151–172.

67. Gadd GM, White C. Bioremediation. Microbial treatment of metal pollution - a working biotechnology? Trends Biotechnol 1993;11:353–359.

68. Goldstein MA, Doi R. Prokaryotic promoters in biotechnology. In: El-Gewely MR (ed) Biotechnology Annual Review, vol 1. Amsterdam: Elsevier, 1995;105–128.

69. Barinaga M. Ribozymes: killing the messenger. Science 1993;262:1512–1514.

70. Kobayashi H, Dorai T, Holland JF, Ohnuma T. Reversal of drug sensitivity in multidrug-resistant tumor cells by an MDR (PGY1) ribozyme. Cancer Res 1994;54:1271–1275.

71. Kashani-Sabet M, Funato T, Florenes VA, Fodstad O, Scanlon KJ. Suppression of the neoplastic phenotype in vivo by an anti-ras ribozyme. Cancer Res 1994;54:900–902.

72. Ouellette RP, Cheremisinoff PN. Essentials of Biotechnology. Ch 4, Chemicals From Plants. Lancaster, Basel: Technomic Publishing Co Inc 1985;57–63.

73. Frey NM. Introduction. In: Fraley RT, Frey NM, Schell J (eds) Genetic Improvements of Agriculturally Important Crops. Current Communication in Molecular Biology. Cold Spring Harbor Laboratory, 1988;1–2.

74. Martin GB, Ganal MW, Tanksley SD. Construction of a yeast artificial chromosome library of tomato and identification of cloned segments linked to two disease resistance loci. Mol Gen Genet 1992;233:25–32.

75. Kochert G. Restriction fragment length polymorphism in plants and its implications. Subcell Biochem 1991;17:167–190.

76. Wang G-L, Wing RA, Paterson AH. PCR amplification form single seeds, facilitating DNA marker-assisted breeding. Nucleic Acid Res 1993;21:2527.

58

77. Moser H, Lee M. RFLP variation and genealogical distance, multivariate distance, heterosis, and genetic variance in oats. Theor Appl Genet 1994;87:947–956.
78. Stuber CW, Lincoln SE, Wolff DW, Helentjaris T, Lander ES. Identification of genetic factors contribution to heterosis in a hybrid from two elite maize inbred lines using molecular markers. Genetics 1992;132:823–839.
79. Negrutiu I, Hinnisdaels S, Cammaerts D, Cherdshewasart W, Gharti-Chhetri CG et al. Plant protoplasts as genetic tool: selectable markers for developmental studies. Int J Dev Biol 1992;36:73–84.
80. Morrison RA, Evans DA, Fan Z. Haploid plants from tissue culture. Application in crop improvement. Subcell Biochem 1991;17:53–72.
81. Evans DA. Somaclonal variation-genetic basis and breeding applications. Trends Genet 1989;5:46–50.
82. Comai L. Impact of plant genetic engineering on foods and nutrition. Annu Rev Nutr 1993;13:191–215.
83. Hooykaas PJ, Schilperort RA. *Agrobacterium* and plant genetic engineering. Plant Mol Biol 1992;19:15–38.
84. Wang Z-u, Takamizo T, Iglesias VA, Osusky M, Nagel J, Potrykus I, Spangenberg G. Transgenic plants of tall fescue (*Festuca Arundinacea* Schreb.) obtained by direct gene transfer to protoplasts. BioTechnology 1992:10:691–696.
85. Vasil V, Castillo AM, Fromm ME, Vasil IK. Herbicide resistant fertile transgenic wheat plants obtained by microprojectile bombardment of regenerable embryogenic callus. BioTechnology 1992:10:667–674.
86. Sanford JC. Biolistic plant transformation. Physiol Plant 1990;79:205–209.
87. Horsch RB. Commercialization of genetically engineered crops. Philos Trans R Soc London Ser B Biol Sci 1993;342:287–291.
88. Baulcombe D. Novel strategies for engineering virus resistance in plants. Curr Opin Biotechnol 1994;5:117–124.
89. Martin GB, Brommonschenkel SH, Chunwongase J, Frary A, Ganal MW et al. Map-based cloning of a protein kinase gene conferring disease resistance in tomato. Science 1993;262:1432.
90. Koziel MG, Beland GL, Bowman C, Carozzi NB, Crenshaw R, Crossland L, Dawson J, Desai N, Hill M, Kadwell S, Launis K, Lewis K, Maddox D, McPherson K, Meghji MR, Merlin E, Rhodes R, Warren GW, Wright M, Evola SV. Field performance of elite transgenic maize plants expressing an insecticidal protein derived from *Bacillus thuringiensis*. BioTechnology 1993;11:194–200.
91. Fraley R. Sustaining the food supply. BioTechnology 1992;10:40–43.
92. Fichtner K, Quick WP, Schulze E-D, Mooney HA, Rodermel SR, Bogorad L, Stitt M. Decreased ribulose-1,5-bisphosphate carboxylase-oxygenase in transgenic tobacco transformed with antisense rbcS V. Relationship between photosynthetic rate, storage ability, biomass allocation and vegetative plant growth at three different nitrogen supplies. Planta 1993;190:1–9.
93. Stitt M. Manipulation of carbohydrate partitioning. Curr Opin Biotechnol 1994;5:137–143.
94. Kinny A. Genetic modification of the storage lipids of plants. Curr Opin Biotechnol 1994;5:144–151.
95. Knauf VC. Progress in the cloning of genes for plant storage lipids biosynthesis. In: Stelow JK (ed) Genetic Engineering, vol 15. Plenum Press 1993;149–164.
96. Savelkoul FH, Van der Poel AF, Tammingara S. The presence and inactivation of trypsin inhibitors, tannins, lectins and amylase inhibitors in legume seeds during germination. A Review. Plant Foods Hum Nutr 1992;42:71–85.
97. Theologis A. Control of ripening. Plant Gene Expression Center, Albany, USA. Curr Opin Biotechnol 1994;5:152–157.
98. Nordvåg BY, Riise HMF, Husby G, Nilsen I, El-Gewely R. Direct use of blood in PCR. In: Sarkar G (ed) Methods in Neurosciences, vol 26, ch 2. San Diego, CA: Academic Press, 1995;15–25.
99. Chehab FF. Molecular diagnostics: past, present and future. Hum Mutat 1993;2:331–337.
100. Nordvåg BY, Ranløv I, Riise HMF, Husby G, El-Gewely R. Retrospective molecular detection of

transthyretin Met 111 mutation in a Danish kindred with familial amyloid cardiomyopathy, using DNA from formalin-fixed and paraffin-embedded tissues. Hum Genet 1993;92:265–268.

101. Martin JB. Molecular genetics in neurology. Ann Neurol 1993;34:757–773.

102. Jaworsky C. The molecular diagnosis of cutaneous infection. J Cutan Pathol 1993;20:508–512.

103. Barnes WM. PCR amplification of up to 35-kb DNA with high fidelity and high yield from λ bacteriophage templates. Proc Natl Acad Sci USA 1994;91:2216–2220.

104. Kolls JK, Nelson S, Summer WR. Recombinant cytokines and pulmonary host defense. Am J Med Sci 1993;306:330–335.

105. Mannucci PM. Modern treatment of hemophilia: from shadows towards the light. Thromb Haemostasis 1993;70:17–23.

106. Helm RE, Gold JP, Rosengart TK, Zelano JA, Isom OW, Kriger KH. Erythropoietin in cardiac surgery. Department of Cardiothoracic Surgery, New York Hospital-Cornell M Center, NY 10021. J Card Surg 1993;8:579–606.

107. Metcalf D. Thrombopoietin - at last. Nature 1994;369:519–520.

108. Gordon MS, Hoffman R. Growth factors affecting human thrombocytopoiesis: potential agents for the treatment of thrombocytopenia. Blood 1992;80:302–307.

109. de Sauvage FJ, Hass PE, Spencer SD, Malloy BE, Gurney AL, Spencer SA, Darbonne WC, Henzel WJ, Wong SC, Kuang W-J, Oles KJ, Hultgren B, Solberg Jr LA, Goeddel DV, Eaton DL. Stimulation of megakaryocytopoiesis and thrombopoiesis by the c-Mpl ligand. Nature 1994;369:533–538.

110. Lok S, Kaushansky K, Holly RD, Kuijper JL, Lofton-Day CE, Oort PJ, Grant FJ, Heipel MD, Burkhead SK, Kramer JM, Bell LA, Sprencher CA, Blumberg H, Johnson R, Prunkard D, Ching AFT, Mathewes SL, Bailey MC, Forstrom JW, Buddle MM, Osborn SG, Evans SJ, Sheppard PO, Presnell SR, O'Hara PJ, Hagen FS, Roth GJ, Foster DC. Cloning and expression of murine thrombopoietin cDNA and stimulation of platelet production in vivo. Nature 1994;369:565–568.

111. Kaushansky K, Lok S, Holly RD, Broudy VC, Lin N, Bailey MC, Forstrom JW, Buddle MM, Oort PJ, Hagen FS, Roth GJ, Papyannopoulou T, Foster DC. Promotion of megakaryocyte progenitor expansion and differentiation by the c-Mpl ligand thrombopoietin. Nature 1994;369:568–571.

112. Wendling F, Maraskovsky E, Debili N, Florindo C, Teepe M, Titeux M, Methia N, Breton-Gorius J, Cosman D, Vainchenker W. c-Mpl ligand is a humoral regulator of megakaryocytopoiesis. Nature 1994;369:571–574.

113. Kolata, G. Blood hormone discovery seen aiding cancer patients. June 17, No 34617 (1994) 1 and 4.

114. Cipolla D, Gonda I, Shire SJ. Characterization of aerosols of human recombinant deoxyribonuclease (rhDNase) generated by yet nebulizers. Pharm Res 1993;11:491–496.

115. Ryser S, Weber M. Genetic Engineering: What is happening at Roche. Basel, Switzerland: Editiones Roche, F. Hoffmann-La Roche Ltd, 1991.

116. Cohn J. The HIV vaccine paradox. Science 1994;264:1072–1074.

117. Committee on infectious diseases, American Academy of Pediatrics. Universal hepatitis B immunization. Pediatrics 1992;89:195–199.

118. Harford N, Cabezon T, Colau B, Delisse AM, Rutgers T, De-Wilde M. Construction and characterization of a Saccharomyces cerevisiae strain (RITA376) expressing hepatitis B surface antigen. Postgrad Med J 1987;63:65–70.

119. Petre J, Van-Wijnendaele F, De-Neys B, Conrath K, Van-Opstal O, Hauser P, Rutgers T, Cabezon T, Capiau C, Harford N. Development of a hepatitis B vaccine from transformed yeast cells. Postgrad Med J 1987;63:73–81.

120. (a) Hilleman MR. Yeast recombinant hepatitis B vaccine. Infection 1987;15:5–9. (b)Committe on Infectious Diseases. Universal hepatitis B immunization. Pedriatics 1992;89:795–800.

121. Payton CD, Scarisbrick DA, Sikotra S, Flower AJ. Vaccination against hepatitis B: comparison of intradermal and intramuscular administration of plasma derived and recombinant vaccines. Epidemiol Infect 1993;110:177–180.

122. Greenberg DP. Pediatric experience with recombinant hepatitis B vaccines and relevant safety and immunogenicity studies. Pediatr Infect Dis J 1993;12:439–445.

123. Yap I, Guan R, Chan SH. Recombinant DNA hepatitis B vaccine containing Pre-S components of the HBV coat protein - a preliminary study on immunogenicity. Vaccine 1992;10:439–442.

124. Pastan I, FitzGerald DJ. Recombinant toxins for cancer treatment. Science 1991;254:1173–1177.

125. Chaudhary VK, Queen C, Junghans RP, Waldmann T, FitzGerald DJ, Pastan I. A recombinant immunotoxin consisting of two antibody variable domains fused to Pseudomonas exotoxin. Nature 1989;339:394–397.

126. Pai LH, Pastan I. The use of immunotoxins for cancer therapy. Eur J Cancer 1993;29A:1606–1609.

127. Wang Q-C, Pai LH, Debinski W, FitzGerald DJ, Pastan I. Preparation of polyethylene glycol modified forms of TGFα-PE40 that are cytotoxic and have a markedly prolonged survival in the circulation of mice. (Abstract) Third International Symposium on Immunotoxins, Orlando, FL, 1992.

128. Talwar GP, Singh O, Pal R, Chatterjee N, Nupadhyay SN, Kaushic C, Garg S, Kaur R, Singh M, Chandrasekhar S, Gupta A. A birth control vaccine is on the horizon for family planning. Ann Med 1993;25:207–212.

129. Fishel SB, Edwards RG, Evans CJ. Human chorionic gonadotropin secreted by preimplantation embryos cultured in vitro. Science 1984;223:816–818.

130. Hearn JP, Gidley-Baird AA, Hodges JK, Summers PM, Wibley GE. Embryonic signals during the preimplantation period in primates. J Reprod Fertil 1983;36(suppl):49–58.

131. Dirnhoffer S, Klieber R, de Leeuw R, Bidart J-M, Merz WE, Wick G, Berger P. Functional and immunological relevance of the COOH-terminal extension of human chorionic gonadotropin β: implications for the WHO control vaccine. FASEB J 1993;7:1381–1385.

132. Wright RM, Suri AK, Kornreich B, Flickinger CJ, Herr JC. Cloning and characterization of the gene coding for the human acrosomal protein SP-10. Biol Reprod 1993;49(2):316–325.

133. Alteri E, Bold G, Cozens R, Faessler A, Klimkait T, Lang M, Lazdins J, Poncioni B, Roesel JL, Schneider P et al. CGP 53437, an orally bioavailable inhibitor or human immunodeficiency virus type 1 protease with potent antiviral activity. Antimicrob Agents Chemother 1993;37:2087–2092.

134. Lang M, Roesel J. HIV-1 protease inhibitors: development, status, and potential role in the treatment of AIDS. Arch Pharm Weinheim 1993;326:921–924.

135. Perno CF, Bergamini A, Pesce CD, Milanese G, Capozzi M, Aquaro S, Thaisrivongs S, Tarpley WG, Zon G, D'Agostini C et al. Inhibition of the protease of human immunodeficiency virus blocks replication and infectivity of the virus in chronically infected macrophages. J Infect Dis 1993;168:1148–1156.

136. Meek TD. Inhibitors of HIV-a protease. J Enzym Inhib 1992;6:65–98.

137. Bugg CE, Carson WM, Montgomery JA. Drugs by design. Sci Am 1993;269:92–98.

138. Goel R, Beard WA, Kumar A, Cases-Finet JR, Strub MP, Stahl SJ, Lewis MS, Bebenek K, Becerra SP, Kunkel TA et al. Structure/function studies of HIV-1(1) reverse transcriptase: dimerization-defective mutant L289K. Biochemistry 1993;32:13012–13018.

139. Mathez D, Schinazi RF, Liotta DC, Leibowitch J. Infectious amplification of wild-type human immunodeficiency virus from patients' lymphocytes and modulation by reverse transcriptase inhibitors in vitro. Antimicrob Agents Chemother 1993;37:2206–2211.

140. Buttice G, Kurkin M. A polyomavirus enhancer A-binding protein-3 site and Ets-2 protein have a major role in the 12–0-tetradecanoylphorbol-13-acetate response of the human stromelysins gene. J Biol Chem 1993;268:7196–7204.

141. Brinckerhoff CE, Sirum-Connolly KL, Karmilowicz MJ, Auble DT. Expression of stromelysin and stromelysin-2 in rabbit and human fibroblasts. Matrix 1992;(Suppl 1):165–175.

142. Ye QZ, Johnson LL, Nordan I, Hupe D, Hupe L. A recombinant human stromelysin catalytic domain identifying tryptophan derivatives as human stromelysin inhibitors. J Med Chem 1994;37:206–209.

143. Griffin LC, Toole JJ, Leung LL. The discovery and characterization of a novel nucleotide-based thrombin inhibitor. Gene 1993;137:25–31.
144. Wiley RA, Rich DH. Peptidomimetics derived from natural products. Med Res Rev 1993;13:327–384.
145. Miceli RM, DeGraaf ME, Fischer HD. Two-stage selection of sequences from a random phage display library delineates both core residues and permitted structural range within an epitope. J Immunol Methods 1994;167:279–287.
146. Koivanen E, Wang B, Ruoslahti E. Isolation of a highly specific ligand for the alpha 5 beta i integrin from a phage display library. J Cell Biol 1994;124:373–380.
147. Roberts D, Guegler K, Winter J. Antibody as a surrogate receptor in the screening of a phage display library. Gene 1994;128:67–69.
148. McLafferty MA, Kent RB, Ladner RC, Markland W. M13 bacteriophage displaying disulfide-constrained microproteins. Gene 1993;128:29–36.
149. DeGraff ME, Miceli RM, Mott JE, Fischer HD. Biochemical diversity in a phage display library of random decapeptides. Gene 1993;128:13–17.
150. Brenner S, Lerner RA. Encoded combinatorial chemistry. Proc Natl Acad Sci USA 1992;89:5381–5383.
151. Scott JK, Smith GP. Searching for peptide ligands with an epitope library. Science 1990;249:386–390.
152. Jayawickreme CK, Graminski GF, Quillan JM, Lerner MR. Creation and functional screening of a multi-use peptide library. Proc Natl Acad Sci USA 1994;91:1614–1618.
153. Salmon SE, Lam KS, Lebl M, Kandola A, Khattri PS, Wade S, Patek M, Kocis P, Krchnak V, Thorpe D et al. Discovery of biologically active peptides in random libraries: solution-phase testing after staged orthogonal release from resin beads. Proc Natl Acad Sci USA 1993;90:11708–11712.
154. Ecker DJ, Vickers TA, Hanecak R, Driver V, Anderson K. Rational screening of oligonucleotide combinatorial libraries for drug discovery. Nucleic Acid Res 1993;21:1853–1856.
155. Nissim A, Hoogenboom HR, Tomlinson IM, Flynn G, Midgley C, Lane D, Winter G. Antibody fragments from a "single pot" phage display library as immunochemical reagents. EMBO J 1994;13:692–698.
156. Alper J. Drug discovery on the assembly line. Science 1994;264:1399–1401.
157. Pease AC, Solas D, Sullivan EJ, Cronin MT, Holms CP, Fodor SPA. Light- generated oligonucleotide arrays for DNA sequence analysis. Proc Natl Acad Sci USA 1994;91:5022–5026.
158. Chen C, Ahlberg-Randall LA, Miller RB, Jones AD, Kurth MJ. "Analogous" organic synthesis of small-compound libraries: validation of combinatorial chemistry in small- molecule synthesis. J Am Chem Soc 1994;116:2661–2662.
159. Jacobs JW, Fodor SPA. Combinatorial chemistry-applications of light-directed chemical synthesis. Trends Biotechnol 1994;12:19–26.
160. Mitani K, Clemens PR, Moseley AB, Caskey CT. Gene transfer therapy for heritable disease: cell and expression targeting. Philos Trans R Soc London Ser B 1993;339:217–224.
161. Woo SLC. Introductory remarks to the review series on gene therapy. Trends Genet 1994;10:111–112.
162. McKusick VA (ed). Mendelian Inheritance and Man. Baltimore, MD: Johns Hopkins University Press, 1991.
163. Morgan RW, Gelb Jr J, Pope CR, Sondermeijer PJ. Efficacy in chickens of a herpes virus of turkeys recombinant vaccine containing the fusion gene of Newcastle disease virus: onset of protection and effect of maternal antibodies. Avian Dis 1993;37:1032–1040.
164. O'Malley Jr BE, Ledley FD. Somatic gene therapy. Methods for the present and future. Arch Otolaryngol Head Neck Surg 1993;119:1100–1107.
165. Atkinson J, Martin R. Mutations to nonsense codon in human genetic disease: implications for gene therapy by nonsense suppressor tRNAs. Nucleic Acids Res 1994;22:1327–1334.

62

166. Hilman BC, Sorensen RU. Management options: SCIDS with adenosine deaminase deficiency. Ann Allergy 1994;72:395–403.
167. Kaptein LC, Einerhand MP, Braakman E, Valerio D, van-Beusechem VW. Bone marrow gene therapy for adenosine deaminase deficiency. Immunodeficiency 1993;4:335–345.
168. Noonan NA, Senner AM. Gene therapy techniques in the treatment of adenosine deaminase-deficiency severe combined immune deficiency syndrome. J Perinat Neonatal Nurs 1994;7:65–78.
169. Setoguchi Y, Jaffe HA, Chu CS, Crystal RG. Intraperitoneal in vivo gene therapy to deliver alpha 1-antitrypsin to the systemic circulation. Am J Respir Cell Mol Biol 1994;10:369–377.
170. Rettinger SD, Kennedy SC, Wu X, Saylors RL, Hafenrichter DG, Flye MW, Ponder KP. Liver-directed gene therapy: quantitative evaluation of promoter elements by using in vivo retroviral transduction. Proc Natl Acad Sci USA 1994;91:1460–1464.
171. Canonico AE, Conary JT, Meyrick BO, Brigham KL. Aerosol and intravenous transfection of human alpha 1-antitrypsin gene to lungs of rabbits. Am J Respir Cell Mol Biol 1994;10:24–29.
172. Kolodka TM, Finegold M, Woo SL. Hepatic gene therapy: efficient retroviral-mediated gene transfer into rat hepatocytes in vivo. Somat Cell Mol Genet 1993;19:491–497.
173. Demarquoy J, Herman GE, Lorenzo I, Trentin J, Beaudet AL, O'Brien WE. Long-term expression of human argininosuccinate synthetase in mice following bone marrow transplantation with retrovirus-transduced hematopoietic stem cells. Hum Gene Ther 1992;3:3–10.
174. Yorifuji T, Wilson RW, Beaudet AL. Retroviral mediated expression of CD18 in normal and deficient human bone marrow progenitor cells. Hum Mol Genet 1993;2:1443–1448.
175. Krauss JC, Bond LM, Todd III RF, Wilson JM. Expression of retroviral transduced human CD18 in murine cells: an vitro model of gene therapy for leukocyte adhesion deficiency. Hum Gene Ther 1991;2:221–228.
176. Welsh MJ, Smith AE, Zabner J, Rich DP, Graham SM, Gregory RJ, Pratt BM, Moscicki RA. Cystic fibrosis gene therapy using an adenovirus vector: in vivo safety an efficacy in nasal epithelium. Hum Gene Ther 1994;5:209–219.
177. Simon RH, Engelhardt JF, Yang Y, Zepeda M, Weber-Pendleton S, Grossman M, Wilson JM. Adenovirus-mediated transfer of the CFTR gene to lung of nonhuman primates toxicity study. Hum Gene Ther 1993;4:771–780.
178. Engelhardt JF, Simon RH, Yang Y, Zepeda M, Weber-Pendleton S, Doranz B, Grossman M, Wilson JM. Adenovirus-mediated transfer of the CFTR gene to lung of nonhuman primates biological efficacy study. Hum Gene Ther 1993;4:759–769.
179. Bout A, Perricaudet M, Baskin G, Imler JL, Scholte BJ, Pavirani A, Valerio D. Lung gene therapy: in vivo adenovirus-mediated gene transfer to rhesus monkey airway epithelium. Hum Gene Ther 1994;5:3–10.
180. Dickson G, Dunckley M. Human dystrophin gene transfer: genetic correction of dystrophin deficiency. Mol Cell Biol Hum Dis Ser 1993;3:283–302.
181. Fardeau M. Cellular therapy and gene therapy: perspectives in neuromuscular pathology. Pathol Biol Paris 1993;41:681–685.
182. Clemens PR, Caskey CT. Gene therapy prospects for duchenne muscular dystrophy. Eur Neurol 1994;34:181–185.
183. Lozier JN, Brinkhous KM. Gene therapy and the hemophilias. J Am Med Assoc 1994;271:47–51.
184. Kay MA, Landen CN, Rothenberg SR, Taylor LA, Leland F, Wiehle S, Fang B, Bellinger D, Finegold M, Thompson AR, Reed M, Brinkhous KM, Woo SLC. In vivo hepatic gene therapy: complete albeit transient correction of factor IX deficiency in hemophilia B dogs. Proc Natl Acad Sci USA 1994;91:2353–2357.
185. Kurachi K, Yao SN. Gene therapy of hemophilia B. Thromb Haemostasis 1993;70:193–197.
186. Thompson AR, Palmer TD, Lynch CM, Miller AD. Gene transfer as an approach to cure patients with hemophilia A or B. Curr Stud Hematol Blood Transf 1991;58:59–62.
187. Occhiodoro T, Hopwood JJ, Morris CP, Anson DS. Correction of alpha-L-fucosidase deficiency in

fucosidosis fibroblasts by retroviral vector-mediated gene transfer. Hum Gene Ther 1992;3:365–369.

188. Krall WJ, Challita PM, Perlmutter LS, Skelton DC, Kohn DB. Cells expressing human glucocerebrosidase from a retroviral vector repopulate macrophages and central nervous system microglia after murine bone marrow transplantation. Blood 1994;83:2737–2748.

189. Aran JM, Gottesman MM, Pastan I. Drug-selected coexpression of human glucocerebrosidase and P-glycoprotein using a bicistronic vector. Proc Natl Acad Sci USA 1994;91:3176–3180.

190. Xu L, Stahl SK, Dave HP, Schiffmann R, Correll PH, Kessler S, Karlsson S. Correction of the enzyme deficiency in hematopoietic cells of Gaucher patients using a clinically acceptable retroviral supernatant transduction protocol. Exp Hematol 1994;22:223–230.

191. Naffakh N, Bohl D, Salvetti A, Moullier P, Danos O, Heard JM. Gene therapy for lysosomal disorders. Nouv Rev Fr Hematol 36(Suppl 1):S11–S16.

192. Naffakh N, Pinset C, Montarras D, Pastoret C, Danos O, Heard JM. Transplantation of adult-derived myoblasts in mice following gene transfer. Neuromuscul Disord 1993;3:413–417.

193. Wolfe JH, Sands MS, Barker JE, Gwynn B, Rowe LB, Vogler CA, Birkenmeier EH. Reversal of pathology in murine mucopolysaccharidosis type VII by somatic cell gene transfer. Nature 1992;360:749–753.

194. Bender MA, Gelinas RE, Miller AD. A majority of mice show long-term expression of a human beta-globin gene after retrovirus transfer into hematopoietic stem cells. Mol Cell Biol 1989;9:1426–1434.

195. Dzierzak EA, Papayannopoulou T, Mulligan RC. Lineage-specific expression of a human beta-globin gene in murine bone marrow transplant recipients reconstituted with retrovirus-transduced stem cells. Nature 1988;331:35–41.

196. Karlsson S, Bodine DM, Perry L, Papayannopoulou T, Nienhuis AW. Expression of the human beta-globin gene following retroviral-mediated transfer into multipotential hematopoietic progenitors of mice. Proc Natl Acad Sci USA 1988;85:6062–6066.

197. Forget BG. The pathophysiology and molecular genetics of beta thalassemia. Mt. Sinai J Med 1993;60:95–103.

198. Anson DS, Bielicki J, Hopwood JJ. Correction of mucopolysaccharidosis type I fibroblasts by retroviral-mediated transfer of the human alpha-L-iduronidase gene. Hum Gene Ther 1992;3:371–379.

199. Kozarsky KF, McKinley DR, Austin LL, Raper SE, Stratford-Perricaudet LD, Wilson JM. In vivo correction of low density lipoprotein receptor deficiency in the Watanabe heritable hyperlipidemic rabbit with recombinant adenoviruses. J Biol Chem 1994;269:13695–13702.

200. Raper SE, Wilson JM. Cell transplantation in liver-directed gene therapy. Cell Transplant 1993;2:381–400.

201. Kozarsky K, Grossman M, Wilson JM. Adenovirus-mediated correction of the genetic defect in hepatocytes from patients with familial hypercholesterolemia. Somat Cell Mol Genet 1993;19:449–458.

202. Morsy MA, Alford EL, Bett A, Graham FL, Caskey CT. Efficient adenoviral-mediated ornithine transcarbamylase expression in deficient mouse and human hepatocytes. J Clin Invest 1993;92:1580–1586.

203. Cristiano RJ, Smith LC, Woo SL. Hepatic gene therapy: adenovirus enhancement of receptor-mediated gene delivery and expression in primary hepatocytes. Proc Natl Acad Sci USA 1993;90:2122–2126.

204. Shedlovsky A, McDonald JD, Symula D, Dove WF. Mouse models of human phenylketonuria. Genetics 1993;134:1205–1210.

205. Liu TJ, Kay MA, Darlington GJ, Woo SL. Reconstitution of enzymatic activity in hepatocytes of phenylalanine hydroxylase-deficient mice. Somat Cell Mol Genet 1992;18:89–96.

206. Cournoyer D, Caskey CT. Gene therapy of the immune system. Annu Rev Immunol 1993;11:297–329.

64

207. Foresman MD, Nelson DM, McIvor RS. Correction of purine nucleoside phosphorylase deficiency by retroviral-mediated gene transfer in mouse S49 T cell lymphoma: a model for gene therapy of T cell immunodeficiency. Hum Gene Ther 1992;3:625–631.
208. Suchi M, Dinur T, Desnick RJ, Gatt S, Pereira L, Gilboa E, Schuchman EH. Retroviral-mediated transfer of the human acid sphingomyelinase cDNA: correction of the metabolic defect in cultured Niemann-Pick disease cells. Proc Natl Acad Sci USA 1992;89:3227–3231.
209. Dinur T, Schuchman EH, Fibach E, Dagan A, Suchi M, Desnick RJ, Gatt S. Toward gene therapy for Niemann-Pick disease (NPD): separation of retrovirally corrected and noncorrected NPD fibroblasts using a novel fluorescent sphingomyelin. Hum Gene Ther 1992;3:633–639.
210. Culver KW, Blaese RM. Gene therapy for cancer. Trends Genet 1994;10:174–178.
211. Culver KW. Clinical applications of gene therapy for cancer. Clin Chem 1994;40:622–628.
212. Sanda MG, Ayyagari SR, Jaffee EM, Epstein JI, Clift SL, Cohen LK, Dranoff G, Pardoll DM, Mulligan RC, Simons JW. Demonstration of a rational strategy for human prostate cancer gene therapy. J Urol 1994;151:622–628.
213. Macri P, Gordon JW. Delayed morbidity and mortality of albumin/SV40 T-antigen transgenic mice after insertion of an alpha-fetoprotein/herpes virus thymidine kinase transgene and treatment with ganciclovir. Hum Gene Ther 1994;5:175–182.
214. Moritz D, Wels W, Mattern J, Groner B. Cytotoxic T lymphocytes with a grafted recognition specificity for ERBB2-expressing tumor cells. Proc Natl Acad Sci USA 1994;91:4318–4322.
215. Friedmann T. Gene therapy of cancer through restoration of tumor-suppressor functions? Cancer 1992;70:1810–1817.
216. Lauret E, Riviere I, Rousseau V, Vieillard V, De-Maeyer-Guignard J, De-Maeyer E. Development of methods for somatic cell gene therapy directed against viral diseases, using retroviral vectors carrying the murine or human interferon-beta. Hum Gene Ther 1993;4:567–577.
217. Chatterjee S, Johnson PR, Wong Jr KK. Dual-target inhibition of HIV-1 in vitro by means of an adeno-associated virus antisense vector. Science 1992;258:1485–1488.
218. Ohno T, Gordon D, San H, Pompili VJ, Imperiale MJ, Nabel GJ, Nabel EG. Gene therapy for vascular smooth muscle cell proliferation after arterial injury. Science 1994;265:781–784.
219. Nabel EG, Pomili VJ, Plautz GE, Nabel GJ. Gene transfer and vascular disease. Cardiovasc Res 1994;28:445–455.
220. Takeshita S, Gal D, Leclerc G, Pickering JG, Riessen R, Weir L, Isner JM. Increased gene expression after liposome-mediated arterial gene transfer associated with intimal smooth muscle cell proliferation. In vitro and in vivo findings in a rabbit model of vascular injury. J Clin Invest 1994;93:652–661.
221. Patel PI. Identification of disease genes and somatic gene therapy: an overview and prospects for the aged. J Gerontol 1993;48:B80–85.
222. Mastrogeli A, Pavirani A, Leccocq JP, Crystal RG. diversity of airway epithelial cell targets for in vivo recombinant adenovirus-mediated gene therapy. J Clin Invest 1993;91:225–234.
223. Setoguchi Y, Jaffe HA, Danel C, Crystal RG. Ex vivo and in vivo gene transfer to the skin using replication-deficient recombinant adenovirus vectors. J Invest Dermatol 1994;102:415–421.
224. Belldegrun A, Tso CL, Sakata T, Duckett T, Brunda MJ, Barsky SH, Chai J, Kaboo R, Lavey RS, McBride WH et al. Human renal carcinoma line transfected with interleukin-2 and/or interferon alpha gene(s): implications for liver cancer vaccines. J Natl Cancer Inst 1993;85:207–216.
225. Tani K. Implantation of genetically manipulated fibroblasts into mice as a model of gene therapy-supplementations of human granulocyte colony-stimulating factor (hG-CSF) and interferon-alpha (IFN-alpha). Hum Cell 1991;4:25–32.
226. Weaver JC. Electroporation: a general phenomenon for manipulating cells and tissues. J Cell Biochem 1993;41:426–435.
227. Matthews KE, Mills GB, Horsfall W, Hack N, Skorecki K, Keating A. Bead transfection: rapid and efficient gene transfer into marrow stromal and other adherent mammalian cells. Exp Hematol 1993;21:697–702.

228. Alton EW, Middleton PG, Caplen NJ, Smith SN, Steel DM, Munkonge FM, Jeffery PK, Geddes DM, Hart SL, Williamson R et al. Non-invasive liposome-mediated gene delivery can correct the ion transport defect in cystic fibrosis mutant mice. Nat Genet 1993;5:135–142.

229. Morgan JE. Cell and gene therapy in Duchenne muscular dystrophy. Hum Gene Ther 1994;5:165–173.

230. Nabel GJ, Nabel EG, Yang Z-Y, Fox BA, Plautz GE, Gao X, Huang L, Shu S, Gordon D, Chang AE. Direct gene transfer with DNA-liposome complexes in melanoma: Expression, biologic activity, and lack of toxicity in humans. Proc Natl Acad Sci USA 1993;90:11307–11311.

231. Chen J, Gamous S, Takayanagi A, Shimizu N. A novel gene delivery system using EGF receptor-mediated endocytosis. FEBS Lett 1994;338:167–169.

232. Ferkol T, Kaetzel CS, Davis PB. Gene transfer into respiratory epithelial cells by targeting the polymeric immunoglobulin receptor. J Clin Invest 1993;92:2394–2400.

233. Ferkol T, Lindberg GL, Chen J, Perales JC, Crawford DR, Ratnoff OD, Hanson RW. Regulation of the phosphoenolpyruvate carboxykinase/human factor IX gene introduced into the livers of adult rats by receptor-mediated gene transfer. FASEB J 1993;7:1081–1091.

234. Harris CE, Agarwal S, Hu P, Wagner E, Curiel DT. Receptor-mediated gene transfer to airway epithelial cells in primary culture. Am J Respir Cell Mol Biol 1993;9:441–447.

235. Rossiter BJ, Stirpe NS, Caskey CT. Report of the MDA Gene Therapy Conference, Tucson, Arizona, September 27–28, 1991. Neurology 1992;42:1413–1418.

236. Gal D, Weir L, Leclerc G, Pickering JG, Hogan J, Isner JM. Direct myocardial transfection in two animal models. Evaluation of parameters affecting gene expression and percutaneous gene delivery. Lab Invest 1993;68:18–25.

237. Walsh CE, Liu JM, Miller JL, Nienhuis AW, Samulski RJ. Gene therapy for human hemoglobinopathies. Proc Soc Exp Biol Med 1993;204:289–300.

238. Flotte TR, Solow R, Owens RA, Afione S, Zeitlin PL, Carter BJ. Gene expression from adeno-associated virus vectors in airway epithelial cells. Am J Respir Cell Mol Biol 1992;7:349–356.

239. Walters L. Human gene therapy: ethics and public policy. Hum Gene Ther 1991;2:115–122.

240. Wivel NA, Walters L. Germ-line gene modification and disease prevention: some medical and ethical perspectives. Science 1993;262:533–538.

241. Carmen IH. Human gene therapy: a biopolitical overview and analysis. Hum Gene Ther 1993;4:187–193.

242. Hennighausen L. The mammary gland as a bioreactor: production of foreign proteins in milk. Protein Expr Purif 1990;1:3–8.

243. Hennighausen L, Ruiz L, Wall R. Transgenic animals production of foreign proteins in milk. Curr Opin Biotechnol 1990;1:74–78.

244. Pittius CW, Hennighausen L, Lee E, Westphal H, Nicols E, Vitale J, Gordon. A milk protein gene promoter directs the expression of human tissue plasminogen activator cDNA to the mammary gland in transgenic mice. Proc Natl Acad Sci USA 1988;85:5874–5878.

245. Janne J, Hyttinen JM, Peura T, Tolvanen M, Alhonen L, Halmekyto M. Transgenic animals as bioproducers of therapeutic proteins. Ann Med 1992;24:273–280.

246. Hansson L, Edlund M, Edlund A, Johansson T, Marklund SL, Fromm S, Stromqvist M, Tornell J. Expression and characterization of biologically active human extracellular superoxide dismutase in milk of transgenic mice. J Biol Chem 1994;269:5358–5363.

247. Wilmut I, Archibald AL, McClenaghan M, Simons JP, Whitelaw CB, Clark AJ. Production of pharmaceutical protein in milk. Experientia 1991;47:905–912.

248. Witelaw CB, Archibald AL, Harris S, McClenaghan M, Simons JP, Clark AJ. Targeting expression to the mammary gland: intronic sequences can enhance the efficiency of gene expression in transgenic mice. Transgenic Res 1991;1:3–13.

249. Wright G, Carver A, Cottom D, Reeves D, Scott A, Simons P, Wilmut I, Garner I, Colman A. High level expression of active human alpha-1-antitrypsin in the milk of transgenic sheep. BioTechnology 1991;9:830–834.

66

250. Denman J, Hayes M, O'Day C, Edmunds T, Bartlett C, Hirani S, Ebert KM, Gordon K, McPherson JM. Transgenic expression of a variant of human tissue-type plasminogen activator in goat milk: purification and characterization of the recombinant enzyme. BioTechnology 1991;9:839–843.
251. Cartè BK. Marine natural products as a source of novel pharmacological agents. Curr Opin Biotechnol 1993;4:275–279.
252. Rinehart KL. Secondary metabolites from marine organisms. Ciba Found Symp 1992;171:236–249.
253. Kaul PN. Drug molecules of marine origin. Prog Drug Res 1990;35:521–557.
254. Scheuer PJ. Some marine ecological phenomena: chemical basis and biomedical potential. Science 1990;248:173–177.
255. Hamamoto T, Takata N, Kudo T, Horikoshi K. Effect of temperature and growth phase on fatty acid composition of the psychrophilic *Vibrio* sp. strain no. 5710. FEMS Microbiol Lett 1994;119:77–82.
256. Trischman JA, Tapiolas DM, Jensen PR, Dwight R, Fenical W, McKee TC, Ireland CM, Stout TJ, Clardy J. Salinamides A and B: anti-inflammatory from a marine streptomycete. J Am Chem Soc 1994;116:757–758.
257. Gross M, Jaenicke R. Proteins under pressure. The influence of high hydrostatic pressure on structure, function and assembly of protein and protein complexes. Eur J Biochem 1994;221:617–630.
258. Welch TJ, Farewell A, Neidhardt FC, Bartlett DH. Stress response of *Escherichia coli* to elevated hydrostatic pressure. J Bacteriol 1993;175:7170–7177.
259. Trischman JA, Tapiolas DM, Jensen PR, Dwight R, Fenical W, McKee TC, Ireland CM, Stout TJ, Clardy J. Salinamides A and B: Anti-inflammatory depsipeptides from a marine streptomycete. J Am Chem Soc 1994;116:757–758.
260. Wiemer DF, Idler DD, Fenical W. Vidalols A and B, new anti-inflammatory bromophenols from the Caribbean marine red alga Vidalia obtusaloba. Experientia 1991;47:851–853.
261. Gil-Turnes MS, Hay ME, Fenical W. Symbiotic marine bacteria chemically defend crustacean embryos from a pathogenic fungus. Science 1989;246:116–118.
262. Burres NS, Sazesh S, Gunawardana GP, Clement JJ. Antitumor activity and nucleic acid binding properties of dercitin, a new acridine alkaloid isolated from a marine Dercitus species sponge. Cancer Res 1989;49:5267–5274.
263. Imamura N, Nishijima M, Adachi K, Sano H. Novel antimycin antibiotics, urauchimycins A and B, produced by marine actinomycete. J Antibiot Tokyo 1993;46:241–246.
264. Loya S, Tal R, Hizi A, Issacs S, Kashman Y, Loya Y. Hexaprenoid hydroquinones, novel inhibitors of the reverse transcriptase of human immunodeficiency virus type 1. J Natl Prod 1993;56:2120–2125.
265. Naruse N, Tenmyo O, Kobaru S, Hatori M, Tomita K, Hamagishi Y, Oki T. New antiviral antibiotics, kistamicins A and B. I. Taxonomy, production, isolation, physico-chemical properties and biological activities. J Antibiot Tokyo 1993;46:1804–1811.
266. Grant-Burgess J, Miyashita H, Sudo H, Matsunaga T. Antibiotic production by the marine photosynthetic bacterium Chromatium purpuratum NKPB 031704: localization of activity of the chromatophores. FEMS Microbiol Lett 1991;68:301–305.
267. Myrnes B, Johansen A. Recovery of lysozyme from scallop waste. Prep Biochem 1994;24:69–80.
268. Sugano Y, Terada I, Arita M, Noma M, Matsumoto T. Purification and characterization of a new agarase from a marine bacterium Vibrio sp. strain JT0107. Appl Environ Microbiol 1993;59:1549–1554.
269. Danaher RJ, Stein DC. Expression of cloned restriction and modification genes, hjaIRM from *Hyphomonas jannaschiana* in *Escherichia coli*. Gene 1990;89:129–132.
270. Mathur EJ, Adams MW, Callen WN, Cline JM. The DNA polymerase gene from the hyperthermophilic marine archaebacterium, *Pyrococcus furiosus*, shows sequence homology with alpha-like DNA polymerases. Nucleic Acids Res 1991;19:6952.

271. Bassler BL, Yu C, Lee YC, Roseman S. Chitin utilization by marine bacteria. Degradation and catabolism of chitin oligosaccharides by *Vibrio furnissii*. J Biol Chem 1991;266:24276–24286.

272. Arnosti C, Repeta DJ. Extracellular enzyme activity in anaerobic bacterial cultures: evidence of pullulanase activity among mesophilic marine bacteria. Appl Environ Microbiol 1994;60:840–846.

273. Paoletti E. Poxvirus recombinant vaccines. Ann NY Acad Sci 1990;590:309–325.

274. Garduno RA, Thornton JC, Kay WW. *Aeromonas salmonicida* grown in vivo. Infect Immun 1993;61:3854–3862.

275. Vaughan LM, Smith PR, Foster TJ. An aromatic-dependent mutant of the fish pathogen *Aeromonas salmonicida* is attenuated in fish and is effective as a live vaccine against the salmonid disease furunculosis. Infect Immun 1993;61:2172–2181.

276. Wong G, Kaattari SL, Christensen JM. Effectiveness of an oral enteric coated *vibrio* vaccine for use in salmonid fish. Immunol Invest 1992;21:353–364.

277. Oberg LA, Wirkkula J, Mourich D, Leong LJ. Bacterially expressed nucleoprotein of infectious hematopoietic necrosis virus augments protective immunity induced by the glycoprotein vaccine in fish. J Virol 1991;65:4486–4489.

278. Xu L, Mourich DV, Engelking HM, Ristow S, Arnzen J, Leong JC. Epitope mapping and characterization of the infectious hematopoietic necrosis virus glycoprotein, using fusion proteins synthesized in Escherichia coli. J Virol 1991;65:1611–1615.

279. Manning DS, Leong JC. Expression in *Escherichia coli* of the large genomic segment of infectious pancreatic necrosis virus. Virology 1990;179:16–25.

280. Thiry M, Lecocq-Xhonneux F, Dheur I, Renard A, de-Kinkelin P. Molecular cloning of the mRNA coding for the G protein of the viral haemorrhagic septicaemia (VHS) of salmonids. Vet Microbiol 1990;23:221–226.

281. Muller F, Lele Z, Varadi L, Menczel L, Orban L. Efficient transient expression system based on square pulse electroporation and in vivo luciferase assay of fertilized fish eggs. FEBS Lett 1993;324:27–32.

282. Ozato K, Wakamatsu Y, Inoue K. Medaka as a model of transgenic fish. Mol Mar Biol Biotechnol 1992;1:346–354.

283. Lu JK, Chen TT, Chrisman CL, Andrisani OM, Dixon JE. Integration, expression and germ-line transmission of foreign growth hormone genes in medaka (*Oryzias latipes*). Mol Mar Biol Biotechnol 1992;1:366–375.

284. Hew CL, Davies PL, Fletcher G. Antifreeze protein gene transfer in Atlantic salmon. Mol Mar Biol Biotechnol 1992;1:309–317.

285. Maclean N, Iyengar A, Rahman A, Sulaiman Z, Penman D. Transgene transmission and expression in rainbow trout and tilapia. Mol Mar Biol Biotechnol 1992;1:355–365.

286. Chen TT, Kight K, Lin CM, Powers DA, Hayat M, Chatakondi N, Ramboux AC, Duncan PL, Dunham RA. Expression and inheritance of RSVLTR-rtGH1 complementary DNA in the transgenic common carp, *Cyprinus carpio*. Mol Mar Biol Biotechnol 1993;2:88–95.

287. Houdebine LM, Chourrout D. Transgenesis in fish. Experientia 1991;15:891–897.

288. Muller F, Ivics Z, Erdelyi F, Papp T, Varadi L, Horvath L, Maclean N. Introducing foreign genes into fish eggs with electroporated sperm as a carrier. Mol Mar Biol Biotechnol 1992;1:276–281.

289. Rahman A, Maclean N. Fish transgene expression by direct injection into fish muscle. Mol Mar Biol Biotechnol 1992;1:286–289.

290. Betancourt OH, Attal J, Theron MC, Puissant C, Houdebine LM. Efficiency of introns from various origins in fish cells. Mol Mar Biol Biotechnol 1993;2:181–188.

291. Zhang PJ, Hayat M, Joyce C, Gonzalez-Villasenor LI, Lin CM, Dunham RA, Chen TT, Powers DA. Gene transfer expression and inheritance of pRSV-rainbow trout-GH cDNA in the common carp, Cyprinus carpio (Linnaeus). Mol Reprod Dev 1990;25:3–13.

292. Inoue K. Expression of reporter genes introduced by microinjection and electroporation in fish embryos and fry. Mol Mar Biol Biotechnol 1992;1:266–270.

293. Powers DA, Hereford L, Cole T, Chen TT, Lin CM, Kight K, Creech K, Dunham RA. Electroporation: a method for transferring genes into the gametes of zebrafish (*Brachydanio rerio*), channel catfish (*Ictalurus punctatus*) and common carp (*Cyprinus carpio*). Mol Mar Biol Biotechnol 1992;1:301–308.
294. Du SJ, Gong Z, Hew CL, Tan CH, Fletcher GL. Development of an all-fish gene cassette for gene transfer in aquaculture. Mol Mar Biol Biotechnol 1992;1:290–300.
295. Bragg JR, Prince RC, Wilkinson JB, Atlas RM. Bioremediation for shoreline cleanup following the 1989 Alaska oil spill. Exxon Company, Houston, TX, USA, 1992.
296. Bragg JR, Prince RC, Harner EJ, Atlas RM. Effectiveness of bioremediation for the Exxon Valdes oil spill. Nature 1994;368:413–418.
297. Swammell RPJ, Head IM. Bioremediation comes of age. Nature 1994;368:396.
298. Zink B, Gerber F, Colombo V, Ryser S. The living microcosm. Basel, Switzerland: Editiones Roche, F. Hoffmann-La Roche AG, 1991.
299. Gimeno CJ, Ljungdahl PO, Styles CA, Fink GR. Unipolar cell divisions in the yeast *S. cerevisiae* lead to filamentous growth: regulation by starvation and RAS. Cell 1992;68:1077–1090.
300. Ryser S, Weber M. Genetic engineering: what happens at Roche. Basel, Switzerland: Editiones Roche, F. Hoffmann-La Roche AG, 1991.

Biotechnology Annual Review Volume 1
M.R. El-Gewely, editor

Application of two-dimensional protein gels in biotechnology

Ruth A. VanBogelen and Eric R. Olson
Department of Biotechnology, Parke-Davis Pharmaceutical Research, Division of Warner-Lambert, Ann Arbor, Michigan, USA

Abstract. The optimal use of biological systems for technologically developed products will not be achieved until biological systems are completely defined in biochemical terms. Two-dimensional polyacrylamide gel electrophoresis, 2-D gels, are contributing to this goal. These gels separate complex mixtures of proteins into individual polypeptide species. The ultimate use of 2-D gels is the construction of cellular 2-D gel databases which identify the proteins on the gels and catalog their responses to different environmental conditions. In addition to these global analyses, many applications for 2-D gels in basic, applied and clinical research have been shown.

Introduction

Over 50% of the mass of most cells is comprised of proteins. The chromosomes of prokaryotic cells are estimated to encode from 1,000 to 5,000 proteins [1] and those of eukaryotic cells encode up to 10 times this number [2]. Proteins provide the basic structural, synthetic and enzymatic functions for the cell. Not all proteins are present and active in every cell. Cells have control mechanisms which regulate the synthesis, the stability and the activity of each of these proteins so that as the environment of a cell varies so does its complement of proteins. Historically these changes have not been monitored simultaneously but rather a single or small number of proteins are studied. This chapter describes a technique, two-dimensional polyacrylamide gel electrophoresis, used for separating and analyzing complex mixtures of proteins. For many applications, both in basic research and applied biotechnology, this technique has an advantage over other separation techniques since it provides the investigator with the opportunity to examine the cells' entire aggregate of protein simultaneously. This ability to see the "forest" and the "tree" often provides valuable insights into the role(s) a particular gene or protein plays in the global strategy of the cell.

Many two-dimensional separation systems for resolving cellular molecules have been described. This chapter discusses a specific 2-D gel technique which electrophoretically (through polyacrylamide) separates polypeptides by their isoelectric point in the first dimension and their molecular mass in the second dimension. This chapter highlights the basic principles of these 2-D gels and gives a sampling of

Address for correspondence: R.A. VanBogelen, Department of Biotechnology, Parke-Davis Pharmaceutical Research, Division of Warner-Lambert, 2800 Plymouth Road, Ann Arbor, MI 48105, USA.

applications in areas ranging from basic research, drug discovery and clinical diagnosis from a diverse array of disciplines. The chapter is divided into four sections:

1. description of 2-D gel technology;
2. identification of proteins on 2-D gels;
3. applications of 2-D gels;
4. construction of cellular protein database using 2-D gels.

The first section is an overview of the 2-D gel method with special emphasis on the following key elements: gel size, isoelectric focusing techniques, SDS-PAGE, protein detection strategies, membrane blotting procedures and image analysis. The second section describes a variety of methods used to correlate the spots on 2-D gels to their corresponding gene. Examples of the kinds of studies which have been carried out using this technique are given in the third section. The fourth section describes the construction and functions of cellular protein databases, perhaps the most powerful application of this technology.

Description of 2-D Gel Technology

O'Farrell first described the separation of complex mixtures of proteins by two-dimensional polyacrylamide gel electrophoresis (2-D gels) [3]. A general description of 2-D gels can be found in most recent biochemistry textbooks (e.g. [4]), detailed protocols are available in many laboratory manuals (e.g. Current Protocols in Molecular Biology [5]) and the apparatus and prepared reagents necessary to run these gels are now available from numerous commercial vendors. All these services have made it possible for individual investigators to utilize this technology without extensive training or expertise. In addition, some research facilities have incorporated a 2-D gel running service into their "core facilities" (e.g. Cold Spring Harbor Laboratories) and many others have at least one laboratory that maintains state-of-the-art equipment and assists other investigators.

The first dimension, isoelectric focusing (IEF), separates proteins based on their isoelectric point (pI) and is accomplished by establishing a pH gradient in acrylamide. Proteins migrate through the pH gradient until they reach the pH at which their net charge is zero whereupon the protein stops and focuses at that point. The second dimension, SDS-PAGE, separates proteins based on their molecular mass. Upon treatment with SDS (in reducing conditions), most proteins dissociate into their polypeptide subunits and are coated with SDS. This protein–SDS complex (mass ratio 1:1.4) consists of the hydrocarbon chains of the detergent in a rigid association with the polypeptide chain leaving negatively charged sulfate ions of the detergent (SDS) exposed to the aqueous medium [6]. The native charge of the polypeptide is overcome and the complex migrates as an anion to the anode. The rate of migration, then, correlates with the molecular mass of the specie.

No single 2-D gel can resolve and detect a cell's complement of polypeptides because of the enormous variation in size, in pI and in abundance. Cellular polypeptides vary in size, from about 1,000 Da to greater than 200,000 Da; in pI, from 3 to

13; and in cellular abundance from 1 molecule per cell to greater than 100,000 molecules per cell. The following paragraphs discuss a number of variables which affect the ability to resolve individual proteins from a complex mixture and also a variety of methods used to detect proteins separated on 2-D gels.

Gel size

There are four standard sizes of 2-D gels in current use, each developed for particular applications. In general larger gels are capable of resolving more individual polypeptide species because a greater physical distance between polypeptides of similar molecular weights and pIs can be achieved. Even with sufficient resolving capacity, the ability to detect a single protein species requires that a sufficient quantity of the protein can be loaded onto the 2-D gels to allow detection of the protein.

1. The mini-gel has the advantage of speed; both dimensions can be run in less than a day and the entire system is available commercially as a kit that requires very little technical skill to run. This size gel is useful for applications that require separation of only a small number of different polypeptides or detection of only the major species of polypeptides. For example, mini gels were used to show that a single mutation in the gene for 16S rRNA in *E. coli* (*rrnE*) caused major changes in the overall protein pattern [7]. The limited resolving power of this size gel eliminates it as a method for global studies of cellular responses and for developing cellular protein databases.

2. The mid-size gel [3] is probably the most widely used format because it allows for a broad spectrum of applications. Although it requires more time than the mini-gel, the mid-size gel can be used for the same applications as well as for global studies of cellular responses. The resolution is sufficient to satisfy most investigators who plan to analyze the gels manually. Soon after he described the technique, O'Farrell used it to study the cellular responses to amino acid limitation in *Escherichia coli*. He observed changes in rates of synthesis of individual proteins as well as heterogeneity in the isoelectric points of proteins, the latter of which he interpreted as an increase in missense errors in translation [8]. Although this method was used initially for several cellular protein databases most investigators have converted to the large or giant gels which allows them to resolve and/or detect more proteins. The apparatus required to run these gels is available commercially from several vendors and comes as a single unit that runs up to four gels per box or as a multiple gel unit that runs 12 gels simultaneously.

3. The large gels were developed in conjunction with a computer-aided image analysis system [9] for the development of cellular protein databases (called the QUEST system). A large gel version of the O'Farrell method has also been described [10] (called Iso-Dalt system). Both versions of larger 2-D gels are still relatively easy to prepare and run; 10–20 gels are usually run simultaneously. Both were also originally described for analytical purposes (maximum of

0.1 mg of total protein per gel), but can be modified for preparative analysis (maximum of 1.0 mg of total protein). Although the preparative gels increase the detection of some proteins (because more protein is loaded), they also decrease the resolution of other proteins because more abundant proteins are present at such high concentration that they mask by spreading onto the more minute proteins. The Iso-Dalt system is commercially available through Large Scale Biology (also distributed by other companies). A modified version of the QUEST gel system is commercially available as the Investigator 2-D system (Millipore Corp.). Quality-controlled prepared reagents are also available for the Investigator 2-D system.

4. The giant gels were developed to allow the detection of proteins present in very minute quantities in complex mixture containing very large number of individual protein species that vary in concentration over a 6-fold range [11]. Because of the size and the large loading capacity, these gels can resolve proteins that are present at very low abundance in the cell. The amount of protein that can be loaded onto these gels is up to 10 mg of total protein [12]. This format is more challenging to run than the other three sizes and the apparatus is not commercially available.

Sample preparation

The method used to extract the proteins from the cell and prepare the proteins for migration in the first dimension is dependent on the cell type being studied. It was originally thought that any SDS in the sample would disrupt the isoelectric focusing, however, it has been recognized that low level of SDS can be tolerated resulting in a significant improvement in the ability to solubilize cellular extracts. Extracts from most animal cells and gram-negative bacteria can be prepared using a similar protocol. Basically, the cells (either concentrated or in suspension) are treated with a buffer containing SDS, DTT and Tris, boiled for 2–5 min, cooled, treated with DNase and RNase, concentrated (if necessary when the protein concentration is too low) and diluted into a buffer containing urea and a nonionic detergent (NP-40 or Triton X are most common). Some tissues, plant cells, yeast and gram-positive bacteria are more difficult to disrupt and require vigorous methods to break up the cell wall prior to the procedure described above. Methods developed for extraction of DNA, RNA or protein (for SDS-PAGE gels) can usually be used to accomplish this. However, two things should be kept in mind. First, aqueous solutions of urea degrade with time or upon heating above 37°C, resulting in the production of cyanate ions [13]. Proteins exposed to these ions are carbamylated and streak (rather than focusing as a spot) in the first dimension. Second, the DNase and RNase treatment is required since the presence of large quantities of DNA and RNA molecules in the extract will negatively affect the integrity and quality of the IEF gel. For example, analytical IEF gels disintegrate if the extract contains large amounts of DNA and RNA. Also, autoradiograms of ^{32}P-labeled cellular extracts not treated have a cloud masking a region of the protein pattern due to ^{32}P-labeled DNA and/or RNA.

First dimension

A prerequisite to fully exploiting 2-D gel electrophoresis is the ability to obtain reproducible protein separations, not only from run to run within a single laboratory, but also between different laboratories. For the first dimension, the primary hindrance to achieving reproducible separations are the carrier ampholytes used to set up the pH gradient. Carrier ampholytes (trade name ampholines) are a mixture of aliphatic polyamino-polycarborylic acids with closely spaced pI values and high conductances [14]. Commercially available ampholytes suffer from poor reproducibility from lot to lot, resulting in very different protein patterns. In 1982 immobilines (acrylamido acids and bases) were introduced as an alternative to carrier ampholytes [15]. Each acrylamido acid or base, with its unique pK value, acts as a buffer to pool sets of proteins. The first generation of these buffers gave inconsistent results [16], however, in 1988 more stable immobilines were produced [16]. Advances in this product line have continued and are commercially available. This method of isoelectric focusing is becoming increasingly popular largely because of its potential for increasing reproducibility in the first dimension [17].

The isoelectric points of cellular proteins range from about 3 to 13. Separation of the rare proteins with pIs less than 4 and greater than 11 is not achieved by the standard 2-D gel methods. The original 2-D gel method (often called an equilibrium gels) will separate proteins with pIs from about 4 to 8. A nonequilibrium gel system (using a broader range of ampholytes) is used to resolve basic proteins (pIs up to 11.5) [18]. Reproducibility of the nonequilibrium gels requires precise control of the total volt-hours since the proteins do not focus in the gel.

Second dimension

Separation of proteins in the second dimension is accomplished by layering the tube gels from the first dimension on top of a slab-type SDS polyacrylamide gel. Differential rates of migration through these gels, based on the molecular weight of the protein, result in the generation of protein "spots", each representing a different polypeptide chain (Fig. 1). In addition to gel size, the acrylamide content and pH of the gel are important parameters influencing the separation in this dimension. The practical limits for the acrylamide content of gels is 7–16%. Below 7% the gel is soft and difficult to manage and above 16% the gels crack when dried. A modified acrylamide was developed, Duracryl (trademark), which improves the handling characteristics of large, thin, gels [19]. The most commonly used acrylamide concentration, 10%, will separate proteins with molecular weights between 10,000 and 200,000 Da, increasing the concentration of acrylamide favors separation of lower molecular weight proteins. Both the O'Farrell [3] and Garrels [9] methods use a single concentration. Use of an acrylamide gradient in the gels extends the separation range and is particularly useful for the mini-gels where the resolving power is severely hampered by size. Anderson's method [10] and the giant gel method [11] also employ gradient gels.

74

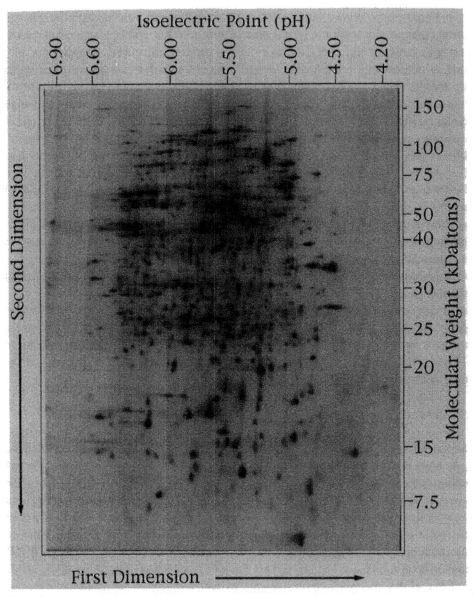

Fig. 1. 2-D gels made from an *Escherichia coli* protein extract. This 2-D gel is a large format gel (apparatus and chemicals from Millipore Corp.). In the first dimensions, pH 4–8 ampholines were used, and in the second dimension the acrylamide concentration was 11.5% and the pH of the trizma was 8.8 (Trizma 8.8 from Sigma Chemical Co.). The proteins in this extract had been pulse labeled with [^{35}S]methionine. The gel image was captured on X-ray film (XAR-5, Kodak Co.), scanned using a camera (Kodak Megaplus Model 1.4) using the BioImage software (Millipore Corp.) and then transferred as a TIFF file to a drawing program (MacDraw Pro on a Macintosh computer) to attach labels. The molecular weight and pI estimates were made based on the locations of certain proteins which have been identified as the protein product of previously characterized genes [34].

The pH of the gel can also be varied to yield a different separation of proteins. For example, at a constant acrylamide concentration of 11.5%, separation of proteins from 6,000 to 100,000 Da is achieved when the pH of the Tris is 8.8 while proteins in the range from 12,000 to 150,000 Da are separated at a pH of 8.5. For optimal reproducibility, the Tris buffer used in the gels should be made by mixing appropriate amounts of Tris base and Tris–HCl or by purchasing Tris that has been pHed. As with SDS-PAGE gels, low molecular weight proteins (less than 10,000 Da) are better resolved using a Tris–Tricine running buffer rather that the Tris–glycine buffer normally used [20].

Protein detection

Proteins separated by 2-D gels can be detected by the same methods used for most other protein separation techniques, including direct (e.g. staining) and indirect (e.g. autoradiography and phosphorimaging of radiolabeled proteins, Western and ligand blotting) methods. Coomassie blue [3] and silver staining [21,22] are the most commonly used staining methods allowing for the detection of very small quantities of a single protein species (10 pg for Coomassie blue and 0.1 pg for silver stain). The sensitivity of these stains varies with the size of the spot. For example, low molecular weight proteins form spots with larger diameters and thus are more difficult to detect at low concentrations. The major drawback of staining is that this detection method does not allow precise quantitation of the amounts of proteins [23]. When high specific activity precursors are used, radiolabeling of proteins is the most sensitive detection method and also allows precise measurement of the amount of individual proteins. Radiolabeled amino acids ([3H], [14C], or [35S]) or their precursors (e.g. [14C] glucose and [35S]SO_4) are commonly used to label cellular proteins in vivo. Radiolabeling proteins also provides a method for measuring rates of synthesis of proteins (pulse-labeling), examining post-translational modifications, and determining the half-lives of proteins (pulse-chase). In order to measure the steady-state levels of individual proteins, the label must be present during a sufficient number of cell generations and the concentration of the chemical form of the isotope must be high enough to ensure that the proportion of label in proteins actually represents their level in the cell. For example, if [35S]methionine is being used to label proteins to determine their steady state level in the cells, then by labeling through four doubling in cell mass, 95% of the protein mass will have been made in the presence of the [35S]methionine but only if the concentration of methionine in the media is sufficiently high to ensure that the [35S]methionine was present during the four generations. These long-term labeling studies or long chase periods in pulse chase studies, however, do not accurately represent the steady state level of proteins with very short half-lives.

Following separation by 2-D gels, radiolabeled proteins are detected by a method suitable for the isotope used. For example, for visualizing the protein pattern in gels containing proteins labeled with [35S] and [32P], autoradiography or phosphorimaging [24] of the dried gel is most convenient. [3H]-Labeled proteins can also be detected by autoradiography if the gel is chemically enhanced with PPO [25] or with commercially available enhancers.

Protein isolation

Due to the high resolving power of 2-D gels the procedure can often be used to isolate a small amount of purified protein for further analytical procedures (e.g. amino acid sequencing) or for use as a reagent (e.g. antibody production). The ability to derive N-terminal sequences from proteins separated on 2-D gels has been exploited for identifying the genes that encode them (i.e. reverse genetics) [26–28]. Two general approaches have been used for isolating proteins, direct extraction from the gel matrix into an aqueous buffer and transfer of the protein to a solid matrix, such as nitrocellulose or nylon membranes. Recently, new types of membranes have been developed for use when the protein will be analyzed by N-terminal microsequencing (Edman degradation on a microscale) or when the amino acid sequence internal to the protein is desired [29].

Image analysis

Obtaining useful information from a gel (or its image) containing thousands of spots relies on the ability of the eye or computer-driven imaging systems to discern "constellations" of spots located in various parts of the image. By aligning constellations from one region of a gel with the same region from another gel, one can easily detect, by eye, qualitative changes in the patterns (appearances, disappearance or changes greater than threefold in the amount of a protein). Changes in the migrations of individual proteins, due to certain post-translational modifications or mutation, can also usually be detected by the human eye provided the sample is co-electrophoresed with the unmodified or wild type proteins. The measurement of more subtle changes in the amounts of individual proteins can be done by two different methods: (1) excising the protein out of the gel and measuring the protein content or radioactivity in the protein [30,31] or (2) optical or radiometric scanning of the gel or gel image. Methods to quantitate the amounts of individual proteins detected from optically scanned X-ray films of gels have been described [9], however they suffer from the requirements of both multiple exposure of each gel (because film has a relatively narrow linear OD range) and use of radiometric standard (to allow the data from multiple exposures to be merged together into a single set of OD for all spots detected on the gel). The recent introduction of phosphorimaging technology has made it possible to quantitate proteins using a single exposure (described in Ref. [24]) and a set of radiometric standards (applied along side the gel) allows the density to be related to dpms.

Optical scanning devices for gels or images of gels, which result in the radioactivity in each spot being converted to a density measurement, include: drum scanners, cameras, document scanner, laser scanners, and phosphorimagers. The simplest software for quantitating the images obtained from these devices require the user to

Table 1. Comparison of methods to analyze maximum number of spots per large size 2-D gel

	Manual spot analysis	Semi-automated spot analysis	Automated spot analysis
Isotope costs	$2,000 (^{14}C and ^{3}H)	$100 (^{35}S)	$100 (^{35}S)
Detection time with film	2 days	30 days (multiple exposures)	30 days (multiple exposures)
Detection time with PI		2 days	2 days
Find spots	Manual (5 min/spot to excise spot from gel)	Manual (1 min/spot to find spot boundaries)	Automated (<2 min/1,200 spots)
Quantitation of spots	Processing of spot for counting	Done in previous step (find spot)	Automated <5 min/ 1,200 spots + 30 min editing
Match spots	Done when finding spot	Manual, average 10–20 min/spot; more if only 1 image is viewed (match for 200 most abundant 2–5 min/ spot)	Match 10–20 spots per set; automated match of 1,000 more; 1–3 hours of editing
Data summary	Manual calculation or enter into computer via counter	Manual or may have some reporting capabilities	Table generated from software
No. of spots analyzed	200	1,200	1,200
Total time to analyze set of 6 gels	3 months; maximum of 200 spots quantitated	1 month; 200 most abundant spots; 12 months; maximum of 1,200 spots quantitated	1 month; maximum of 1,200 spots quantitated

define each spot produced from a protein (i.e. establish boundaries for each spot) and determine the optical density within the defined area. Dedicated software for 2-D gels has been developed independently by several groups (both government-supported and commercial) and offers several powerful features. Some of these programs have automated spot-finding capabilities, will match spots between sets of gels, will identify sets of proteins with shared characteristics (e.g. spots induced by 2–10-fold in one condition compared to another), and even allow information (annotation) to be tagged with the spots. The amount of time and editing required and the type and quality of gels to be analyzed varies greatly among the different programs. In the appendix to this chapter some of the automated software programs associated with 2-D gel databases are referenced. Table 1 gives an example of how improvements in the 2-D technique and developments of automated image analysis techniques have decreased the time required to analyze the gels and increased the number of proteins that can be analyzed.

Identification of Proteins on 2-D Gels

This section describes some of the common methods that have been used for determining the identity of individual proteins on 2-D gels (identity in this sense means to locate the spot on a 2-D gel where a particular protein migrates).

Comigration of purified proteins with cellular extracts

The principle of this method is based on the ability of a previously purified protein to mix with the analogous protein from a total cellular extract due to their identical pIs and molecular weights [10,32]. In effect, the purified protein "dilutes" the protein from the total extract. If the protein from the total extract is radiolabeled, the spot normally observed for this protein on an autoradiogram will appear more diffuse due to this dilution effect. In practice, the purified protein (about 1μg) is diluted into a buffer that is compatible with the first dimension and loaded on the gel with a protein extract from the cells of interest. Identification requires at least three gels; one of the purified protein alone, one with the purified protein and the protein extract, and a third gel of the protein extract alone. About 1μg of purified protein per gel is usually sufficient. The gels are stained with Coomassie blue or silver stain to detect the purified protein and to ascertain its position in the gel. In most cases the purified protein is easily detected (although some do not enter the gel system). This method of identification is only successful about 25% of the time (unpublished observation). The cause of this low success rate has not been thoroughly investigated, however there are several obvious explanations which should be considered. The inability to match up a stained purified protein with a spot on a 2-D gel could be due to the low abundance of the protein in the cellular extract or because one of the species is modified. An example of a protein spot identified using this technique is the RecA protein from *Escherichia coli* [33]. For many years this method of identification was used by cellular protein database investigators. For example, 265 spots in the *E. coli* database have been identified by this method [34].

Immunoblotting

The standard Western blotting techniques used for 1-D gels can be used for 2-D gels, resulting in the ability to identify the content of individual spots following transfer to a solid support (typically nitrocellulose or PVDF membranes) [35]. By using radiolabeled extracts the antibody-tagged spot can easily be aligned with the autoradiogram produced from the same membrane. Alternatively, the total protein pattern can be observed by staining the blotted membrane. An example of how this approach was successfully used is best illustrated with the human amnion cells database where 103 spots were identified using specific antibodies [36]. This approach has also been used to identify similar proteins from different species, for example antibodies to human sera proteins were used to identify the analogous species in cat sera [37]. This is also perhaps the best technique for identifying proteins that exist in more than one form, and, thus migrate to different spots within one gel. For example, separation of human amnion cells extracts on 2-D gels results in four different forms of Elongation factor 2 (EF-2) (two forms are phosphorylated) [36].

Binding assays

The binding of substrate to a protein can often be done on proteins separated by both 2-D and 1-D gels. For example, small GTP-binding proteins were detected by isoprenoid [38]. Even DNA-binding assays have been adapted to 2-D gels. The proteins from a variety of viruses that bind a portion of the $\kappa\beta$ sequence have been identified using a DNA-binding assay [39].

Post-translational modification

Proteins that are post-translationally modified by specific signal transduction pathways can be identified on 2-D gels by binding radiolabeled substrates under conditions specific to that pathway. For example, the *E. coli* chemotaxis proteins, *tsr* and *tar* were identified by this method [40]. These proteins are post-translationally methylated by incubating cells in the appropriate chemotaxis buffer that included [^{3}H]methionine [40]. A fluorogram produced from the 2-D gel made from this cellular extract revealed the methyl-accepting proteins (six spots corresponding to *tar* and eight to *tsr*) [34].

Genetic analysis

Mutations in coding sequences that alter the isoelectric point or molecular weight of the corresponding gene product, or mutations that enhance or decrease expression of a gene, can be invaluable in identifying the location of a protein on 2-D gels. For example, a missense mutation in *mopA*, encoding the GroEL protein from *Escherichia coli* was used to help identify this protein on 2-D gels [41]. The mutant GroEL protein migrated as a more acidic protein so that when an extract prepared from the wild type and mutant where loaded on the same gel, a dumb-bell shaped spot was seen. It has been proposed that 2-D gels could be used to help detect mutations without knowing the identity of the protein spot [42]. The approach would determine over a large number of 2-D gels the variation in relative migration and in abundance of each spot found in the gels. Deviation from this "normal" variation in migration would suggest that a protein spot is the product of a mutant gene and could be confirmed by additional experiments to confirm the heretability of the proposed mutation [42]. Caution must be taken to ensure that changes in the position of proteins or the presence or absence of a protein is a result of the primary mutation and not a secondary effect of the mutation.

For bacterial systems, expression of genes residing on plasmids has been useful for identifying particular proteins. This method is being used to simultaneously identify proteins in *Escherichia coli* [43]; 57 of the proteins identified in the *E. coli* gene-protein database have been identified by this method [34].

Enzyme induction or repression

The enzymatic functions of proteins can be modulated by regulating either their activity or level. Those regulated by changes in the protein level can often be identified on 2-D gels by monitoring the cellular response to inducing and/or repressing conditions. For example, *Escherichia coli* induces a set of proteins when the phosphorus source is not inorganic phosphate. Figure 2 shows two gels made from radiolabeled protein extracts of cells grown either in a media containing inorganic phosphate (left panel) or phosphonate (center panel). The center panel displays the same region in both gels and indicates the positions of two proteins, previously identified by genetic analysis [44], to be induced in cultures grown in phosphonate.

Identifying coding sequences

Algorithms for predicting whether an open reading frame (ORF) identified by DNA sequencing is likely to encode a protein are based on statistical analysis of known genes, and thus are subject to a certain level of uncertainty. An experimental approach to assist in confirming the relationship between an ORF and a gene product is to clone the gene, express the protein, and determine if the product migrates as would be predicted based on its calculated pI and molecular weight. Figure 3A,C illustrates a linear relationship between the isoelectric point and migration in the first dimension and Fig. 3B,D shows a curve fit for the relationship between molecular weight and migration in the second dimension. The isoelectric point and molecular weights were deduced from the amino acid composition inferred from analysis of the DNA sequence (see figure legend for equations used to predict molecular weight and pI). However, for several reasons, this method is insufficient by itself to make a conclusive statement as to whether an ORF encodes a gene product. This approach is being used to identify gene products encoded by ORFs identified by the *E. coli* genome sequencing projects. Briefly, 10–20 kilobase pair chromosomal fragments that are sequenced are cloned into a T7 promoter based expression vector. This vector allows the proteins encoded by the segment to be selectively radiolabeled following induction of the promoter [43]. The resulting products are then matched with protein spots from gels of total cellular proteins and with the ORFs predicted from the DNA sequence. Twenty-eight proteins encoded by ten clones from *E. coli* were identified by this method and 45 more proteins have been mapped to small segments of the chromosome [43].

Microsequencing of proteins from gels

The most popular and one of the most conclusive methods for identifying protein spots on 2-D gels is microsequencing (microscale Edman degradation reaction), a technique that provides the amino acid sequence for 10–20 residues of a peptide [26–29]. These methods often provide enough information to make or confirm an identification [45–48]. Several cellular protein databases rely heavily on this technique to identify the protein spots. For example, 145 proteins found in human kerotinocytes were identified in a single report [49].

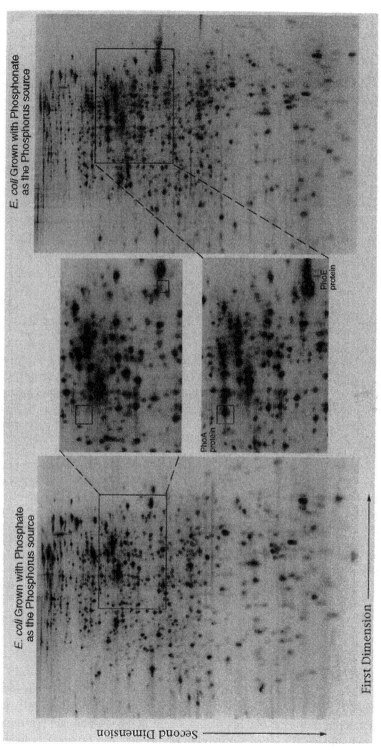

Fig. 2. This figure illustrates a method used to identify individual proteins on 2-D gels. These 2-D gels were made from extracts of *Escherichia coli* cultures grown with phosphate (left panel) or phosphonate (right panel) as the phosphorus source. The gel conditions and image acquisition are the same as described in Fig. 1 except the gel image was captured using a phosphoimager (Molecular Dynamics) and then transferred to the drawing program. The center panels display a portion of the other two gels. The two proteins, PhoA (alkaline phosphatase) and PhoE, boxed in the center panel are known (by genetic and enzymatic analysis) to be repressed in cultures containing phosphate and dramatically induced in cultures containing only phosphonate as the phosphorus source.

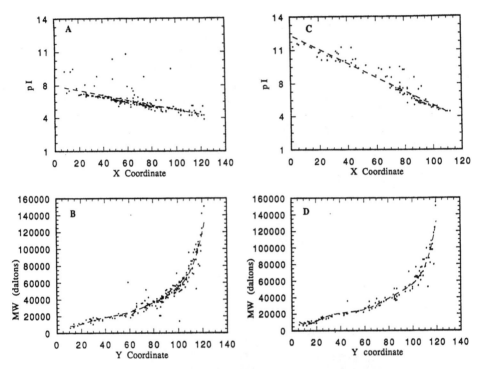

Fig. 3. Relationship between the migration of protein spots in the first and second dimension of 2-D gels and their pI and molecular weight (calculated from their amino acid sequence deduced from the DNA sequence of their gene). The data was taken ECO2DBASE (available through anonymous ftp as described in [34]). A and B plot 233 proteins; C and D plot 132 proteins (using the mid-size gel method described, see Fig. 2A of reference [34]). (A) The relationship between the X-coordinate (migration in the first dimension) and the calculated pI of the protein for a gel separating protein with pIs in the range of 4–7. The equation for the line is; pI = 7.497 – 0.024 (X-coordinate) and R = 0.69. (B) Same 2-D gel as (A) showing the relationship between the Y-coordinate (migration in the second dimension) and the calculated molecular weight (MW) of the protein. The curve fit shown is a sixth order polynomial and the equation is: $MW = 18669 - 2235$ $(Y) + 167$ $(Y^2) - 5.03$ $(Y^3) + 0.0759$ $(Y^4) - 5.6 \times 10^4$ $(Y^5) +$ 0.96×10^6 (Y^6) and R = 0.96. (C) The relationship between the X-coordinate (migration in the first dimension) and the calculated pI of the protein for a mid-size gel separating proteins with pI between 4 and 10 (see Fig. 2B of Ref. [34]). The equation for the line is: pI = 12.22 – 0.068 (X-coordinate) and R = 0.96. (D) Same 2-D gel as (C) showing the relationship between the Y-coordinate (migration in the second dimension) and the calculated molecular weight of the protein. The curve fit shown is a sixth order polynomial and the equation is: $MW = 25558 - 3993$ $(Y) + 304$ $(Y^2) - 9.41$ $(Y^3) + 0.143$ $(Y^4) -$ 1.05×10^3 $(Y^5) + 0.96 \times 10^6$ (Y^6) and R = 0.97.

Peptide mapping

The classical method of peptide mapping using any one of a variety of sequence-specific proteases can be applied to proteins separated by 2-D gels [50]. Briefly, the protein spots are cut out of the gel and applied to a well in the stacking gel portion of an SDS-PAGE gel suitable for resolving peptide fragments. The protease is loaded into the gel and the protein-protease sample is then electrophoresed into the stacking gel, the power is turned off for 30 min or so to allow the protease to work, and then electrophoresis is resumed to separate the resulting fragments [50]. An example of how this approach can be used is illustrated by the identification of two isoenzymes of the *E. coli* lysyl-tRNA synthetase. The two isoenzymes, co-purified by classical chromatography methods, produced two bands on an SDS-PAGE gel and four spots on a 2-D gel. The four spots were excised from gels, treated with a protease (*S. aureus* V8 protease) and the resulting fragments were run on an SDS-PAGE gel. The peptide map showed that three of the four spots were identical [51]. Subsequently, two nearly homologous genes encoding these isoenyzmes were identified [52–55].

The number of different approaches outlined above for analyzing and identifying proteins highlights the versatility of the 2-D gel approach. The application of these procedures run the spectrum from analyzing purified proteins involved in well characterized processes to identifying proteins that might be involved in a physiological response, but for which little prior information has been obtained. The increased use of these techniques, based largely on more classical approaches, is sure to prompt the development of new methods that are more powerful and more closely tailored to take advantage of the strengths inherent in 2-D gel systems.

Applications of 2-D Gels

Protein purification and analysis

The principle uses of 2-D gels in protein purification include isolation of the protein itself, determining the level and nature of contaminants in a protein preparation and ascertaining the nature and/or extent of any covalent modifications. Purifying proteins with this method can solve two central, but related, problems associated with purification: (i) limited amount of material, and (ii) development of a separation strategy. 2-D gels are ideal for certain situations where only a small amount of purified protein is required (e.g. amino terminal sequencing) or when the biological sample containing the protein is in short supply. A nice example illustrating these points was the use of 2-D gels in purifying and characterizing the components of grass pollen thought to be responsible for allergic reactions in humans [56]. Currently, individuals are desensitized to these allergens by exposure, through injection, of crude extracts of pollens, sometimes leading to sensitization to components that were not originally a problem. Using 2-D gels, the components can be separated into individual polypeptides and a partial amino acid sequence of each protein obtained. This information is an important first step in developing a strategy for making large amounts of each antigen via molecular biology techniques. Even in cases in which larger amounts of protein are required (e.g. for antibody production) 2-D gels can be a useful tool. A routine example would be making antibodies in rabbits. Assuming an average size protein, a typical immunization protocol would require from 0.05 to 1 mg of protein per dose (2–3 doses are recommended) [57]. Assuming the gene encoding the protein has been cloned and expressed in a bacterial expression system and that the expression level is modest (5% of the total cell protein), the minimum amount for one dose could be isolated from a single large format gels (20 gels would be required to generate the maximum dosage). If the antibody is to be used in subsequent Western blots caution must be taken to assure that there was not a protein migrating with the protein of interest such that the minor contaminant was masked by the more highly expressed protein.

Modification of proteins can occur either in vivo, as a result of normal cellular processes, or in vitro during the processes of cell disruption, purification or storage. Some of the more common in vivo modifications include phosphorylation, acetylation, glycosylation, farnisylation, methylation and myristilation (many more have been found). Some modifications alter the pI and/or molecular weight of the polypeptide and thus the migration on the 2-D gel, and some can be detected by labeling the proteins with the appropriate precursor. The methods described in the previous section can be used to identify modified forms of proteins. Once extracted from cells, proteins are susceptible to modifications such as oxidation, carbamylation, or proteolytic cleavage due to buffer conditions, etc. In addition, in vivo modifications can be lost (these issues are discussed in detail in Refs. [58–60]). In fact, only about one-fourth of purified proteins comigrate (on 2-D gels) with a protein from the cell extract.

Combining the ability to detect modified proteins with a comparative analysis of cell types or cells grown under different conditions can provide clues for understanding the role of the modification or at least relationships between the modification and the physiological state of the cells. For example, the E. coli ribosomal protein, L12, is acetylated to produce a slightly more acidic protein (called L7). Although the protein is always present at four molecules per ribosome, the degree of modification has been shown to vary significantly with the growth rate of the cells (slower growth rates result in more of the modified form) [61]. Since the translating apparatus of the cell is tightly linked to growth rate, this modification may be a mechanism for controlling ribosome activity. The rationale for the modification has not been determined but is also found in Salmonella typhimurium where this ratio was one method used to estimate the growth rate of the bacteria in macrophages [62]. A similar example has been found in HeLa cells. Three different charge variants of Nuclear factor IV (also called Ku protein) have been identified; the relative abundance of each varies during the cell cycle [63]. The ratio of the forms observed in quiescent cells was the same as was observed when cells were treated with DNA replication inhibitors. Although neither the nature nor the role of the modifications have been identified, the protein's modification appears to be physiologically relevant and a biological marker for cells which are not replicating.

A feature of a large number of eukaryotic and prokaryotic signal transduction pathways is the modification of proteins by phosphorylation/dephosphorylation. The problem of identifying the substrates for the kinases and phosphatases has been approached by separating on 2-D gels whole cell extracts labeled (in vivo) with inorganic phosphate ($^{32}PO_4$) or cellular fractions labeled in vitro with [^{32}P]ATP. An interesting application of this approach was the identification of protein kinase C (PKC) substrates as described by Beckmann et al. [64]. Previous work had revealed a membrane-inserted constitutively activated PKC in bovine brain nuclear membrane. To identify membrane-bound proteins that were phosphorylated by this PKC, nuclear membranes and other subcellular fractions were isolated and proteins were phosphorylated in vitro in the presence or absence of PKC inhibitors. They then looked for proteins that had the following characteristics: (i) phosphorylated proteins found in the nuclear membrane, (ii) proteins that had a decrease in the phosphorylation in the presence of PKC inhibitors, and (iii) protein with an increase in phosphorylation by exogenously added PKC. They found several proteins that met these criteria, picked the two that were most abundant for further analysis, microsequenced the proteins and compared the amino acid sequence to proteins in the databases. Matches were found for both proteins. One protein had previously been determined to be a substrate for PKC. The other protein had a known function but was not known to be regulated by PKC.

Identification of regulatory networks

Often the scope of a response mediated by a particular regulatory protein is not understood when only classical genetic approaches are employed. Witness the number of times, especially in the field of bacterial genetics, in which multiple investigators were studying the effects of a regulatory gene on their particular system with little knowledge that several other pathways were co-regulated. Comparing the pattern of gene expression by 2-D gels in a strain that has a mutated allele of a regulatory gene with that from the wild type strain can reveal the magnitude and complexity of responses mediated by that protein. The number and identification of proteins involved in the heat shock response in *E. coli* was determined through the analysis (on 2-D gels) of a strain that turned out to have a mutation in the gene (*rpoH*) encoding the regulatory factor (Sigma-32) required for activating expression of the target genes [65]. Similarly, this approach has been successful in helping to identify the players involved in the ability of bacteria to adapt to changes in the type and level of nutrients (e.g. carbon, nitrogen, phosphorus), oxygen levels, pH and temperature, as well as responses to exposure to ultraviolet light, oxidizing agents, detergents and antibiotics. Using this approach it has become apparent that for any one of these stimuli there are protein responders unique to that stimulus and also ones that are shared with one or more additional stimuli. It is hard to imagine a better vehicle than 2-D gels for identifying such overlaps. Identifying these overlaps gives one a much more comprehensive view of the regulatory system which in turn helps elucidate the mechanism as well as the role of the response. One *E. coli* protein recognized as being induced by many conditions has been characterized through reverse genetics [48]. At least one function for this protein, named Universal stress protein A, appears to be to coordinate the metabolism of glucose so that intermediate metabolites do not accumulate [66]. This function had been missed by all the classical methods used to study the glucose metabolism in *E. coli*.

Developing genetic tools through comparative studies of different cells

As illustrated in the last few examples, in systems where genetic tools can be easily employed (e.g. bacteria, yeast), 2-D gels can be a valuable supplementary tool. However, in systems where the genetic approach cannot be so easily employed, these gels may be one of only a few tools available for answering specific questions. Analogous to techniques such as Northern hybridization, immunoelectron microscopy or PCR, this approach is useful for doing comparative studies between specific types of cells, whole tissues and fluids from plants and animals. However, while the other techniques mentioned are restricted to examining only the gene or gene product of interest (all require prior identification so that specific probes can be developed), 2-D gels require no prior knowledge about the proteins involved in the process of interest. In fact it is most powerfully used when it serves to identify the components such that subsequent probes can be developed.

For diseases in plants and animals, 2-D gels have been used to identify the factors associated with the disease (e.g. identification of the antigens) or to find proteins that are "markers" of the diseased state (regardless of whether their induction or repression is the cause or consequence of the disease). For example, 2-D gels are being used to identify the antigen portion of immune complexes in blood (the strategy is described in Ref. [67]) in the hope that identifying the antigen will help elucidate the nature of the disease. There are many examples of studies to find protein markers for a diseased cell. In one, six polypeptides markers were found that are strongly up-regulated in diseased (psoriatic) skin cells of human. Further biochemical characterization of these proteins may elucidate their roles in this disease. Studies of other skin diseases should help determine whether the up-regulation of these proteins is unique to psoriatic skin cells or common to all diseased skin cells [68]. Clinical tools exploiting 2-D gel techniques are being developed. Diagnosis of perilymph fistula is difficult because of the lack of a reliable diagnostic tool to identify the fluid in the middle ear as perilymph. 2-D gels were used to compare the protein composition of perilymph with the proteins found in serum and cerebrospinal fluid. A few potential marker proteins were found [69] which can lead to the development of tools (e.g. antibodies) which could be used during surgical procedures to help identify the fluid origin.

In agricultural research, genetic engineering is being used to produce plant varieties which produce yields and better products. 2-D gels are being used to help identify the proteins responsible for some of these more desirable characteristic. A wild relative of tomato is being studied to help identify the genetic determinants which make it more drought resistant than the cultivated varieties of tomatoes [70]. Studies have shown that during a drought, the total protein synthetic capacity of drought sensitive versus drought resistant plants is the same, however, there are dramatic changes in the pattern of proteins seen on 2-D gels during drought conditions suggesting that the more desirable trait could be crossed into the cultivated varieties.

Construction of Cellular Protein Database Using 2-D Gels

A couple of cellular protein databases began immediately after the 2-D gel technique was introduced in 1975. Improvements in the gel techniques and in computer-aided image analysis have made this an attractive approach for cataloging cellular proteins. This chapter summarizes some of the databases that are being built. Investigators are encouraged to search for publications in their area of interest that publish 2-D gels and contact the authors to ask if a database is being established.

The major goal of cellular protein databases is to allow the data obtained from 2-D gels to be accumulated over time and to allow groups of investigators to access and even contribute to the database. For the purpose of this discussion cellular protein database are defined as having these two core features:

1. reference 2-D gel images upon which data is accumulated and
2. a naming or numbering system for the spots on the reference images that allows each spot to be tracked through large numbers of 2-D gels.

Most of the databases also have accessory features that enhance the ease and accessibility of the database including:

1. use of a standardized 2-D gel method that allows other investigators to run gels that match the reference images,
2. use of computer-aided image analysis that accelerates the data acquisition process by automating spot finding and matching functions and
3. use of an electronic version of the database that provides a systematic way to accumulate information about the proteins allowing other investigators (perhaps interested only in the data), to access and use the database.

Most of the databases currently being developed are unique in their design, content and accessibility. Scientific conferences have already been held to discuss issues related to standardizing the various databases [71]. Currently most databases fall into one of two functional types. One type of database is primarily involved in identifying the 2-D gel spots as either a protein with known biochemical or structural function or the product of a particular gene on the chromosome(s). The second type of database concentrates on cataloging the protein content of cells including: the appearance, disappearance and changes in abundance (or synthesis) of spots. Some of the databases described below are being developed for both purposes. Optimal use of these databases requires both functions, and in addition, that it be accessible to the general scientific community and merged with presentational and query types of cellular encyclopedia. A description of published cellular protein databases is given in Appendix A.

Only one bacterial cellular protein database has been published [34] so far although extensive 2-D gel analysis on several other bacterial species has been published. The simplicity of these cells make them an enticing choice for this type of database. The challenge for cellular protein databases in these systems is to catalog the changes in levels of the constitutive proteins, to identify the environments under which other genes are expressed and to find the proteins belonging to each of the many regulatory systems.

Providing a catalog of *what* proteins are produced in response to *which* environment will have many applications in the field of biotechnology. For the production of foreign products in bacterial cells, 2-D gel databases which have cataloged the responses to single stress conditions could provide clues as to how the bacteria are responding to the production of the foreign or unnatural material. Similarly for the pharmaceutical industry the 2-D gel databases can be used as a diagnostic tool to give clues for the mechanism of action of new antibacterial compounds by comparing the cellular response to a new compound to the cellular response to other stress conditions where the cellular target is known. The establishment of protein databases for several bacterial species could be used to compare the cellular response of different species to the same environment, providing valuable clues as to how the species would co-exist. This information would be invaluable for many applications, for example, designing bacterial treatment procedures for sewage plants, landfills and environmental clean-up projects as well as for agricultural applications.

Eukaryotic systems have the capacity to encode up to 10 times more protein species than prokaryotic cells. The 23 human chromosomes are estimated to encode about 50,000 proteins [2]. The protein analysis can still be approached using current 2-D gel methods because most eukaryotic species will be examined one cell type or tissue at a time. Only the very simple eukaryotic organisms (e.g. yeast) will ever be examined on 2-D gels as a whole organism. Individual types of cells or fluids seem to express on average 1,000–5,000 protein species and individual tissues about 5,000. Estimates from the cell types and tissues examined to date suggest that many of the same proteins (housekeeping proteins) are found in all types of cells, although their expression level varies [72]. Several of the human databases are being cross-referenced to aid in the identification of proteins. One database is identifying proteins expressed by five different cell types and cataloging these "common" proteins [73].

The eukaryotic protein databases published so far are for *Drosophila*, mouse, rat and human. In addition to their contributions to basic research efforts (e.g. the Human Genome Project), these databases are being designed and used for disease diagnosis and other clinical applications. The information in each of these databases is presented in Appendix A, but a brief description of the aims and goals of the databases are presented here.

A protein database of *Drosophila* was established to study the changes in expression of proteins during the development of the adult fly from the larvae. The initial study represented the proteins expressed in the wing imaginal discs during metamorphosis in normal and mutant larvae [74].

There are two mouse databases under development. The mouse embryo database [75] is primarily involved in cataloging the changes in protein expression that occur during embryonic development. The aim of the mouse liver database [76] is to provide a directory of liver protein expression, particularly for studies of genetics, biochemistry and toxicology.

Three databases for rat are underway. The original rat database was developed in conjunction with the QUEST software system for analysis of 2-D gels [77,78]. This database used the REF52 (rat embryo fibroblasts) family of normal and virally-transformed rat cell lines and studied different states of proliferation, growth inhibition and growth stimulation. Two rat liver databases have been started recently. One is based on liver tissue samples [79] and the other is based on cultured rat liver cells [80]. Both databases are identifying proteins and looking for changes in levels and synthesis of individual proteins. For the liver tissue database, the major goal of the original publication was the identification of the major protein species, localization of proteins to subcellular organelles and a study of the effects of cholesterol-lowering drugs and a high cholesterol diet on the levels of proteins [79]. The other rat liver database is primarily involved in identifying and classifying proteins [80].

There are several protein databases for human cells. By linking protein spots to DNA sequence and to map positions on the chromosomes, these projects are expected to contribute significantly to the Human Genome Program. One of the very first cellular protein databases was the human plasma database [81]. A second plasma database has also been published [82]. The identifications are cross referenced between the two databases. The second database has as a goal the development of an index of plasma proteins to provide information about disease-associated changes in the protein pattern. The original plasma protein map is widely distributed and has been particularly useful to clinical investigators [81]. A common human protein database is being generated to identify the basic set of "housekeeping" proteins. The goal is to produce a 2-D gel map of spots expressed in many types of human cells and then to identify these protein spots [73]. The human keratinocyte database is an enormous project containing the largest number of identified proteins [72]. In addition to its contribution to the human genome project, this database will also contribute to the understanding of human skin diseases. The MCR-5 human embryonal lung fibroblast protein database is being established to define properties of proteins associated with the transformed state of cells [83]. The human cerebrospinal fluid database maps have so far compared the proteins from "normal" cerebrospinal fluid and to the proteins in cerebrospinal fluid from patients with either schizophrenia or Creutzfeldt–Jakob disease. Goals of this database are to allow the diagnostic classification of disease based on the changes in the proteins pattern of cerebrospinal fluid [84]. The goals of the human liver database are to compare normal and diseased tissue and also to compare the expression level of proteins in the liver with that in other cells and body fluids [85]. The human chorionic villi tissue database is a compilation of proteins consistently seen from samples of human chorionic villi. Different gestational ages are compared and changes in the intensity and isoforms of proteins are cataloged. The objectives of this database are to eventually identify marker proteins for various medical complication of pregnancy [86]. The database of human myocardial proteins plans to provide a global picture of the protein expressed in the human heart in normal and diseased individuals [87]. The human lymphoid protein database is being established to identify differences in the expression of proteins in developing thymocytes and inactive T cells [88]. Of particular interest are the proteins that are cell cycle-regulated. Another contribution of this database is the development of a laboratory information processing system (called LIPS). This system goes well beyond the electronic databases described for the other cellular protein database that focus primarily on the analysis and matching of 2-D gels and accumulation of information on the proteins found on the gels. LIPS combines an image analysis system with an electronic laboratory notebook and establishes relationships among these different types of data [89].

Several new protein databases are introduced each year. Undoubtedly these types of projects will be available for most biological systems in the future. A current development for these databases is to make them accessible to the scientific community through electronic information networks. Some databases are already available through the Internet. For example, the *E. coli* database has been deposited and is updated in the electronic repository at the National Center for Biotechnology Information as ECO2DBASE. Several have been linked to or included in other types of databases (e.g. SWISS-PROT [90]). A new type of protein database, called SWISS-2DPAGE, was initiated recently [92]. It contains genetic, biochemical and physiological data on proteins and includes images of 2-D gels from the cellular protein database. The location of the protein is indicated on the gel image. This database is available through the Internet. A detailed description of how to download this database is included in the publication. This database will be enormously useful for sharing information with other cellular protein databases and also for investigators interested in only small sets of proteins. These initial efforts towards data sharing will hopefully lead to a system allowing both easy access to data and the ability to contribute to these cellular protein databases. The labs engaged in building these databases have always collaborated with other scientists studying the same organism. Many collaborative works have been published (e.g. [45]). In some publications, the database personnel assisted the authors, were acknowledged in the publication and subsequently included the information in the database [91].

These databases are the prototypes to a biological tool of great potential. An analogy could be made with the DNA database. When DNA sequences techniques first emerged in the 1970s, the data (DNA sequence of individual genes) was useful for the analysis of the particular gene. But as more and more DNA sequences for many different organisms became available, databases containing the sequence data were started and many software programs were written to manipulate these data. Now many individuals who never sequenced any DNA spend hours scanning and analyzing the data. Eventually the majority of users of the data in the cellular protein database may never run a 2-D gel or even look at the 2-D images in the databases. These databases will be used like an encyclopedia, to look up data on a protein or set of proteins to find leads for studying specific problems relating to those proteins. Theoreticist will use the data to uncover biological parameters. Systems analysts will use this data to study, for example, biological circuits involved in metabolism or regulation. The cellular protein databases will be a part of the encyclopedia on biological systems that eventually will allow scientists to solve the ultimate question of how cells work.

Concluding Remarks

The optimal use of biological systems for technologically developed products will not be achieved until biological systems are completely defined in biochemical terms. To achieve this ultimate goal in biology a vast number of different scientific disciplines and different techniques will be required. For the analysis of cellular proteins, the 2-D gels discussed in this chapter have applications in many scientific disciplines. In the area of basic science, 2-D gels have been used in thousands of studies of enormous variety including studies focusing on single or small groups of proteins and on whole cells or organisms. In applied and clinical areas of science, 2-D gels have been used to generate better products, to make more specific tools and to aid in diagnosis of disease.

This technique was first introduced nearly 20 years ago. Based on the number of publications which include 2-D gels, it is not out-dated but instead has become increasingly popular. Improvements in the technique and in auxiliary techniques (e.g. computer-aided analysis of images, microsequencing of individual proteins from gels) have made it an even more powerful technique with enormous application in biotechnology.

Appendix A

Summary of the published cellular protein databases. The core and accessory features (discussed in the text) of these types of databases are described and a summary of the information (or types of information) are given. Many of these databases are just being developed and do not have information under all of the categories listed. Each database should be considered a prototype and will contribute to a more uniform style that should emerge in a few years. A description of each category is given at the end of the appendix. The databases are listed as follows:
1. *Escherichia coli*
2. *Drosophila melanogaster* wing marginal discs
3. Mouse embryo
4. Mouse liver tissue
5. Rat embryo fibroblasts (REF-52)
6. Rat liver tissue
7. Rat liver epithelial (RLE)
8. Human keratinocytes
9. Human plasma and red blood cells
10. Human plasma
11. Human embryonial lung fibroblasts (MCR-5)
12. Human amnion cells (AMA database)
13. Human cerebrospinal fluid
14. Human liver tissue
15. Human chorionic villi tissue

16. Human myocardial tissue
17. Human lymphoid
18. Human common proteins

No. 1

Organism/tissue/cell	*Escherichia coli*
Last update	1992 [34] Update and review
Naming system for proteins	Alpha-numeric giving approximate location on gels; X and Y coordinates on reference images give precise location
Reference images	5 images with grids
No. of protein spots in the database	788 protein spots (all listed in table)
No. of identifications	385 spots/ 335 proteins (all listed in tables by protein name, gene name and spot name)
No. of diagnostic experiments included	24 experiments tabulated in electronic version of database (not all protein spots were analyzed in all experiments). Included are: steady state growth in 4 different growth media; steady state growth at 7 different temperatures; shifts to 13 different conditions including environmental and chemical stress conditions.
Other information	Calculated molecular weight and pI from sequence, estimated molecular weight from migration on gel, molecular abundance of some proteins, E.C. number of protein, SWISS-PROT code, GenBank code, location on chromosome, donor of material used in identification, regulator of gene (all known information for each protein spot is listed in the tables)
IEF-SDS gel method	Investigator 2-D apparatus and reagents (Millipore Corp.)
Computer-aided image analysis system	BioImage (Millipore Corp)
Relational database	Macintosh computer using Microsoft Excel; available through Internet at the National Center for Biotechnology Information (Internet address is ncbi.nlm.nih.gov in the directory /ncbi/repository/ECO2DBASE)
Cross-reference to other databases	SWISS-PROT, GenBank, ECD

No. 2

Organism/tissue/cell	*Drosophila melanogaster*/wing marginal discs
Last update	1990 Initial description of database [74,93]
Naming system for proteins	Most abundant protein spots are numbered and spots between these are given that number with a letter
Reference images	Image with each protein spot named
No. of protein spots in the database	1,025 (table list each spot, its MW and its migration in the IEF gel)
No. of identifications	None listed
No. of diagnostic experiments included	One experiment compared to the standard condition. Qualitative and quantitative changes in relative rates of synthesis of proteins in different stages of development
Other information	Molecular weight of proteins estimated from their migration
IEF-SDS gel method	Referenced as O'Farrell method with modifications
Computer-aided image analysis system	
Relational database	
Cross-reference to other databases	

No. 3

Organism/tissue/cell	Mouse embryo
Last update	1992 [75]
Naming system for proteins	Each spot has a name composed of 2 letters (from the reference image) and a number
Reference images	Gel image with a coordinate system
No. of protein spots in the database	
No. of identifications	12 listed in a table
No. of diagnostic experiments included	9 experiments some have multiple time points (selected sets of proteins are presented)
Other information	
Standardized IEF-SDS gel method	Referenced as QUEST system
Computer-aided image analysis system	QUEST
Relational database	
Cross-reference to other databases	

No. 4

Organism/tissue/cell	Mouse liver tissue
Last update	1992 [76]
Naming system for proteins	Computer program assigns spot numbers
Reference images	Gel image; computer map has spot numbers
No. of protein spots in the database	360 proteins (all listed in a table)
No. of identifications	17 listed in a table
No. of diagnostic experiments included	12 experiments (2000 gels) qualitative changes indicated in table
Other information	MW, relative migration in the first dimension, and subcellular location all listed in the table for each spot
IEF-SDS gel method	Iso-Dalt method
Computer-aided image analysis system	Tycho III
Relational database	Oracle
Cross-reference to other databases	

No. 5

Organism/tissue/cell	Rat/embryo fibroblasts (REF-52)
Last update	1989 [77,79]; 1990 [94]; presented as an analysis of the data
Naming system for proteins	Each spot has a name composed of 2 letters (from the reference image) and a number
Reference images	A reference image with coordinates
No. of protein spots in the database	>1,600
No. of identifications	45 protein spots listed in a table
No. of diagnostic experiments included	12 major experiments (79 gels); data on sets of proteins are presented
Other information	Subcellular location, modifications, antibodies, coprecipitation, N-terminal sequence, references (not listed in published database)
Standardized IEF-SDS gel method	QUEST
Computer-aided image analysis system	Quest
Relational database	
Cross-reference to other databases	PIR

No. 6

Organism/tissue/cell	Rat liver tissue
Last update	1991 [79]
Naming system for proteins	Spot numbers assigned by computer program

Reference images	Gel image; computer map with protein spot numbers
No. of protein spots in the database	>1,200
No. of identifications	23 listed in table
No. of diagnostic experiments included	700 IEF-SDS gels matched to reference image including response to drug therapy, and some protein characterization (sampling of data is presented).
Other information	MW and pI given for all protein spots
IEF-SDS gel method	Iso-Dalt (Large Scale Biology)
Computer-aided image analysis system	Kepler
Relational database	Kepler
Cross-reference to other databases	

No. 7

Organism/tissue/cell	Rat liver epithelial (RLE)
Last update	1993 [80]
Naming system for proteins	Spots assigned by computer program.
Reference images	2 gel images and computer maps (not all spot numbers are given)
No. of protein spots in the database	1,100 nucleoplasmic proteins; 850 particulate associated proteins
No. of identifications	10 listed, total number not given in publication
No. of diagnostic experiments included	
Other information	MW, pI, subcellular location, transformation and growth-related information (data not included in any table in publication)
IEF-SDS gel method	Referenced
Computer-aided image analysis system	Elsie 5
Relational database	
Cross-reference to other databases	SWISS-PROT, PIR, PATCHX, NBRF

No. 8

Organism/tissue/cell	Human keratinocytes
Last update	1993 [72] Update and review
Naming system for proteins	Protein spots are numbered through the image analysis software program
Reference images	Two reference images (for equilibrium and non-equilibrium gels) are displayed as enlarged maps with each spot marked with its spot number. The images are divided into sections. Twenty other images locate sets of proteins
No. of protein spots in the database	3,038 protein spots (a table lists the MW and pI for all of these proteins
No. of identifications	763 protein spots identified
No. of diagnostic experiments included	1 experiment comparing normal and psoriatic skin; proteins highly upregulated are indicated on a gel image
Other information	176 spots have been microsequenced with no match in protein databases. Seven images which classify images are included (components of macromolecules, GTP binding proteins, calcium binding proteins, stress proteins, autoantigens, organelle markers and differentiation markers
IEF-SDS gel method	Described in update
Computer-aided image analysis system	PDQuest
Relational database	
Cross-reference to other databases	SWISS-PROT

No. 9

Organism/tissue/cell	Human plasma and red blood cells
Last update	1993 [82]
Naming system for proteins	Spot numbers assigned by computer program
Reference images	Gel image (spot numbers or protein names are given on image)
No. of protein spots in the database	
No. of identifications	10 newly identified; 22 previously identified
No. of diagnostic experiments included	
Other information	The N-terminus of 17 proteins has been determined but not matched to any protein, these proteins have been given SWISS-PROT accession numbers
IEF-SDS gel method	Referenced
Computer-aided image analysis system	Melanie
Relational database	
Linkage to other databases	SWISS-PROT, SWISS-2DPAGE

No. 10

Organism/tissue/cell	Human plasma cells
Last update	1992 [81]
Naming system for proteins	Spot numbers assigned by computer program
Reference images	Gel image; computer map with protein spot numbers
No. of protein spots in the database	About 400 all listed in a table
No. of identifications	49 polypeptides (most are represented by multiple spots and all are listed in the table)
No. of diagnostic experiments included	
Other information	MW and reference to migration in the first dimension (all values given in table)
IEF-SDS gel method	Iso-Dalt (Large Scale Biology)
Computer-aided image analysis system	Tycho
Relational database	Kepler
Linkage to other databases	SWISS-PROT, SWISS-2DPAGE

No. 11

Organism/tissue/cell	Human embryonial lung fibroblasts (MCR-5)
Last update	1990 [83]
Naming system for proteins	Spot numbers assigned by computer
Reference images	Gel image with selected spots; computer map with all protein spot numbers
No. of protein spots in the database	1,895
No. of identifications	By cross-referencing to AMA database
No. of diagnostic experiments included	Comparison of quiescent, proliferating, SV40 transformed MCR-5 fibroblasts cells
Other information	MW and pI (all listed in table)
IEF-SDS gel method	Referenced
Computer-aided image analysis system	PDQuest II
Relational database	AMA cellular protein database
Linkage to other databases	

No. 12

Organism/tissue/cell	Human amnion cells (AMA database)
Last update	1990 [36]
Naming system for proteins	Protein spots numbered by computer program
Reference images	Gel image is labeled with identified proteins; computer map has the numbers of each protein spot

No. of protein spots in the database	3,430
No. of identifications	Many, not totaled in the publication but listed in a series of tables based on method used for identification
No. of diagnostic experiments included	
Other information	Table is given of the annotation fields; included: % of total radioactivity of each spot, antibody against protein, cellular location, levels in fetal human tissues, partial protein sequences, cell-cycle-regulated, protein sensitive to interferon, heat shock protein, annexins and phosphorylated proteins.
IEF-SDS gel method	Referenced
Computer-aided image analysis system	PDQuest
Relational database	
Cross-reference to other databases	

No. 13

Organism/tissue/cell	Human cerebrospinal fluid
Last update	1992 [84]
Naming system for proteins	Spot numbers
Reference images	Enlarged gel images with protein names or spot numbers given
No. of protein spots in the database	931
No. of identifications	31 listed in table
No. of diagnostic experiments included	Other disease associated proteins shown on extra gel images
Other information	MW and charge (not listed in publication)
IEF-SDS gel method	Referenced
Computer-aided image analysis system	GALtool and Advanced Visual Systems software
Relational database	
Cross-reference to other databases	

No. 14

Organism/tissue/cell	Human liver tissue
Last update	1993 [85]
Naming system for proteins	Spot numbers
Reference images	Gel image gives spot number or protein name
No. of protein spots in the database	3,000 spots detected (only identified proteins in the publication)
No. of identifications	65 spots/50 polypeptides
No. of diagnostic experiments included	
Other information	
IEF-SDS gel method	Referenced
Computer-aided image analysis system	Melanie
Relational database	
Cross-reference to other databases	SWISS-2DPAGE, SWISS-PROT

No. 15

Organism/tissue/cell	Human chorionic villi tissue
Last update	1991 [86] (description of database; no data presented)
Naming system for proteins	Spots are lettered and numbered according to the regions on the gel image
Reference images	Gel image is divided into 9 areas, labeled A-I and then spots are given numbers within each area.
No. of protein spots in the database	198
No. of identifications	7 mentioned

No. of diagnostic experiments included   Comparison of 8 week, 12 week and term placenta
Other information   31 types of information collected
IEF-SDS gel method   O'Farrell with modification referenced
Computer-aided image analysis system
Relational database   IBM PC computer and Foxbase + software
Cross-reference to other databases

No. 16

Organism/tissue/cell   Human myocardial tissue
Last update   1992 [87]
Naming system for proteins   Spots listed in publication have manually assigned spot numbers; (spot numbers also assigned by computer)
Reference images   Gel image with spot numbers given
No. of protein spots in the database   1,500 (50 listed in publication)
No. of identifications   46
No. of diagnostic experiments included
Other information   MW and pI
IEF-SDS gel method   Iso-Dalt (Large Scale Biology) and Investigator (Millipore Corp)
Computer-aided image analysis system   PDQuest
Relationship database
Cross-reference to other databases

No. 17

Organism/tissue/cell   Human lymphoid
Last update   1993 [88] (analysis of data in database)
Naming system for proteins
Reference images   Not shown in publication
No. of protein spots in the database
No. of identifications   20 relevant to this analysis were listed
No. of diagnostic experiments included   Several hundred gels entered into database (selected data presented)
Other information   See [91]
IEF-SDS gel method   Referenced
Computer-aided image analysis system   Referenced
Relational database   Referenced
Cross-reference to other databases   See Ali et al., 1992 [89]

No. 18

Organism/tissue/cell   Human common proteins (6 cell lines so far)
Last update   1992 [73]
Naming system for proteins   Computer assigned numbers?
Reference images   Master gel for each cell line, a composite image including only proteins common to all cells lines is "core" of the database
No. of protein spots in the database   450 common protein spots
No. of identifications
No. of diagnostic experiments included   6 cell lines of different germ layers compared
Other information   Subcellular location
IEF-SDS gel method   First dimension is IPG; second dimension is Investigator (Millipore Corp)
Computer-aided image analysis system   BioImage
Relational database
Cross-reference to other databases

Notes

Organism/tissue/cell	The biological organism being studied. For eukaryotic cells the tissue type and the cell type may also be listed. Many of these databases have been named and these names will be given in this section of the description
Last update	Year in which last update was published. For some of the databases, the updates are cumulative and include all of the information the authors have obtained. For other databases, the updates include only the new information or list only some of the database's entries
Naming system for proteins	One of the basic features of any database is that each protein has a unique name or number which allows that protein to be traced through all gels. For groups using computer-aided image analysis systems, the spot numbers assigned to each spot on the standard image is often used as the naming system. Other groups have developed other naming systems, most other naming systems give an approximate location of the spots on the image
Reference images	An image of each type of IEF-SDS gels is given as a reference image. The reference images may indicate the location of each protein spot in the database (using the spot name), it may be overlayed with a grid so that each spot can be given an X and Y coordinate or a computer generated map of the image (labeling the spot numbers) may be used in addition to gel image
No. of protein spots in the database	Some databases list only the protein spots for which some information is known. Others list all of the spots seen on the reference images
No. of identifications	Identification of a spot as a known protein or the product of a gene allows the spot to be cross-reference to information in publications or other databases. These are important for the development and general usage of the database so a sum total of the identification is given (if the number can be extracted from the information given in the database)
No. of diagnostic experiments included	Physiological experiments reveal alterations in the level or rates of synthesis of individual proteins
Other information	This is a catch-all category containing a variety of information about proteins which can be obtained from an IEF-SDS gel (even if the protein spot has not been identified) or which has been extracted from the literature and linked to the protein spot (if the spot is identified)
IEF-SDS gel method	There are many apparatus and chemicals used to run IEF-SDS gels. The databases are most useful if other investigators can run gels which closely match the reference images
Computer-aided image analysis system	Many of the database groups are using software programs to assist in the analysis of the IEF-SDS gels. Many software programs have been developed. Although images can usually be transferred between the difference programs the analysis files associated with the images cannot be transferred. Therefore, investigators interested in accessing a database may want to acquire the same software program as the database developers
Relational database	Databases can be available in electronic form if the data is entered into a database or spreadsheet or if computer software is used in the analysis of gels. These are very

different. The data entered into a database or spreadsheet can be used without the reference images to obtain information about individual proteins or groups of proteins. If the database is only kept as part of the image analysis software program, information about individual proteins or sets of proteins can only be accessed through the reference image and with the use to the software program. This second type of electronic database will limit the number of investigators that can access the information in the database to those who can navigate around the IEF-SDS gel

Cross-reference to other databases | Most of the databases contain an enormous amount of information about proteins. Addition information about many of the identified proteins is available in other types of database (or in other cellular protein databases). Each type of database has an identifier for each protein. Cross-referencing amongst the different databases allows for quick and easy access to additional data

References

1. Neidhardt FC, Ingraham JL, Schaechter M. Physiology of the bacterial cell: a molecular approach. Sunderland, MA: Sinauer Associated, 1990.
2. Celis JE, Gesser B, Pasmussen HH, Madsen P, Leffers H, Dejgaard K, Honore B, Olsen E, Ratz G, Lauridsen JB, Basse B, Mouritzen S, Hellerup M, Andersen A, Walbum E, Celis A, Bauw G, Puype M, Van Damme J, Vandekerckhove J. Comprehensive two-dimensional gel protein databases offer a global approach to the analysis of human cells: the transformed amnion cells (AMA) master database and its link to genome DNA sequence data. Electrophoresis 1990;11:989–1071.
3. O'Farrell PH. High resolution two-dimensional electrophoresis of proteins. J Biol Chem 1975;250:4007–4021.
4. Styer L. Biochemistry, 3rd edn. New York: WH Freeman, 1988.
5. Adams LD. In: Ausubel FM, Brent R, Kingston RE, Moore DD, Seidman JG, Smith JA, Struhl K (eds) Current Protocols in Molecular Biology. Greene/Wiley, 1993;10.4.1–10.4.13.
6. Leninger AL. Biochemistry, 2nd edn. New York: Worth, 1975;p.180.
7. Prescott CD, Dahlberg AE. A single base change at 726 in 16S rRNA radically alters the pattern of proteins synthesized in vivo. EMBO J 1990;9:289–294.
8. O'Farrell PH. The suppression of defective translation by ppGpp and its role in the stringent response. Cell 1978;14:545–557.
9. Garrels JI. Two-dimensional gel electrophoresis and computer analysis of proteins synthesized by clonal cell lines. J Biol Chem 1979;254:7961–7977.
10. Anderson NL, Anderson NG. High resolution two-dimensional electrophoresis of human plasma proteins. Proc Natl Acad Sci USA 1977;74:5421–5425.
11. Young DA, Voris BP, Maytin EV, Colbert RA. Very-high-resolution two-dimensional electrophoretic separation of proteins on giant gels. Methods Enzymol 1983;91:190–214.
12. Young DA. Advantages of separations on giant two-dimensional gels for detection of physiologically relevant changes in the expression of protein gene-products. Clin Chem 1984;30:2104–2108.
13. Zazra JJ, Szklaruk B. Tetramethlurea as a protein denaturing agent in gel electrophoresis. Electrophoresis 1987;8:331–332.
14. Cann JR, Stimpson DI. Isoelectric focusing of interacting systems, I. Carrier ampholyte-induced macromolecular isomerization. Biophys Chem 1977;7:103–114.
15. Bjellqvist BEK, Righetti PG, Gorg A, Westermeir R, Postel W. Isoelectric focusing in immobi-

lized pH gradients: principle, methodology and some applications. J Biochem Biophys Methods 1982;6:317–339.

16. Chiari M, Righetti PG. The Immobiline family: from "vacuum" to "plenum" chemistry. Electrophoresis 1992;13:187–191.

17. Bjellqvist, Passquali C, Ravier F, Sanchez J-C, Hochstrasser D. A nonlinear wide-range immobilized pH gradient for two-dimensional electrophoresis and its definition in a relevant pH scale. Electrophoresis 1993;14:1357–1365.

18. O'Farrell PZ, Goodman HM , O'Farrell PH. High resolution two dimensional electrophoresis of basic as well as acidic proteins. Cell 1977;1133–1142.

19. Patton WF, Lopez MF, Barry P, Skea WM. A mechanically strong matrix for protein electrophoresis with enhanced silver staining properties. BioTechniques 1992;12:580–585.

20. Schagger H, von Jagow G. Tricine-sodium dodecyl sulfate-polyacrylamide gel electrophoresis for the separation of proteins in the range from 1 to 100 kDa. Anal Biochem 1987;166:368–379.

21. Switzer RC, Merril CR, Shifrin S. A highly sensitive silver stain for detecting proteins and peptides in polyacrylamide gels. Anal Biochem 1979;98:231–237.

22. Rabilloud T. Mechanisms of protein silver staining in polyacrylamide: a 10 year synthesis. Electrophoresis 1990;11:785.

23. Gersten DM, Pamagli LS, Johnson DA, Rodriguez LV. The use of radioactive bacteriophage proteins as X-Y markers for silver-stained two-dimensional gels and quantification of the patterns. Electrophoresis 1992;13:1387–1392.

24. Patterson SD, Latter GI. Evaluation of storage phospho imaging for quantitative analysis of 2-D gels using the Quest II system. BioComputing 1993;15:1076–1083.

25. Bonner WM, Laskey RA. A film detection method for tritium-labeled proteins and nucleic acids in polyacrylamide gels. Eur J Biochem 1974;46:83–88.

26. Vandekerckhove J, Bauw G, Puype M, VanDamme J, Van Montagu M. Protein-blotting on polybrene-coated glass-fiber sheets. A basis for acid hydrolysis and gas-phase sequencing of picomole quantities of proteins. Eur J Biochem 1985;152:9–19.

27. Aebersold RH, Teplow DB, Hood LE, Kent SBH. Electroblotting onto activated glass. High efficiency preparation of proteins from analytical sodium dodecyl sulfate-polyacrylamide gels for direct sequence analysis. J Biol Chem 1986;261:4229–4238.

28. Bauw G, Rasmussen HH, Van Den Bulcke M, Van Damme J, Puype M, Gesser B, Celis JE. Two-dimensional gel electrophoresis, protein electroblotting and microsequencing: a direct link between proteins and genes. Electrophoresis 1990;11:528–536.

29. Patterson SD, Hess D, Yungwirth T, Aebersold R. High-yield recovery of electroblotted proteins and cleavage fragments from a cationic polyvinylidene fluoride-based membrane. Anal Biochem 1992;202:193–203.

30. Lemaux PG, Herendeen SL, Bloch PL, Neidhardt FC. Transient rates of synthesis of individual polypeptides in *E. coli* following temperature shifts. Cell 1978;13:427–434.

31. VanBogelen RA, Kelley PM, Neidhardt FC. Differential induction of heat shock, SOS, and oxidation stress regulons and accumulation of nucleotides in *Escherichia coli*. J Bacteriol 1987;169:26–32.

32. Pedersen S, Bloch PL, Reeh S, Neidhardt FC. Patterns of protein synthesis in *E. coli*: a catalog of the amount of 140 individual proteins at different growth rates. Cell 1978;14:179–190.

33. Gudas LJ, Mount DW. Identification for the *recA* (*tif*) gene product of *Escherichia coli*. Proc Natl Acad Sci USA 1977;74:5280–5284.

34. VanBogelen RA, Sankar P, Clark RL, Bogan JA, Neidhardt FC. The gene-protein database of *Escherichia coli*: Edition 5. Electrophoresis 1992;13:1014–1054.

35. Galleagher S, Winston SE, Fuller SA, Hurrell JGR. In: Ausubel FM, Brent R, Kingston RE, Moore DD, Seidman JG, Smith JA, Struhl K (eds.) Current Protocols in Molecular Biology. Greene/Wiley, 1993;10.8.1–10.8.17.

36. Celis JE, Gesser B, Rasmussen HH, Madsen P, Leffers H, Dejgaard K, Honore B, Olsen E, Ratz G,

Lauridsen JB, Basse B, Mouritzen S, Hellerup M, Andersen A, Walbum E, Celis A, Bauw G, Puype M, Van Damme J, Vandekerckhove J. Comprehensive two-dimensional gel protein databases offer a global approach to the analysis of human cells: the transformed amnion cells (AMA) master database and its link to genome DNA sequence data. Electrophoresis 1990;11:989–1071.

37. Miller I, Gemeiner M. Two-dimensional electrophoresis of cat sera: protein identification by cross reacting antibodies against human serum proteins. Electrophoresis 1992;13:450–453.

38. Maltese WA, Sheridan KM, Repko EM, Erdman RA. Post-translational modification of low molecular mass GTP-binding proteins by isoprenoid. J Biol Chem 1990;265:2148–2155.

39. Phares W, Franzo BR, Herr W. The kB enhancer motifs in human immunodeficiency virus type 1 and simian virus 40 recognize different binding activities in human jurkat and H9 T cells: evidence for NF-kB-independent activation of the kB motif. J Virol 1992;66:7490–7498.

40. Engstrom P, Hazelbauer GL. Multiple methylation of methyl-accepting chemotaxis proteins during adaption of E. coli to chemical stimuli. Cell 1980;20:165–167.

41. Neidhardt FC, Phillips TA, VanBogelen RA, Smith MW, Georgalis Y, Subramanian AR. Identity of the B56.5 protein, the A-protein and the groE gene product of Escherichia coli. J Bacteriol 1981;145:513–520.

42. Taylor J, Giometti CS. Use of principal components analysis for mutation detection with two-dimensional electrophoresis protein separations. Electrophoresis 1992;13:162–168.

43. Sankar P, Hutton ME, VanBogelen RA, Clark RL, Neidhardt FC. Expression analysis of cloned chromosomal segments of Escherichia coli. J Bacteriol 1993;175:5145–5152.

44. Wackett LP, Wanner BL, Venditti CP, Walsh CT, Involvement of the phosphate regulon and the psiD locus in carbon-phosphorus lyase activity in Escherichia coli. J Bacteriol 1987;169:1753–1756.

45. Vandekerckhove J, Bauw G, Vancompernolle K, Honore B, Celis J. Comparative two-dimensional gel analysis and microsequencing identifies Gelsolin as one of the most prominent downregulated markers of transformed human fibroblast and epithelial cells. J Cell Biol 1990;111:95–102.

46. Hanash SM, Strahler JR, Neel JV, Hailat N, Melhem R, Keim D, Zhu XX, Wagner D, Gage DA. Highly resolving two-dimensional gels for protein sequencing. Proc Natl Acad Sci USA 1991;88:5709–5713.

47. Aeborsold R, Leavitt J. Sequence analysis of proteins separated by polyacrylamide gel electrophoresis: towards an integrated protein database. Electrophoresis 1990;11:517–527.

48. Nystrom T, Neidhardt FC. Cloning, mapping and nucleotide sequence of a gene encoding a universal stress protein in Escherichia coli. Mol Microbiol 1992;6:3187–3198.

49. Rasmussen HH, Van Damme J, Puype M, Gesser B, Celis JE, Vandekerckhove J. Microsequencing of 145 proteins recorded in human two-dimensional gel protein database of normal epidermal keratinocytes. Electrophoresis 1992;13:960–969.

50. Cleveland DW, Fischer SG, Krischner MN, Laemmli UK. Peptide mapping by limited proteolysis in sodium dodecyl sulfate and analysis by gel electrophoresis. J Biol Chem 1977;252:1102–1106.

51. Hirshfield IN, Bloch PL, VanBogelen RA, Neidhardt FC. Multiple forms of lysyl-transfer ribonucleic acid synthetase in Escherichia coli. J Bacteriol 1981;146:345–351.

52. VanBogelen RA, Vaughn V, Neidhardt FC. Gene for heat-inducible lysyl-tRNA synthetase (lysU) maps near cadA in Escherichia coli. J Bacteriol 1983;153:1066–1068.

53. Clark RL, Neidhardt FC. Roles of two lysyl-tRNA synthetases of Escherichia coli: analysis of nucleotide sequence and mutant behavior. J Bacteriol 1990;172:3237–3243.

54. Kawakami K, Jonsson YH, Bjork GR, Ikeda H, Nakamura Y. Chromosomal location and structure of the operon encoding peptide-chain-release factor 2 of Escherichia coli. Proc Natl Acad Sci USA 1988;85:5620–5624.

55. Gampel MA, Tzagoloff A. Homology of aspartyl and lysyl-tRNA synthetases, Proc Natl Acad Sci USA 1989;86:6023–6027.

56. Ekramoddoullah AKM. Two-dimensional gel electrophoretic analyses of Kentucky bluegrass and rye grass pollen allergens. Int Arch Allergy Appl Immunol 1990;93:371–377.

57. Harlow E, Lane D. Antibodies, a laboratory manual. Cold Spring Harbor, NY: Cold Spring Harbor Laboratory, 1988.

58. Seifter S, Englard S. In: Deutscher MP (ed.) Methods in Enzymology, vol 182. San Diego, CA: Academic Press, 1990;626–645.

59. Wold F, Moldave K. Methods in Enzymology, vol 106. San Diego, CA: Academic Press.

60. Wold F, Moldave K. Methods in Enzymology, vol 107. San Diego, CA: Academic Press.

61. Pedersen S, Bloch PL, Reeh S, Neidhardt FC. Patterns of protein synthesis in E. coli: a catalog of the amount of 140 individual proteins at different growth rates. Cell 1977;14:179–190.

62. Abshire KZ, Neidhardt FC. Growth rate paradox of *Salmonella typhimurium* within host macrophages. J Bacteriol 1993;175:3744–3748.

63. Stuiver MH, Celis JE, van der Vliet PC. Identification of nuclear factor IV/Ku autoantigen in a human 2D-gel protein database. FEBS Lett 1991;282:189–192.

64. Beckmann R, Buchner K, Jungblut PR, Eckerskorn C, Weise C, Hilbert R, Hucho F. Nuclear substrates of Protein Kinase C. Eur J Biochem 1992;210:45–51.

65. Neidhardt FC, VanBogelen RA. Positive regulatory gene for temperature-controlled proteins in *Escherichia coli*. Biochem Biophys Res Comm 1981;100:894–900.

66. Nystrom T, Neidhardt FC. Isolation and properties of a mutant of *Escherichia coli* with and insertional inactivation of the uspA gene, which encodes a universal stress protein. J Bacteriol 1993;175:3949–3956.

67. Fried R, Wiederkehr F, Vonderschmitt DJ. Immune complexes in blood: a new strategy for the analysis of their antigen portion. Electrophoresis 1993;13:718–719.

68. Celis JE, Crüger D Kiil J, Lauridsen JB, Ratz G, Basse B, Celis A. Identification of a group of proteins that are strongly up-regulated in total epidermal keratinocytes from psoriatic skin. FEBS Lett 1990;262:159–164.

69. Paugh DR, Teliam SA, Disher MJ, Identification of perilymph proteins by two-dimensional gel electrophoresis. Onolaryngol: Head Neck Surg 1991;104:517–525.

70. Chen R-D, Tabaeizadeh Z. Expression and molecular cloning of drought-induced genes in the wild tomato Lycopersicon chilense. Biochem Cell Biol 1992;70:199–206.

71. Simpson RJ, Tsugita A, Celis JE, Garrels JI, Mewes HW. Workshop on two-dimensional gel protein databases. Electrophoresis 1992;13:1055–1066.

72. Celis JE, Rasmussen HH, Olsen E, Madsen P, Leffers H, Honore B, Dejgaard K, Gromov P, Hoffmann HJ, Nielsen M. The human keratinocyte two-dimensional gel protein database: update 1993. Electrophoresis 1993;14:1091–1198.

72a. Celis JE, Rasmussen HH, Madsen P, Leffers H, Honoré B, Dejgaard K, Gesser B, Olsen E, Gromov P, Hoffman HJ, Nieslsen M, Andersen AH, Walbum E, Kjærgaard I, Puype M, Van Damme J, Vandekerckhove J. The human keratinocyte two-dimensional gel protein database (update 1992): towards an integrated approach to the study of cell proliferation, differentiation and skin disease. Electrophoresis 1992;13:893–959.

73. Burggaraf D, Andersson K, Eckerskorn C, Lottsplich F. Toward a two-dimensional database of common human proteins. Electrophoresis 1992;13:729–732.

74. Santaren JF. Towards establishing a protein database of Drosophila. Electrophoresis 1990;11:254–267.

75. Latham KE, Garrels JI, Chang C, Solter D. Analysis of embryonic mouse development: construction of a high-resolution, two-dimensional gel protein database. Appl Theor Electrophoresis 1992;2:163–170.

76. Giometti CS, Taylor J, Tollaksen SL. Mouse liver protein database: a catalog of proteins detected by two-dimensional gel electrophoresis. Electrophoresis 1992;13: 970–991.

77. Garrels JI, Franza BR, The REF52 protein database. J Biol Chem 1989;264:5283–5298.

78. Garrels JI, Franza BR, Transformation-sensitive and growth-related changes of protein synthesis in REF52 cells. J Biol Chem 1989;264:5299–5312.

79. Anderson NL, Esquer-Blasco R, Hofmann J-P, Anderson NG. A two-dimensional gel database of

rat liver proteins useful in gene regulation and drug effects studies. Electrophoresis 1991;12:907–930.

80. Wirth PJ, Luo L-D, Fujimoto Y, Bisgaard HC, Olson AD. The rat epithelial (RLE) cell protein database. Electrophoresis 1991;12:931–954.

81. Anderson NL, Anderson NG. A two-dimensional gel database of human plasma proteins. Electrophoresis 1991;12:883–906.

82. Hughes GJ, Frutiger S, Paquat N, Ravier F, Pasquali C, Sanchez J-C, James R, Tissot J-D, Bjellqvist B, Hochstrasser DF. Plasma protein map: update by microsequencing. Electrophoresis 1992;13:707–714.

83. Celis JE, Dejgaard K, Madsen P, Leffers H, Gesser B, Honore B, Rasmussen HH, Olsen E, Lauridsen JB, Ratz G, Mouritzen S, Basse B, Hellerup M, Celis A, Puype M, Van Damme J, Vandekerckhove J. The MCR-5 human embryonal lung fibroblast two-dimensional gel cellular protein database: quantitative identification of polypeptides whose relative abundance differs between quiescent, proliferating and SV40 transformed cells. Electrophoresis 1990;11:1072–1113.

84. Yun M, Wu W, Hood L, Harrington M. Human cerebrospinal fluid protein database: edition 1992. Electrophoresis 1992;13:1002–1013.

85. Hochstrasser D F, Frutiger S, Paquet N, Bairoch A, Ravier F, Pasquali C, Sanchez J-C, Tissot J-D, Bjellqvist B, Vargas R, Appel RD, Hughes GJ. Human liver protein map: a reference database established by microsequening and gel comparison. Electrophoresis 1993;13:992–1001.

86. Kornienko YA, Shishkin SS, Kaurov BA, Proteins of human chorionic villi tissue: construction of a proteins map based upon fractionation by two-dimensional gel electrophoresis. Biomed Sci 1991;2:590–594.

87. Baker CS, Corbett JM, May AJ, Yacoub MH, Dunn MJ. A human myocardial two-dimensional electrophoresis database: protein characterization by microsequencing and immunoblotting. Electrophoresis 1992;13:723–726.

88. Hanash SM, Strahler JR, Chuan Y, Kuick R, Teichroew D, Neel JV, Hailat N, Keim DR, Gratiot-Deans J, Ungar D, Melhem R, Zhu XX, Andrews P, Lottspeich F, Eckerskorn C, Chu E, Ali I, Fox DA, Richardson BC, Turka LA. Database analysis of protein expression patterns during T-cell ontogeny and activation. Proc Natl Acad Sci USA 1993;90:3314–3318.

89. Ali I, Chan Y, Kuick R, Teichroew D, Hanash SM. Implementation of a two-dimensional electrophoresis related laboratory information processing system: database aspects. Electrophoresis 1992;12:747–761.

90. Bairoch A, Boeckmann, B. The SWISS-PROT protein sequence data bank recent developments. Nucleic Acids Res 1993;21:3093–3096.

91. Ernsting BR, Atkinson MR, Ninfa AJ, Matthews RG. Characterization of the regulon controlled by the leucine-responsive regulatory protein in *Escherichia coli*. J Bacteriol 1992;174:1109–1118.

92. Appel RD, Sanchez J-C, Bairoch A, Golaz O, Miu M, Vargas JR, Hochstrasser D F. SWISS-2-DPAGE: a database of two-dimensional gel electrophoresis images. Electrophoresis 1993;14:1232–1238.

93. Assigeo R, Santaren JF. High resolution two-dimensional gel analysis of proteins in the central nervous system of larvae of *Drosophila melanogaster*. Electrophoresis 1992;13:321–328.

94. Garrels JI, Franza BR, Chang C, Latter G. Quantitative exploration of the REF52-protein database: cluster analysis reveals the major protein expression profiles in response to growth regulation, serum stimulation, and viral transformation. Electrophoresis 1990;11:1114–1130.

Biotechnology Annual Review Volume 1
M.R. El-Gewely, editor

Prokaryotic promoters in biotechnology

Marc A. Goldstein[1] and Roy H. Doi[2]
[1]*Section of Plant Biology and* [2]*Section of Molecular and Cellular Biology, University of California, Davis, California, USA*

Abstract. The basic properties of prokaryotic promoters and the promotor region are described with special emphasis on promoters that are found in *Escherichia coli* and *Bacillus subtilis*. Promoters recognized by major and minor forms of RNA polymerase holoenzymes are compared for their specificities and differences. Both natural and hybrid promoters that have been constructed for purposes of efficient and regulated transcription are discussed in terms of their utility. Since promoter regions contain sequences that are recognized not only by RNA polymerase but by positive and negative regulatory factors that regulate expression from promoters, the functions and properties of these promoter regions are also described. The current utility and the future prospects of the prokaryotic promoters in expressing heterologous genes for biotechnology purposes are discussed.

Introduction

The expression of prokaryotic genes is a complex and highly specific process. It encompasses a wide variety of phenomena including transcription, translation, post-translational modifications, protein folding and assembly, protein interactions, protein targeting, and a secretion process for extracellular proteins. At each of these steps there are specific regulatory signals, control factors, and enzymes and proteins that allow the efficient expression of the gene. Under "normal" growth conditions gene expression results in the production of a protein that is present at the correct time, at the proper amount, and at the correct intracellular or extracellular site. This review focuses on a very narrow aspect of this complex process. We concentrate on describing the properties of prokaryotic promoters and promoter regions (the promoter region contains the promoter and other regulatory sites that may overlap or be adjacent to the promoter) that effect the desired expression of a gene and how these properties might be exploited for biotechnology purposes.

Strategies for Selection and Use of Promoters

The strategy for using a specific promoter will depend on one of several reasons for expressing the gene. Do you want to express the gene constitutively and at the high-

Address for correspondence: R.H. Doi, Section of Molecular and Cellular Biology, University of California, Davis, CA 95616, USA.

est rate possible during growth of the host? Do you want to express the gene after growth of the cell has ceased and for a long period in the stationary phase? Do you want to quickly induce the expression of the gene under certain physiological and environmental conditions? Do you want the highest yield possible of the protein product? Do you want expression of the gene under conditions that might ordinarily cause repression of the gene? Do you want to produce reasonably high quantities of the protein in its native state and not as an inclusion body? Do you want to produce extracellular proteins that will not overtax the host's secretion system? These and other factors would determine the particular promoter and physiological conditions that one may want to use for the expression of homologous and/or heterologous genes.

Further considerations include the goal of expressing a gene under conditions that would not be detrimental to the growth or viability of the host. In these cases the promoter would be prevented from being expressed until the cell population has reached a high enough density that the production of even a "toxic" product will be sufficient in spite of the death of the host cells during its production. Other mechanisms including making a nontoxic fusion protein are beyond the scope of this discussion.

Thus, the selection of a promoter would be dependent on a number of factors that take into consideration not only the expression of a gene, but the consequences of gene expression to the cell.

General Requirements for Prokaryotic Gene Transcription

Promoter properties

Strictly speaking a "promoter" is considered as the binding site for RNA polymerase and as the initiation site for transcription. The promoter is, however, regulated by a number of regulatory factors whose DNA binding sites at times are adjacent to or even overlap the promoter, and thus a broader term for this region could be the "promoter region". We attempt to distinguish between these terms during our discussion.

Major promoters

The major promoters would be classified as those promoters that are recognized by the major prokaryotic RNA polymerase holoenzymes. The major promoters would include those promoters that are recognized, for instance, by the *Escherichia coli* sigma-70 or *Bacillus subtilis* sigma-A RNA polymerase holoenzymes. These holoenzymes are considered as "major" since the sigma factors for these enzymes occur in much higher quantities than the minor sigma factors, and these forms of the holoenzyme recognize the promoters of genes that are generally expressed during rapid growth of the cell and are usually not repressed by the presence of glucose in

Table 1. Reasons for promoter selection

Maximum expression during growth
Maximum expression during stationary phase
Constitutive expression of a gene
Induced expression of a gene
Controlled expression of a gene
Moderate expression of a gene
Temporal expression of a gene

---

the medium. These major promoters have been studied extensively and their properties have been well established. Some of the key features of these promoters include the following:

a. A conserved hexamer sequence at the −10 and −35 positions of the promoter (Fig. 1). When a promoter contains both consensus −10 (TATAAT) and −35 (TTGACA) sequences, it is usually a very efficient, constitutive promoter. Very few natural major promoters have both consensus hexamer sequences. Most hexamer sequences in fact do not contain the consensus sequence. Some promoters have a poor homology to the consensus sequence at the −35 region and usually these promoters require ancillary factors to be maximally expressed. There is evidence in some rare cases that the −35 region is not required for promoter recognition and expression. Interestingly, no two exact prokaryotic promoter sequences have been found, indicating that the RNA polymerase is able to bind to DNA sequences that are heterogeneous.

b. A third recognition element, UP. Another recognition element in bacterial promoters is a DNA sequence rich in A-T located upstream of the −35 region of promoters. This region, UP, is involved in stimulating transcription from these promoters and is recognized by the α-subunit of the RNA polymerase core [1]. Mutations in the carboxy-terminal region of the α-subunit prevent stimulation of promoters containing the UP module. In addition purified α binds directly to UP indicating that there is direct interaction between UP and the α subunit of RNA polymerase core. Since the same α mutations also block activation by some transcription activators, there may be a relationship between the regulatory mechanisms of positive control elements and the UP.

c. A typical 17 base spacer region between the −10 and −35 hexamer sequences. When the spacer region is shorter or longer by even one base pair, the efficiency of the promoter can be reduced drastically. There are, however, natural promoters that have spacers that are either shorter or longer than 17 base pairs.

d. A discriminator region between the −10 to +1 region. This region which can be either A-T or G-C rich can be either activated or repressed, respectively, during amino acid starvation and is related to the stringent/relaxed condition of the cell. The promoter with a discriminator region controls the expression of ribosomal structural genes and amino acid biosynthesis genes.

The best evidence for the interaction of RNA polymerase holoenzyme to the promoter site has been the discovery of mutations in the −35 and −10 regions of the promoter that can be suppressed by mutations in the cognate sigma factor. In the case of *E. coli* major holoenzyme, it was shown that a mutation in a phage promoter at the 3rd position of the −35 hexamer in which TTGACA was changed to TTAACA reduced the promoter activity significantly, but this loss in activity was reversed by a mutation in the sigma-70 in which an Arg residue at position 588 was changed to a His residue [2]. In a related experiment a change in the *lac* promoter at the −34 position reduced its activity which was reversed by a different mutation in the sigma-70 [3]. In a similar experiment in *Bacillus subtilis*, a G-C transition mutation to A-T at position −13 of a "down" promoter was suppressed to an active level by a mutation in *sigH* that caused a change from Thr to Ile at position 100 of the sigma H factor [4]. Thus these experiments indicate that the −35 position of the major promoter is recognized by region 4 of the sigma-70 factor [2,3] and that the −10 region is recognized by region 2 of sigma factors [4].

Minor promoters

The minor promoters are those promoters that are recognized by so-called minor forms of RNA polymerase holoenzymes. These minor holoenzymes are minor only in the sense that the sigma factors that bind to the core enzyme occur in much smaller quantities than the major sigma factors. Many of the holoenzymes occur in significant quantities only under special conditions such as starvation, heat shock, sporulation, and chemotaxis. These minor holoenzymes express genes whose products can cope with or counteract the conditions arising from stressful environmental situations.

The properties of the minor sigma factors from *E. coli* and *B. subtilis* are summarized in Table 2. The promoter specificity of the holoenzymes under control of the

Table 2. Prokaryotic sigma factors

Holoenzymes	−35	−10	References
E. coli			
$E\sigma^{70}$	TTGACA	TATAAT	[6]
$E\sigma^{38}$	TTGACA	TATAAT	[53]
$E\sigma^{32}$	TNTCNCCCTTGAA	CCCCATTTA	[54]
$E\sigma^{28}$	TAAA	GCCGATAA	[55]
$E\sigma^{54}$	CTGGYAYR	TTGCA	[56,57]
$E\sigma^{24}$	GAACTT	TCTGA	[58]
B. subtilis			
$E\sigma^{A}$	TTGACA	TATAAT	[59]
$E\sigma^{B}$	AGGTTTAA	GGGTAT	[60]
$E\sigma^{C}$	AAATC	TANTGNTT	[61]
$E\sigma^{D}$	CTAAA	CCGATAT	[62]
$E\sigma^{E}$	GAANAANT	CATATTNT	[63]
$E\sigma^{F}$	TGCATG	ATAATA	[64]
$E\sigma^{G}$	YGHATR	CAHWHTAH	[65]
$E\sigma^{H}$	GCAGGANTT	GAATT	[66]
$E\sigma^{K}$	AC	CATA---TA	[67]

minor sigma factors is different from that of the major holoenzymes. This is obvious from the −10 and −35 conserved consensus sequences that are recognized by these minor RNA polymerase holoenzymes. These conserved sequences are significantly different from those utilized by the major holoenzymes.

Although most of the minor holoenzymes recognize promoters with conserved sequences at the −10 and −35 regions, the holoenzymes evolved to recognize genes involved in nitrogen metabolism recognize a sequence that is conserved more at the −12 and −24 sequences.

Regulatory Transcription Signals in the Promoter Region

There are generally two types of regulatory transcription signals in the promoter region: inhibitory sites (operators) nd activator sites. The operators in many cases overlap the promoter site, and t'.e activator sites, which are target sites for activating proteins, generally are adjacent to the −35 region of the promoter (Fig. 3). However, in activation and repression two regulatory operator or activator sites may be present, one that is adjacent to the promoter and the other may be quite distant from the promoter. When both sites are bound by a regulatory protein, the regulatory proteins may bind to each other causing the DNA to form a loop. This looping process may either repress or activate the promoter (Fig. 2).

Repressors appear to inhibit transcription by either steric hindrance preventing the binding of RNA polymerase to the promoter site or by preventing the transition of the closed state of the promoter to the open state.

Activation on the other hand allows either a tighter binding of the RNA polymerase to the promoter or facilitates the transition of the closed state to the open state.

Transcription terminator sites

Although transcription termination is not part of the promoter function, a brief description of transcription termination may be appropriate. Usually transcription termination is signaled at the 3' end of the gene by rho dependent or rho independent transcription termination signals. The transcription termination mechanism stops further synthesis of the RNA transcript and causes the release of RNA polymerase and the newly synthesized transcript from the DNA template.

However, for many prokaryotic genes and operons there are attenuation signals [5] which are in fact termination signals. These attentuation sites usually lie between the promoter and the open reading frame (ORF) of the gene and their function as termination signals depends on the physiological state of the cell and its environment. If the external medium contains sufficient levels of the product of the gene or genes in a biosynthetic operon, then the attenuation site acts as a terminator site and transcription is terminated prior to the open read frame. However, if a required product is not present in the medium, the attenuation site is by-passed to allow the completion of transcription of the gene(s). The regulation of attenuation occurs at the transcription/translation level, since the conversion of the termination state to the continued transcription state is dependent, in most cases of attenuation, on whether the mRNA is translated or not at the attenuation site. In some cases the level of an aminoacyl-tRNA is also a regulatory signal at the attenuation site.

Properties of a Useful Promoter and Promoter Region

The following elements would comprise the ideal properties of a promoter that would be useful for biotechnology:
a.   the promoter would be efficient during growth and/or stationary phase;
b.   the expression of the promoter would be easily controllable, either by media composition, inducer compounds, or physico-chemical conditions such as temperature, pH, or ionic concentrations;
c.   the design of the promoter region would allow the ready attachment and expression of a foreign gene. This would necessitate having a suitable ribosome binding site (RBS) and an insertion site that could accommodate the insertion or attachment of foreign genes.

	−35	−10	+1	Ref.:
lac	CCCCAGGCT<u>TTACAC</u>TTTATGCTTCCGGCTCG<u>TATGTT</u>GTGTGG<u>A</u>ATT			(11)
lac UV5	CCCCAGGCT<u>TTACAC</u>TTTATGCTTCCGGCTCG<u>TATAa</u>TGTGTGG<u>A</u>ATT			(11)
trp	AATGAGCTG<u>TTGACA</u>ATTAATCATCGAACTAG<u>TTAAC</u>TAGTACGC<u>A</u>AG			(11)
trc	AATGAGCTG<u>TTGACA</u>ATTAATCATCCGGCTCG<u>TATAAT</u>GTGTGG<u>A</u>ATT			(33)
tac	AATGAGCTG<u>TTGACA</u>ATTAATCATC GGCTCG<u>TATAAT</u>GTGTGG<u>A</u>ATT			(13)
P₁	TGGCGGTG<u>TTGACA</u>TAAATACCACTGGCGGT<u>GATACT</u>GAGCAC<u>A</u>TCAG			(24)
P_r	CGTGCGTG<u>TTGACT</u>ATTTTACCTCTGGCGGT<u>GATAAT</u>GGTTGC<u>A</u>TGTA			(24)
T7		TAATACGACTCACTAT<u>G</u>GGAGA		(17)

Fig. 1. Aligned sequences from assorted bacterial and phage promoters. The −10 and −35 regions, and the transcriptional start sites designated "+1," are noted by underlines. The lac UV5 promoter has lower case letters indicating the 2-base pair alteration in the −10 region that brings it closer to the consensus sequence. The T7 promoter, which is recognized by a phage RNA polymerase, consists of a unique 23-base pair sequence, and thus does not have a −10 or −35 region underlined.

A. Arabinose Absent:

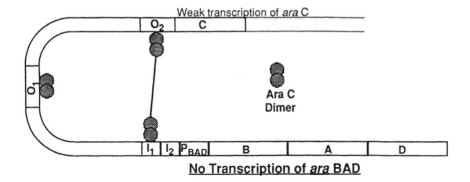

B. Arabinose Present:

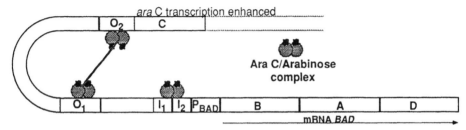

Fig. 2. DNA looping in the arabinose operon [9]. (A) Without arabinose present, AraC dimers at the I_1 site bind to AraC dimers at the O_2 site. The result is weak transcription of the *ara*C repressor gene and essentially no transcription of the *ara*BAD genes. (B) When arabinose is present, it forms complexes with the AraC dimers. The end result is that the dimers at the O_1 and O_2 sites bind one another, and another dimer pair interacts with both the I_1 and I_2 sites. This opens up transcription from of the *ara*-BAD genes, and enhances transcription of the *ara*C repressor gene.

Escherichia coli Promoters Used for Biotechnology

Perhaps the most thoroughly characterized promoters are those that are operative in *Escherichia coli*. In this section, the structure of *E. coli* promoters is examined, with an emphasis on those promoters most useful for biotechnology. Included here will be not only promoters native to *E. coli*, but genetically engineered variants and promoters from phages that use *E. coli* as a host.

The study of *E. coli* promoters began with the sequencing of numerous *E. coli* genes. Early methods of cloning genes typically involved creating a genomic library by the ligation of restriction-digested genomic DNA into a suitable phage or plasmid vector, followed by the probing of this genomic library with either a DNA or antibody probe. This method of cloning is likely to yield the genomic sequences flanking the gene, including in many cases the promoter region. Indeed, by 1983, more than

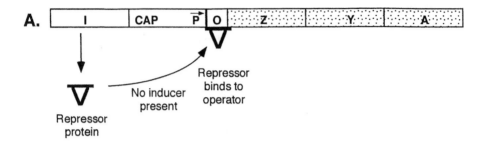

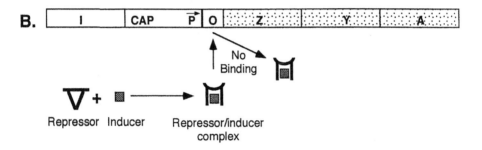

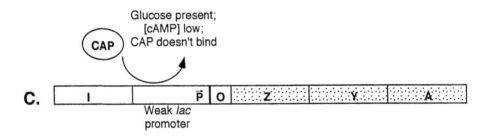

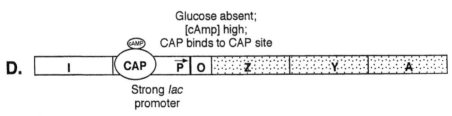

Fig. 3. Negative and positive control of expression of the *lac* operon [9,10]. (A) In the absence of inducer, the *LacI* repressor binds to the *lac* promoter operator site, blocking transcription by RNA polymerase by stearic hindrance. (B) When inducer is present, it binds to *LacI*, altering its conformation and preventing its interaction with the operator site. RNA polymerase can now begin transcription of the *lac* operon, although the transcriptional level is low unless CAP protein binds (see below). (C) When glucose is present, the cAMP concentration is low. CAP protein does not bind to the CAP site, and RNA polymerase recognizes the *lac* promoter only weakly. (D) When glucose is not present, the increased cAMP concentration favors the formation of a CAP–cAMP complex. This complex interacts with the *lac* promoter in such a way that it strongly enhances the *lac* promoter strength.

150 *E. coli* promoter sequences had been reported. The first comprehensive review [6] of these sequences investigated the base content at each position of the region upstream of the transcription start site. It was found that two regions upstream of the transcription start site were particularly important for promoter activity. These regions, designated the "−35" and "−10" regions, formed the primary functional parts of the promoter. The consensus sequence of the −35 region was found to be "tcTTGACat," where the most highly conserved bases are designated by upper case, and the less conserved bases are denoted by lower case. Similarly, the −10 region was found to contain the sequence "TAtAaT." This "Pribnow box" exhibited a particularly high level of conservation. This early review also tallied the results of a variety of promoter-area mutations that had been made. Mutations that brought the sequence closer to the consensus promoter were found to result in an increase in that gene's expression level, while alterations that changed consensus bases to other bases resulted in a decrease in the strength of the promoter. Thus, promoter strength was found to be directly related to the degree of conformation to the consensus sequence. As described below, several researchers have created promoters stronger than those found naturally by genetically engineering existing promoters to more closely match the consensus sequence. Figure 1 is an alignment of some of the promoters most commonly used. All of these promoters conform closely to the strong promoter consensus sequence. Although those listed in the figure are the ones typically used for biotechnological purposes, other promoter types are known to exist, but are inactive except under very special circumstances, heat shock promoters being a well characterized example.

Inducibility

The most common biotechnological use of a promoter is to direct synthesis of a heterologous protein. However, a strong promoter is not necessarily biotechnologically useful. The ability to turn off the promoter is as important as the promoter's strength. Promoters that are constitutively "on" are not generally used for protein expression in *E. coli* because of difficulties in the creation and maintenance of the plasmid construct. Two important reasons for this are that (i) production of heterologous protein in high yield is very draining to the resources of the cell, and (ii) many genes encode "toxic proteins". These factors can result in a strong selection against cells carrying such plasmid constructs, and may be the cause of plasmid instability or even the inability to transform cells successfully with the ligated plasmid construct. The strongest promoters may result in the accumulation of heterologous protein to more than 50% of the total cell protein [7]. This reduces the cell's ability to produce its own proteins needed for growth. The phenomenon of foreign proteins' toxicity is well known, even though it is still poorly understood. Some general rules have been established, however, that can predict the toxicity of given proteins. As an example, DNA-binding proteins tend to be highly toxic when overexpressed in *E. coli*, as are proteins that disrupt vital functions such as electron transport or membrane integrity. These two factors, toxicity and the draining of cell resources, tend to select against cells carrying plasmids with genes following constitutively-on strong promoters.

To get around this problem, protein expression is generally done by using an inducible promoter. Such promoters remain off until a specific signal is delivered to the cell. Under control of this signal, the promoter is turned on only when the expression of the gene is desired. This enables the researcher to keep the gene turned off during the creation of the plasmid construct and the establishment of the plasmid in the desired host cell type. Only when the cells are in the growth media at a sufficient density is the signal given, inducing the gene and beginning the production and accumulation of the target protein. The period of induction varies from system to system and with the growth conditions used; typically 1–12 h of induced growth time is used.

Lac Promoter – A Model System for Negative and Positive Control

As described below, there are other factors that affect promoter strength than simply the DNA sequence. Nearby regions may contain regions of sequence that can dramatically increase or decrease the strength of the promoter. Perhaps the most studied promoter region is that responsible for controlling expression from the *lac* operon. As this promoter was among the first used as a tool for biotechnology, and is still commonly used today, the *lac* promoter (P_{lac}) region is described first as a model system.

The *lac* operon consists of several genes responsible for the enzymatic degradation of lactose into sugars that can enter the main glycolytic pathways. The first gene in the operon, the *lacZ* gene, encodes a β-galactosidase that converts the disaccharide lactose into the monosaccharides glucose and galactose. Next is the *lacY* gene, which is a permease, and helps bring lactose into the *E. coli* cell. The final gene, for the *lacA* transacetylase, has an unclear role in the utilization of lactose by the cell.

Directly upstream of the first of these genes, *lacZ*, lies the *lac* promoter. This promoter is regulated in *E. coli* by two methods, one which decreases and one which increases transcription from the promoter. A diagram of the *lac* operon is shown in Fig. 3. The first of these regulatory systems is the binding of the *lac* I gene product (the *lac* repressor) to a site called the operator, just downstream from the *lac* promoter. The *lacI* gene is a divergently-transcribed gene, with its own promoter, located even further upstream. In the absence of inducer, the *lac* repressor protein binds tightly to the operator site, blocking transcription, and shutting off the *lac* operon (Fig. 3A). When there is lactose present, a small amount of it is converted to the true inducer allolactose [8] (via a transgalactosidation activity of the β-galactosidase enzyme, present in trace amounts even when the operon is "off"), which binds to the *lac* repressor, preventing its interaction with the *lac* promoter, and allowing RNA polymerase to bind and begin transcription from the promoter (Fig. 3B).

The second level of transcriptional regulation of this promoter lies with the catabolite-activator protein (CAP). This protein, which has effects both on the lactose and arabinose operons [9], acts to increase the level of transcription under conditions of glucose starvation. When the cells are starved for glucose, high levels of cyclic adenosine monophosphate (cAMP) are synthesized. Some of this cAMP interacts with CAP, and allows its interaction with a region upstream of the *lac* promoter, called the CAP site. Alone, the lac promoter is very weak (Fig. 3C); when CAP-cAMP binds to the CAP site, however, the promoter strength increases about 50-fold (Fig. 3D) [10], resulting in substantial production of the *lac* operon gene products.

Biotechnological uses of the lac promoter region

The natural purpose in *E. coli* of the *lac* repressor system is to prevent the cell from wasting its valuable resources producing substantial amounts of lactose-utilization enzymes when the substrate is not present. As glucose is a preferable carbon source to lactose, the CAP system allows for *lac* operon expression only when glucose is not present. This precise control system has been turned to many researchers' advantage through the use of modern molecular biology. The *lac* promoter region, the first well-understood inducible promoter, has been used to control the synthesis of cloned genes, as discussed below.

In addition, the *lac* system is used as a cloning aid. The most common plasmid and phage vectors used in cloning today include the pUC series, the pBluescript series (StrataGene, La Jolla, CA), and the phages derived from m13. While they have been used and improved since the 1980s, the basic concept of these vectors remains unchanged. In addition to the ampicillin resistance gene on the plasmids, each of these vectors contains an origin of replication, a multiple cloning site (MCS) containing a variety of unique restriction sites, and a portion of the *lac* operon consisting of the promoter and a portion of the *lacZ* gene sufficient to provide β-galactosidase activity. The MCS is positioned at the beginning of the *lacZ* gene, in order to disrupt *lacZ* expression in the event that a fragment of DNA is cloned into the MCS. Thus, when ligations are transformed into cells, and plated on agar plates containing an inducer of the *lac* promoter, such as isopropyl-β-D-thiogalactoside (IPTG), the promoter is turned on and cells harboring plasmids without inserts create the β-galactosidase enzyme. The plates used typically contain a chromogenic compound such as X-gal, which is cleaved by the enzyme to create a blue dye. Thus, colonies of cells without an insertion are blue, while the colonies containing inserts will be colorless. This allows for easy screening of large numbers of colonies, reducing the labor involved in determining which ones might be harboring a desired clone. Figure 4 illustrates the use of so-called "blue/white" selection.

Another advantage of the placement of P_{lac} next to the MCS of a vector is that some expression of a gene placed there might be possible. Thus, cloning into pUC or similar vectors can be desirable for other methods of screening or selection. This method has been used to identify clones carrying endoglucanases, such as those from cellulolytic bacteria. Whole genomic DNA from *Clostridium cellulovorans*, cleaved by restriction enzymes and cloned into the pUC MCS and plated on IPTG plates has resulted in colonies that produce a diffusible endoglucanase, which was capable of

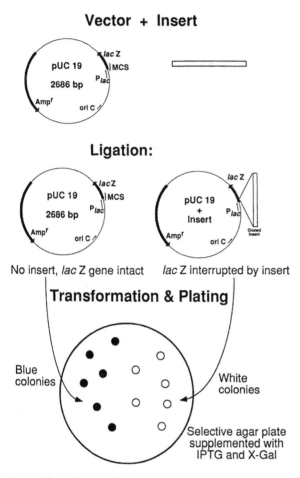

Fig. 4. Use of the *lacZ* gene for selection of plasmid clones bearing insert. Inserts are cloned into a multiple cloning site (MCS), which is located at the beginning of the *lacZ* alpha fragment gene. Insertion of a piece of foreign DNA at this site disrupts the transcription of an intact *lacZ* gene, preventing production of LacZ β-galactosidase in these clones. Colonies harboring clones without insert-disrupted *lacZ* are able to convert a substrate (X-gal) into a blue dye; clones with inserts result in white colonies. IPTG is included in the media to insure that the *lac* promoter is activated.

degradation of a carboxymethyl-cellulose substrate also present in the agar [11]. By suitable staining techniques, these rare colonies were able to be identified and isolated. This application of the promoter system relies on coincidence to insure that the clone is in the proper orientation for the promoter to work, and that there is not so much intervening sequence between the promoter adjacent to the end of the cloned fragment and the beginning of the gene. However, this approach has proven useful in a variety of cases, and is still used today.

Many improvements have been made to the P_{lac} system, by producing mutant and hybrid promoters that are much stronger, with the advantage that higher expression levels are possible with such systems. However, increases in the promoter's maximum activity level have resulted in increases in the basal, or "off" signal strength. To counteract this, the ability of the repressor to turn off the promoter has been improved. The introduction of the *lacI*q up mutation to the weak, constitutive *lacI* promoter [12], a one base-pair change which results in much higher expression of the repressor, has allowed strong promoter vectors to be constructed with reduced fear of adverse effects from "leaky" promoters. In addition, transcription termination signals, placed downstream of the gene cloning site, reduce the negative effects on cell survival of the very strong promoters [13]. By this method, promoters so strong as to be lethal to the *E. coli* cell can be tolerated. Discussed in a later section are several useful promoter systems that are derived from the *lac* promoter.

The trp Promoter

Although the *lac* promoter region was characterized and understood very early, its signal strength even when fully de-repressed and induced is insufficient to make it attractive for the heterologous expression of cloned genes [14]. The search for stronger inducible promoters soon lead to the use of the promoter for the tryptophan operon (*trp*). This operon encodes the genes necessary for biosynthesis of the amino acid tryptophan, and is induced by tryptophan starvation [15]. This promoter is strong, and can lead to the accumulation of the heterologous protein to levels of up to 30% of the total cellular protein. Examples of proteins expressed with the help of so-called *trp* vectors include human proteins such as growth hormones, prolactin, and interleukin [16].

The use of the *trp* promoter has been simplified by the creation of a variety of cloning vectors with conveniently placed restriction sites adjacent to the promoter. All of them allow for easy cloning of the gene of interest with the translation initiation codon a proper distance from the ribosome binding site. These vectors typically are low copy number plasmids, as a high copy number plasmid would risk saturation of the limited number of *trp* repressor molecules present in the bacterial cell, resulting in a high basal activity level of the *trp* promoter.

Although this promoter has been used to express numerous proteins, it does have several drawbacks which have led many researchers to pass it by in favor of other promoters. Studies have shown that the induction ratio is poor, only about 50-fold [17,18], even on pBR322-based (low copy number) vectors. The *lac* promoter described above, although not as strong, has twenty times this range, with an induction ratio of 1,000×. Thus, the "leaky" *trp* promoter would not be a wise choice for use in the expression of genes that are toxic to *E. coli*; in fact it might be difficult to establish the plasmid construct.

Another drawback of the *trp* promoter are the conditions needed to induce it. While the *lac* promoter is easily turned on by the addition of IPTG to the growth media, the *trp* promoter is induced by the removal of tryptophan from the growth media. A minimal medium lacking tryptophan is inoculated with up to 5% of an overnight rich broth culture. After a period of about 2 h at 37°C, the pool of trypto-phan has been depleted, and the *trp* promoters have been fully induced. Indole-3-acrylic acid is then added to the media [19], to speed the synthesis of tryptophan needed to express the heterologous protein as well as the proteins needed for survival of the host cells. As the media is not a rich one, cell growth and the accumulation of the target protein are slow. Expression is then continued for several hours to over-night, with the peak of expression being dependent on factors related to the individ-ual protein of interest and host strain background. Because of the high target protein expression levels possible with this vector, the proteins sometimes form inclusion bodies, which concentrate the proteins in small, insoluble granules within the cell. The majority of cell proteins are soluble cytosolic proteins [20,21]; inclusion body proteins can be purified away from the bulk of the contaminants by centrifuga-tion [22]. These inclusion bodies can often then be subjected to a denaturation/ re-naturation step to refold the protein into an active, or at least soluble, conformation [20,21].

Hybrid Promoters

Although natural *E. coli* promoters have been used to express heterologous proteins, researchers quickly saw the potential for improved, recombinant promoters based upon the *lac* and *trp* promoter regions [23]. Early work recognized the importance of the −35 and −10 regions of the promoter structure [4] and began to characterize the interactions of RNA polymerases with these regions [24–26]. The next logical step was the creation and characterization of hybrid promoters by the construction of clones having the −35 region of one promoter and the −10 region of another. Mu-tants with single base-pair changes and alterations of the −35 to −10 spacing were also made, to investigate the effects of such changes. These studies showed that the P_{trp} −35 region in combination with the P_{lac} −10 region, separated by a 16-base pair spacing, resulted in a promoter stronger than either promoter alone [23]. In addition, when a construct bearing the "UV5" mutation [27] in the *lac* region was created, the promoter (designated P_{tac}) had very high induced strength (3 times the fully derepressed *trp* promoter strength, and 10 times the *lac* UV5 promoter strength [28]), while maintaining *lacI* repression, and thus inducibility, found in the native *lac* pro-moter region. A very similar promoter, designated P_{trc}, differs in that the consensus spacing of 17 base pairs is used to separate the −10 and −35 hexamers [29]. Both of these promoters have similar strength [30].

As these promoter region constructs contain the *lac* operator region instead of the *trp* repressor binding site, induction is easily accomplished by the addition of IPTG to the growth media. Easy induction conditions, coupled with high promoter strength, have led to the commercial development of vectors bearing these promot-ers. Examples of vectors based on these promoters that are commonly used for high-level heterologous protein expression include the pTRC99A vector (Pharmacia, Pis-cataway, NJ), which as its name suggests uses the P_{trc} promoter, as well as the pMAL vectors (New England Biolabs, Beverly, MA) which employ the P_{tac} promoter to direct expression of fusion proteins. Quite a few proteins have been expressed by these and similar vector systems [31], and the number is still growing rapidly, given the popularity of these promoters for heterologous protein expression. The vectors on which these promoters reside typically include the *lac* I^Q allele to maintain tight control of the promoter, a multiple cloning site for easy insertion of the target gene, and an RNA polymerase transcription termination signal downstream. Together, these factors create powerful expression vectors that can be used to express large amounts of most target proteins.

Phage Promoters

Although the promoters described above are often used for protein expression, there are situations in which the target gene product is too toxic for the plasmid to be stably maintained in *E. coli*, even with the strong repression afforded by the presence of the *lac* I^Q allele. All of the strong natural or recombinant *E. coli* promoters previously described have a significant basal activity, even when not induced. To get around this problem, expression systems bypassing the native *E. coli* RNA polymerase have been designed [7,32,33].

Many phage using *E. coli* as a host produce their own proprietary RNA polymerases, which recognize promoter sequences distinct from those found in *E. coli* promoters. As a result, plasmids that place a phage promoter upstream from even the most toxic genes can be created and maintained in *E. coli*. Without the phage RNA polymerase, the promoter is completely inactive. Once the desired plasmid construct is created it can be transformed into a strain of *E. coli* that contains the (typically inducible) gene for the phage polymerase. In this expression strain, the phage promoter becomes active, leading to the production of the target protein. Even in cases of highly toxic genes, significant protein expression can often be obtained by this method.

By far the most popular phage promoter used for this purpose is the one derived from the T7 phage. The T7 RNA polymerase is composed of a single polypeptide of 100,000 kDa, which recognizes a specific 23-base pair stretch of DNA that overlaps the RNA transcription start site [34,35]. In addition to the high specificity of the polymerase for its own peculiar promoters, it has a processivity of roughly five times that of its *E. coli* counterpart [7], which makes it seem ideal for expression of specific cloned genes. Because of these factors, the P_{T7} has become the standard promoter of choice for many researchers. There are many vectors which employ this promoter, including general cloning vectors such as the pBluescript series (Stratagene, La Jolla, CA), but for the purpose of protein expression the most thoroughly described and used are those described by Studier et al. [7,32], now available commercially (Novagen, Madison, WI). This series of plasmid vectors has been named the pET series, and now comprises almost two dozen variations including transcription vectors, translation vectors, protein fusions of several kinds, as well as other refinements. However, the basic concept behind all of them is simple, and will be described in a generic fashion in this section.

The pET vectors place the P_{T7} directly upstream of a Shine-Dalgarno ribosome-binding site, in turn directly upstream from a cloning site such as Nco I (recognition site: C ↓ CATGG) in the case of the commonly-used pET-3d. There is a strong transcription terminator downstream of a second cloning site, to which the 3' end of the gene is anchored. The gene of interest is ligated into these sites such that the ATG start codon of the target gene is in frame with the original ATG of the Nco I restriction site. Although many genes, particularly eukaryotic ones, naturally have an Nco I site around their start codon, for those genes without such a site it is a simple task to use site-directed mutagenesis to engineer one into place. One drawback to this is that the Nco I site demands that the second codon begin with a G. As a survey of eukaryotic translation initiation sites showed a bias towards G as the first base of the second codon [29,36,37], this will be significant in relatively few situations. In cases where this would be inconvenient, an alternate version of the pET vector may be used, or the Nco I cohesive overhang can be blunted with mung bean nuclease, leaving a blunt-end to which the gene can be ligated.

Once the desired plasmid construct has been created, it is transformed into the E. coli strain to be used for protein expression. The T7 RNA polymerase can be provided by one of two methods: expression from a λ phage lysogen (e.g. λDE3) which contains the T7 RNA polymerase gene under P_{lac} control, or by infection with phage λCE6, a lytic phage which provides constitutive expression of the phage polymerase. The bacterial culture is grown to an OD_{600} around 0.5–0.8 [20,21], with the optimal density depending on the host strain and target gene, followed by induction of gene. This is accomplished in λDE3 lysogens by the addition to the growth media of IPTG, which induces expression of T7 RNA polymerase from the lysogen. The use of phage CE6 is beneficial for the expression of genes whose products are so toxic that even the basal "leaky" T7 RNA polymerase from a lysogen host results in enough protein synthesis to cause problems. This method allows for complete withholding of the phage polymerase until the moment the culture is safely at the desired cell density. DE3-lysogenic host strains may also provide for constitutive expression of varying levels of lysozyme, a natural inhibitor of T7 RNA polymerase [7]. These strains have the designation pLysE and pLysS, denoting the lysozyme gene encoded on a plasmid. The plasmid is from a different incompatibility group than those that are commonly used as protein expression vectors [7], so can be maintained without trouble in the same host cells. The lysozyme produced acts to reduce the basal level of active T7 RNA polymerase in the cells, usually without significantly affecting the induced level.

As the target gene expression requires prior expression of the phage polymerase, there is a lag of an hour or so before significant target protein synthesis is seen. After this lag, however, accumulation of the target protein is rapid, due to the highly active phage RNA polymerase/promoter combination, and by 3 or 4 h post-induction, the levels of target protein are likely to have peaked. In cases where the target gene product exerts toxic effects on the cells, it is often advantageous to reduce the temperature of expression or to reduce the pre-harvest time interval. Aliquots of cells taken over a 6-h period of expression may be helpful in determining the optimum expression period for a given gene. It is also common for the proportion of the expressed protein present in the soluble phase to vary with time or temperature; in cases where either specifically soluble or inclusion-body proteins are desired, sonication of the protein samples at each time point can reveal the location of the desired peptide. An example of successful use of the T7 phage promoter in the vector pET-3d is shown in Fig. 5. In this case, the cellulose binding domain (CBD) from the *C. cellulovorans* CbpA was expressed at a level of about 100 mg/l cell culture. The CBD was purified via cellulose affinity, resulting in over 70 mg/l final purified protein [21].

There are other phage promoters that are useful to molecular biologists and thus for biotechnological purposes. In particular are the T3 promoter, and the λP_l and P_r promoters [14,38]. The former is often used for the creation of strand-specific RNA probes. Several plasmids, such as the pBluescript series, have this promoter facing into one side of the plasmid's MCS. On the opposite side of the MCS, there will typically be a T7 promoter facing in, whose promoter is distinct. Strand specific probes can be synthesized by the addition in vitro of either polymerase.

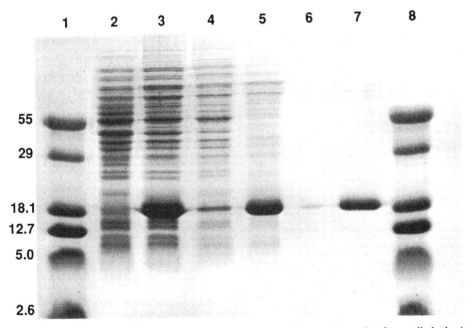

Fig. 5. Expression and purification of the CBD protein [21]. Whole-cell proteins from cells harboring pET-3d (lane 2), whole-cell proteins from cells harboring pET-CBD (lane 3), cytosolic fraction from lysed pET-CBD cells (lane 4), guanidine–HCl fraction from lysed pET-CBD cells (lane 5), final wash of cellulose pellet (lane 6), and purified CBD protein (lane 7) were loaded on a 15% acrylamide gel. Each lane was loaded with an equal percentage of the total protein from each fraction, with the exception of lane 6, which was a 10× concentrate. Prestained molecular mass markers (lanes 1 and 8) have mobilities of approximately 2.6, 5, 12.7, 18.1, 29, and 55 kDa.

The λP_l and P_r promoters are useful for heterologous protein expression. As early as 1983, vectors such as pCQV2 were reported that allowed for convenient cloning of a gene of interest downstream from the phage promoter [14]. Examples of proteins successfully expressed with this vector include β-galactosidase, t-antigen, and a viral protein kinase [39]. Accumulation of target protein of up to 10% has been reported, although the level of course depends on factors specific to the individual gene. This vector encodes the CI857 mutant of the *cI* repressor gene. This temperature-sensitive repressor maintains a tight control on protein synthesis at low temperature (~32°C), but at 42°C, denatures and no longer inhibits transcription by *E. coli* polymerase at the phage promoter. This temperature-dependent control of transcription is very convenient and suggests that the promoters controlled thus would be useful in industrial settings, for large-scale protein synthesis. Where many other inducible systems rely on the addition to the growth media of costly IPTG to accomplish the induction, the ability of this system to be turned on simply by an increase in temperature make it attractive, despite the relatively low yield of protein per volume of cell culture.

Another approach to the use of the λP_l for protein expression has been incorporated in the pTrxFus vector (Invitrogen, San Diego, CA). This unique vector uses the tightly regulated λP_l promoter to direct expression of a thioredoxin fusion protein in *E. coli* strains whose genome has been engineered to carry the gene for the λcI repressor under control of the *trp* promoter. Cloning and growth of the cells is accomplished under tryptophan starvation conditions, resulting in activity at the *trp* promoter and thus high levels of λcI repressor protein which shuts off expression from the P_l promoter. Expression from the vector is accomplished by switching the cells to rich media, which results in the *trp* promoter being shut off, and rapidly decreasing concentration of λ repressor. Activity at the P_l promoter quickly turns on, resulting in substantial yields of the fusion protein under its control.

Bacillus subtilis Promoters

In this section we describe a few promoters that have been utilized to express foreign genes in *B. subtilis*. Although the subject is outside of the purview of this review, the *B. subtilis* promoter systems can be utilized in conjunction with signal peptides for the secretion of foreign proteins from protease deficient *B. subtilis* cells [40]. Thus there is an inherent interest in the development and utilization of *B. subtilis* promoters. The disadvantage of utilizing *B. subtilis* promoters is that very few inducible promoters have been well characterized. To overcome this problem, the development of hybrid promoters and promoter regions is described.

Pspac promoter

The Pspac promoter represents an elegant example of a genetically engineered promoter that is efficient and controllable and capable of expressing foreign genes [41,42]. This hybrid promoter is a derivative of a promoter of the *B. subtilis* phage SPO1 and the operator site of the *E. coli lac* gene. The controlled expression of Pspac is dependent on the simultaneous presence of the *lac* operator that lies between the SPO1 promoter and the *B. subtilis* ribosome binding site and the *E. coli* repressor gene, *lacI*, that was engineered to be expressed in *B. subtilis*. This was done by substituting the *B. licheniformis* promoter and RBS of the penicillinase gene for the *E. coli lacI* gene promoter and RBS. The presence of the *lac* repressor in *B. subtilis*, in the absence of a *lac* inducer, prevented the expression of the hybrid Pspac promoter. The addition of isopropyl-β-thiogalactoside (ITPG) allowed expression from the Pspac promoter.

This whole construct of the Pspac promoter followed by a *lac* operator site, ribosome binding site, initiation codon and a foreign gene and the *lac* repressor gene was inserted into a plasmid shuttle vector capable of replicating in both *E. coli* and *B. subtilis*. With these constructs it was found that the promoter was indeed regulated by the *lac* operator/repressor system and that foreign gene expression could be controlled by the presence and absence of the inducer for this system.

SacB expression system

The *sacB* system can be utilized for the expression of foreign genes in *B. subtilis*. *sacB* is one of a number of genes involved in the metabolism of sucrose. The expression of *sacB*, the structural gene for extracellular levansucrase, is induced by the addition of sucrose to the medium. The regulation of expression of *sacB* is controlled by an adjacent *sacR* regulatory region consisting of a constitutive promoter followed by a stem-and-loop sequence that acts as a terminator in the absence of inducer. When sucrose is present a positive regulatory gene, *sacY*, is expressed. SacY is an anti-terminator protein that allows transcription to proceed from the promoter through the *sacB* sequence to form an active mRNA. In addition the expression of *sacB* is controlled by a number of other regulatory factors such as *sacQ* [43], *sacU* [44], *sacV* [45] and *prtR* [46] that can further stimulate *sacB* expression at the transcription level by up to 100-fold.

One *sacB* system utilized the *sacR* region of *B. subtilis* that controlled the expression of a reporter gene, *xylE* (catechol 2,3-dioxygenase or C230) in the presence (pA51S) and absence (pA51) of the *sacY* gene on a suitable plasmid vector [47]. The expression of this system in the absence or presence of *sacY* was inducible with sucrose. A further modification of this system utilized a constitutive synthetic phage T5 promoter in place of the constitutive promoter present in *sacR* and this construct also had the *sacY* gene present (pAMB12T5S) or absent (pAMB12T5) on the vector. Both of these systems were also found to be induced in the presence of sucrose.

Another modification of the *sacB* system utilized the complete *sacR* region including the constitutive promoter-regulatory region and signal peptide fused to a reporter gene, TEM β-lactamase (Bla),on a pUB110-derived plasmid, pWB [48]. Another plasmid was constructed (pQL-1) to contain not only the sacB system of pWB but also the *sacQ* gene transcribed from a *B. subtilis* promoter P43 [48]. The expression of Bla was induced in both pWB and pQL-1 by the presence of 2% sucrose in the medium. However, the presence of the *sacQ* gene (pQL-1) increased the synthesis of Bla by 17-fold over the construct lacking *sacQ* (pWB). Thus a high copy number of the regulatory *sacQ* gene made a dramatic difference in the expression level. The important role of other regulatory factors was illustrated when a host cell containing a single copy of the *sacU^h* mutation in its chromosome was able to enhance the expression of *sacB* expression as much as a high copy number of the sacQ gene [48]. Thus with *B. subtilis* not only the promoter and promoter region are important factors for gene expression, but it is also necessary to consider the numerous *trans* acting regulatory factors that affect total gene expression.

This inducible *sacB* system in conjunction with a *B. subtilis* strain deficient in six extracellular proteases has been used to produce β-lactamase [49] and streptokinase [50].

P43 promoter

If expression of a gene is required during growth of the *B. subtilis* in a rich medium, the P43 is a very useful promoter [51]. The P43 promoter is a strong constitutive promoter that is expressed during growth and the early stages of sporulation in the presence of high levels of glucose in an enriched medium. It has been used to produce a functional single chain antidigoxin antibody [52] and SacQ [48].

Future Uses of Prokaryotic Promoters in Biotechnology

Prokaryotic promoters can play an important role in biotechnology particularly in expressing those genes whose products can be made in their mature active forms and in large quantities in prokaryotic hosts. In many cases this limits the products to prokaryotic proteins. However, eukaryotic proteins and peptides can also be synthesized in prokaryotes and still be functional albeit at a reduced level.

The inducible promoters are useful in those cases where the product of the gene may be toxic to the host or when the production of a protein at a particular growth stage is desirable. When foreign proteins are synthesized at a very high rate from highly efficient promoters, the rapid accumulation of the protein may cause it to form inclusion bodies or aggregates. There are advantages and disadvantages when this occurs. The advantages are that formation of inclusion bodies reduces the possibility of proteolytic degradation of the product by intracellular proteases and that inclusion bodies are readily purified by centrifugation methods that enrich the desired proteins. The disadvantages are that the protein is probably "inactive" and that it will have to be denatured gently and renatured to obtain an active form of the protein. The renaturation may not be efficient and thus there would be a low yield of the desired product.

In the case of extracellular proteins, the promoter region of choice can be modified to have the promoter followed by a signal peptide sequence and a cloning site to insert a sequence for the extracellular protein. This strategy is useful for *B. subtilis*, since many proteins that are synthesized intracellularly tend to be proteolyzed by intracellular proteases. With the availability of a *B. subtilis* host deficient in six extracellular proteases [49], the secreted protein can be isolated in its native form. On the other hand with *E. coli*, some proteins secreted into the periplasmic space are degraded by proteases found in the periplasm and the removal of signal peptides may allow the successful accumulation of the mature protein in the cytoplasm. Therefore the strategy of expressing and producing proteins will depend not only on the promoter, but on the host organism.

In the case of *E. coli*, the major problem is that too rapid synthesis of a foreign protein appears to tax the capability of the system to correctly fold the protein into its proper soluble or active conformation. Usually protein inclusion bodies are formed which is useful in protecting the protein from proteolytic digestion, but then it is necessary to denature and renature the protein to its active form. This could be difficult and the yield of the active protein could be very low. Thus a better understanding of the system for correct and rapid folding of the nascent protein is required.

The use of *B. subtilis* hosts lacking extracellular proteases is promising, since secreted proteins can accumulate even slowly without being substantially degraded. The major problem remaining is that even with the use of *B. subtilis* promoters and signal peptides, not all eukaryotic proteins are secreted efficiently. A better understanding of the secretion mechanism, protein translocation, and protein folding in this and other organisms is necessary to construct an optimal host expression system.

References

1. Ross W, Gosink KK, Salomon J, Igarashi K, Zou C, Ishihama A, Severinov K, Gourse RL. A third recognition element in bacterial promoters: DNA binding by the a subunit of RNA polymerase. Science 1993;262:1407–1413.
2. Gardella T, Noyle H, Susskind MM. A mutant *Escherichia coli* sigma-70 subunit of RNA polymerase with altered promoter specificity. J Mol Biol 1989;206:579–590.
3. Siegele DA, Hu JC, Walter WA, Gross CA. Altered promoter recognition by mutant forms of sigma 70 subunit of *Escherichia coli* RNA polymerase. J Mol Biol 1989;206:591–604.
4. Zuber P, Healy J, Carter HLI, Cutting S, Moran CPJ, Losick R. Mutation changing the specificity of an RNA polymerase sigma factor. J Mol Biol 1989;206:605–614.
5. Yanofsky C. Transcription attenuation. J Biol Chem 1988;263:609–612.
6. Hawley DK, McClure WR. Compilation and analysis of *Escherichia coli* promoter DNA sequences. Nucleic Acids Res. 1983;11:2237–2255.
7. Studier FW, Rosenberg AH, Dunn JJ, Dubendorff JW. Use of T7 RNA polymerase to direct expression of cloned genes. Methods Enzymol 1990;185:60–89.
8. Muller-Hill B, Rickenberg HV, Wallerfels K. Specificity of the induction of the enzymes of the *lac* operon in *Escherichia coli*. J Mol Biol 1964;10:303–318.
9. Darnell J, Lodish H, Baltimore D (eds). Molecular Cell Biology. New York: Scientific American, 1986;287–288.

126

10. Beckwith J, Grodzicker T, Arditti R. Evidence for two sites in the *lac* promoter region. J Mol Biol 1972;69:155–160.

11. Shoseyov O, Hamamoto T, Foong F, Doi RH. Cloning of *Clostridium cellulovorans* endo-1,4-beta-glucanase genes. Biochem Biophys Res Commun 1990;169:667–672.

12. Calos MP. DNA sequence for a low-level promoter of the *lac* repressor gene and an 'up' mutation. Nature 1978;274:762–765.

13. Gentz R, Langner A, Chang ACY, Cohen SN, Bujard H. Cloning and analysis of strong promoters is made possible by the downstream placement of a RNA termination signal. Proc Natl Acad Sci USA 1981;78:4936–4940.

14. Queen C. A vector that uses phage signals for efficient synthesis of proteins in *Escherichia coli*. Mol Appl Gen 1983;2:1–10.

15. Yanofsky C, Platt T, Crawford IP, Nichols BP, Christie GE, Horowitz H, VanCeemput M, Wu AM. The complete nucleotide sequence of the tryptophan operon of *Escherichia coli*. Nucleic Acids Res. 1981;9:6647–6668.

16. Yansura DG, Henner DJ. Use of *Escherichia coli trp* promoter for direct expression of proteins. Methods Enzymol 1990;185:54–60.

17. Bogosian G, Somerville RL. Analysis *in vivo* of factors affecting the control of transcription initiation at promoters containing target sites for *trp* repressor. Mol Gen Genet 1984;193:110–118.

18. Hallewell RA, Emtage S. Plasmid vectors containing the tryptophan operon promoter suitable for efficient regulated expression of foreign genes. Gene 1980;9:27–47.

19. Doolittle WF, Yanofsky C. Mutants of *Escherichia coli* with an altered tryptophanyl-transfer ribonucleic acid synthetase. J Bacteriol 1968;95:1283–1294.

20. Chang BY, Doi RH. Overproduction, purification and characterization of *Bacillus subtilis* RNA polymerase sigma A factor. J Bacteriol 1990;172:3257–3263.

21. Goldstein MA, Takagi M, Hashida S, Shoseyov O, Doi RH, Segel IH. Characterization of the cellulose-binding domain of the *Clostridium cellulovorans* cellulose-binding protein A. J Bacteriol 1993;175:5762–5768.

22. Shire SJ, Bock L, Ogez J, Builder S, Kleid D, Moore DM. Purification and immunogenicity of fusion VP1 protein of foot and mouth disease virus. Biochemistry 1984;23:6474–6480.

23. Russell DR, Bennett GN. Construction and analysis of in vivo activity of *E. coli* promoter hybrids and promoter mutants that alter the -35 to -10 spacing. Gene 1982;20:231–243.

24. Rosenberg M, Court D. Regulatory sequences involved in the promotion and termination of RNA transcription. Annu Rev Genet 1979;13:319–353.

25. Siebenlist U, Simpson RB, Gilbert W. *E. coli* RNA polymerase interacts homologously with two different promoters. Cell 1980;20:269–281.

26. Wells RD, Goodman TC, Hillen W, Horn GT, Klein RD, Larson JE, Müller UR, Nevendorf SK, Panayotatos N, Stirdivant SM. DNA structure and gene regulation. In: Chen W (Ed.) Progress on Nucleic Acids Research and Molecular Biology. New York: Academic Press, 1980:168–267.

27. Carpousis AJ, Stefano JE, Gralla JD. 5′ nucleotide heterogeneity and altered initiation of transcription at mutant *lac* promoters. J Mol Biol 1982;157:619–633.

28. De Boer HA, Comstock LJ, Vasser M. The *tac* promoter: a functional hybrid derived from the *trp* and *lac* promoters. Proc Natl Acad Sci USA 1983;80:21–25.

29. Amann E, Ochs B and Abel K-J. Tightly regulated *tac* promoters useful for the expression of unfused and fused proteins in *Escherichia coli*. Gene 1988;69:301–315.

30. Mulligan ME, Brosius J, Clure WR. Characterization in vitro of the effect of spacer length on the activity of *Escherichia coli* RNA polymerase at the *tac* promoter. J Biol Chem 1985;260:3529–3538.

31. di Guan C, Li P, Riggs PD, Inouye H. Vectors that facilitate the expression and purification of foreign peptides in *Escherichia coli* by fusion to maltose-binding protein. Gene 1987;67:21–30.

32. Studier FW, Moffat BA. Use of bacteriophage T7 RNA polymerase to direct selective high-level expression of cloned genes. J Mol Biol 1986;189:113–130.

33. Tabor S, Richardson CC. A bacteriophage T7 RNA polymerase/promoter system for controlled exclusive expression of specific genes. Proc Natl Acad Sci USA 1985;82:1074–1078.

34. Davanloo P, Rosenberg AH, Dunn JJ, Studier FW. Cloning and expression of the gene for bacteriophage T7 RNA polymerase. Proc Natl Acad Sci USA 1984;81:2035–2039.

35. Dunn JJ, Studier FW. Complete nucleotide sequence of bacteriophage T7 DNA and the locations of T7 genetic elements. J Mol Biol 1983;166:447–535.

36. Kozak, M.: Compilation and analysis of sequences upstream from the translational start site in eukaryotic mRNAs. Nucleic Acids Res. 1984;12:857–872.

37. Kozak M. An analysis of $5'$-noncoding sequences from 699 vertebrate messenger RNAs. Nucleic Acids Res. 1987;15:8125–8148.

38. Remaut E, Stanssens P, Fiers W. Plasmid vectors for high-efficiency expression controlled by the Pl promoter of coliphage lambda. Gene 1981;15:81–93.

39. Wang JYJ, Queen C, Baltimore D. Expression of an Abelson murine leukemia virus-encoded protein in *Escherichia coli* causes extensive phosphorylation of tyrosine residues. J Biol Chem 1982;257:13181–13184.

40. Doi RH, Wong SL, Kawamura F. Potential use of *Bacillus subtilis* for secretion and production of foreign proteins. Trends Biotechnol 1986;4:232–235.

41. Yansura DG, Henner DJ. Development of an inducible promoter for controlled gene expression in *Bacillus subtilis*. In: Ganesan AT, Hoch JA (eds) Genetics and Biotechnology of *Bacilli*. Orlando: Academic Press, 1984;249–263.

42. Yansura DG,. Henner DJ. Use of the *Escherichia coli lac* repressor and operator to control gene expression in *Bacillus subtilis*. Proc Natl Acad Sci USA 1984;81:439–443.

43. Tomioka N, Honjo M, Funakoshi K, Manabe K, Akaoka A, Mita I, Furatini Y. Cloning, sequencing and some properties of a novel *Bacillus amyloliquefaciens* gene involved in the increase of extracellular protease activities. J Biotechnol 1985;3:85–96.

44. Lepesant JA, Kunst F, Pascal M, Lepesant-Kejzlarova J, Steinmetz M, Dedonder R. Chromosomal location of mutations affecting sucrose metabolism in *Bacillus subtilis* Marburg. Gen Genet 1972;118:135–160.

45. Martin I, Debarbouille M, Klier A, Rapaport G. Identification of a new locus, *sacV*, involved in the regulation of levansucrase synthesis in *Bacillus subtilis*. FEMS Microbiol Lett 1987;44:39–43.

46. Nagami Y, Tanaka T. Molecular cloning and nucleotide sequence of a DNA fragment from *Bacillus natto* that enhances production of extracellular proteases and levansucrase in *Bacillus subtilis*. J Bacteriol 1986;166:20–28.

47. Zukowski M, Miller L, Cogswell P, Chen K. Inducible expression system based on sucrose metabolism genes of *Bacillus subtilis*. In: Ganesan AT, Hoch JA (eds) Genetics and Biotechnology of *Bacilli*. San Diego, CA: Academic Press, 1988;17–22.

48. Wong S-L. Development of an inducible and enhancible expression and secretion system in *Bacillus subtilis*. Gene 1989;83:215–223.

49. Wu X-C, Lee W, Tran L, Wong S-L. Engineering a *Bacillus* expression-secretion system with a strain deficient in six extracellular proteases. J Bacteriol 1994;173:4952–4958.

50. Wong S-L, Ye R-Q, Nathoo S. Engineering and production of streptokinase in a *Bacillus subtilis* expression-secretion system. Appl Environ Microbiol 1994;60:517–523.

51. Wang P-Z, Doi RH. Overlapping promoters transcribed by *Bacillus subtilis* σ^{55} and σ^{37} RNA polymerase holoenzymes during growth and stationary phases. J Biol Chem 1984;259:8619–8625.

52. Wu X-C, Ng SC, Near RI, Wong S-L. Efficient production of a functional single chain antidigoxin antibody via an engineered *Bacillus subtilis* expression-secretion system. Bio/Technology 1993;11:71–76.

53. Tanaka K, Takayanagi Y, Fujita N, Ishihama A, Takahashi H. Heterogeneity of the principle s factor in *Escherichia coli*: the *rpoS* gene product, σ^{38}, is a second principal factor of RNA polymerase in stationary phase *Escherichia coli*. Proc Natl Acad Sci USA 1993;90:3511–3515.

54. Cowing DW, Bardwell JCA, Craig EA, Woolford C, Hendrix RW, Gross CA. Consensus sequence for *Escherichia coli* heat shock gene promoters. Proc Natl Acad Sci USA 1985;82:2679–2683.

55. Arnosti DN, Chamberlin MJ. Secondary sigma factor controls transcription of flagellar and chemotaxis genes in *Escherichia coli*. Proc Natl Acad Sci USA 1989;86:830–834.

56. Ausubel FM. Regulation of nitrogen fixation genes. Cell 1984;37:5–6.

57. Hunt TP, Magasanik B. Transcription of glnA by purified Escherichia coli components: core RNA polymerase and the products of glnF, glnG and gln L. Proc Natl Acad Sci USA 1985;82:8453–8457.

58. Erickson JW, Gross CA. Identification of the sigma E subunit of *Escherichia coli* RNA polymerase - a second alternative sigma factor involved in high temperature gene expression. Genes Dev 1989;3:1462–1471.

59. Moran CP, Lang N, Legrice FFJ, Lee G, Stephens M, Sonenshein AL, Pero J, Losick R. Nucleotide sequences that signal the initiation of transcription and translation in *Bacillus subtilis*. Mol Gen Genet 1982;186:339–346.

60. Tatti KM, Moran CPJ. Promoter recognition by σ^{37} RNA polymerase from *Bacillus subtilis*. J Mol Biol 1984;175:285–297.

61. Johnson WC, Moran CPJ, Losick R. Two RNA polymerase sigma factors from *Bacillus subtilis* discriminate between overlapping promoters for a developmentally regulated gene. Nature 1983;302:800–804.

62. Gilman MZ, Wiggs JL, Chamberlin MJ. Nucleotide sequences of two *Bacillus subtilis* promoters used by *Bacillus subtilis* sigma-28 RNA polymerase. Nucleic Acids Res. 1981;9:5991–6000.

63. Rather PN, Hay RE, Ray GL, Haldenwang WG, Moran CPJ. Nucleotide sequences that define promoters that are used by *Bacillus subtilis* sigma-29 RNA polymerase. J Mol Biol 1986;192:557–565.

64. Sun D, Fajardo-Cavazos P, Sussman MD, Tovar-Rojo F, Cabrera-Martinez R-M, Setlow P. Effects of chromosome location of *Bacillus subtilis* forespore genes on their spo gene dependence and transcription by $E\sigma^{F}$: identification of features of good $E\sigma^{F}$-dependent promoters. J Bacteriol 1991;173:7867–7874.

65. Nicholson WL, Sun D, Setlow B, Setlow P. Promoter specificity by sigma-G containing RNA polymerase from sporulating cells of *Bacillus subtilis*: identification of a group of fore-spore specific promoters. J Bacteriol 1989;171:2708–2718.

66. Carter HL, Wang L-F, Doi RH, Moran CPJ. rpoD operon promoter used by sigma-H RNA polymerase in *Bacillus subtilis*. J Bacteriol 1988;170:1617–1621.

67. Zheng L, Halberg R, Roels S, Ichikawa H, Kroos L, Losick R. Sporulation regulatory protein GerE from *Bacillus subtilis* binds to and can activate or repress transcription from promoters for mother cell specific genes. J Mol Biol 1992;226:1037–1050.

Biotechnology Annual Review Volume 1
M.R. El-Gewely, editor

Comparative methods for identifying functional domains in protein sequences

Steven Henikoff
Howard Hughes Medical Institute, Fred Hutchinson Cancer Research Center, Seattle, Washington, USA

Abstract. This chapter reviews the different approaches that have been applied to the problem of protein motif identification. Several methods are available for finding motifs within protein families, for clustering databases to identify family relationships and for searching databases that consist of motif representations. With the rapid expansion of sequence databases, which currently appear to represent most protein families, these methods are becoming increasingly important for interpretation of molecular sequence information.

Introduction

The detection of amino acid sequence homology between a newly identified sequence and previously determined sequences is often the most important clue to the function of a gene of interest. In fact, the "homology search" has become so important to molecular biology and biotechnology that this is often the major reason for sequencing a gene. Extreme manifestations of this view are the many limited cDNA sequencing projects in which random coding sequences representing mRNAs from an organism or a tissue are acquired at the maximum possible rate with the hope that some of the sequences will have informative database matches, leading to further study [1]. Because homology searching is so important, there has been a recent expansion in the number and kinds of tools available for identifying homology. The traditional searching approach, in which a sequence is compared to the current databanks of sequences, remains the most important method for identifying homology, with several recent practical improvements [2–6]. Since this approach locates evolutionarily conserved protein segments, it can also be used to identify functional domains, which are typically highly conserved. However, other methods have been introduced in recent years whereby functional domains are identified directly, rather than inferred on the basis of sequence database search results. These direct methods for functional domain identification are the subject of this chapter. These methods are valuable for guiding site-directed mutagenesis and protein design, especially in the absence of a known structure. Furthermore, these

Address for correspondence: S. Henikoff, Howard Hughes Medical Institute, Fred Hutchinson Cancer Research Center, Seattle, WA 98104, USA. Tel.: +1 206 667 4515; Fax: +11 206 667 5889; Internet: steveh@howard.fhcrc.org.

conserved regions typically provide candidate DNA sequences for PCR-based or hybridization-based approaches to homolog isolation. Finally, accurate identification of a functional domain can guide a choice of peptide for structure determination using two-dimensional NMR, a method that is currently limited to small peptides, about the size of functional domains.

Motif Representation

The term "domain" is widely used to describe a separately folded portion of a protein. Domains are readily identified once the 3D structure of a protein is known. However it is far more common that there is no model structure available for a sequence, in which case a domain cannot be directly identified. Also, the concept of a separately folded domain is too restrictive, since critical stretches of residues within a protein fold, such as those at the active site of an enzyme, are often of interest. Therefore, the more inclusive term "motif" will be used to describe a sequence segment consisting of a contiguous stretch of residues or residue types, sometimes with limited gaps (i.e. deletions or insertions), shared by a group of related proteins and conserved for function. Even this description of motif is too restrictive, because it is not necessary that the group of proteins be related. So, for example, the helix-turn-helix DNA-binding motif is found in many otherwise unrelated groups of proteins that are thought to have converged to form a similar 20-residue structure with weak sequence constraints [7]. The issue of whether or not any particular motif reflects common ancestry or convergence can be difficult to resolve in cases of short, weak similarities, so that the term "group" (rather than "family" or "superfamily") will be used to describe a collection of sequences that share a motif. The terms "signature" and "pattern" are often synonyms for motif, although here the term "pattern" will be used in a more restricted sense. Motifs represent "local" as opposed to "global" features of a protein, which can contain multiple motifs.

A simple representation of a motif is a consensus sequence extracted from a multiply aligned set of sequence segments (Fig. 1a). A consensus is a string of amino acids (plus "x" for any non-consensus residue that intervenes), given some arbitrary criteria for whether or not a residue is conserved. An example of a consensus sequence is shown in Fig. 1b for selected homeodomain proteins. Note that to make a consensus, information is necessarily discarded, since only a single amino acid can be represented. For example, in the last position of the homeodomain alignment, R [for arginine] is present in 16/35 sequences and is considered the consensus residue, however, K [for lysine] is present in 10/35 sequences and is not represented. A consensus such as this can be used to detect homology in the context of a database search, in which case it is searched exactly as is a sequence, but it should be more representative of the alignment than any single sequence.

A simple pattern is potentially a more informative representation of a motif than a consensus, since multiple amino acids are allowed for every position. For example, using the PROSITE convention for representation of patterns [8], the occurrence of

a)   BLOCK   MATRIX
                                                                                  COLUMN

                        1   2   3   4
                        0   0   0   0
HKN1_MAIZE   281 SWWDQHYKWPYPSETQKVALAESTGLDLKQINNWFINQRKRHWK   A   0
HM1D_DROAN   345 TLEKSFERQKYLSVQERQELAHKLDLSDCQVKTWYQNRRTKWMR   C   0
HMAB_DROME   399 ELEKEFLFNAYVSKQKRWELARNLQLTERQVKIWFQNRRMKNKK   D   0
HMBC_DROME   111 ELEQHFLQGRYLTAPRLADLSAKLALGTAQVKIWFKNRRRRHKI   E   0
HMCU_DROME   1759 ALRLAFALDPYPNVGTIEFLANELGLATRTITNWFHNHRMRLKQ   F   3
HMES_DROME   405 KLEHAFESNQYVVGAERKALAQNLNLSETQVKVWFQNRRTKHKR   G   0
HMH2_DROME   301 GLEIQFQQQKYITKPDRRKLAARLNLTDAQVKVWFQNRRMKWRH   H   3
HMIX_XENLA   110 ILEQFFQTNMYPDIHHREELARHIYIPESRIQVWFQNRRAKVRR   I   3
HMM3_CAEEL   223 VLNEMFSNTPKPSKHARAKLALETGLSMRVIQVWFQNRRSKERR   K 29
HMN1_DROME   559 SLENKFKTTRYLSVCERLNLALSLSLTETQVKIWFQNRRTKWKK   L   0
HMOC_DROME   87 VLEALFGKTRYPDIFMREEVALKINLPESRVQVWFKNRRAKCRQ   M   0
HMRO_DROME   205 RLEVEFHRNEYLSRSRRFELAETLRLTETQIKIWFQNRRAKDKR   N   0
HMTI_DROME   315 ELECRFRLKKYLTGAEREIIAQKLNLSATQVKIWFQNRRYKSKR   P   0
HOX1_HALRO   659 GLEKSFQSQKYVAKPERRKLADALSLTDAQVKIWFQNRRMKWRQ   Q   9
ISL1_RAT   195 TLRTCYAANPRPDALMKEQLVEMTGLSPRVIRVWFQNKRCKDKK   R 46
MTA2_YEAST   146 SWFAKNIENPYLDTKGLENLMKNTSLSRIQIKNWVSNRRRKEKT   S   0
OCT3_MOUSE   237 SLETMFLKCPKPSLQQITHIANQLGLEKDVVRVWFCNRRQKGKR   T   6
PHO2_YEAST   91 VLKRKFEINPTPSLVERKKISDLIGMPEKNVRIWFQNRRAKLRK   V   0
TTF1_RAT   175 ELERRFKQQKYLSAPEREHLASMIHLTPTQVKIWFQNHRYKMKR   W   0
CF1A_DROME   357 ALEQHFHKQPKPSAQEITSLADSLQLEKEVVRVWFCNRRQKEKR   Y   3
HMEV_DROME   84 RLEKEFYKENYVSRPRRCELAAQLNLPESTIKVWFQNRRMKDKR
HMAN_DROME   311 ELEKEFHFNRYLTRRRRIEIAHALCLTERQIKIWFQNRRMKWKK
HMCA_DROME   257 ELEKEYCTSRYITIRRKSELAQTLSLSERQVKIWFQNRRAKERT
HMEN_DROVI   500 RLKREFNENRYLTERRRQQLSSELGLNEAQIKIWFQNKRAKIKK
HMGD_DROME   199 ALERIFARTQYPDVYTREELAQSTGLTEARVQVWFSNRRARLRK
HMP1_HUMAN   228 ALERHFGEQNKPSSQEIMRMAEELNLEKEVVRVWFCNRRQREKR
HMX1_HUMAN   180 ALERKFRQKQYLSIAERAEFSSSLSLTETQVKIWFQNRRAKAKR
HMZ1_DROME   104 ELENEFKSNMYLYRTRRIEIAQRLSLCERQVKIWFQNRRMKFKK
HNFA_HUMAN   234 EECNRAECIQRGVSPSQAQGLGSNLVTEVRVYNWFANRRKEEAF
HXDB_MOUSE   265 ELEREFFFNVYINKEKRLQLSRMLNLTDRQVKIWFQNRRMKEKK
HXDC_MOUSE   225 ELENEFLVNEFINRQKRKELSNRLNLSDQQVKIWFQNRRMKKKR
IPOU_DROME   305 SLEAYFAVQPRPSGEKIAAIAEKLDLKKNVRVWFCNQRQKQKR
MTA0_YEAST   84 FLEQVFRRKQSLNSKEKEEVAKKCGITPLQVRVWVCNMRIKLKY
OCT1_HUMAN   393 ALEKSFLENQKPTSEEITMIADQLNMEKEVIRVWFCNRRQKEKR
PAX6_HUMAN   224 ALEKEFERTHYPDVFARERLAAKIDLPEARIQVWFSNRRAKWRR

b) Consensus: ELExEFxxNPYLSxxEREELAxxLGLTERQVKIWFQNRRMKEKR

c) Simple Pattern: LxxxxxLxxxxIRxWxxxxxxxxR
                                 I   I   VK   K
                                 V   V   Q
                                 M   M
                                 F
                                 Y

d) Covering: LpxxdxxxxxcxxxrxxxfjxxfxcxxxranxWfxNxRxn

Fig. 1. Different representations of a homeodomain, using selected sequences. (a) Block derived from
the 112 full-length sequences in PROSITE v. 11.0 using the PROTOMAT system, pruned from the
entry in BLOCKS v. 6.2 [BL00027] to 35 segments such that no segment is more than 80% identical to
any other. Although the homeodomain is 60–61 amino acids wide, PROTOMAT default parameters
yield an ungapped block of 44 amino acids. The starting position for each sequence is indicated after
the SWISS-PROT name. A single column of a scoring matrix derived from the last column of the block
is to the right. (b) Consensus sequence derived by taking the most frequent residue at every position. An
"x" indicates that no residue is represented in more than 20% of the sequences at that position. (c)
PROSITE pattern detects 12 false positives and misses 8 true positives in a search of SWISS-PROT 26.
(d) Covering produced by the PIMA program [25] for the largest subset of 19 homeodomains, where
p = [EQKR], d = [FWY], c = [IVLM], r = [HNDEQKRST], f = [CIVLMFWY], j = [STAGP], a = [IV],
and n = [KR].

either R or K is represented as [RK]. In this way, any desired minimum frequency of a residue in a position of a multiple sequence alignment can be represented. The pattern shown in Fig. 1c, [LIVMFY]-x(5)-[LIVM]-x(4)-[IV]-[RKQ]-x-W-x(8)-[RK], is one in which nearly all residues that occur in a given position are represented up to a maximum of six different residues, with "x" representing more degenerate positions. While patterns are potentially more informative than consensus sequences because substitutions are allowed, the desire that nearly all sequences in a group have the pattern leads to the arbitrary omission of otherwise strong consensus residues. For example, at position 2 or position 35 of the block shown in Fig. 1a, a single residue is present in 32/35 sequences. Still, not all sequences in a group necessarily contain all residues in the pattern; for example, the first sequence in the block shown in Fig. 1a [HKN1_MAIZE] conflicts with the pattern at two of the six residues. Patterns have been used in attempts to detect homology by counting any aligned residue that matches one of the occurrences in the pattern at that position. So, for example, at the last position in the pattern, either R or K would be counted as a match.

Patterns need not be limited to the 20 amino acids. For example, a "covering" is an extended pattern representing an entire alignment that uses a hierarchy of amino acid classes [9]. Classes are composed of residues with similar properties, such that any amino acid belonging to a class at a position in the alignment is given a special symbol to represent the degeneracy (Fig. 1d). In this way, a covering is a consensus in which residue classes are allowed in addition to single residues. Coverings have been used to search for homology in the same way that sequences are used in searches, since the amino acid class symbols can be scored in the same way as the letters representing the 20 amino acids.

The above motif representations are like sequences in that they can be expressed as strings of single characters, one per position. However, it is possible to use the multiple sequence alignment in an uncondensed form, that is, as an array in which each position is represented as a column with as many rows as there are sequences in the alignment. When the array represents highly conserved regions, typically without gaps, it is referred to as a block (Fig. 1a). To search for homology, the alignment or block is not used directly. Rather, it is first converted to a second array, a position-specific scoring matrix, where each position in the alignment corresponds to a column of the matrix, an example of which is shown to the right of Fig. 1. A matrix of this type can be simple, with 20 rows corresponding to the 20 amino acids, or complex, with a 21st row for a position-specific penalty for inserting a gap, and a 22nd row to penalize extending the gap, in which case it is typically called a "profile" [10]. Each position in the matrix is represented by a score based on how frequently a residue appears at that position in the alignment. So for the first column of the matrix in Fig. 1, the value for E is 29%, representing the frequency of E in this position of the alignment. In a search, the segment to be scored is aligned with the matrix, and each amino acid residue in the segment is given the score found in the aligned column for that amino acid. In this way, no positional information from the alignment is discarded. The matrix shown provides an especially simple scoring scheme in which a score can be interpreted as a fractional match (×100%). However, matrix scores are typically adjusted to account for nonrandom expected occurrences of residues in real proteins [10,11], and can be weighted to reduce the contribution of more closely related sequences in the multiple sequence alignment [12,13]. Profiles can be searched against sequence databanks using a dynamic programming algorithm for searching sequences to allow the scoring of gaps [10]. Matrices representing blocks do not score variable gaps, so it is feasible to exhaustively score all possible segments the width of the block in a search, where the number of segments scored is about the same as the number of amino acids in the sequence databank [14].

Scoring matrices are general tables in which any position-specific property can be represented. Structural properties can be represented by specifying position-specific structural environments rather than amino acids [15]. Several variations of this general approach have been applied to the problem of predicting structural similarity, such as using contacts between residues rather than environments to model protein folds [16]. These methods have been tested on proteins that align structurally over their full length, such as actins and HSP70 proteins, but have not been applied to individual motifs.

While a profile or scoring matrix is quite general, it can be further generalized by specifying additional parameters. Hidden Markov models are constructed by training on known family members in such a way that an alignment results. While there is evidence that this generalization can sometimes provide improved detection of distant relationships, the added parameters apparently necessitate a very large training set, on the order of hundreds of sequences [17–19].

Finding Motifs Within Groups

Several different strategies have been used to detect motifs. These are listed in Table 1 along with particular examples of programs described in more detail below.

Given a set of proteins that can be aligned one to another in pairs, methods are available for multiple sequence alignments using a string comparison algorithm, such as dynamic programming [20]. While simultaneous multiple sequence alignment methods are adequate for a few, or even several sequences of average length [21], they soon lead to impossible time and space computational requirements. As a result, multiple sequence alignment is frequently carried out using hierarchical clustering methods in order to reduce the n-dimensional problem into a series of two-dimensional ones [22–25]. These methods have proven effective in providing good alignments for many protein families, and often these alignments are used for manual identification and extraction of motifs. Unlike simultaneous methods, however, hierarchical methods are not guaranteed to find the best alignment given a set of parameters: the possibilities are limited by the order in which sequence pairs are chosen, and the optimal alignment might be dependent on information not obtained by successive pairwise comparisons. Hierarchical methods do well when sequences

Table 1. Summary of methods for finding motifs within groups

Method	Advantages	Disadvantages
String comparison		
Simultaneous alignments	Optimal based on scores used	Is computationally explosive
	OK when uniformly diverged	Not designed for motifs
Hierarchical clustering	Can be applied to many sequences	Depends on pairwise alignments
	OK when differentially diverged	Not designed for motifs
Hidden Markov models	Can be applied to many sequences	Model needed
	Can deduce parameters during modeling	Not designed for motifs
Motif-based, consecutive identities		
MALIGN	Looks for shared information	Need consecutive identities
		Need anchors
GENALIGN	Looks for shared information	Is computationally expensive
		Need some consecutive identities
Pairwise, not consecutive identities		
FIL-LOG, pab, MACAW	Exhaustive	Depends on pairwise alignments
	Not limited to identities	Is computationally expensive
SCR finder	Looks for shared information	Need anchors [or]
		Is computationally explosive
Motif-based, not consecutive identities		
Spaced triplets	Looks for shared information	Some identities needed
MSC	Looks for shared information	Model needed
		Is computationally expensive
Gibbs sampler	Looks for shared information	Model needed

are differentially diverged, so that closely related sequences can be aligned first and their consensus used to aid in alignment of more distantly related sequences. Both simultaneous and hierarchical methods are generally effective when the sequences being aligned are related to one another throughout their lengths, having acquired substitutions and gaps during divergence from a common ancestor. However, they are less effective when similarity is confined to only segments of each protein. For example, the homeobox proteins consist of many different families that are grouped together because of a single shared domain (Fig. 1). Outside of the highly conserved homeodomain, there is often no similarity at all between many members of this group, so that any alignment for flanking segments is a spurious one. This problem is potentially addressable by choosing appropriate thresholds to prevent spurious alignments, or by editing sequences prior to a final alignment.

Hidden Markov models also provide multiple sequence alignments of distantly related sequences. A potential advantage is that all parameters in the model can be deduced during training. This method appears best suited for sequences related throughout their lengths. To detect and align motifs, the width of the motif model must be specified in advance, necessitating repeated runs at different widths in the expected range [18].

Other alignment methods for finding motifs look for them directly by searching for segments in common among proteins in a group to form blocks. In some cases the detection of candidate blocks is followed by full or partial multiple alignment in which blocks are strung together and gaps inserted [26,27]. These methods are potentially more sensitive in cases where similarity is limited to just a small portion of the sequences in the group. Methods that detect motifs first rather than extracting them from multiple sequence alignments take advantage of the general observation that even distant motifs are rarely gapped, and when they are, the gaps are restricted in length. However, the general problem of examining all possible motifs is computationally explosive, even for a single motif length, since the number of possible motifs increases with sequence length (m) and number of sequences (n) as m^n. However, effective methods have been devised in which only a subset of possible motifs are examined; some of these have computational requirements that increase only as $m \times n$. This can be accomplished by specifying a list of possible motifs for initial examination, then exhaustively scanning all sequences for occurrences of each one. An occurrence of a possible motif in some or all of the sequences provides a candidate that can be examined for other features to determine whether it is likely to be real.

An early example of an exhaustive search method, MALIGN [28], examines a group of sequences for identical segments of two or more amino acids. These identical segments become potential anchors for a multiple alignment by determining which of the segments are in the same order for all of the sequences. By choosing weights and gap penalties, an optimal alignment is obtained as the one with the maximum score. Since only identities are scored, the multiple sequence alignment consists of ordered identical segments separated by unaligned regions. However, distant relationships rarely have segments of consecutive identities shared by all sequences in a group, so that this method is not one that easily generalizes to large and/or distantly related groups. An improvement in this method, GENALIGN [29], includes a sequence clustering algorithm that makes possible detection of block alignments in which consecutive identities are shared by only some of the segments. That is, if segments A and B share a high-scoring match not shared by C, and segments A and C share a high-scoring match not shared by B, then all three segments are alignable. This approach leads to an explosion in the number of possible alignments when the length of consecutive identities is small and the number of sequences is large. Therefore, a lower length threshold is set in a first pass, and the alignments found are used as "anchors". The threshold is then reduced and the region between anchors is examined for new alignments. This "telescoping" procedure can be continued down to alignments based on dipeptide matches. An alternative to telescoping involves (1) finding the locations of all shared dipeptides, (2) drawing a directed graph with shared dipeptides as nodes connected by arcs when the nodes are in the same order for all of the sequences, and (3) finding a best path through the graph that maximizes the score [30].

The exhaustive search for possible motifs need not be based on consecutive identities. Another approach is to carry out pairwise comparisons between sequences and then ask which high-scoring segments are in common for most or all of the sequences in the group. Multiple alignment of these segments leads to a block. However, determining which segments belong in a block can be challenging, and different methods have been reported. The FIL-LOG algorithm accepts segments that are in common for a minimum number of pairwise comparisons [26]. *pab* accepts only segments that are "connected" in the sense that every segment must be represented in at least two high-scoring pairs with other segments in the block [31]. MACAW asks which segment pairs exceed a threshold score, then combines those with shared segments into a block [32]. The extent of a block is limited by requiring that each column have some minimum level of homogeneity, although adjacent blocks with the same number of residues between them are fused, thus allowing single blocks to include both conserved and diverged positions.

The pairwise nature of the above methods present the same potential drawback as for hierarchical multiple sequence alignment programs, that information in common for all of the sequences might not be represented in the pairwise alignments. In addition, the number of pairwise comparisons needed is n^2 for n sequences. Simultaneous methods for finding motifs can potentially avoid these problems. MALIGN is an example of a simultaneous method, but is severely limited by the requirement that all sequences share consecutive identities. This limitation can be circumvented by allowing any shared motif to become an anchor. In the SCR finder method [33], sequences in the group are examined for the presence of any motif of predetermined length, where the definition of a motif can be broadened to include non-identical but similar segments. In one variation, the highest scoring shared motifs in the same order for all sequences become anchors. This method is sensitive to occasional misalignments, which will occur with increasing frequency as sequences become more numerous and more distant. A more satisfactory alternative, to consider all high-scoring shared motifs, becomes computationally infeasible for more than a few sequences.

Intermediate between approaches that require consecutive identities and those that allow any similar region to serve as anchors are methods that examine all sequences in a group for spaced identities [34,35]. One implementation of this approach, MOTIF [35], examines all sequences for the presence of spaced triplets of the form $aa_1 \, d_1 \, aa_2 \, d_2 \, aa_3$ where d_1 and d_2 are fixed distances between the amino acids. So Ala-Ala-Ala is one triplet, Ala-x-Ala-Ala is another, and Val-x[16]-Ala-x[7]-Cys is another. An exhaustive search is carried out for all such triplets in the full set of sequences using all combinations of d_1 and d_2 out to a reasonable maximum distance (about 20). The rationale is that true motifs will typically include one or more sets of spaced triplets in all of the sequences in the group. Since some true motifs do not contain aa_1, aa_2, and aa_3 in the full set of sequences, the number of sequences required to contain a triplet (the "significance level") can be reduced. In such cases, the block containing the triplet is scanned along each of the sequences that lack the triplet to find the best segment based on maximizing an overall score for the block. Each sequence is then rescanned to maximize the score.

Other approaches avoid limiting the motif search to a predetermined list without becoming computationally explosive by detecting motif "seeds" that occur in as few as two sequences, then asking whether any of these seeds can mature to include other sequences in the group. For example, the MSC program considers all possible segment pairs of specified length between two sequences as candidate patterns, and then examines these pairs by aligning them with all possible segments from other sequences in the group [36]. The rationale is that real patterns will strengthen with iterative addition of new members, whereas false starts will fail to improve. Another method, designed for nucleotide sequences but applicable to amino acid sequences, builds up solutions in a similar manner, but uses a measure of statistical significance to ascertain alignment quality [37]. A different method that optimizes a measure of statistical significance is a "Gibbs sampling" strategy. Starting with a block of specified width, the sampler aligns random positions within all but one sequence [38]. Sampling consists of choosing a segment from the remaining sequence at random, where the probability of being chosen is proportional to how well the segment conforms with the block. Other sequences are then sampled in the same way to further improve the significance of the alignment. Resampling continues until no further improvement is seen.

While "seed" methods can theoretically find motifs that are missed by exhaustive search methods because the true motif was not represented in the list, they risk getting trapped in local optima, a problem that cannot occur for exhaustive search methods by definition. However, this does not appear to be a problem in practice for current implementations (G. Stormo, personal communication; S. Henikoff and J.G. Henikoff, unpublished results). A potentially more serious problem is that seed methods might require a model that specifies the number of motifs sought and their widths. Because of the quite different challenges that these two general approaches face, it is likely that seed methods and exhaustive search methods will each have different strengths and weaknesses. Therefore, comprehensive evaluations using realistic data are crucial.

Database Clustering Methods for Identifying Groups

Each of the above methods generally assumes that all sequences in the group are related, i.e. that there are motifs to find. However, the converse is also true, that the presence of one or more motifs shared among sequences is evidence for membership in a group. Therefore, the identification of motifs shared by a subset of sequences can form the basis for carving out groups from a protein sequence database using a sequence clustering algorithm. A database of n sequences is searched using every sequence in it as query, leading to an $n \times n$ array of search results. The major challenge is to determine criteria that will highlight real groups and ignore spurious ones in this array. One database clustering method based on identifying motifs starts with a simple consecutive matching criterion to provide candidate anchors for full multiple sequence alignment [9]. The hierarchical alignment algorithm aligns the first pair and constructs a covering, then adds the next pair, and so on, leading to a covering representing the full group. This approach was successful in obtaining coverings for sequences as distant as the alpha and beta hemoglobins (43% identical). Another database clustering method also begins with simple consecutive matching criteria to identify groups that share unique peptide words [39]. The words are then ordered using a dynamic programming algorithm and refined into blocks representing the group. A more elaborate approach is to initially cluster sequences based on dipeptide match frequencies [40], then refine and analyze the clusters using statistical methods [41]. The resulting statistics are then used to design a suitable neural net for clustering the database [42].

Other database clustering methods based on motif identification use segment similarity criteria rather than exact matches in the $n \times n$ search. A potential advantage is that amino acid substitution tables that underlie segment similarity measures are more informative and so better able to detect initial homology than identical matches [43,4]. One method [44] carries out the $n \times n$ search using BLAST3, a program that generates three-segment blocks in the context of a single-sequence database search [3]. BLAST3 carries out a standard search for local alignment, then examines the "twilight zone" of high-scoring segment pairs for high-scoring three-way alignments involving two database segments that align with the same query segment. Starting with three-way alignments that exceed a high threshold of chance expectation, those that overlap and exceed a threshold number of matches are clustered to provide a group for construction of a block [44]. Blocks consisting of segments of low compositional complexity, such as runs of a single amino acid, are avoided by requiring that the composition of segments approach that of the database as a whole. An important simplification in this approach is that lengthy pairwise matches or two matches distant from one another along the lengths of the sequences eliminate the matches from consideration, thus avoiding the detection of globally homologous relationships. In this way, the method can focus on single local motifs such as the homeodomain (Fig. 1), which was in fact one of the larger groups found when this method was implemented. In addition to known motifs, two previously unrecognized candidate groups were identified in this study. Another method, Megaclassification [45], uses the BLAST searching program [2], which provides a list of high-scoring segment pairs rank ordered according to decreasing statistical significance. Each hit is extended if possible allowing gaps, and the resulting assembled hit is compared to its nearest "neighbors" based on similarities identified in the $n \times n$ search. Groups of neighbors can represent both global and local alignments, from which the underlying motifs can potentially be extracted.

Clustering methods have applications beyond motif detection and homology searching per se. The widely used Entrez program from the National Center for Biotechnology Information includes a neighboring function using a data structure that was created using a BLAST-based clustering algorithm. A similar clustering approach was employed in the identification of conserved regions that cross phylum boundaries [46].

Motif Databases

There are now several examples in which motif-finding methods have been applied to provide a database of motifs which can be used to classify new sequences with respect to the possible presence of a motif. Each of the database clustering approaches can potentially provide an associated database of multiple alignments or motif representations, and in fact a database of coverings has been publicly available for sequence classification since 1990 [9]. In addition, there are publicly-available databases of sequence segments, patterns, blocks and profiles, some of which can be searched for sequence similarity. These compilations are ultimately based on human decisions as to the composition of groups rather than on algorithmic criteria. Humans have a considerable advantage over machines in deciding on group membership in that not all relevant information concerning a protein is present in the sequence. For example, no available clustering algorithm takes into account the fact that G-protein coupled receptors function similarly in signal transduction pathways. However, since highly motivated biologists have succeeded in accurately classifying proteins belonging to this very diverse and important group, it seems worthwhile to take advantage of this enormous reservoir of expert knowledge. Disadvantages of relying on human decision-making for grouping proteins include the inevitable lag between entry of a sequence into a databank and its appropriate annotation as a group member, and the concern that some groups have escaped detection [44,46].

The PROSITE database of patterns [8] is probably the most widely-used motif database, consisting of nearly 1,000 different patterns representing about 40% of the sequences in SWISS-PROT [47], a fully annotated nonredundant protein databank with about 30,000 separate entries. PROSITE patterns are derived manually starting with published reviews that provide candidate groups. Conserved regions are noted, and are especially favored for determination of a pattern if thought to be important for function, such as enzyme active sites. A short pattern of up to five residues is chosen, and this is used to search SWISS-PROT, noting whether or not false positives or potentially new true positives appear. The pattern is then extended and adjusted to eliminate false positives. An especially valuable feature of PROSITE is the informative annotation for each group. Updates of PROSITE are scheduled at 6-month intervals to coincide with even-numbered releases of SWISS-PROT.

The SDOMAIN database consists of motif segments derived from protein databanks with manual intervention [48,49]. The form of an SDOMAIN entry is simply an annotated sequence segment, with more than 34,000 entries in the latest database. This database can be searched using conventional sequence searching programs such as BLAST and BLAST3. The rationale behind searching SDOMAIN is that the background level of spurious similarities is reduced by eliminating regions that are not conserved in evolution. SDOMAIN entries are cross-referenced to PROSITE to aid in evaluation of a suspected similarity.

The BLOCKS database is generated by the automated PROTOMAT system [50]. Groups are obtained from the PROSITE database, although the PROSITE patterns are not used in any way. The current database consists of 2679 individual blocks derived from 698 nonoverlapping PROSITE groups. PROTOMAT (1) extracts SWISS-PROT sequences listed in a PROSITE group, (2) applies the MOTIF [35] spaced-triplet detection program in an automated mode for empirical determination of significance level, (3) merges and extends MOTIF-generated alignments to obtain high-scoring candidate blocks and (4) considers blocks as nodes in a directed graph to determine a best path of blocks in which all blocks are in the same order in all sequences. In some cases a minority of sequences in a PROSITE group are too dissimilar from the other sequences to end up in a best path, and these are excluded. PROTOMAT finds accurate blocks for many of the groups in the PROSITE database that are too diverged to be aligned by automated string alignment methods, such as the G-protein coupled receptors.

The HSSP database consists of multisequence alignments and accompanying profiles in which 2D and 3D structural information have been used to guide sequence alignment [51]. Database entries are available as separate files, one per group, each corresponding to one of the entries in the PDB database of 3D structures. In a recent database there were 1,532 files representing about 24% of SWISS-PROT. It should be noted that because of the redundancy in PDB, there is considerable overlap between groups. To reduce this redundancy, a selected subset is available in which no PDB sequence is more than 25% identical to any other over a stretch of 80 residues. In addition to HSSP, two other databases provide multiple sequence alignments based on aligning 3D structures [52,53].

Searching Databases of Motifs

Originally, motif representations were used to search sequence databases [54], an application that is still widely used. However, with the availability of motif databases, searches can be reversed whereby a sequence is used to query a motif database; detection of a motif potentially classifies it into the group represented by the motif. In such cases, the same matching or scoring algorithm can be used in either direction. Since a motif database is typically much smaller than a sequence databank, such reverse searches can be very fast.

Motifs that are represented as sequence segments can be searched using any algorithm designed for searching sequence databases. This applies to consensus sequence queries of sequence databanks, and to coverings and SDOMAIN segments either as queries or as database compilations, although for coverings a special scoring table including additional symbols is needed. In other cases, searches are carried out using programs designed to compare a motif representation and a sequence.

Simple patterns can be searched using generic text-matching algorithms; perhaps as a result, there are 20 different programs available for searching PROSITE patterns [8]. High speed is a feature of such searches, typically requiring seconds to search PROSITE and a minute or so to search a pattern against a protein databank. Beyond speed, there is no reason to use simple patterns for homology detection, as evidence is lacking that simple patterns can detect true homologies that other methods cannot. This should not be surprising, since any simple pattern is a special case of a more complex representation, such as a profile. In contrast to searches using sequences, consensus sequences, coverings or matrices, simple pattern searches report only a yes/no answer, which does not provide any guide for distinguishing true from false positives and true from false negatives. There are some searching programs for simple patterns that provide scoring systems claimed to alleviate this problem, e.g. [55], but comparative evaluations are lacking. Simple patterns provide a useful heuristic representation of motifs, but nothing more.

Blocks can also be searched by pattern matching. In the case of the clustered database based on unique peptide words, a sequence is aligned with a block and a match between a residue in the sequence and any residue in the aligned column of the block is considered a match [39]. If the fraction of matches exceeds a threshold, the block is scored as a hit. While this procedure led to an excessive number of false positives when single block hits were allowed, adequate specificity was achieved by limiting search reports to hits consisting of multiple blocks.

Blocks can be searched by conversion to scoring matrices, where each column of a block becomes a column of a matrix. In contrast to pattern-matching methods, which provide only yes/no answers, matrix-based searches provide scores that can be used to infer homology. Programs are available for searching a block or multiple blocks versus a sequence database and for searching a sequence versus a database of blocks [11,56,57,58]. In these programs, a block is first converted to a scoring matrix "on-the-fly", then every segment in the block is aligned with every possible segment of equal length in the sequence, and then the alignment is scored using values from the matrix. These computations are carried out with sufficient rapidity that a desktop computer can carry out a search of typical block against SWISS-PROT 27 in about 10 min, and a search of a typical sequence against the BLOCKS 7.0 database in about 1 min. High speed is a consequence of not scoring gaps, whereas for profiles that allow gaps, gap scoring requires a string-comparison algorithm which is more computationally intensive than exhaustive comparisons without gaps. Although gaps are not scored directly in searches involving blocks, it has proven useful to measure global information by evaluating hits in which multiple blocks from a group obtain high scores when aligned in order along a sequence [59].

Profiles have been used to detect or confirm distant relationships in searches of sequence databanks using the PROFILESEARCH program [10], although at present the best database of profiles [HSSP] is not available in searchable form.

The clustered database produced by Megaclassification is different from those described above in that it consists of sequence classifications based on clustering, and is not based on multiple sequence alignments per se [45]. Therefore, searching this database involves sequence comparison using BLAST, followed by classification of the query depending on the group membership of significant hits.

Future Directions

Two different analyses argue that sequence representatives of most protein groups are already known, so that protein classification, as opposed to discovery of new motifs, will be an important future endeavor. One estimate, that there are only about 1,000 different protein folds, comes from dividing the number of families with known structures (120) by the fraction of database sequences judged to be homologous to these known structures (1/4) and by the fraction of database sequences that have identified homologs (1/3) [60]. A different type of estimate, that there are only about 900 ancient protein families (Green et al., 1993), is based on the surprising finding that protein sequences sampled from organisms in different phyla only rarely identify ancient families that were not previously known. To explain why most sequences still do not fall into existing ancient groups, one must suppose that either these sequences represent more modern inventions, or that current methods for sequence classification are insufficiently sensitive. The latter possibility is supported by several findings in which structurally similar proteins lack sequence similarity (e.g. [61]). Clearly, there is considerable room for incremental improvements in methods for sequence classification. For example, improvements in the sensitivity of matrix searching have been obtained by sequence weighting and improved position weighting of profile scores [12,13]. It is less clear that current methods for sequence classification are hampered by difficulties in identifying motifs within a group using current methods. A fully automated implementation of the spaced-triplet method [50] has been able to identify motifs successfully for even the most challenging groups compiled in PROSITE [8]. Similar results have been obtained using a Gibbs sampling strategy [32] modified and extended to allow for full automation (S. Henikoff and J. Henikoff, unpublished results).

Beyond improvements in existing systems, the question arises whether there are promising approaches to motif identification and sequence classification that will become standard in the future. Structure-based profiles are the first that come to mind. It might be that when there are representatives from most of the families and domains in the structure database, it will be possible to classify sequences based on structural profiles more reliably than based on sequence information. However, recent studies do not provide grounds for optimism. For example, it has been pointed out that structural environment classes are largely proxies for residue features [16]. When "pure" structural information is used, general performance in sequence classification falls far below what can be obtained using sequence-based methods. Furthermore, in the few cases in which structure-based methods do detect homology that sequence-based methods do not, such as for HSP70 and actin, it appears that shared compositional features varying over long sequence stretches underlie detection [62]. In contrast, current sequence-based methods depend on highly informative local motifs. One wonders whether a purely sequence-based method could be designed to do as well or better in these cases by paying attention to compositional features such as hydrophobicity and amphipathy.

Part of the uncertainty concerning the future of structure-based methods stems from a need for realistic and comprehensive head-to-head evaluations of current approaches, a problem for sequence analysis methodology in general [63–65]. A method might have theoretical advantages, but these might not be of great importance in dealing with real problems. Sometimes, examples are selected to illustrate the strengths of a new method without discussion of important limitations. For instance, the Gibbs sampling strategy [38] was illustrated with three difficult examples, but none of these involved finding multiple motifs of different optimal widths, which is the case for the large majority of protein groups. Further work is necessary before this very promising approach can be evaluated in the context of real-world problems. It should also be realized that instances can always be found in which even a poor performer will shine, perhaps even by chance. For example, in comprehensive comparisons of amino acid substitution matrices, the worst-performing matrix overall occasionally did better in a particular search than all of the others [65]. In a search, however, one would always choose to use the best matrix overall; occasionally it will not be the best choice, but there is no way of knowing that a priori. Likewise, it makes no sense to use a structure-based method in searching a new sequence unless it works better in general, and not just in a few selected cases. With improvements in purely sequence-based methods accumulating rapidly, there is reason to be skeptical that structure-based methods using published approaches will have more than supplemental value. New ideas are needed for exploiting the richness of structural information to identify motifs in protein sequence.

While alignment is currently the most successful means of determining similarity between proteins and for classifying proteins into groups, it is not the only one possible. For example, 20×20 displays of dipeptide counts have been used to characterize sequences [40]. Related sequences often lie close together in the 400-dimensional space of dipeptide counts, so that the distance between points can be used as a measure of sequence similarity. While this approach has been applied to clustering databases and classifying sequences, it does not appear to be applicable to subsequences representing motifs. More progress has been made using neural network approaches to sequence classification [41,42,66,67]. An attractive feature of neural nets is that complicated relationships between segments and between positions that might be idiosyncratic for a group can be represented. A potential disadvantage of this approach is that a large number of examples are usually required to train a net to avoid overspecificity. Therefore, especially large and diverse groups would seem to be excellent candidates. In one study, a net was trained on sequences belonging to different groups and then was asked to classify test set sequences into these groups or no group at all [66]. Inputs consisted of variable-length strings of amino acid identities or of classes, and outputs consisted of 620 groups represented in the test set. In a different approach, a net was trained on a multiple sequence alignment in order to search a database for new group members [67]. Here, inputs for training consisted of overlapping 12-residue segments of sequences from a multiple alignment, in which gap residues were represented as random amino acids. Searching performance of the resulting nets in querying a protein database was compared to that of profiles constructed from the training set of sequences. For the three most diverse families tested, a net proved to be more sensitive than a profile in finding true positive members of the test set. It is interesting that the net appears to discover conserved motifs, whereas a profile representing a multiple sequence alignment searches for both diverged and conserved regions. This suggests that scoring matrices derived from conserved blocks [50] or those in which low information regions and gaps are ignored [12] might give better performance in database searches than profiles derived from full multiple sequence alignments.

In conclusion, the rapid growth of sequence databanks and the likelihood that representatives from nearly all protein groups will be soon be known focuses attention on classifying within groups rather than on identifying new ones. At the same time, the percentage of these groups for which one or more 3D structures are known will increase, so that the identification of motifs in protein sequences will become increasingly important for inferring structure and function. Several different methods already exist for classifying sequences into known groups, and no doubt new methods will soon be introduced. Regardless of the method used, present experience suggests that methods sensitive to concentrated motifs will be more successful than those sensitive to weaker information spread out over a multiple sequence alignment. In considering what approach is best, it must be kept in mind that the goal is accurate discrimination between true positives and true negatives and not detection sensitivity per se. So, the challenge is not so much finding idiosyncratic features in protein groups, but rather deciding on which ones add useful information and which ones only increase the noise. Structure-based methods, for all the attention that has been paid to them, have not yet been proven successful in real-world situations, perhaps because they are not effective at discriminating motifs. Sequence-based methods continue to be improved upon and appear to hold the most future promise for group classification of sequences. Which methods represent improvements and which do not can only be judged by comprehensive, as opposed to anecdotal, evaluation.

Acknowledgements

I thank Jorja Henikoff, Denise Clark and Gary Stormo for helpful comments on the manuscript. Sequence analysis work in my lab is supported by grant GM29009 from the National Institutes of Health.

References

1. Adams MD, Kelley JM, Gocayne JD, Dubnick M, Polymeropoulos MH, Xiao H, Merril CR, Wu

A, Olde B, Moreno RF, Kerlavage AR, McCombie WR, Venter JC. Complementary DNA sequencing: expressed sequence tags and human genome project. Science 1991;252:1651–1656.

2. Altschul SF, Gish W, Miller W, Myers EW, Lipman DJ. Basic local alignment search tool. J Mol Biol 1990;215:403–410.

3. Altschul SF, Lipman DJ. Protein database searches for multiple alignments. Proc Natl Acad Sci USA 1990;87:5509–5513.

4. Henikoff S, Henikoff JG. Amino acid substitution matrices from protein blocks. Proc Natl Acad Sci USA 1992;89:10915–10919.

5. Claverie JM, States DJ. Information enhancement methods for large scale sequence analysis. Comput Chem 1993;17:191–201.

6. Wootton JC, Federhen S. Statistics of local complexity in amino acid sequences and sequence databases. Comput Chem 1993;17:149–163.

7. Dodd IB, Egan JB. Improved detection of helix-turn-helix DNA-binding motifs in protein sequences. Nucleic Acids Res. 1990;18:5019–5026.

8. Bairoch A. PROSITE: a dictionary of sites and patterns in proteins. Nucleic Acids Res. 1992;20:2013–2018.

9. Smith RF, Smith TF. Automatic generation of primary sequence patterns from sets of related protein sequences. Proc Natl Acad Sci USA 1990;87:118–122.

10. Gribskov M, McLachlan AD, Eisenberg D. Profile analysis: detection of distantly related proteins. Proc Natl Acad Sci USA 1987;84:4355–4358.

11. Wallace JC, Henikoff S. PATMAT: a searching and extraction program for sequence, pattern, and block queries and databases. CABIOS 1992;8:249–254.

12. Thompson JD, Higgins DG, Gibson TJ. Improved sensitivity of profile searches through the use of sequence weights and gap excision. CABIOS 1994;10:19–29.

13. Luthy R, Xenarios I, Bucher P. Improving the sensitivity of the sequence profile method. Protein Sci 1994;3:139–146.

14. Henikoff S, Wallace JC, Brown JP. Finding protein similarities with nucleotide sequence databases. Methods Enzymol 1990;183:111–132.

15. Bowie JU, Luthy R, Eisenberg D. A method to identify protein sequences that fold into a known three-dimensional structure. Science 1991;253:164–170.

16. Ouzounis C, Sander C, Scharf M, Schneider R. Prediction of protein structure by evaluation of sequence-structure fitness. Aligning sequences to contact profiles derived from three-dimensional structures. J Mol Biol 1993;232:805–825.

17. Stultz CM, White JV, Smith TF. Structural analysis based on state-space modeling. Protein Sci 1993;2:305–314.

18. Krogh A, Brown M, Mian IS, Sjolander K, Haussler D. Hidden Markov models in computational biology. J Mol Biol 1994;235:1501–1531.

19. Baldi P, Chauvin Y, Hunkapiller T, McClure MA. Hidden Markov models of biological primary sequence information. Proc Natl Acad Sci USA 1994;91:1059–1063.

20. Pearson WR, Miller W. Dynamic programming algorithms for biological sequence comparison. Methods Enzymol 1992;210:575–601.

21. Lipman DJ, Altschul SF, Kececioglu JD. A tool for multiple sequence alignment. Proc Natl Acad Sci USA 1989;86:4412–4415.

22. Feng DF, Doolittle RF. Progressive sequence alignment as a prerequisite to correct phylogenetic trees. J Mol Evol 1987;25:351–360.

23. Barton GJ, Sternberg MJ. A strategy for the rapid multiple alignment of protein sequences. J Mol Biol 1987;198:327–337.

24. Higgins DG, Sharp PM. CLUSTAL: a package for performing multiple sequence alignment on a microcomputer. Gene 1988;73:237–244.

25. Smith RF, Smith TF. Pattern-induced multi-sequence alignment (PIMA) algorithm employing

secondary structure-dependent gap penalties for use in comparative protein modelling. Protein Eng 1992;5:35–41.

26. Vingron M, Argos P. Motif recognition and alignment for many sequences by comparison of dot-matrices. J Mol Biol 1991;218:33–43.

27. Miller W. Building multiple alignments from pairwise alignments. CABIOS 1993;9:169–172.

28. Sobel E, Martinez HM. A multiple sequence alignment program. Nucleic Acids Res 1986;14:363–374.

29. Martinez HM. A flexible multiple sequence alignment program. Nucleic Acids Res 1988;16:1683–1691.

30. Vingron M, Argos P. A fast and sensitive multiple sequence alignment algorithm. CABIOS 1989;5:115–121.

31. Boguski MS, Hardison RC, Schwartz S, Miller W. Analysis of conserved domains and sequence motifs in cellular regulatory proteins and locus control regions using new software tools for multiple alignment and visualization. New Biol. 1991;4:247–260.

32. Schuler GD, Altschul SF, Lipman DJ. A workbench for multiple alignment construction and analysis. Proteins 1991;9:180–190.

33. Depiereux E, Feytmans E. Simultaneous and multivariate alignment of protein sequences: correspondence between physicochemical profiles and structurally conserved regions (SCR). Protein Eng 1991;4:603–613.

34. Posfai J, Bhagwat AS, Posfai G, Roberts RJ. Predictive motifs derived from cytosine methyltransferases. Nucleic Acids Res 1989;17:2421–2435.

35. Smith HO, Annau TM, Chandrasegaran S. Finding sequence motifs in groups of functionally related proteins. Proc Natl Acad Sci USA 1990;87:826–830.

36. Bacon DJ, Anderson WF. Multiple sequence alignment. J Mol Biol 1986;191:153–161.

37. Stormo GD, Hartzell GW III. Identifying protein binding sites from unaligned DNA fragments. Proc Natl Acad Sci USA 1989;86:1183–1187.

38. Lawrence CE, Altschul SF, Boguski MS, Liu JS, Neuwald AF, Wootton, JC. Detecting subtle sequence signals: a Gibbs sampling strategy for multiple alignment. Science 1993;262:208–214.

39. Ogiwara A, Uchiyama I, Seto Y, Kanehisa M. Construction of a dictionary of sequence motifs that characterize groups of related proteins. Protein Eng 1992;5:479–488.

40. van Heel M. A new family of powerful multivariate statistical sequence analysis techniques. J Mol Biol 1991;220:877–887.

41. Ferran EA, Pflugfelder B. A hybrid method to cluster protein sequences based on statistics and artificial neural networks. CABIOS 1993;9:671–680.

42. Ferran EA, Ferrara P. Clustering proteins into families using artificial neural networks. CABIOS 1992;8:39–44.

43. Dayhoff M. Atlas of protein sequence and structure, vol 5, suppl 3. Washington DC: National Biomedical Research Foundation, 1978;345–358.

44. Sheridan RP, Venkataraghavan R. A systematic search for protein signature sequences. Proteins 1992;14:16–28.

45. Harris N, Hunter L, States D. Megaclassification: discovering motifs in massive datastreams. In: Tenth National Conference on Artificial Intelligence. San Jose: AAAI Press, 1992;837–842.

46. Green P, Lipman D, Hillier L, Waterston R, States D, Claverie J-M. Ancient conserved regions in new gene sequences and the protein databases. Science 1993;259:1711–1716.

47. Bairoch A, Boeckmann B. The SWISS-PROT protein sequence data bank. Nucleic Acids Res. 1992;20:2019–2022.

48. Pongor S, Skerl V, Cserzo M, Hatsagi Z, Simon G, Bevilacqua V. The SBASE domain library - a collection of annotated protein segments. Protein Eng 1993a;6:391–395.

49. Pongor S, Skerl V, Cserzo M, Hatsagi Z, Simon G, Bevilacqua V. The SBASE protein domain library, release 2.0: a collection of annotated protein sequence segments. Nucleic Acids Res 1993b;21:3111–3115.

50. Henikoff S, Henikoff JG. Automated assembly of protein blocks for database searching. Nucleic Acids Res 1991;19:6565–6572.
51. Sander C, Schneider R. Database of homology-derived protein structures and the structural meaning of sequence alignment. Proteins 1991;9:56–68.
52. Pascarella S, Argos P. A data bank merging related protein structures and sequences. Protein Eng 1992;5:121–137.
53. Holm L, Ouzounis C, Sander C, Tuparev G, Vriend G. A database of protein structure families with common folding motifs. Protein Sci 1992;1:1691–1698.
54. Brenner S. Phosphotransferase sequence homology. Nature 1987;329:21.
55. Sternberg MJE. PROMOT: a FORTRAN program to scan protein sequences against a library of known motifs. CABIOS 1991;7:257–260.
56. Fuchs R. MacPattern: protein pattern searching on the Apple Macintosh. CABIOS 1991;7:105–106.
57. Fuchs R: Block searches on VAX and Alpha computer systems. CABIOS 1993;9:587–591.
58. Geourjon C, Deleage G. Interactive and graphic coupling between multiple alignments, secondary structure predictions and motif/pattern scanning into proteins. CABIOS 1993;9:87–91.
59. Henikoff S, Henikoff JG. Protein family classification based on searching a database of blocks. Genomics 1994;19:97–107.
60. Chothia C. One thousand families for the molecular biologist. Nature 1992;357:543–544.
61. Kraulis PJ. Similarity of protein G and ubiquitin. Nature 1991;254:581–582.
62. Blundell TL, Johnson MS. Catching a common fold. Protein Sci 1993;2:877–883.
63. Pearson WR. Searching protein sequence libraries: comparison of the sensitivity and selectivity of the Smith-Waterman and FASTA algorithms. Genomics 1991;11:635–650.
64. Fickett JW, Tung C-S. Assessment of protein coding measures. Nucleic Acids Res. 1992;20:6441–6450.
65. Henikoff S, Henikoff JG. Performance evaluation of amino acid substitution matrices. Proteins 1993;17:49–61.
66. Wu CH, Whitson G, McLarty J, Ermongkonchai A, Chang T-C. Protein classification artificial neural system. Protein Sci 1992;1:667–677.
67. Frishman D, Argos P. Recognition of distantly related protein sequences using conserved motifs and neural networks. J Mol Biol 1992;228:951–962.

Biotechnology Annual Review Volume 1
M.R. El-Gewely, editor

Peptide and protein display on the surface of filamentous bacteriophage

Franco Felici, Alessandra Luzzago, Paolo Monaci, Alfredo Nicosia, Maurizio Sollazzo and Cinzia Traboni

IRBM (Istituto di Biologia Molecolare P. Angeletti), Rome, Italy

Abstract. The isolation of ligands that bind biologically relevant molecules is fundamental to the understanding of biological processes and to the search for therapeutics. Filamentous phage can be used to display foreign peptides and proteins in physical association with their DNA coding sequences. Repertoires larger than 10^8 phage clones expressing different peptide sequences can be prepared using molecular genetic techniques. The strategies utilizing this technology promise to provide not only new binding and possibly catalytic activities, but also lead structures for the development of new drugs and vaccines.

Introduction

Most biological events depend on interactions between molecules. The identification, cloning and expression of their relative genes are crucial to understanding the molecular mechanism of these interactions as well as to producing compounds of therapeutical value. In recent years the generation of large "repertoires" of chemically and biologically synthesized molecules has provided a novel, powerful approach to the identification of unknown ligands or variants of already known molecules, with the addition of new and more favorable properties. One of the distinctive features of these molecular repertoires is that they are solvent exposed, which allows them to be affinity selected using a given ligate. In this way vast numbers of different structures (in theory, 10^{12} or more) can be rapidly surveyed.

Over the past 3 or 4 years, molecular biology tools have been used to construct a series of ligand libraries [1–7], most of which exploit filamentous bacteriophage as molecular vectors. The closely related filamentous phage M13, f1, fd [8,9] are made up of a proteic envelope, constituted by several copies of five different proteins (pIII, pVI, pVII, pVIII and pIX) and a single-stranded DNA molecule (contained in such envelope) carrying the phage genetic information. When exogenous DNA is cloned in frame at the 5′-end of the gene encoding for one of the two capsid proteins pIII or pVIII, the corresponding fusion product is displayed on the surface of the virion. These bacteriophage infect *Escherichia coli* and grow without the lysis of the host cells, giving very high titers of ligand displaying phage (up to 10^{12} particles/ml of

---

Address for correspondence: F. Felici, IRBM, via Pontina Km. 30.600, 00040 Pomezia, Rome, Italy.
Fax: +39 6 91093 225; E-mail: felici@irbm.it.

culture supernatant). Phage displayed ligands thus offer the additional advantage over most of the other biological or chemical repertoires, of being physically associated with their genetic information which can be easily rescued and propagated by infection of competent bacteria. This allows the individual selected leads to be identified and characterized, even when only tiny amounts are recovered in the selection procedure (in theory, even at the level of a single molecule).

The above selection strategy has been applied to many different ligand/ligate systems. Phage display of proteins or protein domains has been widely adopted to identify mutants with increased affinity for their target or with different binding specificity. Large collections of antibody fragments have been expressed on the surface of the phage and successfully screened with different antigens. Finally, libraries of short peptides of random sequence have been displayed on phage and screened with antibodies as well as with nonantibody molecules, leading to the identification of new ligands which do not necessarily resemble the natural ones, but display similar binding specificity.

An overview of the published work in the main areas of the phage display technology is presented in this review, with particular emphasis on those aspects which can be of more general interest for further biological, biotechnological and pharmaceutical applications.

Biology of filamentous bacteriophage

Filamentous bacteriophage M13, f1 and fd belong to a larger family of bacteriophages infecting different bacterial species [8,9] and have been extensively studied and utilized in molecular biology for many years. M13, f1 and fd phage infect *E. coli* male strains (either F+ or Hfr), as their receptor on the bacterial surface is the tip of the sex pilus encoded by the F episome.

The complete nucleotide sequence of their single-stranded genomic DNA has been determined [10–13] and many vectors deriving from these phage genomes have been described to allow cloning and sequencing of DNA inserts [14–16]. The genomic DNA is around 6,400 nucleotides long and codes for ten proteins, three of which are required for phage DNA synthesis (the products of genes II, V and X), two, encoded by genes I and IV, serve phage morphogenesis and five are virion structural proteins (those encoded by genes III, VI, VII, VIII and IX). All the genes are very close (when not overlapping) in the genome. An "intergenic region" is located between genes II and IV which does not code for any protein, but contains essential information for phage incapsidation and the (+) and (−) origins of replication (Fig. 1).

At the beginning of the infection cycle, the single-stranded DNA is injected into the cell and the complementary (−) strand is immediately synthesized by host enzymes (this process does not require phage-encoded proteins). Then a pool of double stranded genomic DNA (replicative form) is produced with the involvement of the gene II product and phage genes are expressed. Finally, a progeny of single-stranded (+) DNA is synthesized for which both gene II and V products (pII and pV) are re-

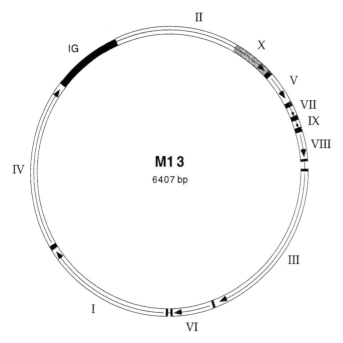

Fig. 1. Map of the M13 genome; the (+) strand is represented. The genes and their orientation are indicated (the shaded part indicates the genes II and X overlapping region). The intergenic region (IG) is in black.

quired. The newly synthesized (+) strand DNA is then covered with pV, which is a single-stranded DNA binding protein, and the phage assembly and production takes place at the membrane level (in adhesion zones between the inner and outer membranes of the host cells [17]) with the participation of the proteins encoded by genes I and IV and several host factors [9,18]. The single-stranded DNA is extruded through the membrane and covered by the capsid proteins. The products of structural genes VII and IX (pVII and pIX) are incorporated at the end of the virion which is extruded first and gene V proteins are removed from the DNA and shed back in the cytoplasm, being replaced by molecules of the major coat protein (the product of gene VIII, pVIII). The products of genes III and VI (pIII and pVI) are incorporated at the other end of the viral particle and the assembled phage is released in the medium.

All the above biological processes act without the lysis or death of the host cell. The major phenotypic effect of phage infection and production is a slowing down of bacterial growth rate, which causes the formation of plaques (consisting of slowly doubling phage-infected cells) on a bacterial lawn of noninfected cells.

Structure of phage particles

Phage particles are about 1–2 µm long and about 6–7 nm in diameter, and consist of a circular, single-stranded DNA molecule, encapsidated in a protein envelope constituted by about 2,700 copies per virion of the major coat protein VIII.

No high resolution three-dimensional structures of filamentous phage particles are available, nevertheless X-ray crystallographic and NMR studies have shown that pVIII consist of a single, slightly distorted, α-helix (45 amino acids long) with an N-terminal disordered extension of five residues [19–22], and that its copies in the phage capsid are tightly packed to form a protein tube containing the DNA. The carboxy-terminus of pVIII, which contains many basic residues, faces the nucleic acid, the hydrophobic core favors the formation of the protein stack, and the N-terminus (rich in acidic residues) is exposed to the solvent.

Three to five copies of each of the minor coat proteins are localized at the two tips of the filament. pVII and pIX (33 and 32 amino acids long, respectively) are localized at the end which emerges first from the bacterial surface, pIII and pVI (406 and 112 amino acids long, respectively) at the other end. In electronic micrographs of phage particles the adsorption apparatus appears as knobs at one end of each filament, each knob is mainly formed by a monomer of pIII [23].

Fusion Vectors

Two of the capsid proteins, pIII and pVIII, have been reported as suitable for expression of foreign peptide sequences [24–27]. In most cases the insertion takes place at (or near) the amino-terminus of the protein molecule. The large number of copies of pVIII on a single phage particle contrasts with the 5 peptides which are the maximum that can be presented by the pIII protein. This constitutes an important difference between the two kinds of fusions, making them suitable for different purposes: selection of low affinity or high affinity binders, respectively. In addition, display of large peptides have been more efficiently achieved as pIII fusions.

Two main types of phage-derived vectors have been developed: the first one is used to obtain recombinant phage particles where all copies of one capsid protein are modified (one-gene system), while the second can be utilized to produce chimerical virions bearing both wild type and recombinant versions of the capsid protein (two-gene system, Fig. 2).

One-gene system

This first type of vector has been described mainly for pIII, whose biological function does not seem in most cases to be significantly affected by N-terminal fusions. In contrast, attempts to insert specific peptides longer than 6–10 amino acids in all the copies of the major coat protein pVIII have failed, probably due to defects in phage assembly [27].

Smith [28] was the first to show that pIII could be used to display peptides on the surface of filamentous phage, utilizing a naturally occurring BamHI restriction endonuclease site near the middle of the protein sequence. A family of phage molecular vectors has been described by Parmley and Smith [29]; in these vectors foreign DNA fragments can be inserted into the region just downstream the signal peptide of gene

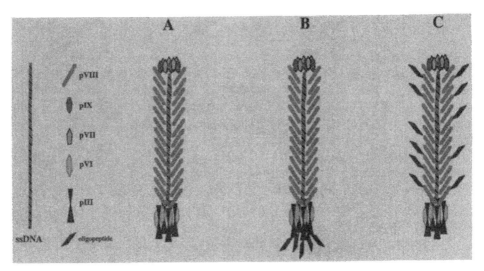

Fig. 2. (A) Schematic representation of a filamentous phage. (B) Representation of peptide display in pIII obtained through the one gene-system (all the pIII copies are modified). (C) Oligopeptide fusion to pVIII (obtained through the two-gene system) resulting in the assembly of a chimerical phage.

III in phage derivative of fd-tet [30], which also carry the gene for resistance to tetracycline and can thus be propagated like a plasmid. Recently, many more vectors carrying unique restriction endonuclease recognition sites in the amino-terminal region of pIII have been developed [31].

Two-gene system

Vectors for two-gene systems utilize a concomitant expression of recombinant and wild-type molecules of the structural protein pIII or pVIII. This can be accomplished by constructing a tandem repeat of the corresponding gene in the phage genome (one of the two copies is modified) or by utilizing a phagemid system. In this last system recombinant genes III or VIII are present in an expression plasmid containing the origin of replication of the phage [32]. Superinfection of bacteria containing such plasmids results in the release of hybrid phage in the supernatant, with both recombinant and wild type coat proteins present on the phage coat. Moreover, part of this phage population will contain the plasmid DNA coding for the recombinant coat protein.

With the two-gene system, phage particles displaying recombinant peptide sequence in pVIII can be constructed, or pIII fusions can be obtained that carry inserts or deletions that significantly affect the coat protein function (required for phage infectivity). By modulating the level of expression of the recombinant copy of the gene, monovalent display of the foreign peptide can be achieved.

Several variations and developments in the use of recombinant pIII and pVIII expression systems for peptide and protein display have appeared during the last years, some of which are mentioned during the present review.

Peptide Display and Peptide Libraries

Phage display of single peptide sequences has been reported by several groups. In 1985, Smith [28] cloned in the gene III a DNA sequence encoding for the EcoRI endonuclease and showed that the corresponding protein sequence was accessible to interaction with a peptide-specific antibody. In 1989, Ilyichev and co-workers [33] first showed that pVIII proteins carrying short amino-terminal insertions can assemble into functional viral capsids.

The construction of large libraries of random peptides fused to pIII was first reported in 1990 [2,34,35], 1 year after the first library of random peptides displayed on pVIII was described [36]. In the few years since then, many research groups have become involved in phage peptide library construction and screening with many different ligates (Table 1).

Some peptide libraries display the peptide in a linear fashion, others have more complex and/or constrained inserts. Constrained libraries in particular, could lead to higher affinity phage binders being identified: as the peptide is presented in a more rigid (constrained) context, the number of conformations it assumes should be much more limited than all those possible with a linear flexible peptide. This has the disadvantage of restricting the overall number of possible conformations, but presenting more copies of each single conformation, thus contributing to increased affinity, although the data reported until now do not provide definitive conclusions about this.

As our knowledge of these mechanisms is still incomplete, in order to increase the possibilities of selecting specific ligands for a given ligate (whether an antibody molecule or another protein), it is always advisable to screen several different phage peptide libraries. Affinity maturation of the selected phage (and therefore the construction of a "secondary" library) could then lead to the best binders being identified.

Immunological Applications of Peptide Libraries on Phage

Mapping of linear epitopes

Over the past few years several examples of identification of antigenic determinants utilizing phage peptide libraries have been described.

Monoclonal antibodies (mAbs) directed against short linear peptide epitopes of a protein antigen have been used as a model system by several groups to select phage bearing peptides specifically recognized by the ligate. In most cases the peptide sequence exposed by the affinity selected clones revealed a sequence similarity with the original epitope.

Scott and Smith [2] reported the construction of a 200 million-clone hexapeptide library fused to the amino-terminal region of pIII. A subset of this library was used to select clones binding to two mAbs recognizing the sequence DFLEKI, both as a linear peptide and as part of the protein myohemerythrin. The authors were able to

Table 1. Phage peptide libraries

Authors	Fusion protein	Displaying system	Genetic marker	Random positions	N-terminus of fusion protein[a]
Scott and Smith [2]	pIII	One-gene	Tet	6aa	A-DGA(6aa)GAAGA-ETVE
Devlin et al. [34]	pIII	One-gene	Amp	15aa	AE-(15aa)PPPPPAE-TVE
Cwirla et al. [35]	pIII	One-gene	Tet	6aa	(6aa)GG-TVE
Felici et al. [36]	pVIII	Two-gene	Amp	9aa	AEG-EF(9aa)-DPAK
Christian et al. [43]	pIII	One-gene	Tet	10aa	A-DVA(10aa)AASGA-ETVE
O'Neil et al. [61]	pIII	One-gene	Kan	6aa	(6aa)GGG-AETVE
O'Neil et al. [61]	pIII	One-gene	Kan	6aa	AE-C(6aa)CGG-TVE
Keller et al. [55]	pIII	One-gene	Tet	15aa	A-DGA(15aa)GAAGA-ETVE
McLafferty et al. [38]	pIII	One-gene	Amp	6aa	AE-G(1aa)C(4aa)C(1aa)SYIEGRIVE-TVE
Luzzago et al. [48]	pVIII	Two-gene	Amp	9aa	AEG-EFC(9aa)CG-DPAK
Kay et al. [68]	pIII	One-gene	–	36aa	S(S/R)(18aa)PG(18aa)SRPAR-TVE
Hammer et al. [63]	pIII	One-gene	Amp	9aa	AE-LGGGG(9aa)GGGGVP-
Blond-Elguindi et al. [66]	pIII	One-gene	Tet	8aa	(8aa)ASGSA-
Blond-Elguindi et al. [66]	pIII	One-gene	Tet	12aa	(12aa)ASGSA-
Matthews and Wells [73]	pIII	Two-gene	Amp	5aa	GPGG(5aa)GGPG-
Matthews and Wells [73]	pIII	Two-gene	Amp	5aa	GPAA(5aa)AAPG-
Jellis et al. [40]	pIII	One-gene	Tet	20aa	A-DGA(20aa)GAAGA-ETVE
Miceli et al. [39]	pIII	One-gene	Tet	10aa	A-DASSGA(10aa)SALSGSGA-ETVE
Koivunen et al. [59]	pIII	One-gene	Tet	7aa	A-DGAC(7aa)CGAAGA-ETVE

Examples of different phage peptide libraries constructed in various laboratories as fusions to pIII or pVIII.
[a]The predicted amino acidic sequences of the N-termini of the resulting fusion proteins are shown.

isolate many independent clones resembling the original epitope and containing the consensus sequence DFL.

Felici et al. [36] screened a pVIII based nonapeptide library with a mAb raised against the peptide VQGEESNDK (residues 163–171 of human Interleukin 1-beta protein). This mAb is able to recognize both the synthetic peptide and the entire protein. After two rounds of selections, several independent phage binding to the mAb were isolated, all of them displaying peptides with the consensus sequence SND/E, which is also contained in the nonapeptide used to elicit the antibody, therefore suggesting which peptide sequence is critical for the mAb binding.

Another example of antigenic mimicking by random peptide libraries, where the original epitope is represented by a continuous peptide sequence, has been described by Cwirla et al. [35]. A hexapeptide library displayed at the amino-terminus of pIII was screened with the mAb 3E7, which recognizes the sequence YGGF of beta-endorphin. All 51 affinity selected clones analyzed displayed Y as the first residue, with a majority displaying G in the second position. Most of the clones analyzed showed low affinity for the mAb. By using low 3E7 Fab concentration during affinity selection, the same group was able to select high affinity phage with sequence similarity (all of them with YG in the first two positions), most of them closely resembling the peptide YGGFL which binds with high affinity to the mAb [37].

mAb 3E7 was also used by McLafferty et al. to screen a constrained library [38]. The 18-amino acid residue displayed at the amino-terminus of pIII contained a pair of two fixed cysteine residues, and six variegated residues, four between the cysteines and one either side of them. The selected phage showed sequence similarity with those previously obtained by Dower's group, although with a more stringent sequence specificity, as to be expected in a constrained library.

A different strategy has been used by Miceli et al. [39] to define the epitope of the anti-FLAG octapeptide M2 mAb. The affinity selection of a first small decapeptide library in pIII allowed the core motif to be identified. A second library was constructed with randomized residues in positions different from the putative core, and used to select a permissible range of flanking residues within the epitope.

Jellis et al. screened a 20 amino acid random library in pIII with a HIV-1 isotype MN envelope reactive mAb, using two different panning procedures [40]. The consensus sequence which was obtained agreed with that previously determined by testing several synthetic peptides derived from natural HIV-1 isolates.

In all the works described above, the mAbs utilized for affinity selection were known to bind short linear peptide sequences.

The first demonstration of the validity of random peptide libraries in precisely mapping a continuous native epitope has been reported by Stephen and Lane [41], through screening a pIII-based hexapeptide library with a mAb recognizing a conformation-dependent epitope occurring in p53 mutants, and not reacting with the correctly folded wild type p53. All the selected clones shared the sequence RHSVV/I, with RHSV representing the minimal requirement for antibody binding. On the basis of the consensus sequence the authors could also predict a specific cross-reaction between the mAb and another protein, the *Xenopus* transcription factor TFIIIA, also containing the motif RHSVV.

The precise mapping of a continuous epitope and the prediction of a potentially cross-reactive determinant defined by random peptide libraries has also been accomplished by du Plessis et al. [42]. Once again the consensus sequence obtained by the selected clones resembles a sequence found in the native original antigen.

There are, however, some exceptions to this rule; one is represented by peptides that mimic conformational epitopes, which is discussed later in this chapter. Examples of peptide sequences corresponding to inserts of phage clones selected with mAbs recognizing linear epitopes are listed in Table 2.

The identification of phage mimotopes that do not share any sequence homology with a continuous epitope of the natural ligand has also been reported [2,43]. Christian et al. [43] screened a decapeptide library with a mAb recognizing 10 residues on the V3 loop of gp120. Two common motifs were identified through phage sequence analyses, indicating that selection had occurred; nevertheless, neither of the two consensus sequences were related to the sequence of gp120 epitope.

Monoclonal antibodies directed against native proteins are useful tools for structural and functional studies. Defining the key residue requirement for antibody binding has therefore important implications in elucidating relationships between protein structure and function. Yayon et al. [44], screened a pIII hexapeptide library with two mAbs which affect the binding of basic fibroblast growth factor to its receptor. The results obtained suggest that the two mAbs recognize a distinct, although similar epitope. Sequence comparison between bFGF and the consensus sequence obtained from isolated specific phage identify two possible homologous and continuous regions of bFGF as candidates for the mAbs binding. Synthetic peptides derived from phage sequence and from the identified bFGF regions block bFGF-receptor interaction, suggesting that these sequences contribute in binding to the receptor.

Malik et al. [45] mapped a surface-exposed epitope of the Na^+, K^+ ATPase by screening a 15-amino acid library in pIII with a mAb recognizing a region previously characterized with the synthetic peptide HLLVMKGAPER and located within the ATP binding domain. Despite its binding location, the mAb does not affect enzyme activity, in contrast with the enzyme activity inhibition obtained with a polyclonal antibody raised against the sequence GAPER, included in the longer peptide. The consensus sequence obtained by the mAb binding phage included clones with five continuous residues PRHLL, perfectly matching the original antigen. Analyzing the consensus sequence together with species-specificity mapping led to the complete identification of the epitope, including those amino acid residues not contained in the mAb reacting synthetic peptide and located in a region adjacent but separate from the GAPER sequence.

Mapping discontinuous epitopes

The utilization of overlapping synthetic peptides to generate information about

Table 2. Phage displayed peptides mimicking linear epitopes

	mAb				
	PAb 240 [2]	α-IL1β [36]	3-E7 [35]	M2 [39]	M33 [41]
Antigen	Myohemerythrin	h-Interleukin 1β	β-Endorphin	FLAG peptide	p 53
Epitope	**DFLEKI**	**VQGEESNDK**	**YGGF**	**DYKDDDDK**	**FRHSVV**
Selected phage inserts	**DFL**E**RI**	**L**F**DRESNDW**	**YGGF**PD	**K**S**DYKEQDRT**	**W**RHSVV
	DFLEQL	**DK**SSNDS**IA**	**YGGL**GL	**EDDYKNEDIK**	**L**RHSVV
	DFLEML	R**DLHSNDVV**	**YGGL**GI	V**NDYKAPDTR**	**YRHSV**I
	DFLEWL	**SNDGVWAIP**	**YGGL**GR	**L**I**DYKCRDST**	**LRHSV**I
	DFLHFI	**Y**T**DSNERTT**	**YGGL**NV	**LSDYKGADRI**	
	DFLVQL	**YRDSNEHQP**	**YGGL**RA	**GLDYKDGDPC**	
	DFMEW**L**		**YGGL**EM	VSS**YKDDDQI**	
	CEFLEC		**YGGI**AV	**FE**M**YKDLDTR**	
	CRFLEC		**YGGI**AS	**I**I**GYKDIDGS**	

Sequences of inserts from phage isolated with five different mAbs recognizing linear epitopes. Amino acidic sequences deduced from DNA sequence are reported; the residues matching the sequence of the original epitope are in indicated in bold.

epitope structures is sometimes misleading and cannot be used when the epitope is discontinuous. On the basis of the known structurally defined epitopes, we can expect the majority of B cell epitopes to be of a conformational type, with crucial residues for antibody binding that are close upon protein folding, whilst very far apart in the primary structure.

The evidence that a peptide displayed on phage is able to mimic a discontinuous epitope is still very limited, and in all the cases published it was very difficult, if not impossible, to identify homology between the selected phage clones and the antigen. The first demonstration that a linear peptide library could be used to isolate phage mimicking assembled epitopes was provided by Felici et al. [46], by screening a nonapeptide library at the amino-terminus of pVIII with 1B7 mAb that recognizes a discontinuous epitope of *Bordetella pertussis* toxin. Several independent phage clones were isolated and proved to interact with the mAb antigen binding site by competition experiment with the original antigen. In this case it was not possible to precisely identify the residues constituting the epitope by sequence inspection of the isolated phage clones, nor to identify a precise consensus sequence among different clones.

Balass et al. [47] used a mAb recognizing the ligand-binding site of nicotinic acetylcholine receptor (AcChoR) to screen a hexapeptide library displayed on pIII. The antibody recognizes a conformation-dependent epitope, since it reacts with the native protein and not with the denatured AcChoR, nor with synthetic peptides derived from AcChoR sequence. Selected phage allowed the consensus sequence not present in the continuous amino acid sequence of the receptor to be identified. A synthetic peptide derived from one of the selected sequences was shown to compete for the binding of the mAb to the receptor and able to inhibit the effect of the mAb on the receptor in in vivo functional studies.

Although the two reported examples show that phage mimicking conformational epitopes can be isolated, in neither case was it possible to correlate the peptide sequences obtained by phage selection with the protein regions involved in the mAb binding.

Mapping of a discontinuous epitope has been reported by Luzzago et al. [48] through screening a cysteine-constrained nonapeptide library in pVIII with the mAb H107, which recognizes recombinant human H ferritin whose three-dimensional structure is known. The isolated phage clones could be grouped into two different consensus sequences. Comparing the phage sequences with residues of H ferritin exposed to the solvent in the assembled molecule identified two different ferritin regions that are distant in the primary sequence, but close together in the folded molecule. Both phage consensus sequences contained amino acid residues spanning the "discontinuity" between the two regions (Fig. 3).

Single point mutation of H ferritin in these regions abolished binding to the mAb H107, without affecting the correct ferritin folding, as proved by these mutants being recognized by a set of different mAbs.

In this latter case the epitope mapping was possible only by combining the phage sequence information with antigen structural data.

```
                                    YPYGSTFGL
       YDGSYRWAP                    RLPGSAFTY
       YDAQHHWTR                    HIPGSIFHL
       YNAFLNWVP                    RTTGSVFHL
                                    RWQGSAFAY
```

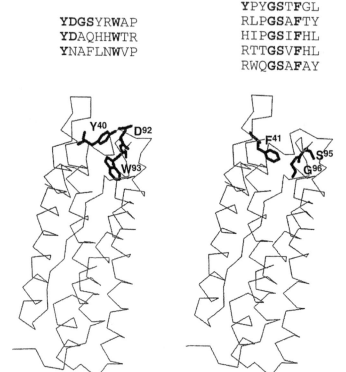

Fig. 3. Human H ferritin subunit backbone with the side-chains (in bold) of the amino acids matching the two consensus sequences. The sequences of the inserts of phage clones binding to the mAb H107 are shown in the upper part of the figure.

Phage-displayed peptides have been reported as not only capable to mimic complex proteic epitopes like discontinuous ones, but also nonproteinaceous antigens, like carbohydrates. The mAb B3 reacts with a carbohydrate antigen (Lewis^Y), and was used by Hoess et al. [49] to screen a linear and a constrained libraries displayed at the amino-terminus of pIII. From the nonconstrained library, phage clones specifically binding the mAb and exposing the amino acid sequence APWLYGPA were isolated. Mutagenesis studies indicated the tetrapeptide PWLY as the critical core for the mAb binding, and the authors suggest that tyrosine and tryptophan may play a role in mimicking sugars.

Phage as immunogens

In 1988, de la Cruz et al. [50] expressed repeat regions derived from the circumsporozoite protein of the human malaria parasite *Plasmodium falciparum* on phage as pIII N-terminal fusions in a polyvalent fashion; recombinant phage were shown to be both antigenic and immunogenic in rabbits and some mouse strains. The phage display system was therefore proposed as an alternative approach to that of antigen chemical synthesis. Compared with other expression systems it has the advantage of producing large quantities of antigen at minimal cost and with no need for extensive purification of the recombinant protein.

Displaying multiple copies of an antigenic peptide was forecast as increasing its immunogenicity. With this idea in mind, different groups generated phage particles displaying multiple copies of the foreign peptide as fusion to the major coat protein pVIII.

Greenwood et al. [51] showed that phage displaying two major antigenic determinants of the circumsporozoite protein of the human malaria parasite *Plasmodium falciparum* were much more immunogenic when expressed in multiple copies as pVIII fusions in a two-gene system, than when expressed at the N-terminus of pIII.

In a follow up to the earlier work, the same group showed that Abs raised in this manner were highly specific against the peptide insert [52]. Phage particles were shown to perform a "carrier function" that obviates the need for conjugation of peptides to carrier molecules or co-injection of adjuvant. Furthermore, the immune response was found to be T-cell dependent, as the phage particles themselves were able to recruit T-cell help, and to undergo class switching from IgM to IgG.

Minenkova et al. [53] inserted a small antigenic determinant of the HIV-1 p17 gag protein at the N-terminus of pVIII, generating a recombinant virion in which all copies of pVIII display the foreign peptide. The recombinant phage raised Abs in rabbits that reacted in Western blots with the p17 protein and its precursor polyprotein, p55.

In line with the above findings, injection without adjuvant of hIL3 expressed on phage as pIII fusion was shown by Gram et al. [54] to raise a specific immune response in mice.

Immunogenic mimicking

Phage isolated from peptide libraries by selection with mAbs appear in most cases to be good antigenic mimotopes, for their ability to specifically interact with the mAb antigen-binding site. In some cases, selected peptides exposed on phage have been reported to be also capable of immunogenically mimicking the original antigen.

By screening a pIII based 15-residue library, Keller et al. [55] identified a series of peptides that bind strongly to the human mAb 447–52D, a neutralizing, broadly reactive antibody that recognizes the GPXR (X = any amino acid) motif of HIV gp120 V3 loop. Almost all the phage contained GPXR as consensus sequence. A peptide derived from one of these sequences was used to immunize rabbits and produced elevated titers of specific antibodies, capable of neutralizing HIV-1 variants.

More recently, antigenic mimotopes of the surface antigen of the human hepatitis B virus (HBsAg) have been selected by an anti-HBsAg mAb from a library of random peptides expressed as pIII fusions [56]. One of the selected phage was shown to elicit in mice a specific immune response directed against the HBsAg.

In both reported cases the mAbs utilized for the selections recognized a continuous epitope. A different result was obtained by Felici et al. [46] when phage mimotopes of a discontinuous epitope of *Bordetella pertussis* toxin were used to immunize mice. Phage elicited a good immune response directed against the peptide sequences displayed on the phage, but no response against the original antigen was detected, indicating that antigenic and immunogenic mimicking are not overlapping properties in every case.

Selection with immune sera

As reported in the previous sections, a large body of information has been accumulated to corroborate the conclusion that libraries of small random peptides displayed on phage are sufficiently "complete" to contain ligands for almost any given antibody. On this basis phage peptide libraries were considered a good source of epitopes to explore a new strategy for discovering disease-related epitopes: screening random peptide libraries on phage using human sera from patients.

To this end Dybwad et al. [57] used immunoglobulins from different rheumatoid arthritis (RA) patients and from healthy individuals in subsequent cycles of enrichment and depletion of a nonapeptide library displayed on pVIII. Phage were identified which showed a percentage of positivity in ELISA when assayed with RA sera higher than that observed with the sera from healthy individuals.

Further and more convincing evidence of the identification of disease-specific epitopes on phage (phagotopes) using human sera came from the work by Folgori et al. [58]. These authors screened a cysteine-flanked nonapeptide library based on the pVIII display system with human sera from individuals immunized against HBsAg. By a three step procedure which avails itself only of clinically characterized sera from immune and nonimmune individuals, four different phagotopes were identified. By competition experiments, three of these phage were shown to display variants of the same epitope, while the fourth mimicked a different one. The selected phagotopes were shown to react with the majority of sera from immunized individuals (80%), but not with sera from a nonimmune population. The same phagotopes were also shown to be immunogenic mimics of the natural antigen, since they were able to elicit the production of Abs against HBsAg when injected into different animals.

These results open new perspectives for the application of phage display technology in the field of diagnosis and vaccine development. In principle it is feasible that the above strategy can be used to identify specific diagnostic reagents for a given disease, even in the absence of any information about the pathological agent and/or its antigens. Besides having a diagnostic value, disease-related phagotopes could well be useful tools to identify unknown etiologic agents. Furthermore, in those cases where protective Abs are known to exist, a new generation of vaccines could be developed by preparing a "cocktail" of phagotopes selected with these Abs.

Selection of Peptide Libraries with Nonantibody Molecules

As reported in the previous section, affinity selection of phage peptide libraries with mAbs as ligates can be used to map and/or identify epitopes relevant to antibody/ antigen structural studies, disease diagnosis and immunization. A further application of phage display technology is the screening of phage peptide libraries with non-antibody ligates, aimed at identifying peptide mimics of unknown natural targets, as well as new ligands with improved and/or "ad hoc" modified binding properties. Work published so far can be classified according to the molecular nature of the ligate used for the selection and that of its natural target.

Selection of peptides mimicking surfaces of protein/protein interactions

By panning a hexapeptide library based on the pIII display system with $\alpha_5\beta_1$ integrin purified from human placenta, Koivunen et al. [59] selected a family of peptides bearing a common RGD consensus sequence, present in an exposed loop of the 10th type III repeat of $\alpha_5\beta_1$ natural target fibronectin. Synthetic peptides reproducing the inserts of the selected phage were able to inhibit $\alpha_5\beta_1$/fibronectin binding. The selected sequences displayed broad binding specificity in that they were able to recognize other related integrins ($\alpha_v\beta_1$, $\alpha_v\beta_5$, $\alpha_v\beta_3$).

In accordance with previous results showing improved receptor recognition by cyclization of RGD containing synthetic peptides, screening the library in more stringent conditions led to the selection of cysteine-flanked RGD peptides showing a 10-fold higher binding affinity, presumably due to disulfide induced looping of the ligand sequence. Consistent with these data, subsequent screening of a cysteine-flanked heptapeptide library led to the isolation of high affinity and more selective $\alpha_5\beta_1$-binding peptides, which needed to be presented in a cyclic form to retain full binding activity [60]. Similar conclusions were reached by O'Neil et al. [61] who screened a cysteine-flanked hexapeptide library displayed on pIII, using another member of the integrin family ($\alpha_{IIb}\beta_3$). Synthetic peptides reproducing the sequences of the selected phage were able to inhibit binding of fibrinogen to platelets, as well as aggregation of agonized platelet. Oxidation of these peptides resulted in a 2- to 20-fold increase of their IC_{50} in the anti-aggregatory assay.

Both groups also identified bio-active peptides displaying sequence variations from the consensus NGR and KGD, respectively, as well as a high affinity peptide bearing no resemblance to the RGD consensus (CRRETAWAC; [60]), indicating that, as has been observed with antibodies (Abs), also for screening with proteins other than Abs the peptide display approach can identify unexpected, novel ligands.

The possibility of selecting specific binders bearing little or no resemblance to the natural ligand from phage peptide libraries was clearly shown by Smith et al. [62]. These authors used a 104-amino acid long fragment of bovine pancreatic ribonuclease (S-protein) to screen a pIII-displayed hexapeptide library. Several phage showing the common sequence motif (F/Y)NF(E/V)(I/V)(L/V) were selected. Despite the lack of any significant similarity between this sequence and that of the natural target (S-peptide), one of the selected fusion phage bearing the sequence YNFEVL was able to interfere with S-protein/S-peptide interaction.

As in the case of mAb selected mimotopes, phage peptide library screening can also be successfully employed to generate information on the sequence requirements for binding other classes of proteins to their natural targets.

Since this technology allows large numbers of ligand variants, from which it is possible to derive a consensus, to be identified rapidly and inexpensively, it is particularly suitable in those cases where the ligate interacts with different ligands that share no obvious sequence similarity.

For this reason, Hammer et al. [63,64] chose this strategy to map promiscuous and allele-specific anchor residues in the peptides binding to MHC II molecules. Several distinct pIII-fused nonapeptides were selected by screening an M13 library of nonapeptides with three DR1 allelic variants. Sequence comparison analysis of each of the three selected peptide pools identified a "core" of six amino acids in which positions 1 and 4 were practically invariant, while the amino acid at position 6 was different in the various peptides binding to the three MHC II alleles. On this basis, the residue at position 6 was suggested as conferring allelic specificity. Binding assays performed with the three DR1 molecules and a series of peptides sharing the same motif, but displaying DR-specific residues at position 6, confirmed this hypothesis [64]; recently, crystallographic data have provided support to these findings [65].

A similar case of semi-promiscuous recognition is represented by the endoplasmic reticulum (ER) chaperonin BiP which mediates translocation of newly synthesized polypeptides across the ER membrane. The role of BiP as a chaperone derives from its ability to interact with a wide variety of unrelated nascent polypeptides with a marked hydrophobicity. Selection and screening of pIII-displayed octa- and dodecapeptide libraries with purified murine BiP [66] yielded a population of phage sharing the common heptameric motif Hy(W/X)HyXHyXHy (Hy = large hydrophobic, W = tryptophan, X = any amino acid). The alternating pattern of hydrophobic residues displayed by the identified consensus fits the model that peptides are bound in an extended conformation with large hydrophobic and/or aromatic side chains lying on one side and promoting stable interactions with corresponding pockets in the binding cleft of the BiP molecule. The selected peptides were able to bind BiP and to stimulate its ATPase activity with an efficiency comparable to that of naturally occurring polypeptides (μM range). Because of the large number of peptides selected, the authors were able to develop a scoring system for accurately predicting BiP binding sites in synthetic and naturally occurring polypeptides.

Selection of peptides mimicking nonproteinaceous ligands

Random peptide libraries displayed on phage can be operationally defined as "all purpose libraries" as they can be used as a source of ligands for any given ligate. Perhaps the most extreme example of this is the selection of peptides on phage which mimic nonproteinaceous molecules.

In 1990 Devlin and co-workers [34] first addressed the issue of whether peptides on phage can mimic nonproteic structures by screening a pentadecameric library fused to pIII using the biotin-binding protein streptavidin, a molecule with no previous known affinity for peptides. Phage encoding for nine different streptavidin-binding peptides were identified; all isolates shared the tripeptide sequence HPQ (with occasional substitutions of methionine or asparagine for glutamine) and were efficiently competed by biotin for their binding to streptavidin. Crystal structure determination of the streptavidin/peptide complex confirmed that this tripeptide is important in stabilizing the complex as it forms a hydrogen-bonded network including the side chains of the HPQ sequence [67]. However, in spite of the fact that selected tripeptide and biotin occupy roughly the same location in the hydrophobic binding pocket, there is little atom-for-atom correspondence between the two ligands. The evidence of additional interactions between residues at both sides of the HPQ segment and the streptavidin surface, would also suggest that there is a substantial difference in the mode of recognition between peptidic and nonpeptidic streptavidin ligands. The same HPQ consensus tripeptide was identified by Kay et al. [68] in the screening of a pIII displayed library containing larger peptides (38 amino acids). These authors also isolated streptavidin-binding phage bearing novel, unrelated sequences.

Further demonstration that peptidic molecules can substitute for nonproteinaceous ligands comes from the work by Oldenburg et al. [69] and Scott et al. [70] who used the lectin concanavalin A (Con A) in phage peptide library screening. Both groups identified different families of Con A-binding phage, with the most tightly binding ones bearing the sequence YPY. Both phage-borne peptides and synthetic peptides containing this consensus sequence were able to bind Con A and to compete for the natural ligand methyl α-D-mannopyranoside (MeαMan). The MeαMan binding to Con A involves a series of hydrogen bonds between the ligand oxygens and five protein residues, as well as van der Waals interactions between the sugar ring and two tyrosine residues (Y^{12} and Y^{100}) on the ligate. It has been suggested that a similar hydrophobic stacking of the tyrosine side chains of the selected peptide against Y^{12} or Y^{100} in the binding site of Con A might be responsible for most of the complex stabilization. However, the low affinity of the tripeptide YPY (>10 mM [69]) indicates that other interactions contribute to the binding energy of the selected mimics. Surprisingly, peptide ligands did not bind to closely-related MeαMan-binding lectins [70]. As in the previous case, it is probable that binding of the peptide to Con A involves more than the residues known to contact MeαMan, thereby generating new properties that are not present in the natural lectin-binding molecule.

These findings indicate that drug design of nonpeptidic molecules derived from crystal structure data of the peptide complex is difficult, since specific stereochemical features of the natural ligand cannot be easily derived from analysis of affinity selected peptide/ligate interactions. Nonetheless it is clear that mimicking nonproteinaceous ligands (i.e. carbohydrates) can be achieved through peptide sequences. This constitutes a unique and essential feature for those projects where the nature of the ligand is either unknown or known to contain nonproteinaceous elements which are important for ligate recognition.

Furthermore, it appears that, at least in some cases, peptide mimics selected from phage peptide libraries can show higher selectivity than the natural targets. This property is particularly useful in the development of receptor antagonists which need to discriminate between different, but ligand-related molecules.

Selection of peptides binding to nonproteinaceous ligates

Work reported so far on phage library screening as a novel approach to the identification of ligands is mainly concerned with the use of Abs and other proteins as ligates. Results obtained by Saggio and Laufer [71] have expanded the range of application of phage display technology to the search for ligands of small nonproteinaceous molecules. Biotin-binding phage were affinity selected from a cysteine-flanked nonapeptide library displayed on pVIII [48]. These phage all shared the common consensus sequence WXPPF(K/R), which does not show any significant homology with other known biotin-binding proteins even if a tryptophan residue, known to be required for the activity of some of these proteins, is always found in the inserts of the selected phage. A 20 amino acid-long synthetic peptide containing one of the selected sequences was shown to retain binding activity. This peptide represents the shortest nonavidin biotin-binding sequence identified to date and displays comparable affinity to the shortest avidin fragment which was shown to recognize biotin (50 μM and 100 μM, respectively) [71,72].

This first successful example of random peptide library screening with nonproteinaceous molecules lends further support to these kinds of molecular repertoires being a source of artificial ligands for many different kinds of ligates.

Selection of peptide substrate for enzymatic activity

The identification of phage displayed peptide mimics of nonAb ligands will certainly prove an invaluable tool to the development of competitive inhibitors for several biomolecules. A particularly effective way of employing phage peptide libraries in this direction has been implemented by Matthews and Wells to determine sensitive and resistant protease substrate sequences [73]. These authors constructed a library of fusion proteins containing the human growth hormone (hGH) linked to a five amino acid-long random sequence and followed by a truncated pIII gene. Fusion proteins were displayed on the surface of the phage and bound to immobilized hGH-binding protein. After treatment with protease, released phage providing good substrates, and bound phage expressing protease-resistant peptides were recovered. Using this so-called "phage substrate" strategy, the specificity of a variant of the bacterial subtilisin (BPN'-H64A) and of factor Xa were investigated. BPN'-H64A is a mutant which cleaves substrates containing a histidine residue in P2 or P1' position by a substrate-assisted catalytic mechanism. After multiple rounds of selection and cleavage with BPN'-H64A, all sensitive clones displayed a sort of consensus represented by a histidine, almost invariably located at position 2 or 4 of the random sequence, preferentially flanked by hydrophobic residues (Y, M, L and T). This is also in accordance with the finding that hydrophobic residues (particularly Y, M, L and F) are preferred at the P1 position of subtilisin. No prolines or cysteines, known to be poor substrates, were found among the sensitive sequences. On the contrary, the resistant clones did not exhibit a recognizable consensus and in many cases they displayed more than one proline.

The same approach was used to determine the specificity of the blood-clotting protease factor X_a. Also in this case enriched sensitive sequences were found to contain residues important for factor X_a cleavage (namely P1 arginine), with many sequences matching previously known factor X_a substrates.

Expression of Protein Domains on Phage

An additional, important and fast growing application of fusion phage technology is the display of protein domains on the surface of the phage. A number of advantages promoted the use of this expression system:
1. protein domains expressed by phage retain at least partial native folding and activity, and their stability and solubility can be enhanced;
2. phage-displayed fusion proteins are solvent exposed and can be selectively enriched by affinity purification;
3. it is possible to generate vast numbers of mutant proteins and, through a proper selection strategy, rapidly identify mutants with the desired phenotype;
4. phage-borne proteins are secreted in culture supernatant from which they can be easily purified;
5. as reported in the previous sections, phage displayed peptides/proteins are highly immunogenic, and thus do not require conjugation to carrier moieties.

Limitations to the display of protein on the surface of phage derive mainly from whether or not the foreign protein can be secreted through the *E. coli* membrane.

Hormone and inhibitor phage

The possibilities opened up by the application of this technology to protein engineering and structure-function studies has drawn the attention of many laboratories, especially in the pharmaceutical field. In 1990, Wells and co-workers [3] pioneered this field by expressing the 22 kDa human growth hormone (hGH) as a fusion to the C-terminal domain of pIII to analyze more conveniently protein-receptor interactions. The resulting hGH-pIII fusion was properly folded and hGH-expressing phage could be specifically enriched upon affinity selection with hGH-receptor.

The two-gene system adopted in this case produced phage which rarely contained more than one copy of the fusion protein per virion particle. This strongly reduces the interference of the foreign moiety on phage assembly and infectivity, allowing propagation of phage that display large and properly folded proteins. In addition, monovalent phage display eliminates the avidity effects in the selection process, enabling high affinity variants to be identified.

Using this system, hGH variants with up to eight-fold higher affinity for hGH receptor were selected from a library in which several positions were randomized [74]. Combinations of these mutants produced an hGH derivative with 30-fold higher affinity than wild type. The same hGH variants exhibiting an increased affinity for hGH receptor showed a 1,000-fold less affinity for the prolactin receptor, which shares with hGH a set of overlapping residues modulating the hormone binding.

An extension of these examples is the selection for molecules with novel binding specificities, as reported for the bovine pancreatic trypsin inhibitor (BPTI). BPTI was displayed on phage both as pVIII and as pIII fusion [75–77] and a limited library of variants has been generated by randomizing a five-residue region known to interact with human neutrophil elastase (HNE). From this library it was possible to identify variants that bound HNE with 10^6-fold higher affinity than that of the wild type molecule: this inhibitor is 50 times more potent than the most powerful one derived so far from structure-based design.

In this case a multicopy display system was used: the negligible effect of avidity in isolating high affinity mutants can be ascribed to a substantial degradation of BPTI on phage or to stringent HNE concentrations used in affinity cycles.

The secretion pathway of phage coat proteins has been exploited to express an active form of the 42 kDa human plasminogen activator inhibitor 1 (PAI-1), a member of the serine protease inhibitor family [78]. The native protein is produced as an inactive polypeptide after prolonged cytoplasmic synthesis in *E. coli*, from which activity can be recovered upon a denaturation/renaturation process. In contrast, routing of the PAI-1 to the periplasm of *E. coli* and its subsequent monovalent display as a pIII-fusion on phage particle produced a biologically active inhibitor. Phage-expressed PAI-1 specifically bound to specific polyclonal Abs and mAb, and retained its capacity to form equimolar complexes with its target serine protease tissue-type plasminogen activator.

Monovalent displaying of PAI-1 on phage has allowed the construction of a vast library of variants, for the isolation of mutants defective in their interaction with other components of the homeostatic system.

Very recently, the B domain of protein A of *Staphylococcus aureus* has been expressed on phage surface as pIII-fusion [79]. Protein A is a well characterized immunoglobulin binding cell wall component of *S. aureus*, which is widely used for antibody purification because of its specificity, stability and low cost. It contains five independent binding domains, each of which possesses affinities for the Fc region of many types of Abs. In particular, the 58-amino acid long B domain displays micromolar binding to human IgG.

Phage displaying the protein A domain selectively and specifically interact with immobilized IgG molecules. As highlighted in the previous cases, phage surface display libraries can be constructed by mutagenesis of this construct. Massive screening of mutant forms of protein A for alteration in binding and elution properties is anticipated to improve utility of protein A as a purification "handle" and in therapeutic applications.

Enzymes on phage

The first example of an active enzyme expressed on the surface of phage was reported by Chiswell and co-workers [80]. They expressed an *E. coli* alkaline phosphatase (AP) as pIII-fusion using a polyvalent display system in which the expression of the 47 kDa was accompanied by a 40–70% degradation of fusions, depending on the preparation. Phage particles expressing the enzyme were enriched in comparison to wild type phage by chromatography using an immobilized inhibitor of the enzyme. The phage-borne enzyme had a catalytic activity comparable to that of the free enzyme, but showed reduced, although qualitatively similar, kinetic properties. This decrease can be ascribed to the fact that AP functions as a dimeric complex and association of AP-fusions is influenced by their being "anchored" to the phage particle.

More recently, Light and Lerner expressed an active AP enzyme as pVIII-fusion on phage harboring a pIII-fused Fab fragment [81]. The use of such AP-expressing phage allows a simpler and faster ELISA screening of Fab libraries expressed on pIII, eliminating the need for a secondary conjugated anti-phage antibody. No data about the enzymatic activity of this phage-linked enzyme, nor an estimation of the number of AP molecules per phage were reported, but the data recorded stand for a poor activity/expression of the AP-fusion. This was the first example of different proteins being displayed on the same phage framework, which has many potential applications, such as the creation of new, specific enzyme delivery systems.

Results indicating that the enzyme being attached to the phage surface does not block catalysis promoted the idea that the display of enzymes (as well as catalytic Abs) on phage would make it possible to generate and screen vast libraries of mutants. In this way, tailored selection strategies could help improve catalytic efficiency, for example by sorting phage on transition state columns. Similarly, engineering enzymes with novel substrate/inhibitor specificities could become possible.

With the aim of developing new protease specificities and novel catalytic properties, Craik and co-workers at UCSF expressed the serine protease trypsin on phage both as pIII and pVIII fusion [82]. In the latter case, they directly expressed the fusion protein within an M13 vector, avoiding technical complexities of helper phage/phagemid system. Trypsin on phage were hydrolytically active, showed kinetic properties similar to the wild type trypsin, and could be captured by immobilized inhibitor and specifically eluted. This indicated a correct formation of intramolecular disulfide bonds and a proper folding of the enzyme, despite its fusion to the N-terminus of the coat proteins. Interestingly, an endogenous *E. coli* inhibitor, ecotin, associated and co-purified with trypsin-phage.

An elegant example of the use of phage display technology to create new enzyme specificities was presented by the recent reports about the expression on phage of a zinc finger structure, a common eukaryotic DNA-binding motif [83]. The three zinc fingers of the Zif268 protein were expressed on the surface of phage as pIII fusions and a library of variants was prepared by randomizing critical positions in the first finger. Affinity selection using DNA binding sites altered in the region recognized by the first finger, identified of mutants with new DNA-binding specificities and provided information about the interaction of the protein with DNA.

Toxin on phage

Swimmer et al. [84] expressed on phage the B chain of ricin (ricin B). Ricin is a heterodimeric toxin, comprised of a cytotoxic A chain and a galactose-binding B chain (32 kDa). The B chain has a dual function: it attaches holotoxin to the cell binding galactosides moieties on cell surface and facilitates penetration of cell membranes by the toxic A chain.

Ricin has been extensively used in the production of immunotoxins. However, selectivity of immuno-conjugates is hampered by the unspecific binding to cells mediated by the B chain. Mutations of ricin B which abrogate sugar-binding activity without interfering with its translocation function would eliminate the lack of cell specificity of the immunotoxins preserving their cytotoxicity.

Ricin B, as well as the two independent sugar-binding domains in which it can be dissected, were displayed in a functional conformation and bound to their specific ligand when expressed as a pIII-fusion on the surface of phage. Stable expression of the active lectin on the surface of phage made it possible to efficiently select mutants of the ricin B chain or its sugar-binding domains with altered binding activities from a vast randomly generated library.

Cytokine and receptor on phage

The improved solubility and easy purification procedure of fusion proteins expressed on phage was exploited by Zenke and co-workers [54], who expressed the human interleukin 3 (hIL-3) on the phage surface as pIII-fusion. Cytokine production in *E. coli* is in fact often hampered by their being trapped in inclusion bodies, with consequent loss of folding and biological activity. They demonstrated that hIL-3 on phage is predominantly displayed in its native conformation and is able to specifically promote the proliferation of an hIL-3 dependent cell line. Recombinant phage directly used as immunogen, elicited an anti-hIL-3 response in mice, and neutralizing mAb directed against native hIL-3 could be established from these mice with a high frequency.

The display of protein on phage has also emerged as a powerful technique for the study of receptor-ligand interactions. The first indication that a biologically active receptor can be produced on the phage was provided by two reports [85,86] describing the expression of functional domains of the human high-affinity receptor (FcεRI) for immunoglobulin E (IgE). FcεRI is a multisubunit complex that participates in IgE-dependent activation of mast cells and basophils in the allergic reaction. The IgE-binding portion of the receptor is confined to the extra cellular domains of the α subunit.

The human FcεRI α-subunit extracytoplasmic portion has been displayed on the surface of phage as fusion with the C-terminal part of pIII, using a monovalent display system. The phage-borne α receptor was able to interact with IgE, indicating that a proper folding of the receptor moiety has occurred.

The individual α-subunit ectodomains or a chimerical rat-human α-subunit fragment were also displayed on phage. In accordance with previous observations derived from different structural contexts, the C-terminal ectodomain of the α subunit was sufficient for interaction with IgE, although with a lower binding affinity than the comparably displayed two-domain fragment. This system offers the opportunity to identify in great detail residues involved in IgE binding.

Development of phage display of proteins

Phage display can potentially be utilized to engineer proteins, even if they are not directly expressed on the phage coat. An example of this is the association of the immunoglobulin heavy chain displayed on phage with the periplasmically expressed light chain (or vice versa) to yield functional heavy and light chain complexes (discussed in the next section). The co-expression of the endogenous *E. coli* inhibitor ecotin with phage-borne trypsin [82] is another general indicator of the potential of phage display technology.

More recently, a system which allows the display of proteins on phage via their association with a partner fused to pIII has been developed [87]. Jun and Fos are two transcription activators which heterodimerize through a defined short peptide domain known as "leucine zipper". The Jun domain has been expressed as a fusion protein with pIII and, from the same phagemid, the leucine zipper of Fos was co-expressed as an N-terminal fusion peptide to a foreign protein. Using this strategy, the resulting Fos-fusion protein associated with the Jun-expressing phage particles. To avoid exchange between phage of Fos-fusion products and subsequent loss of the appropriate genetic information, a covalent link was provided by cysteines engineered at the extremities of each leucine zipper. Enzymes or antigens covalently attached to the phage coat by the modified Jun/Fos molecules were shown to be biologically active.

This system has been developed to allow the expression and screening of cDNA libraries on phage. So far, the use of phage for cDNA library screening was prevented by the fact that cDNA gene products cannot be directly expressed as fusion proteins to the N-terminus of the viral coat proteins, due to transcription stop sites present at the 3'-end of nontranslated regions in eukaryotic cDNA: the Jun/Fos mediated display overcomes this limitation. Several additional useful applications can be indicated for this system, such as the screening and analysis of homo- or heterodimeric molecules.

Phage Display of Antibody Fragments

Given the fact that phage-display technology has revolutionized the engineering of antibody combining sites, we decided to dedicate a specific section to this topic. In this section we attempt to survey the advances made in this rapidly moving field,

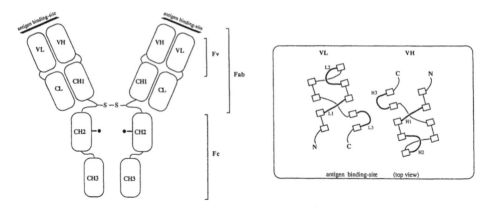

Fig. 4. The basic structure of an IgG molecule. There are four domains on the heavy chains (VH, CH1, CH2, CH3) and two on the light chains (VL, CL). Carbohydrates are indicated (●) in CH2 domains. Papain cleaves the molecule at the hinge, releasing two Fab units and one Fc unit. Fv indicates the variable domains. The top view of the antigen-binding site (right panel) is adapted from Chothia et al. [148]. H1, H2, H3 and L1, L2, L3 correspond to the hypervariable regions defined as Chothia and Lesk [92].

focusing on the phage display of Ab fragments and of their repertoires. A more detailed review dedicated entirely to this subject has been published recently by Winter et al. [88].

To perceive the enormous progress made in this field, it is crucial to comprehend the structure of Ab molecules, particularly their modular array of structural domains (Fig. 4). This feature has a counterpart in the genetic organization of immunoglobulin (Ig) loci [89], which has made it possible to produce Abs on demand. The work of Edelman and co-workers was a milestone in our current understanding of the structure of Igs; since then, much progress has been made regarding not only the structure but also the genetics and the functional properties of Igs (for review, see Padlan [90]). Abs are tetrameric molecules consisting of two identical heavy (H) chains joined to two identical light (L) chains by disulfide bonds and held together by inter-heavy-chain disulfide linkages. Both H and L chains have a variable (V) and a constant (C) region. Kabat and Wu [91] pointed out the existence of discrete areas of variability within Ab V regions and introduced the notion of hypervariable (HV) or complementarity-determining regions (CDR). Variable regions are organized as β-strands interconnected by HV loops. Analysis of Abs of known atomic structure clarified the relationship between sequence and spatial structure [92,93] of HV loops: with the exception of the loop formed by the CDR3 of the H chain, binding site loops have a restricted number of functional conformations or "canonical structures" determined by size and sites of key residues.

The idea of using Igs for passive therapy began with the pioneering work of Behring and Kitasato in 1890, and since this initial stage, Igs and their clinical use have undergone important developments. Ideally, Igs for therapy should be of human origin to prevent eliciting undesired anti-isotypic and idiotypic responses and the induction of serum sickness (for review on history of serotherapy, see [94]). The advent of hybridoma [95] raised a legitimate hope for the production of human hybridomas that secrete Abs of desired specificity. Historically, researchers have accepted the disadvantages of murine mAbs because of the impracticability of working with mAbs of human origin. However, the advantages of human mAbs, particularly their lack of immunogenicity, make them very attractive for diagnostic and therapeutic applications, although human hybridoma are unfortunately characterized by instability and poor secretion. The need for homogeneous sources of Abs with single antigen specificities therefore has promoted the search for alternative techniques to produce Abs useful in human therapy. Today, advances such as the phage display of Ab repertoires makes it possible to overcome many of these difficulties.

Phage display of immunoglobulins from natural repertoires

Since the first demonstration that it was possible to display functional Ab fragments on the surface of f1 [96–100], the perspective of constructing a "synthetic human immune system" came about [101]. The display of human repertoires of Ab fragments offers a new way of making Abs with pre-defined specificities. In fact the in vitro process of display and selection of Ab fragments on filamentous phage can actually mimic the in vivo immune selection [102] leading to Abs with improved features.

Functional Ab fragments have been displayed on phage fused to pVIII phage protein [103–105] or, more frequently, fused to pIII protein; moreover, these fragments can be expressed in the form of Fab or scFv fragments, that is V_L and V_H joined together by means of a polypeptide tether (for a review of different cloning strategies, see [88]). Recombinant Abs can be isolated from repertoires of Ab genes of immunized or nonimmunized donors (in vivo rearranged Abs), from libraries of germline genes (in vitro rearranged) or from Ab gene repertoire synthetic libraries.

The construction of human Ab repertoire libraries from immunized donors has been particularly fruitful for the selection of Abs against causative agents of a wide range of important human pathologies. Some groups, in particular those of Burton and Barbas at the Scripps Research Institute, by following this strategy and using PBLs obtained by leukapheresis or lymphocytes from bone marrow, have been able to isolate a large panel of recombinant Abs. Some of the agents for which Abs have been selected by this methodology are: *Clostridium tetani* toxin [106,107]; hepatitis B virus [108]; type 1 human immunodeficiency virus [109–111]; herpes simplex 1 and 2 [112]; respiratory syncytial virus [113,114]; human cytomegalovirus, varicella zoster virus, rubella virus [115]. Some of the Fabs have been shown to be effective in neutralizing viruses, as in the case of HIV-1, RSV, and HSV. The use of immune libraries has also led to the identification of Abs produced in nonviral diseases as in the case of autoimmune diseases in human and mice [116,117].

The main advantages of utilizing nonimmunized libraries are related to the concept of "universal" libraries, since it is possible to produce Abs with different specificities from a single library. On the other hand, the screening of mouse [104,118] and human [4] repertoires of in vivo rearranged V-genes from nonimmunized individuals for Abs binding to specific antigens has sometimes led to results inferior to those obtained with the same antigen by screening an immunized donor library, in terms both of quantity and "quality" (affinity) of selected Abs. The results can be particularly discouraging when Ab fragments derived from IgG isotypes are sought, whereas specific IgM-derived V region repertoires can be readily selected, albeit with low binding affinity; this is expected as IgM have not undergone affinity maturation in vivo. Nevertheless, both from in vivo [4] and in vitro (germline genes [119]) rearranged libraries, it has also been possible to isolate Abs with micromolar range affinities. Using repertoires of about 10^7–10^8 different V regions, it was shown that even Abs specific for self-antigens [120,121], including intracellular proteins [122], could be rescued.

Semisynthetic and synthetic libraries

A more effective alternative to nonimmunized libraries is the construction of semisynthetic or synthetic libraries, that is of libraries in which part or all of the naturally occurring CDR coding regions are substituted by random sequences. Barbas et al. have shown that libraries of human Abs containing synthetic random CDR can be utilized to select Abs which bind novel antigens [123,124]; other examples of how this strategy can be utilized are the selection of Abs that bind Con A [125], or self-antigens as human integrins, where the recombinant Ab mimics natural ligands, with very high K_D, 10^{-10} M [126]; also catalytic activities can be selected from these semisynthetic repertoires [127].

Recently, it was proposed that much larger repertoires could be constructed by exploiting the lox P/Cre system [128]. The strategy was carried out entirely in vitro, using V-gene segments as building blocks, by cloning most of the H and L chain segments used in vivo fused to synthetic CDR3 regions. Bacteria harboring a plasmid encoded H chain repertoire are infected with a phage encoded L chain repertoire. The recombination of the two chains is mediated by the Cre enzyme at the specific Lox P sites. With this strategy a combinatorial library approaching 10^{11} members has been constructed, from which a large panel of Ab fragments with nM or even pM K_D have been isolated [129].

Other developments of phage display of antibody fragments

There is some evidence that at least one of the two chains of recombinant Abs selected by the combinatorial approach corresponds to the one used in vivo against the corresponding antigen [115], and comparative studies with antigen-selected or non selected B-cells as a source of RNA for recombinant Ab construction [130] suggest that libraries from selected cells reflect at least in part the natural Ab repertoire of the memory compartment. Likewise, the only approach that ensures the cloning of Abs retaining the original chain pairing is the use of either cell lines [130–

133a–c] or, even in a mixed B cell population, of the "in cell PCR" technique [134] in which PCR amplifications occur inside the Ab producing cells.

Strategies to optimize affinity for targeting antigens while minimizing cross-reactivity have been developed, including error-prone PCR random mutagenesis [104], chain-shuffling [102], codon-based mutagenesis [135], enzymatic inverse PCR mutagenesis [136], computer-assisted "parsimonious mutagenesis" [137].

A highly effective strategy to humanize mouse Abs was developed by Hoogenboom and co-workers named Epitope Imprinted Selection (EIS) [138,139]. The strategy is based on chain shuffling whereby the mouse V regions are substituted stepwise, with the "best fitted" V_L and V_H human counterparts and affinity-selected with the antigen.

Phage displayed Abs have been selected from libraries not only with purified proteins as antigens but also using transfected intact bacterial [140] and eukaryotic cells [116] expressing foreign proteins as well as human cells expressing endogenous surface molecules [121].

Other developments to overcome the inherent delivery problems of large molecules have also been proposed. Fragments smaller than Fv, such as single domain Abs [141], may have special applications, e.g. imaging in vivo , targeting of immunotoxins [142], where small size may be useful for tissue penetration. Some of the problems observed with single domain Abs may be overcome by using camel Abs [143]. Camels can mount effective immune response using only V_H repertoires, as they appear to lack functional L chains. Amino acid differences in V_H interface and the lack of C_H1 domain explains how these molecules can be secreted and remain monomeric. Their CDR3 is longer than mouse and human counterparts thus, greater diversity can be achieved despite the in vivo usage of a single V domain. Smaller sub-domain fragments such as the minibody, which embodies three β-strands from each of the two β-sheets of the V_H domain along with the exposed hypervariable H1 and H2 regions [144,145a], can serve as a platform for displaying discontinuous structural epitopes. From a repertoire of minibodies [145b] a variant that binds human interleukin-6 with micromolar dissociation constant and with antagonistic properties was affinity-selected. This system also provides a strategy to readily identify "Minimal Recognition Units" and for developing Ig-like peptidomimetics [146,147].

The ability to isolate and manipulate antibody genes to produce Abs on demand brings a powerful tool to the fight against many diseases including cancer and leukemia, allergy, sepsis and AIDS. Pharmacokinetic, pre-clinical and clinical data where human mAbs are being used including radioimmune-detection therapy of cancer, the neutralizing of HIV 1 virus, RSV, HSV, CMV, retinitis and other autoimmune and viral targets are showing promising results. The ongoing technological revolution in Abs engineering using phage display brings ever closer the practical reality of readily available, target-specific Abs for diagnostic and therapeutic purposes.

176

Acknowledgement

We thank Ms. Janet Clench for revising the manuscript.

References

1. Huse WD, Sastry L, Iverson SA, Kang AS, Alting-Mees M, Burton DR, Benkovic SJ, Lerner RA. Generation of a large combinatorial library of the immunoglobulin repertoire in phage lambda. Science 1989;46:1275–1281.
2. Scott JK, Smith GP. Searching for peptide ligands with an epitope library. Science 1990;249:386–390.
3. Bass S, Greene R, Wells JA. Hormone phage: an enrichment method for variant proteins with altered binding properties. Proteins Struct Funct Genet 1990;8:309–314.
4. Marks JD, Hoogenboom HR, Bonnert TP, McCafferty J, Griffiths AD, Winter G. By-passing immunization. Human antibodies from V-gene libraries displayed on phage. J Mol Biol 1991;222: 581–597.
5. Cull MG, Miller JF, Schatz PJ. Screening for receptor ligands using large libraries of peptides linked to the C terminus of the lac repressor. Proc Natl Acad Sci USA 1992;89:1865–1869.
6. Birnbaum S, Mosbach K. Peptide screening. Curr Opin Biotechnol 1992;3:49–54.
7. Hoess RH. Phage display of peptides and protein domains. Curr Opin Struct Biol 1993;3:572–579.
8. Rasched I, Oberer E. Ff coliphages: structural and functional relationships. Microbiol Rev 1986;50:401–427.
9. Model P, Russel M. In: Calendar E (ed) The Bacteriophages 2. New York: Plenum, 1988;375–456.
10. Beck E, Sommer R, Auerswald EA, Kurz C, Zink, B, Osterburg G, Schaller H, Sugimoto K, Sugisaki H, Okamoto T, Takanami M. Nucleotide sequence of bacteriophage fd DNA. Nucleic Acids Res 1978;5:4495–4503.
11. van Wezenbeck PMGF, Hulsebos TJM, Schoenmakers JGG. Nucleotide sequence of the filamentous bacteriophage M13 genome: comparison with phage fd. Gene 1980;11:129–148.
12. Beck E, Zink B. Nucleotide sequence and genome organisation of filamentous bacteriophages f1 and fd. Gene 1981;16:35–58.
13. Hill DF, Petersen GB. Nucleotide sequence of bacteriophage f1 DNA. J Virol 1982;44:32–46.
14. Zinder ND, Boeke JD. The filamentous phage (Ff) as vectors for recombinant DNA - a review. Gene 1982;19:1–10.
15. Smith GP. In: Rodriquez RL, Denhardt DT (eds) Vectors: a Survey of Molecular Cloning Vectors and their Uses. Boston: Butterworth, 1987;61–83.
16. Viera J, Messing J. Production of single-stranded plasmid DNA. Methods Enzymol 1987;153:3–11.
17. Lopez J, Webster RE. Assembly site of bacteriophage f1 corresponds to adhesion zones between the inner and outer membranes of the host cell. J Bacteriol 1985;163:1270–1274.
18. Russel M. Protein-protein interactions during filamentous phage assembly. J Mol Biol 1993;231: 689–697.
19. Opella SJ, Stewart PL, Valentine KG. Protein structure by solid-state NMR spectroscopy. Quart Rev Biophys 1987;19:7–49.
20. Colnago LA, Valentine KG, Opella SJ. Dynamics of fd coat protein in the bacteriophage. Biochemistry 1987;26:847–854.
21. Marvin DA. Model building studies of *Inovirua*: genetic variations on a geometric theme. Int J Biol Macromol 1990;12:125–138.
22. Glucksman MJ, Bhattacharjee S, Makowski L. Three-dimensional structure of a cloning vector. X-ray diffraction studies of filamentous bacteriophage M13 at 7 Å resolution. J Mol Biol 1992;226: 455–470.

23. Gray CW, Brown RS, Marvin DA. Adsorption complex of filamentous fd virus. J Mol Biol 1981; 146:621–627.
24. Smith GP. Surface presentation of protein epitopes using bacteriophage expression systems. Curr Opin Biotechnol 1991;2:668–673.
25. Cesareni G. Peptide display on filamentous phage capsids. A new powerful tool to study protein-ligand interaction. FEBS Lett 1992;307:66–70.
26. Smith GP. Surface display and peptide libraries. Gene 1993;128:1–2.
27. Makowski L. Phage display: structure, assembly and engineering of filamentous bacteriophage M13. Curr Opin Struct Biol 1994;4:225–230.
28. Smith GP. Filamentous fusion phage: novel expression vectors that display cloned antigens on the virion surface. Science 1985;228:1315–1317.
29. Parmley SF, Smith GP. Antibody-selectable filamentous fd phage vectors: affinity purification of target genes. Gene 1988;73:305–318.
30. Zacher III AN, Stock CA, Golden II JW, Smith GP. A new filamentous phage cloning vector: fd-tet. Gene 1980;9:127–140.
31. Smith GP, Scott JK. Libraries of peptides and proteins displayed on filamentous phage. Methods Enzymol 1993;217:228–257.
32. Dente L, Cesareni G, Cortese R. pEMBL: a new family of single stranded plasmids. Nucleic Acids Res 1983;11:1645–1655.
33. Ilyichev AA, Minenkova OO, Tat'kov SI, Karpyshev NN, Eroshkin AM, Petrenko VA, Sandachshiev LS. Production of a viable variant of the M13 phage with a foreign peptide inserted into the basic coat protein. Dokl Acad Nauk USSR 1989;307:481–483.
34. Devlin JJ, Panganiban LC, Devlin PE. Random peptide libraries: a source of specific protein binding molecules. Science 1990;249:404–406.
35. Cwirla SE, Peters EA, Barrett RW, Dower WJ. Peptides on phage: a vast library of peptides for identifying ligands. Proc Natl Acad Sci USA 1990;87:6378–6382.
36. Felici F, Castagnoli L, Musacchio A, Jappelli R, Cesareni G. Selection of antibody ligands from a large library of oligopeptides expressed on a multivalent exposition vector. J Mol Biol 1991;222: 301–310.
37. Barrett RW, Cwirla SE, Ackerman MS, Olson AM, Peters EA, Dower WJ. Selective enrichment and characterization of high affinity ligands from collections of random peptides on filamentous phage. Anal Biochem 1992;204:357–364.
38. McLafferty MA, Kent RB, Ladner RC, Markland W. M13 bacteriophage displaying disulfide-constrained microproteins. Gene 1993;128:29–36.
39. Miceli RM, DeGraaf ME, Fischer HD. Two-stage selection of sequences from a random phage display library delineates both core residues and permitted structural range within an epitope. J Immunol Methods 1994;167:279–287.
40. Jellis CL, Cradick TJ, Rennert P, Salinas P, Boyd J, Amirault T, Gray GS. Defining critical residues in the epitope for a HIV-neutralizing monoclonal antibody using phage display and peptide array technologies. Gene 1993;137:63–68.
41. Stephen CW, Lane DP. Mutant Conformation of p53. Precise epitope mapping using a filamentous phage epitope library. J Mol Biol 1992;225:577–583.
42. du Plessis DH, Wang LF, Jordan FA, Eaton BT. Fine mapping of a continuous epitope on VP7 of bluetongue virus using overlapping synthetic peptides and a random epitope library. Virology 1994;198:346–349.
43. Christian RB, Zuckermann RN, Kerr JM, Wang L, Malcolm BA. Simplified methods for construction, assessment and rapid screening of peptide libraries in bacteriophage. J Mol Biol 1992;227: 711–718.
44. Yayon A, Aviezer D, Safran M, Gross JL, Heldman Y, Cabilly S, Givol D, Katchalski-Katzir E. Isolation of peptides that inhibit binding of basic fibroblast growth factor to its receptor from a random phage-epitope library. Proc Natl Acad Sci USA 1993;90:10643–10647.

45. Malik B, Jamieson GA, Ball WJ. Identification of the amino acids comprising a surface-exposed epitope within the nucleotide-binding domain of the Na^+, K^+-ATPase using a random peptide library. Protein Sci 1993;2:2103–2111.
46. Felici F, Luzzago A, Folgori A, Cortese R. Mimicking of discontinuous epitopes by phage-displayed peptides, II. Selection of clones recognized by a protective monoclonal antibody against the *Bordetella pertussis* toxin from phage peptide libraries. Gene 1993;128:21–27.
47. Balass M, Heldman Y, Cabilly S, Givol D, Katchalski-Katzir E, Fuchs S. Identification of a hexapeptide that mimics a conformation-dependent binding site of acetylcholine receptor by use of a phage-epitope library. Proc Natl Acad Sci USA 1993;90:10638–10642.
48. Luzzago A, Felici F, Tramontano A, Pessi A, Cortese R. Mimicking of discontinuous epitopes by phage-displayed peptides, I. Epitope mapping of human H ferritin using a phage library of constrained peptides. Gene 1993;128:51–57.
49. Hoess R, Brinkmann U, Handel T, Pastan I. Identification of a peptide which binds to the carbohydrate-specific monoclonal antibody B3. Gene 1993;128:43–49.
50. de la Cruz VF, Lal AA, McCutchan TF. Immunogenicity and epitope mapping of foreign sequences via genetically engineered filamentous phage. J Biol Chem 1988;263:4318–4322.
51. Greenwood J, Willis AE, Perham RN. Multiple display of foreign peptides on a filamentous bacteriophage. Peptides from *Plasmodium falciparum* circumsporozoite protein as antigens. J Mol Biol 1991;220:821–827.
52. Willis AE, Perham RN, Wraith D. Immunological properties of foreign peptides in multiple display on a filamentous bacteriophage. Gene 1993;128:79–83.
53. Minenkova OO, Ilyichev AA, Kishchenko GP, Petrenko VA. Design of specific immunogens using filamentous phage as the carrier. Gene 1993;128:85–88.
54. Gram H, Strittmatter U, Lorenz M, Glück D, Zenke G. Phage display as a rapid gene expression system: production of bioactive cytokine-phage and generation of neutralizing monoclonal antibodies. J Immunol Methods 1993;161:169–176.
55. Keller PM, Arnold BA, Shaw AR, Tolman RL, van Middlesworth F, Bondy S, Rusiecki VK, Koenig S, Zolla-Pazner S, Conard P, Emini EA, Conley AJ. Identification of HIV vaccine candidate peptides by screening random phage epitope libraries. Virology 1993;193:709–716.
56. Motti C, Nuzzo M, Meola, A, Galfrè, G, Felici F, Cortese R, Nicosia A, Monaci P. Recognition by human sera and immunogenicity of HBsAg mimotopes selected from an M13 phage display library. Gene 1994;146:191–198.
57. Dybwad A, Førre Ø, Kjeldsen-Kragh J, Natwig JB, Sioud M. Identification of new B cell epitopes in the sera of rheumatoid arthritis patients using a random nonapeptide phage library. Eur J Immunol 1993;23:3189–3193.
58. Folgori A, Tafi R, Meola A, Felici F, Galfrè G, Cortese R, Monaci P, Nicosia A. A general strategy to identify mimotopes of pathological antigens using only random peptide libraries and human sera. EMBO J 1994;13:2236–2243.
59. Koivunen E, Gay DA, Ruoslahti E. Selection of peptides binding to the a_5b_1 integrin from phage display library. J Biol Chem 1993;268:20205–20210.
60. Koivunen E, Wang B, Ruoslahti E. Isolation of a highly specific ligand for the a_5b_1 integrin from a phage display library. J Cell Biol 1994;124:373–380.
61. O'Neil KT, Hoess RH, Jackson SA, Ramachandran NS, Mousa SA, DeGrado WF. Identification of novel peptide antagonists for GPIIb/IIIa from a conformationally constrained phage peptide library. Proteins Struct Funct Genet 1992;14:509–515.
62. Smith GP, Schultz DA, Ladbury JE. A ribonuclease S-peptide antagonist discovered with a bacteriophage display library. Gene 1993;128:37–42.
63. Hammer J, Takacs B, Sinigaglia F. Identification of a motif for HLA-DR1 binding peptides using M13 display libraries. J Exp Med 1992;176:1007–1013.
64. Hammer J, Valsasnini P, Tolba K, Bolin D, Higelin J, Takacs B, Sinigaglia F. Promiscuous and allele-specific anchors in HLA-DR-binding peptides. Cell 1993;74:197–203.

65. Stern LJ, Brown JH, Jardetzky TS, Gorga JC, Urban RG, Strominger JL, Wiley DC. Crystal structure of the human class II MHC protein HLA-DR1 complexed with an influenza virus peptide. Nature 1994;368:215–221.

66. Blond-Elguindi S, Cwirla SE, Dower WJ, Lipshutz RJ, Sprang SR, Sambrook JF, Gething MJH. Affinity panning of a library of peptides displayed on bacteriophages reveals the binding specificity of BiP. Cell 1993;75:717–728.

67. Weber PC, Pantoliano MW, Thompson D. Crystal structure and ligand-binding studies of a screened peptide complexed with streptavidin. Biochemistry 1992;31:9350–9354.

68. Kay BK, Adey NB, He Y-S, Manfredi JP, Mataragnon AH, Fowlkes DM. An M13 phage library displaying random 38-amino-acid peptides as a source of novel sequences with affinity to selected targets. Gene 1993;128:59–65.

69. Oldenburg KR, Loganathan D, Goldstein IJ, Schultz PG, Gallop MA. Peptide ligands for a sugar-binding protein isolated from a random peptide library. Proc Natl Acad Sci USA 1992;89:5393–5397.

70. Scott JK, Loganathan D, Easley RB, Gong X, Goldstein IJ. A family of concanavalin A-binding peptides from a hexapeptide epitope library. Proc Natl Acad Sci USA 1992;89:5398–5402.

71. Saggio I, Laufer R. Biotin binders selected from a random peptide library expressed on phage. Biochem J 1993;293:613–616.

72. Hiller Y, Bayer EA, Wilchek M. Studies on the biotin-binding site of avidin. Minimized fragments that bind biotin. Biochem J 1991;278:573–585.

73. Matthews DJ, Wells JA. Substrate phage: selection of protease substrates by monovalent phage display. Science 1993;260:1113–1116.

74. Lowman HB, Bass SH, Simpson N, Wells JA. Selecting high-affinity binding proteins by monovalent phage display. Biochemistry 1991;30:10832–10838.

75. Markland A, Roberts BL, Saxena MJ, Guterman SK, Ladner RC. Design construction and function of a multicopy display vector using fusions to the major coat protein of bacteriophage M13. Gene 1991;109:13–19.

76. Roberts BL, Markland W, Ley AC, Kent RB, White DW, Guterman SK, Ladner RC. Directed evolution of a protein: selection of potent neutrophil elastase inhibitors displayed on M13 fusion phage. Proc Natl Acad Sci USA 1992;89:2429–2433.

77. Roberts BL, Markland W, Siranosian K, Saxena MJ, Guterman SK, Ladner RC. Protease inhibitor display M13 phage: selection of high-affinity neutrophil elastase inhibitors. Gene 1992;121: 9–15.

78. Pannekoek H, van Meijer M, Schleef RR, Loskutoff DJ, Barbas III CF. Functional display of human plasminogen-activator inhibitor 1 (PAI-l) on phages: novel perspectives for structure-function analysis by error-prone DNA synthesis. Gene 1993;128:135–140.

79. Djojonegoro BM, Benedik MJ, Willson RC. Bacteriophage surface display of an immunoglobulin-binding domain of *Staphylococcus aureus* protein A. Bio/Technology 1994;12:169–172.

80. McCafferty J, Jackson RH, Chiswell DJ. Phage-enzymes: expression and affinity chromatography of functional alkaline phosphatase on the surface of bacteriophage. Protein Eng 1991;4:955–961.

81. Light J, Lerner R. Phophabs: antibody-alkaline phosphatase conjugates for one step ELISA's without immunization. Bioorg Med Chem Lett 1992;2:1073–1078.

82. Corey DR, Shiau AK, Yang Q, Janowski BA, Craik CS. Trypsin display on the surface of bacteriophage. Gene 1993;128:129–134.

83. Rebar EJ, Pabo CO. Zinc finger phage: affinity selection of fingers with new DNA-binding specificities. Science 1994;263:671–673.

84. Swimmer C, Lehar SM, McCafferty J, Chiswell DJ, Blätter WA, Guild BC. Phage display of ricin B chain and its single binding domains: system for screening galactose-binding mutants. Proc Natl Acad Sci USA 1992;89:3756–3760.

85. Robertson MW. Phage and *Escherichia coli* expression of the human high affinity immunoglobulin E receptor a-subunit ectodomain. J Biol Chem 1993;268:12736–12743.

86. Scarselli E, Esposito G, Traboni C. Receptor phage. Display of functional domains of the human high affinity IgE receptor on the M13 phage surface. FEBS Lett 1993;329:223–226.
87. Crameri R, Suter M. Display of biologically active proteins on the surface of filamentous phages: a cDNA cloning system for selection of functional gene products linked to the genetic information responsible for their production. Gene 1993;137:69–75.
88. Winter G, Griffiths AD, Hawkins RE, Hoogenboom HR. Making antibodies by phage display technology. Annu Rev Immunol 1994;12:433–455.
89. Rathbun G, Berman J, Yancopoulos G, Alt FW. In: Honjo T, Alt FW, Rabbitts TH (eds) Immuno-globulin Genes. New York: Academic Press, 1989;63–90.
90. Padlan EA. Anatomy of the antibody molecule. Mol Immunol 1994;31:169–217.
91. Kabat EA, Wu TT. Attempts to locate complementarity-determining residues in the variable positions ol light and heavy chains. Ann NY Acad Sci 1971;190:382–393.
92. Chothia C, Lesk AM. Canonical structures for the hypervariable regions of immunoglobulins. J Mol Biol 1987;196:901–917.
93. Tramontano A, Chothia C, Lesk AM. Framework residue 71 is a major determinant of the position and conformation of the second hypervariable region in the V_H domains of immunoglobulins. J Mol Biol 1990;215:175–182.
94. Silverstein AM. In: Paul WE (ed) Fundamental Immunology, 3rd edn. New York: Raven Press, 1993;43–74.
95. Kohler G, Milstein C. Continuous cultures of fused cells secreting antibody of predefined specificity. Nature 1975;265:495–497.
96. McCafferty J, Griffiths AD, Winter G, Chiswell DJ. Phage antibodies: filamentous phage displaying antibody variable domains. Nature 1990;348:552–554.
97. Hoogenboom HR, Griffiths AD, Johnson KS, Chiswell DJ, Hudson P, Winter G. Multi-subunit proteins on the surface of filamentous phage: methodologies for displaying antibody (Fab) heavy and light chains. Nucleic Acids Res 1991;19:4133–4137.
98. Barbas III CF, Kang AS, Lerner RA, Benkovic SJ. Assembly of combinatorial antibody libraries on phage surfaces: the gene III site. Proc Natl Acad Sci USA 1991;88:7978–7982.
99. Breitling F, Dübel S, Seehaus T, Klewinghaus I, Little M. A surface expression vector for antibody screening. Gene 1991;104:147–153.
100. Garrard LJ, Yang M, O'Connell MP, Kelley RF, Henner DJ. Fab assembly and enrichment in a monovalent phage display system. Bio/Technology 1991;9:1373–1377.
101. Winter G, Milstein C. Man-made antibodies. Nature 1991;349:293–299.
102. Marks JD, Griffiths AD, Malmqvist M, Clackson TP, Bye JM, Winter G. By-passing immunization: building high affinity human antibodies by chain shuffling. Bio/Technology 1992;10:779–783.
103. Kang AS, Barbas III CF, Janda KD, Benkovic SJ, Lerner RA. Linkage of recognition and replication functions by assembling combinatorial Fab libraries along phage surfaces. Proc Natl Acad Sci USA 1991;88:4363–4366.
104. Gram H, Marconi L-A, Barbas III CF, Collet TA, Lerner RA, Kang AS. In vitro selection and affinity maturation of antibodies from a naive combinatorial immunoglobulin library. Proc Natl Acad Sci USA 1992;89:3576–3580.
105. Huse WD, Stinchcombe TJ, Glaser SM, Starr L, MacLean M, Hellström KE, Hellström I, Yelton DE. Application of a filamentous phage pVIII fusion protein system suitable for efficient production, screening, and mutagenesis of F(ab) antibody fragments. J Immunol 1992;149:3914–3920.
106. Persson MAA, Caothien RH, Burton DR. Generation of diverse high-affinity human monoclonal antibodies by repertoire cloning. Proc Natl Acad Sci USA 1991;88:2432–2436.
107. Hogrefe HH, Mullinax RL, Lovejoy AE, Hay BN, Sorge JA. A bacteriophage lambda vector for the cloning and expression of immunoglobulin Fab fragments on the surface of filamentous phage. Gene 1993;128:119–126.
108. Zebedee SL, Barbas III CF, Hom YL, Caothien RH, Graff R, DeGraw J, Pyati J, LaPolla R, Burton

DR, Lerner RA, Thornton GB. Human combinatorial antibody libraries to hepatitis B surface antigen. Proc Natl Acad Sci USA 1992;89:3175–3179.

109. Burton DR, Barbas III CF, Persson MAA, Koenig S, Chanock RM, Lerner RA. A large array of human monoclonal antibodies to type 1 human immunodeficiency virus from combinatorial libraries of asymptomatic seropositive individuals. Proc Natl Acad Sci USA 1991;88:10134–10137.

110. Barbas III CF, Björling E, Chiodi F, Dunlop N, Cababa D, Jones TM, Zebedee SL, Persson MAA, Nara PL, Noorby E, Burton D. Recombinant human Fab fragments neutralize human type 1 immunodeficiency virus in vitro. Proc Natl Acad Sci USA 1992;89:9339–9343.

111. Barbas III CF, Collet ATA, Amberg W, Roben P, Binley JM, Hoekstra D, Cababa D, Jones TM, Williamson RA, Pilington GR, Haigwood NL, Cabezas E, Satterthwait AC, Sanz I, Burton DR. Molecular profile of an antibody response to HIV-1 as probed by combinatorial libraries. J Mol Biol 1993;230:812–823.

112. Burioni R, Williamson A, Sanna PP, Bloom FE, Burton DR. Recombinant human Fab to glycoprotein D neutralizes infectivity and prevents cell-to-cell transmission of herpex simplex viruses 1 and 2 in vitro. Proc Natl Acad Sci USA 1994;91:355–359.

113. Barbas III CF, Crowe Jr JE, Cababa D, Jones TM, Zebedee SL, Murphy BR, Chanock RM, Burton DR. Human monoclonal Fab fragments derived from a combinatorial library bind to respiratory syncytial virus F glycoprotein and neutralize infectivity. Proc Natl Acad Sci USA 1992;89:10164–10168.

114. Crowe JE, Murphy BR, Chanock RM, Williamson RA, Barbas III CF, Burton DR: Recombinant human respiratory syncytial virus (RSV) monoclonal antibody Fab is effective therapeutically when introduced directly into lungs of RSV-infected mice. Proc Natl Acad Sci. USA 1994;91: 1386–1390.

115. Williamson RA, Burioni R, Sanna PP, Partridge LJ, Barbas III CF, Burton DR. Human monoclonal antibodies against a plethora of viral pathogens from single combinatorial libraries. Proc Natl Acad Sci USA 1993;90:4141–4145.

116. Portolano S, McLachan SM, Rapoport B. High affinity, thyroid-specific human autoantibodies displayed on the surface of filamentous phage use V genes similar to other autoantibodies. J Immunol 1993;151:2839–2851.

117. Calcutt MJ, Kremer MT, Giblin MF, Quinn TP, Deutscher SL. Isolation and characterization of nucleic acid-binding antibody fragments from autoimmune mice-derived bacteriophage display libraries. Gene 1993;137:77–83.

118. Clackson T, Hoogenboom HR, Griffiths AD, Winter G. Making antibody fragments using phage display libraries. Nature 1991;352:624–628.

119. Hoogenboom HR, Winter G. By-passing immunisation human antibodies from synthetic repertoires of germline V_H gene segments rearranged in vitro. J Mol Biol 1992;227:381–388.

120. Griffiths AD, Malmqvist M, Marks JD, Bye JM, Embleton MJ, McCafferty J, Baier M, Holliger KP, Gorik BD, Hughes-Jones NC, Hoogenboom HR, Winter G. Human anti-self antibodies with high specificity from phage display libraries. EMBO J 1993;12:725–734.

121. Marks JD, Ouwehand WH, Bye JM, Finnern R, Gorick BD, Voak D, Thorpe SJ, Hughes-Jones NC, Winter G. Human antibody fragments specific for human blood group antigens from a phage display library. Bio/Technology 1993;11:1145–1149.

122. Nissim A, Hoogenboom HR, Tomlinson IM, Flynn G, Midgley C, Lane D, Winter G. Antibody fragments from a 'single pot' phage display library as immunochemical reagents. EMBO J 1994; 13:692–698.

123. Barbas III CF, Bain JD, Hoekstra DM, Lerner RA. Semisynthetic combinatorial antibody libraries: a chemical solution to the diversity problem. Proc Natl Acad Sci USA 1992;89:4457–4461.

124. Barbas III CF, Amberg W, Simoncsits A, Jones TM, Lerner RA: Selection of human anti-hapten antibodies from semisynthetic libraries. Gene 1993;137:57–62.

125. Akamatsu Y, Cole MS, Tso JY, Tsurushita N. Construction of a human Ig combinatorial library from genomic V segments and synthetic CDR3 fragments. J Immunol 1993;151:4651–4659.

126. Barbas III CF, Languino LR, Smith JW. High-affinity self-reactive human antibodies by design and selection: targeting the integrin ligand binding site. Proc Natl Acad Sci USA 1993;90:10003–10007.

127. Janda KD, Lo CHL, Li T, Barbas III CF, Wirsching P, Lerner RA. Direct selection for a catalytic mechanism for combinatorial antibody libraries. Proc Natl Acad Sci USA 1994;91:2532–2536.

128. Waterhouse P, Griffiths AD, Johnson KS, Winter G. Combinatorial infection and in vivo recombination: a strategy for making large phage antibody repertoires. Nucleic Acids Res 1993;21:2265–2266.

129. Griffiths AD, Williams SC, Hartley O, Tomlison IM, Waterhouse P, Crosby WL, Kontermann RE, Jones PT, Low NM, Allison TJ, Prospero TD, Hoogenboom HR, Nissim H, Cox JPL, Harrison JL, Zaccolo M, Gherardi E, Winter G. Isolation of high affinity human antibodies directly from large synthetic repertoires. EMBO J 1994;13:3245–3260.

130. Hawkins RE, Winter G. Cell selection strategies from making antibodies from variable gene libraries: trapping the memory pool. Eur J Immunol 1992;22:867–870.

131. Chang CN, Landolfi NF, Queen C. Expression of antibody Fab domains on bacteriophage surfaces potential use for antibody selection. J Immunol 1991;147:3610–3614.

132. George AJT, Titus JA, Jost CR, Kurucz I, Perez P, Andrew SM, Nicholls PJ, Huston JS, Segal DM. Redirection of T cell-mediated cytotoxicity by a recombinant single-chain Fv molecule. J Immunol 1994;152:1802–1811.

133. (a) Jiang W, Bonnert TP, Venugopal K, Gould EA. A single chain antibody fragment expressed in bacteria neutralizes Tick-borne flavivirus. Virology 1994;200:21–28. (b) Esposito G, Scarselli E, Traboni C. Phage display of human antibody against *Clostridium tetani* toxin. Gene 1994;148: 167–168. (c) Esposito G, Scarselli E, Cerino A, Mondelli MU, LaMonica N, Traboni C. A human antibody specific for hepatitis C virus core protein: expression in a bacterial system and characterization. Gene 1995 (in press).

134. Embleton MJ, Gorochov G, Jones PT, Winter G. In-cell PCR from mRNA: amplifying and linking the rearranged immunoglobulin heavy and light chain V-genes within single cells. Nucleic Acids Res 1992;20:3831–3837.

135. Glaser SM, Yelton DE, Huse WD. Antibody engineering by codon-based mutagenesis in a filamentous phage vector system. J Immunol 1992;149:3903–3913.

136. Stemmer WPC, Morris SK, Kautzer CR, Wilson BS. Increased antibody expression from *Escherichia coli* through wobble-base library mutagenesis by enzymatic inverse PCR. Gene 1993; 123:1–7.

137. Balint RF, Larrick JW. Antibody engineering by parsimonious mutagenesis. Gene 1993;137:109–118.

138. Johnson KS, Chiswell DJ: Human antibody engineering. Curr Opin Struct Biol 1993;3:564–571.

139. Figini M, Marks JD, Winter G, Griffiths AD. Diversifying antibody binding site: template directed selection using repertoires of Fab fragments assembled on phage *in vitro*. J Mol Biol 1994;239: 68–78.

140. Bradbury A, Persic L, Werge T, Cattaneo A. Use of living columns to select specific phage antibodies. Bio/Technology 1993;11:1565–1569.

141. Ward SE, Güssow D, Griffiths AD, Jones PT, Winter G. Binding activities of a repertoire of single immunoglobulin variable domains secreted from *Escherichia coli*. Nature 1989;341:544–546.

142. Brinkmann U, Reiter Y, Jung S-H, Byungkook L, Pastan I. A recombinant immunotoxin containing a disulfide-stabilized Fv fragment. Proc Natl Acad Sci USA 1993;90:7538–7542.

143. Hamers-Casterman C, Atarhouch T, Muyldermans S, Robinson G, Hamers C, Bajyana Songa E, Bendahaman N, Hamers R. Naturally occurring antibodies devoid of light chains. Nature 1993; 363:446–448.

144. Pessi A, Bianchi E, Crameri A, Venturini S, Tramontano A, Sollazzo M. A designed metal-binding protein with a novel fold. Nature 1993;362:367–369.

145. (a) Venturini S, Martin F, Sollazzo M. Phage display of the minibody: a β-scaffold for the selec-

tion of conformationally-constrained peptides. Prot Peptide Lett 1994;1:70–75. (b) Martin F, Toniatti C, Salvati AL, Venturini S, Ciliberto G, Cortese R, Sollazzo M. The affinity-selection of a minibody polypeptide inhibitor of human interleukin-6. EMBO J 1994;13:5303–5309.

146. Saragovi HU, Greene MI, Chrusciel RA, Kahn M. Loops and secondary structure mimetics: development and applications in basic science and rational drug design. Bio/Technology 1992;10:773–778.

147. Taub R, Greene MI. Functional validation of ligand mimicry by anti-receptor antibodies: structural and therapeutic implications. Biochemistry 1992;31:7431–7435.

148. Chothia C, Lesk AM, Gherardi E, Tomlinson IM, Walter G, Marks JD, Llewelyn MB, Winter G. Structural repertoire of the human V_H segments. J Mol Biol 1992;227:799–817.

Biotechnology Annual Review Volume 1
M.R. El-Gewely, editor

Aptamers as potential nucleic acid pharmaceuticals

Andrew D. Ellington and Richard Conrad

Department of Chemistry, Indiana University, Bloomington, Indiana, USA

Abstract. In vitro selection is a technique for the isolation of nucleic acid ligands that can bind to proteins with high affinity and specificity, and has potential applications in the development of new pharmaceuticals. This review summarizes the protein targets that have successfully elicited nucleic acid binding species (also known as "aptamers") and explores examples of how they might be developed for clinical use. In particular, the use of aptamers for the alleviation of blood clotting and the treatment of AIDS are considered.

Introduction

While nucleic acids are generally thought of as informational macro molecules, they can also assume structures that can specifically interact with, and potentially inhibit the function of, particular proteins or other metabolites. The ability of antishape (as opposed to antisense) nucleic acids to alter cellular metabolism was originally observed during the development of techniques for the overexpression of particular DNA or RNA molecules. For example, repressor binding sites on multicopy plasmids were often found to derepress the chromosomal copy of a regulated gene by soaking up most of the repressor molecules in a cell. More recently, antishape nucleic acids have been purposefully engineered to inhibit protein function. For example, the Tat protein of HIV-1 normally binds to its cognate RNA element, TAR, in order to activate transcription of the viral genome. However, when the TAR element is separated from the viral genome and overexpressed it acts as an effective decoy for the Tat protein and almost completely inhibits viral replication [1].

Since specific protein–nucleic acid interactions govern a large portion of cellular and viral metabolism, the use of antishape nucleic acid "decoys" to inhibit or augment the function of nucleic acid binding proteins can in theory be used as a quite general strategy for the treatment of disease. Furthermore, nucleic acid decoys and inhibitors need not be limited to nucleic acid binding proteins alone. A detailed examination of the NMR and crystal structures of known protein–nucleic acid complexes suggests that RNA and DNA molecules can assume shapes and present surfaces that should be capable of recognizing virtually any protein structure. Many different types of interactions between individual nucleotides and individual amino acids are possible; for example, homeodomain proteins use residues such as glu-

---

Address for correspondence: A.D. Ellington, Department of Chemistry, Indiana University, Bloomington, IN 47405, USA.

tamine to make specific contacts with regulatory sequences [2], HIV Tat interacts with RNA via blocks of arginines [3], and ss DNA binding proteins contact DNA via tyrosine and tryptophan residues [4,5]. Thus, although nucleic acid binding proteins often contain patches of basic amino acids that can partially neutralize the polyanionic backbone of RNA or DNA, they may not otherwise differ significantly from the surfaces of proteins that do not normally bind cellular nucleic acids.

Unfortunately, it is not currently possible to rationally design nucleic acid shapes or surfaces that can specifically interact with a given protein. The detailed interactions between protein and nucleic acid surfaces are known for only a few complexes, and, thus, there is a paucity of relevant information that might be used to construct new interfaces. In addition, even if it were possible to predict what molecular shapes would likely interact with one another, a feat that is difficult even for well-studied protein–protein complexes, it might still prove difficult to determine which nucleic acid sequences would map to a desired shape. While the secondary structures of nucleic acids can be predicted with some accuracy (excluding structures such as pseudoknots), the rules that govern the folding of secondary structure into tertiary structure remain largely unknown. Physical and evolutionary data have sometimes been used to successfully model the three-dimensional structures of nucleic acids [6], but the de novo design of a particular structure is still well beyond the ability of modelling programs.

An efficient alternative to the design of protein–nucleic acid interfaces is their selection. Natural selection has already proven to be particularly adept at generating double-stranded DNA and single-stranded RNA shapes that bind to (or, conversely, be bound by) proteins. Most natural, functional nucleic acid molecules, such as operator sequences and tRNAs, probably originally arose by the accretion or rearrangement of random sequence blocks and have since been finely sculpted by mutation during the course of evolution. This same basic strategy can now be mimicked in vitro, using enzymes rather than whole cells for the preferential amplification of beneficial genotypes.

Nucleic acids that have been selected in vitro to bind to and inhibit a particular target molecule can potentially be used as pharmaceuticals in vivo. However, as is the case with antisense nucleic acids, it is currently quite difficult to either stabilize antishape nucleic acids in serum (for extracellular targets) and/or to introduce them directly into cells (for intracellular targets). A number of technologies are being evaluated for stabilization and delivery, but none are yet in common clinical use. Nonetheless, the development of antishape nucleic acid pharmaceuticals continues unabated and is driven by two essential factors.

First, in many instances it is much easier to find nucleic acid binding species (also known as aptamers) than conventional organic drug leads. In the simplest cases, decoys can be immediately fashioned from wild-type binding sequences and sites. However, there is no reason to expect that natural nucleic acids will necessarily make the best decoys. Although some protein–ligand interactions, such as avidin-biotin, are probably as tight as evolution can make them, many natural protein–nucleic acid complexes have equilibrium dissociation constants that are purposefully weaker than they could be, in order to promote facile dissociation for replicative or regulatory reasons. Moreover, protein binding sites on natural nucleic acids are often constrained by factors that may be unrelated to their ability to bind to a protein. For example, an overlapping amino acid coding sequence will necessarily restrict the choice of bases in a protein binding site. Thus, even when an obvious nucleic acid decoy exists, it may be worthwhile to use selection to search for alternate or unnatural binding sequences. Of course, when no natural nucleic acid ligand for a protein is known selection must be used to generate decoys. The fact that the protocols for the isolation of aptamers from random sequence pools are relatively straightforward and that the pitfalls are relatively simple to avoid augurs well for the adoption of these techniques to generate drug leads.

Second, aptamers identified by in vitro selection can bind their targets with extremely high affinities and specificities, and can in most cases effectively compete with natural ligands for binding. For example, as we shall see, aptamers to nucleic acid binding proteins have been found to bind from one to several orders of magnitude better than corresponding wild-type nucleic acid ligands. More impressively, an aptamer selected to bind to a small pharmaceutical, theophylline, recognizes its cognate substrate much more specifically than monoclonal antibodies generated for the same purpose [7].

This review focuses on these advantages by providing a complete summary of the wide range of proteins that have successfully elicited aptamers. Wherever possible, the affinities and specificities of the selected molecules for their targets are emphasized. In addition, in order to give some feel for how these molecules may eventually proceed to clinical use, several in-depth examples are provided of how selected nucleic acid decoys and inhibitors might function both extra- and intracellularly.

In Vitro Selection

In vitro selection can be thought of as a tool for directly querying molecular targets regarding what nucleic acid shapes and sequences they can bind. However, while this tool is relatively simple to master, it must be intelligently applied to ensure that meaningful nucleic acid "answers" are derived. The dialogue between target and pool can break down in a number of ways; for example, a problem that sometimes crops up during in vitro selections is the accidental amplification of random nucleic acid contaminants, rather than those molecules that have truly bound to a target. Similarly, while in principle virtually any soluble or immobilized protein or small molecule can be used as a selection target, and a wide range of enzyme inhibitors or protein decoys can potentially be generated, in practice, experimental success with different targets is largely a function of how the selection "questions" are initially posed. Some simple rules and procedures are codified here and will hopefully serve as a guide for expanding the range of targets and experiments that are being attempted in both academia and industry (for a more technical review, see Ref. [8]).

Overall, the principles that guide the in vitro selection of aptamers are similar to those that govern the natural selection of biopolymers. First, a pool of heritable variation is generated. In general, single-stranded DNA oligonucleotides that contain a core of random sequence flanked by constant regions necessary for enzymatic manipulation are synthesized chemically or biochemically. Second, the nucleic acid population is winnowed by extracting functional sequences. For example, sequences that fold into shapes that are complementary to a target molecule can be extracted by immobilizing the target, allowing the pool to equilibrate with the target, and then washing away all species that fail to bind either tightly or specifically. Finally, the survivors (and, in the case of selection for binding function, it truly is survival of the "fittest") are amplified. We consider each portion of this schema in turn.

The design and construction of the initial random sequence nucleic acid pool will in large part control what aptamers can eventually be recovered from a selection. The variables that govern design include what type of nucleic acid is used, how long and complex the pool will be, how the pool should be synthesized, and precisely what flanking sequences should be present. Obviously, RNA binding proteins may prefer RNA aptamers, while double-stranded DNA binding proteins may prefer double-stranded DNA aptamers. However, it should be noted that in many cases it will be possible to recover almost any combination of either RNAs or DNAs that can bind to RNA binding proteins, DNA binding proteins, or proteins that are generally not thought to bind nucleic acids at all. In general, however, an aptamer sequence is specific to the type of pool that it was derived from; selected DNA sequences will not bind if represented as RNA and vice versa [9].

Depending on the length of sequence that is considered to be necessary for function, the size of the random sequence core in a nucleic acid library can be correspondingly varied. In those cases where there is a natural nucleic acid ligand for a given protein target, for example, a rough estimate of the likely size of potential aptamer sequences can be garnered. For example, it was known that a short RNA stem-loop could bind to T4 DNA polymerase. Therefore, Tuerk and Gold [10] selected new binding sequences from a library that spanned the 8 loop positions of the wild-type molecule. Similarly, viral regulatory proteins containing arginine-rich motifs (ARMs), such as the Rev protein of HIV-1 or the Rex protein of HTLV-1, bind to RNA molecules that contain short helical elements interspersed with unpaired bases, such as internal loops and bulges, and libraries that can be used to select RNA aptamers that will bind to this class of proteins have been synthesized. One oligonucleotide library, 79.9, that was prepared for these targets contained only a short (12–18 base) random sequence tract, and displayed this tract within the context of a defined secondary structure, a stem-internal loop-stem (Fig. 1a) [11]. Since the range of possible shapes that can be assumed by the 79.9 random sequence region are constrained by the flanking loops and, hence, more restricted than they would be with a longer, unconstrained random tract, it is possible that selections using this library might miss novel binding motifs. However, the 79.9 library does allow motifs and shapes that are similar to known viral RNA binding sites to be quickly, efficiently, and completely scanned by in vitro selection. The shorter random sequence tract also may be able to stretch over the surface of a target, secondary structural analyses and chemical protection studies have revealed that some of the aptamers derived from the 169.1 pool sometimes have an extremely elongated shape; these shapes would not have been found in more limited forays into sequence space. However, it should be noted that this pool contains only around the 10 to the 13th members, a tiny fraction of all possible oligonucleotide sequences of length 120 (4 to the 120th = 10 to the 72nd). However, incomplete coverage of sequence in the longer random sequence pool in no way means that it will be more difficult to search this pool for functional sequences. Again, the primary a priori consideration in this respect is how many bases are likely necessary for binding to a given target. Many selected aptamers have been shown to contain only 15–25 essential nucleotides, and individual subsegments of the 120-mer library will easily span all possible 25-mers (4 to the 25th = 10 to the 15th). In addition, this pool should fold into shape libraries that contain most known natural nucleic acid ligands, up to and including tRNAs and domains of ribosomal RNAs. In fact, Schuster and co-workers [12] have demonstrated using a simple folding algorithm that even relatively limited sets of nucleic acid sequences can completely cover secondary structural space, making it unlikely that any but the rarest shapes will be underrepresented in a random sequence pool. This theoretical prediction has been largely borne out: aptamers selected from random sequence pools represent almost all possible secondary structural classes, including stem loops, internal loops, bulges, and even quadruplexes.

Once a nucleic acid library has been constructed, a selection protocol must be decided on. In general, separating nucleic acid shapes that can interact productively with a given target from the remaining random sequence pool requires immobilization of either the target molecule or the nucleic acid–target complex. There are a variety of techniques that can (and have) been used for immobilization, and the choice of a particular immobilization technique is largely dependent on experimental prejudices as to what the likely affinity of an aptamer for the target will be. For example, if the equilibrium dissociation constant for a complex between a nucleic acid binding protein and its natural ligand is on the order of 1 nM, then it is likely that selected aptamer sequences will also bind with K_d values of 1 nM or less. In this instance, an extremely stringent selection protocol that can eliminate from consideration all aptamers whose K_d values are 10 nM or more should be used. Modified cellulose filters capture proteins but not nucleic acids and can be used to sieve nucleic acids that form stable complexes with a protein target in solution from the remaining mixture of unbound random sequence molecules. After protein–nucleic acid complexes are captured on the filter, weakly bound species can be washed away with from tighter binders that do not dissociate from protein molecules as quickly. Since the volume of wash is generally extremely large compared with the volume of the filter itself, only aptamers with very low dissociation constants are retained following even limited washing, and these species can be subsequently eluted from the filter and amplified. However, one problem with this method is that not all proteins bind efficiently to modified cellulose filters. Therefore, each potential protein target must therefore first be tested to ensure that it can be captured by the filter and in turn capture aptamers.

Another difficulty arises from the preferential enrichment of aptamers that bind solely to the filters. In particular, purine-rich sequences in single stranded regions have been found to bind quite well to cellulose-acetate/-nitrate filters. Therefore, depending on the stringency of the selection, more filter-binding sequences than protein-binding sequences may be isolated in a given round, and over the course of several rounds can quickly take over a selection. Once filter-binding sequences have been preferentially enriched, they are extremely difficult to remove from a selection, since the concentration of the unintended target, the cellulose filter, will always be extremely high relative to any protein target that can be introduced into solution. In order to avoid the accumulation of filter-binding molecular "parasites", random sequence pools should be passed over a modified cellulose filter in the absence of protein at each round. Alternatively, a different selection modality can be used to try to eliminate filter-binding sequences; for example, aptamers can be selected by looking for a decrease in electrophoretic mobility on a native polyacrylamide gel in the presence of a protein (gel-shift or gel-retardation). This method has proven useful for either the de novo selection of aptamers from a random sequence population or for the purification of protein-binding aptamers away from filter-binding aptamers. Despite potential problems, the simplicity of filtration techniques generally still makes them the method of choice for the iterative isolation of aptamers that form complexes whose dissociation constants are nanomolar or lower. Since it is never clear whether or not a protein that does not normally bind a cellular nucleic acid ligand will or will not have a patch of amino acids somewhere on its surface that will prove to be chemically complementary to some shape in a nucleic acid library, filtration selections should almost always be carried out before trying other selection protocols.

In those cases where filtration either does not work or yields filter-binding sequences that cannot be separated form protein-dependent aptamers, affinity chromatography can be used as a selection method. The important parameters for aptamer isolation by affinity chromatography include the concentration of column ligand (which will almost always be in excess over the number of nucleic acid species applied), the number of column volumes of wash, and the choice of procedure for elution (generally either denaturation of the aptamer from the column or affinity elution with an appropriate ligand or set of ligands). By varying these parameters, affinity chromatography can yield nucleic acids binding species that can form complexes with their cognate targets that have K_d values of from 100 μM down to 1 nM. Unfortunately, column-binding nucleic acids can be selected as readily as filter-binding nucleic acids. To avoid accumulating column-binding molecular parasites, nucleic acid pools should always be passed over an unliganded or blocked pre-column prior to being applied to the selection column containing the immobilized target ligand.

Following the actual sieving of functional from non-functional species, amplification can occur either in vitro or in vivo. In vitro amplification encompasses a variety of techniques, including PCR (for either single- or double-stranded DNA pools); reverse transcription, followed by PCR (RT/PCR; for RNA pools); or a retroviral-like replication cycle that relies on the combined activities of reverse transcriptase and RNA polymerase (TAS or 3SR; for RNA or single-stranded DNA pools). In vivo amplifications can be carried out by transforming double-stranded DNA libraries mounted in plasmids or viral vectors into cells. In vivo amplifications are often coupled to in vivo selections for function (for example, identifying a double-stranded DNA sequence that can serve as a promoter for an essential gene).

Since any one round of selection is usually not completely efficient in removing nucleic acids that bind either poorly or not at all to a target ligand, multiple cycles of selection and amplification are frequently required to enrich the selected population in the tightest binding aptamers. Depending on the type and stringency of a selection, anywhere from two to fifteen cycles may be warranted. The total time involved can be anywhere from several weeks to several months. To speed this process, the stringency of a given round can be increased so that fewer nonspecifically binding species are retained. There are several possible ways to increase the intrinsic stringency of a selection: (a) the amount of target used in a given round can be decreased, forcing increased competition between aptamers for available binding sites on the target. This method is most amenable to either filtration or gel-shift selections, but is not really applicable to affinity chromatography protocols, where ligand is typically in excess; (b) the number of wash volumes that are applied prior to the elution of aptamers from either filters or affinity columns can be increased; (c) the buffer conditions used for selection can be altered to favor tighter-binding variants. For example, increasing the ionic strength of a buffer may promote the retention of species that form better hydrogen bonds and salt bridges with a ligand; and, finally, (d) a natural ligand that competes for a known binding site on the target can be introduced into a selection. While this tactic is similar in some ways to (a), reducing the amount of available target, it has the advantage both of allowing easily measurable (as opposed to vanishingly small) quantities of target to continue to be used and of establishing an affinity "benchmark" that other aptamers have to compete with. However, it should be noted that all methods that reduce the amount of ligand-dependent "signal" will simultaneously decrease the ratio of ligand-dependent to nonspecific binding. Thus, while any of these methods can be used to encourage the selection of high-affinity aptamers, they may also promote the selection of filter-binding (or column-binding) contaminants.

Extracellular Protein Targets

Given that the selection of antishape nucleic acids is technically quite feasible, a chief problem that remains with using aptamers as therapeutic reagents is, of course, that nucleic acids are difficult to deliver to the interiors of cells. While transfection of plasmids or viral vectors can be used to deliver genes that encode either nucleic acids or proteins to cells in tissue culture, these methods are not yet clinically practical. Therefore, for the time being it will be necessary to focus on targets that are immediately accessible to nucleic acid pharmaceuticals delivered via the vasculature, lymphatic system, or other readily accessible physiological space. Such targets will by and large be extracellular proteins. Paradoxically, the best pharmacological targets are those which are not known to bind nucleic acids by virtue of their localization. Nonetheless, to the extent that nucleic acid shape libraries in fact contain a range of shapes that can be recognized by more than the basic patches on nucleic acid binding proteins, it should be possible to isolate aptamers that can bind to cell surface proteins.

This conjecture has already been shown to be true. Jellinek et al. [13] have selected RNAs that can bind to basic fibroblast growth factor (bFGF) from a RNA library that spanned 30 random sequence positions. The aptamers isolated in this selection fell into two apparent structural classes: Motif 1 was a short stem-loop structure with a side bulge, while Motif 2 was a stem capped by a large loop. In both cases, both stem and loop structures contained invariant residues, indicating that binding to the growth factor receptor was distributed over the surface of the RNA shape. The best members from each class of motifs bind to the receptor in vitro with affinities of, respectively, around 3 and 50 nM. More importantly, the RNA aptamers compete with the polyanionic compound heparin, a known antagonist of the receptor, and themselves act as antagonists of bFGF binding to both low- and high-affinity cell surface receptors. These results raise the possibility that RNAs might be used directly to block signal transduction pathways with high specificity by interacting with particular cell surface receptors.

Other extracellular targets that RNAs can productively interact with include proteins involved in cell–cell interactions and the immune response, such as antibody molecules. Tsai et al. [14] selected an RNA molecule that could bind to an antiserum raised against a specific 13 amino acid peptide. The selected aptamers were from a N10 library couched in the loop region of a stem-loop structure, and could specifically block the interaction of the peptide with the antibody. This result is particularly interesting, since the original peptide, MASMTGGQQMGRC, is not particularly basic (it contains only one arginine residue). Hence, even a simple RNA sequence and structural motif seems to be capable of mimicking surfaces presented by proteins (or peptides). Moreover, the selected RNA stem loop can be recognized when flanked by different sequences. This stem loop can therefore potentially be used as a nucleic acid epitope "tag", raising the possibility that entirely new techniques for the detection of specific RNA sequences and structures both in vitro and in vivo can be developed.

The Example of Thrombin

The best current example of how aptamers might be used to modify human physiology by interacting with an extracellular protein target comes from researchers at Gilead. Thrombin, one of many proteins involved in the blood-clotting cascade, is a particularly good extracellular target for nucleic acid pharmaceuticals. Aptamers that inhibit thrombin function can potentially prevent reocclusion of blood vessels following coronary artery enzymatic thrombolysis or angioplasty [15]. Guided by these considerations, Bock et al. [16] generated a single-stranded DNA aptamer against thrombin. These authors' specific application of the general techniques we have described was as follows: a DNA oligonucleotide consisting of a core of 60 random sequence positions flanked by two 18 base constant sequence priming sites was chemically synthesized. This DNA pool was amplified in a polymerase chain reaction in which one of the two PCR primers was biotinylated. Following amplification, the duplex DNA library was bound to an avidin column and the non-biotinylated strand was eluted by denaturation with base. For the actual selection, thrombin (a glycoprotein) was immobilized on a concanavalin A (conA) affinity column, the single-stranded DNA library was passed over the conA/thrombin column, and then the column was washed with several column volumes of buffer to remove nonspecifically bound DNAs. Both thrombin and thrombin–nucleic acid complexes were specifically eluted from the column with a competitor sugar, α-methyl mannoside. To guard against selecting DNAs that might interact with the affinity matrix itself, the pool was pre-selected by passing it over an unliganded conA column. After each round of selection, the winnowed pool was reamplified using one biotinylated and one nonbiotinylated primer, and single-stranded DNA was prepared as before for further selection. After five cycles of selection and amplification, the pool was assayed for binding to thrombin, ovalbumin and human fibrinogen using a nitrocellular filter immobilization assay. Of these three potential targets, only thrombin bound to the selected aptamer pool. Individual members of the pool were isolated by cloning, and their relative and absolute binding affinities determined. Each aptamer was found to bind specifically to thrombin with a K_d of around 200 nM. Sequence analysis of 32 clones from the final round of selection showed that while no two aptamers were exactly alike, almost all of the high-affinity binders contained a consensus sequence that was a variably spaced repeat of two similar motifs, GGTTGG ... GGNTGG.

Single-stranded DNA from one thrombin-binding clone, as well as oligonucleotides comprising both the minimal consensus repeat (GGTTGG) and an extended consensus that contained both copies of the repeat (GGTTGGTGTGGTTGG) were tested for their ability to inhibit clotting in a standard assay. All showed an increase in clotting time relative to nonselected same-size DNA controls. Interestingly, the single-repeat hexamer was only effective at concentrations approaching $20\,\mu$M, while the double-repeat pentadecamer and the 96-nt full-size clone from which it was derived were effective at over 200-fold lower concentrations. In fact, the full-sized clone has an IC_{50} of 25 nM, and the consensus 15-mer is at least twice as effective as this clone.

The consensus sequence for the thrombin-binding aptamers has no potential Watson–Crick base pairs, but is reminiscent of telomeres, which also have repetitive poly G/poly T motifs. In a recent group of papers, the research groups of Philip Bolton [17,18] and Juli Feigon [19,20] have examined the structure of the consensus 15-mer using NMR. Both of these groups determined that the aptamer structure is stabilized by potassium, and that increasing the potassium concentration from 5 mM (the conditions under which the aptamer was selected) to 100 mM concurrently increases the melting temperature from 38°C to 51°C. The increased stability in potassium and position 1 of guanosine to carbonyl at position 6, lone pair on nitrogen at position 7 to amino group at position 2. The orientation of guanosine residues alternates between syn and anti when moving between connected residues: G1 is syn, while G2 is anti. The orientation also alternates when going around the guanosines that comprise an individual tetrad: G1 is syn, G6 is anti, G10 is syn, G15 is anti. Thus, each tetrad has an internal dyad symmetry, and this symmetry is flipped over between tetrads. As is the case with double helices, the superficial symmetry of the quadruplex breaks down when examined in more detail. For example, the G2-G5-G11-G14 quartet appears to be much more planar than the G1-G6-G10-G15 quartet. Similarly, a feature of the quadruplex that is not intuitively obvious is that it has two wide and two narrow grooves. The narrow grooves happen to be those that are terminated in TT loops, and the skewing of symmetry in the quadruplex may be due at least in part to the fact that these Ts may be paired with one another and that the 3' member of each doublet (residues T4 and T13) is stacked over a neighboring G quartet.

Recently, Padmanabhan et al. [21] have determined a crystallographic structure for the aptamer–thrombin complex. The basic structure indicated by NMR was verified – two G quartets with one more planar than the other and T4 and T13 folded over the "G box" and potentially base paired – but the strand polarity was reversed. The net effect of this reversal is to switch the location of the wide and narrow grooves identified in the NMR work [20], and to make the G1-G6-G10-G15 quartet the more planar of the two. Also, the TT loops seem to be in slightly different positions in the bound aptamer; pairing between T4 and T13 is shifted from the O4 carbonyl to the O2 carbonyl and T3, T9, and T12 are now flipped out away from the quadruplex box, with T3 and T12 extended towards solvent. The differences between the crystal structure of the complex and the NMR solution structure of the aptamer may reflect the differences in the molecules and methods used for study, or may imply that conformational changes occur on binding.

Prior to delineating the structure of the complex, the portion of thrombin that was bound by the thrombin aptamer had been determined by genetic and biochemical experiments [22]. The aptamer did not inhibit the ability of thrombin to cleave the amide bond of a chromogenic substrate, and, hence, it was unlikely that the aptamer bound close to the active site of the enzyme. The actual site of action was found to be an anion-binding exosite that is known to be an important determinant for recognition of substrates (such as fibrinogen), cofactors, and inhibitors. For example, the aptamer competes for binding with a known substrate for this site, thrombomodulin. Similarly, mutations at this site eliminate the anti-clotting effect of the aptamer. The crystal structure further clarified these findings: each thrombin molecule actually bound two aptamers, one in the anion-binding exosite and one in a separate binding site that can also be occupied by the polyanionic ligand heparin. Exosite interactions are with the portion of the quadruplex spanned by the TGT loop; specifically, G8 and G10 phosphate oxygens form ion pairs with basic amino acids and the thymine moieties T7 and T9 are involved in hydrophobic interactions.

It is unclear what the relative affinities of the two classes of sites are for the thrombin aptamer, although it is likely that inhibition is in fact due to binding in the exosite.

Intracellular Protein Targets

In vitro selection has also proven to be an extremely adept method for the identification of decoys for intracellular nucleic acid binding proteins. Interestingly, however, artificial selection can be used to isolate sequences and structures that often have higher affinities and better specificities for a given protein target than the wild-type sequences that nucleic acid binding proteins have co-evolved with.

Some of the earliest selection experiments attempted to find double-stranded DNA molecules that could specifically interact with DNA binding proteins. For example, Oliphant et al. [23] synthesized a degenerate double-stranded DNA library that contained 23 random sequence positions, and selected molecules from this mix that could bind to the yeast transcription factor GCN4 immobilized on an affinity column. After only one pass through the column, the authors found that most of the retained DNAs contained a common sequence, TGA(C/G)TCA, that matched the known yeast GCN4 consensus binding site. While this one-pass selection was extremely efficient, Oliphant et al. noted that the isolated sequence population could have been amplified and carried through further rounds of selection had this been necessary.

Many other researchers have also employed in vitro selection both to define the sites of other regulatory proteins and to delve into the biochemistry of these proteins. In this respect, in vitro selection methods have proven especially useful for the initial identification of potential pharmacological targets. For example, double-stranded DNA binding sites for the yeast transcription factor RAP1 were recently identified by using a gel-shift assay to isolate binding motifs from a DNA library that contained 13 random sequence positions [24]. Similar techniques have previously been employed to identify the consensus binding sequences for other transcription factors, such as MCM1 [25], SRF [26], E2A homodimers [27], and E2A heterodimers with MyoD [28].

In addition to quickly and efficiently identifying a dominant binding motif for a given protein, in vitro genetic selections can also provide insights into the range of sequences that can be recognized by a given protein, allowing new or nonconsensus binding sites to be ascertained. For example, while Graham and Chambers [24] identified the core of the RAP1 consensus sequence, flanking sequences were not as precisely determined and could admit to several degenerate binding solutions. Since numerous different RAP1 binding sites could exist in the yeast genome, the selection data were used as an informational template for identifying new RAP1 binding sites. A direct comparison of each individual aptamer sequence with the entire EMBL + GenBank sequence data base yielded 160 matches. Since the selection data also gives some indication of where departures from consensus may critically affect binding, many of these putative binding sites were immediately culled from the list, leaving 102 newly identified potential binding sites for RAP1 within yeast promoters. Similarly, the GCN4 selection not only picked out the consensus binding site, but also revealed strong sequence preferences in the regions flanking the consensus site, and suggested that the GCN4 homodimer might bind with a measurable asymmetry. Again, while these nuances might have been picked up by examining many different natural GCN4 binding sites, the diverse contexts in which natural binding sites are found (positioned differently in different genes with different regulatory features) might have significantly confounded the analysis. However, it was relatively simple to identify additional, weak consensus elements in the in vitro selected molecules, since all of the GCN4 binding site variants were presented within the same sequence and structural context and shared the same evolutionary history. These experiments provide an apt practical contrast to some of the arguments that we have advanced for using pools with long random sequence tracts to initiate selections: short pools can restrict travels through sequence space to the point where subtle functional variations and covariations can be appraised, while long pools can be used to more quickly identify the most important aspects of functional sequence and structural motifs. Finally, in addition to defining where a nucleic acid binding protein may interact with a nucleic acid, binding site selections can also aid in determining how a protein binds and, thus, may illuminate the biology of regulatory factors. Blackwell and Weintraub [28] identified consensus binding motifs for several basic helix-loop-helix proteins, including MyoD homodimers, and MyoD heterodimers with E12 and E47. Interestingly, these results, especially when contrasted with results from selections that targeted E12 and E47 homodimers [27], suggest that the binding sites for heterodimers can be viewed as combinations of homodimer "half sites", and that different proteins can be combined with one another to create new binding specificities. Taken together, these results may make it possible to predict the previously unknown binding sites of heterodimeric helix-loop-helix transcription factors.

RNA–protein interactions are also important for gene expression and regulation, and an obvious extension of allowing DNA binding proteins to search through shape libraries of double-stranded DNA shapes for high-affinity sites is to allow RNA binding proteins to search through random sequence RNA shape libraries. A number of potential RNA targets have been identified using in vitro selection, and the selected sequences and shapes have proven valuable in deciphering the web of interactions between cellular RNAs and their cognate recognition proteins (for a review, see Ref. [29]). For example, RNA binding proteins in eukaryotic cells are primarily members of the RNA Recognition Motif (RRM) family. Proteins in this family interact with RNA molecules via conserved arginine and aromatic amino acids presented in the context of a β-sheet structure. It has proven possible to select RNA sequences that can interact with a number of proteins in this family. Keene and co-workers used a N10 stem-loop library (the same library that was used to select the antiantibody aptamer) to search for RNAs that could bind to the U1 snRNP-A protein, a member of the RRM family [30]. The selected sequences generally mirrored the wild-type element, both in terms of its consensus sequence and in terms of the distribution of known phylogenetic variants. In a series of experiments whose results are reminiscent of the GCN4 selection with double-stranded DNA libraries, Levine et al. used in vitro selection to define potential binding sites for Hel-N1, a neuron-specific RNA binding protein, in the 3' untranslated regions of neural RNAs [31]. A variety of specific poly U-containing sequences were extracted from a random sequence pool by HelN1, and these motifs could be matched up with sequences in oncoprotein and lymophokine mRNAs. Finally, Tuerk and Gold [10] used in vitro selection (which they termed SELEX) to identify sequence motifs that could bind to T4 DNA polymerase, a protein that is regulated by a small RNA ligand. These authors constructed an RNA library that displayed 8 random sequence positions in the context of a stable, constant sequence stem-loop, similar to the wild-type ligand, and sieved this library using a filter-binding technique. Unsurprisingly, they recovered the wild-type sequence motif; surprisingly, however, they also recovered a very different sequence motif that bound as well as the wild type.

Proteins select RNAs from random sequence libraries based primarily on affinity for particular sequences, as in the examples cited above, but can also choose those shapes they will best interact with. In an experiment designed to show that in vitro selection can yield answers more quickly and efficiently than laborious site-directed mutagenesis methods, Schneider et al. [32] isolated ligands for the R17 bacteriophage coat protein from an RNA pool that contained a core of 32 completely random sequence positions. The R17 bacteriophage uses the coat protein as a translation repressor for its replicase mRNA, and the region of the R17 genomic RNA that binds to the coat protein has been determined experimentally to be a 7 bp hairpin loop with a single bulged base. Further, through an elegant set of experiments involving many separate oligonucleotide constructions, Uhlenbeck and his collaborators had previously determined that canonical base-pairing in the RNA stem, a preference for purine in the bulge, and an ANYA tetraloop closure were important for coat protein binding [33]. Using a filter-binding selection, Schneider et al. [32] carried out 11 rounds of selection and amplification to enrich the random sequence population in R17 binding species, and then cloned and sequenced ligands. Of 38 isolated aptamers that could bind the coat protein, 36 could be represented as stem-loop structures and essentially confirmed the structural features originally identified by multiple site-directed mutations. In a similar series of experiments designed to elaborate on what sequence sets could form recognizable tRNA structures, Uhlenbeck and his colleagues randomized either 10 or 14 phylogenetically conserved positions in E. coli tRNA(phe) and selected molecules that met two functional criteria: first, molecules were selected on the basis of their ability to bind to the cognate tRNA synthetase, and then those tRNAs that were charged by the synthetase were further selected on the basis of their ability to bind to elongation factor Tu–GTP complexes (which normally carry and load charged tRNAs onto the ribosome). Since the pockets on both the synthetase and the elongation factor are unlikely to accept (and charge) molecules that do not resemble tRNA, this experiment in effect asks how many and what sorts of sequence changes can be accepted in a native tRNA structure. In the sequences that were recovered, many of the positions were the same as in their wild-type counterparts. However, a number of positions also changed to alternate combinations of bases, once again indicating that there may exist multiple sequence solutions that map to a given shape, even when the shape is a naturally evolved biomolecule. In addition, several new combinations of bases involved in tertiary structural interactions were recovered, indicating that a common nucleic acid shape (the tRNA "L") can be buttressed in multiple ways.

Not only are selected RNA sequences sometimes different from those of wild-type binding sites, but selected RNA shapes also do not necessarily resemble the structures of the cognate ligands of RNA binding proteins. Tuerk et al. [32] used a N32 library and filter-binding to isolate RNAs that could bind to HIV reverse transcriptase (RT). After nine rounds of selection and amplification, a population of aptamers that were retained on nitrocellulose filters only in the presence of RT was isolated. Sequence analysis of individual clones from this population revealed a consensus primary and secondary structure for the aptamers: RT-binding RNAs apparently form a pseudoknot containing a defined sequence stem, connecting loops of one and three unpaired bases, and a second paired stem whose precise sequence was relatively unimportant. This pseudoknot structure was somewhat surprising in view of the fact that the normal substrate for HIV (and other) RTs is a paired complex between a tRNA and the viral genome. Although some tRNA mimics found at the end of various viral RNAs have pseudoknot structures, it is likely that the HIV RT aptamer does not recreate known protein–RNA interactions but instead fortuitously fits the active site of the enzyme. Further support for this interpretation comes from the fact that the RT aptamer cannot itself serve as a primer for reverse transcription. In fact, the aptamer stoichiometrically inhibits HIV RT in a standard assay that utilized a DNA primer. The affinity and specificity of the RT aptamer were quite good: the K_d value for some of the best selected variants was ca. 5 nM, and the RNAs were incapable of inhibiting other reverse transcriptases, such as those from Moloney murine leukemia virus and avian myeloblastosis virus.

The Example of Rev

The human immunodeficiency virus regulates its replication through a cascade of complex interactions between viral RNA binding proteins and the RNA genome. In particular, like many other viruses, HIV has genes that are expressed both early and late in its life cycle and the orderly progression between the synthesis of these sets of genes is governed in large part by the regulatory protein Rev. Rev is a short (116 amino acids) highly basic protein that tetramerizes and binds to a specific site on the viral mRNA, the Rev responsive element (RRE). In the early phase of cellular infection, little Rev is present, and the RRE is spliced out of messages to create transcripts that encode early proteins, including Rev itself. As Rev accumulates in cells, it binds to the RRE and facilitates the transport of unspliced RNAs out of the nucleus. The unspliced messages encode structural proteins, such as Gag, and allow new viral particles to be synthesized. There is some indication that Rev may be explicitly involved in the long latent periods between HIV infection and progression to AIDS, and that a critical concentration or threshold of Rev may be necessary for viral reactivation [34,35]. The mechanistic basis for the observed "threshold effect" may be the cooperative assembly of multiple Rev molecules onto the RNA [36].

In vitro selection was originally used to define the sequences and structures that comprised the primary Rev binding in the RRE, the Rev binding element (RBE). Starting from a 66 nucleotide template that included most of the sequences known to be important for Rev responsiveness, Bartel et al. [37] designed and synthesized a doped random sequence pool that had, on average, 65% of a wild-type base, 10% of each non-wild-type base, and 5% deletions. Using a filter-binding selection, elements in the mutagenized population that could interact tightly and specifically with Rev were isolated after only three rounds of selection. Alignment of the sequences of the Rev-binding aptamers revealed a core of nine residues that were always identical to the wild-type element. In addition, there were 11 nearby residues that were conserved in 24 out of 29 clones examined. These functionally important bases could be folded into a stem-internal loop-stem secondary structure. This structure had not previously been proposed for this portion of the RRE; the selection results gave the first accurate picture of what the viral RNA looked like. To test whether the short stretch of conserved residues was in fact all that was necessary for Rev-binding the stem-internal loop stem structure was synthesized in isolation as a 30 base RNA. This RNA was able to bind Rev as well as the full-length wild-type RRE. More importantly, the sequence covariations identified in vitro could promote the Rev recognition in cell culture as well, and could also facilitate the transport of mRNA molecules from the nucleus.

Since the limited selection carried out by Bartel et al. found some RNA molecules that could bind Rev better than the wild-type element, it seemed likely that even better aptamers could be identified by more thorough searches of sequence space. Therefore, we synthesized RNA pools (such as 79.9, Fig. 1a) that completely randomized portions of the stem-internal loop-stem sequence found in the RBE [38]. In vitro selections with these pools were similar to those carried out by Bartel et al., except that we routinely included the wild-type RRE sequence as a competitive benchmark. After only three to four rounds of selection and amplification, our pools bound Rev, on average, 2–3-fold better than the wild-type competitor. Individual aptamers were cloned and sequenced, and their sequences compared. A pool that contained only ten random sequence positions (76.6; corresponding to the internal loop region of the RBE) yielded sequences that were similar to the wild-type motif. The best of these sequences contained a pentuple substitution, including the A:A homopurine pairing originally identified by Bartel, and bound Rev threefold better than the wild-type RRE. On the other hand, the pool (79.9) that contained up to 18 random sequence positions, including both internal loop and flanking sequences, yielded sequences that bore a partial similarity to the wild-type element, but also contained novel subsequences and substructures. The best of these sequences had only four bases that were similar to the original wild-type element, and bound Rev up to ten-fold better than the wild-type RRE (Fig. 3). Similar results were obtained

202

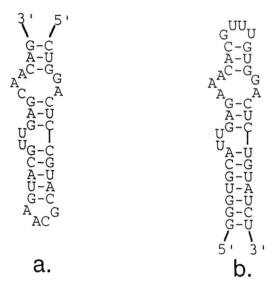

Fig. 3. Rev-binding aptamers selected by the (a) Ellington and (b) Tuerk (Nexagen) groups. Loop ends are switched to align homologous regions.

from an N32 library by a group from the company NeXagen [39,40]. It can be argued that the motif that was independently discovered by these two groups is the best possible Rev-binding sequence of this length. First, the 18-mer library used by Giver et al. [38] completely spanned sequence space; all possible stem-internal loop-stem molecules competed with one another for Rev-binding. Second, further mutagenesis and selection of the Rev-binding motif shown in Fig. 3 by Jensen et al. [41] did not discover any sequence changes that significantly improved binding. However, several bases in the motif were seen to coordinately change during this selection, perhaps indicating that selection can more finely discriminate between molecules with similar binding abilities than can functional assays.

The data derived from in vitro selections have proven useful not only in developing inhibitory aptamers, but have also aided in modelling the three-dimensional structure of the RBE [42]. All modelling experiments were carried out in collaboration with Bob Cedergren and Fabrice Leclerc at the University of Montreal.

Typically, in a natural molecular phylogeny, the identification of simultaneous, concerted sequence substitutions will facilitate the identification of Watson–Crick base pairs and the placement of A-helical stems within an RNA secondary structure. For example, if a G at position 4 in an RNA molecule and a C at position 44 simultaneously change to, respectively, U and A this may indicate that these residues are involved in a Watson–Crick paired stem. However, while sequence analysis of high-affinity RNA ligands for Rev revealed multiple sequence covariations between residues that formally correspond to positions in the wild-type RBE [38], very few of these were simple Watson–Crick substitutions. Instead, the sequence covariations found in the internal loop of the RBE were quite complex: for example, in our selections G48 and G71 frequently changed in concert to A48 and A71, and, less frequently to C48 and A71. When the A:A double substitution is introduced into short RBE their affinity for Rev actually increases. Similarly, multiple, different non-random sequence sets were found to occupy positions 50, 68, and 69. It was possible to use these various covariation results in two different ways to construct a molecular model of the wild-type RBE.

First, it was assumed that the residues involved in each sequence covariation directly interacted with one another. Thus, position 48 was assumed to lie adjacent to position 71, and positions 50, 68, and 69 were assumed to be in contact with one another. At first glance, these restrictions do not appear to delimit the universe of possible structures, since these residues necessarily lie across from one another within the internal loop region of the RBE. However, other viral RNAs involved in gene regulation, such as the TAR element of HIV-1, have been found to assume baroque structures in which bulged regions fold-back to form triple-base paired pockets for arginine recognition [43]. Thus, the finding that residues "paired" at various positions along the length of the internal loop allowed modelling efforts to be focused on structures that were essentially extended nucleic acid helices with unusual pairings and kinks.

Second, the types of sequence covariations that were observed in some cases indicated what types of base pairings might feasibly be introduced into the internal loop of the RBE. For example, while the G:G and A:A pairings between positions 48 and 71 were obviously not Watson–Crick base pairs, Bartel et al. [37] had previously hypothesized that these covariations might represent a particular type of non-Watson–Crick interaction: G:G and A:A interactions can be drawn as isosteric homopurine pairings in which only the hydrogen bond donors and acceptors change. Our results provided further credence for this model; the C:A pairing that was identified in some of the aptamers to Rev could be modeled as an additional base pair that was different from but isosteric with the postulated G:G and A:A pairings. Thus, in this instance, in vitro selection data not only constrained positions 48 and 71 to be near one another, but also limited the ways in which residues at these positions could be juxtaposed in space. These non-Watson–Crick interactions had previously been missed by more conventional mutational analyses [44].

These results and others that were garnered from selection data were used by the program MC-SYM [6,45] to generate initial structural models for the Rev-binding element. As outlined above, distance constraints and base pairing models delimited the range of possible bond angles for individual nucleotide residues in the RBE. The actual bond angles used in the model were in turn selected from a conformational database derived from all known nucleic acid structures. The conformations of some residues, such as those found in a perfectly paired A-helical region, were severely constrained, whereas other residues, such as the bulge at U72, were much less well constrained. Residues were built serially into the initial model; each residue that was added further restricted the conformational choices available for those residues that followed. A traceback routine eliminated those combinations of conformations that were implausible. For example, in constructing the internal loop structure, an inherent constraint was that opposing RNA strands should be able to bridge the distance between flanking helices in a way that was geometrically feasible: sets of conformations that sent the strands in opposite directions in space, for example, were deemed a priori implausible.

Several initial MC-SYM models (or "scripts") were further optimized by energy minimization. To ensure that the resultant structures were not trapped in local conformational minima, the models were also subjected to simulated annealing, a technique that in essence kinetically excites a given modeled structure and frees it to explore the conformers adjacent to it in conformational space. Following energy minimization and simulated annealing most of the initial models converged to a single structure (Fig. 4). The consensus model for the RBE immediately suggests how it might interact with the Rev protein. The RNA recognition region of Rev spans positions 34–50 and is known to be α-helical [46]. A normal A-helical structure cannot readily accommodate the Rev34–50 α-helix, but the internal loop structure of the RBE has a significantly widened major groove. The opening of the groove is due in large part to non-Watson–Crick pairings (such as G47:G71 in the wild-type element) and to a series of unusual interactions, including a network of bifurcated hydrogen

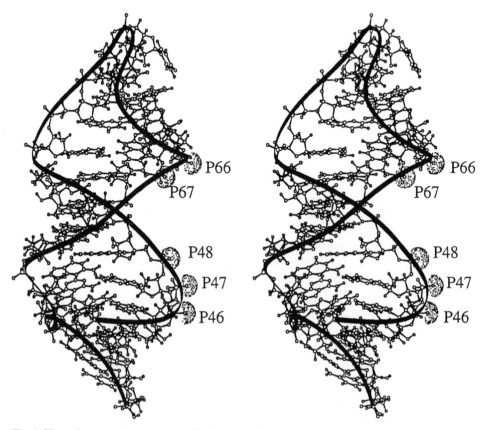

Fig. 4. Three dimensional structural model of the Rev-binding element.

bonds, between residues at positions 50, 68, and 69. Both of these features were originally identified by our in vitro genetic selection, but were subsequently refined and optimized during the course of model building. Of course, this model represents a prediction that must be confirmed by structural analysis. In this respect, it is gratifying to note that Jamie Williamson and his co-workers have recently published an NMR structure for the RBE that agrees in many details with our model [47].

Determining the sequence and structure of the RBE has allowed us to go beyond the physical data, in that it can be easily manipulated to predict how ligands such as peptides and antibiotics can interact with a structured RNA molecule. RNA binding pockets that are recognized by a compound or family of compounds can immediately lead to the design of novel drug leads. Certain aminoglycoside antibiotics, such as neomycin B and tobramycin, have been shown to bind to the RBE of HIV-1, and to disrupt Rev–RBE interactions [48]. As was the case with ribosomal RNA and Group I introns, the binding seems to be highly specific; neomycin B protects particular bases within the RBE from chemical modification, and a battery of other aminoglycosides do not similarly disrupt Rev–RBE interactions. Since the structure of the RBE had already been modeled using distance and conformational constraints supplied by in vitro selection [42], it was possible to dock the antibiotic structure and to identify potential contacts with the RNA. A similar docking experiment had previously been carried out with a model of a short peptide fragment of Rev known to specifically bind the RBE; a comparison of the two models revealed that most of the residues that were hypothesized to be contacted by the peptide were similarly contacted by the antibiotic. Taken together, the experimental and modelling results strongly suggest that the antibiotic is recognizing a particular site on the RBE and is making specific contacts with both bases and phosphates. Based on our model of the interactions between neomycin B and the RBE, we are in the process of designing and synthesizing antibiotic derivatives that are predicted to bind with even greater avidity and specificity.

Therapeutic Considerations

Any aptamer that has therapeutic potential must be realistically assessed regarding whether it is suitable for clinical application. As was the case with the thrombin aptamer, it must be first be tested for its ability to inhibit its target in an in vitro assay at levels that would not be prohibitive in terms of material or expense. To make this task simpler, aptamers are usually pared down to a functional core sequence or structure. For example, the HIV RT and Rev-binding aptamers [11] can be reduced to sequences of only 30 bases or so. Once a minimal, economical target compound has been produced, it must then be tested in serum or, better yet, in an appropriate animal model to get an idea of its physiological half-life. In general, natural nucleic acids will have extremely short half-lives in serum or in vivo. This barrier to their use as pharmaceuticals can be overcome in one of several ways: (a) by focusing on applications that require only a short burst of activity; (b) by altering the composition of the aptamer so that it will have a longer half-life in vivo (primarily for extra-cellular targets); or (c) by developing a delivery system that will stabilize the aptamer (primarily for intracellular targets).

Some therapeutic applications may well be suitable for a short-lived aptamer, such as providing a pulse of clotting inhibition. For example, Li et al. [15] have tested the thrombin aptamer for efficacy against clot formation. These authors first demonstrated the ability of the aptamer to inhibit platelet aggregation in platelet-rich plasma, and then examined whether it could also inhibit thrombin which was already immobilized on clots. This was done by immersing washed clots from recalcified citrated plasma in plasma that had been treated with either nothing (negative control), heparin, a synthetic thrombin inhibitor (PPACK), or the aptamer. The activity of thrombin in plasma was then measured by the release of fibrinopeptide A from the clots. Of the various inhibitors used, the aptamer was found to have much greater efficacy than heparin, although not as much as PPACK at physiologically maintainable concentrations. To assess the potential clinical relevance of these findings, the researchers used ex vivo assays in which a dissected rabbit artery was subjected to balloon angioplasty, and then perfused with whole human blood at known flow rates. Two assays of thrombin activity were employed to determine the efficacy of the inhibitory aptamer: platelet aggregation in the injured region, and fibrinopeptide A levels in the perfused blood. Both assays demonstrated that aptamer treatment was effective, while heparin, even at 10 times the clinically achievable concentration, was not. Encouragingly, the amount of inhibition was about the same as that seen with high-dose PPACK treatments. However, other studies [49] have shown that the aptamer has a half-life of only about 100 s in monkeys, so applications related to wounding or angioplasty would presumably require constant administration.

Fortunately, some clinical applications require only a brief inhibition of thrombin, and the short half-life of the aptamer in serum is actually of unique benefit to these treatments. For example, cardiopulmonary bypass operations require short-term use of anticoagulants. While heparin is often employed, some patients are resistant to it and in others cases it can induce thrombosis. The thrombin aptamer can potentially be used as a heparin substitute in these cases. To determine the pharmacokinetic properties of this oligo, DeAnda et al. [50] carried out a cardiopulmonary bypass study using dogs. At a 0.5 mg/kg per min dose the activated clotting time rose from 106 s to >1500 s. The half-life of the aptamer during the operation rose from 1.9 min to 7.7 min, perhaps suggesting that the heart and/or lungs play a role in clearance of the nucleic acid pharmaceutical. The coagulation profile of the infused animals returned to normal within 5 min of stopping aptamer infusion, and postoperative bleeding was minimal. Thus, the selected oligo appears to be ideally suited for some types of bypass operations.

However, most pharmaceutical applications are not quite as delimited either in terms of time or material and will require repeated doses of a compound; the number of doses will in turn be related to the half-life of the drug in the body. Thus, stabilization of aptamer structures is essential for their efficient and economical use as pharmaceuticals against extracellular targets. There are two chief routes to stabilization: the inclusion of modified residues or structures within nucleic acids during in vitro selection, or the chemical or structural modification of an aptamer following in vitro selection.

An astounding variety of modified nucleotides have been synthesized, primarily as potential anti-viral compounds. However, it is not known, by and large, which of these might be effectively used by polymerases such as Taq DNA polymerase, T7 RNA polymerase, or reverse transcriptase. Therefore, we first focus on those analogues that are known to be accepted by polymerases, and that could potentially be included in selections. For example, Aurup et al. [51] have produced full-length tRNA transcripts in which the 2' hydroxyl groups of pyrimidine ribonucleotides were replaced with either fluoro or amino groups. While the incorporation of these analogues by T7 RNA polymerase was from 2- to 20-fold slower than was the case for natural nucleotides, they were nonetheless able to get reasonable yields of product using a synthetic DNA template. Moreover, these 2' substituted analogues have been shown to drastically increase the stability of nucleic acids that include them; for example, Pieken et al. [52] synthesized hammerhead ribozymes in which all of the pyrimidines were completely replaced with either 2' amino or 2' fluoro pyrimidines. Although the activity of the resultant ribozymes is reduced, the stability of these RNAs in rabbit serum is increased over 1,200-fold. Similarly, Heidenreich and Eckstein [53] made hammerheads that contained various combinations of 2' fluoro pyrimidines and tested their resistance to digestion in cell culture supernatants. While transcripts that contained only 2' fluoro CTP were digested as fast as unmodified RNAs, when both 2' fluoro CTP (fC) and 2' fluoro UTP (fU) were included the RNAs were "stabilized considerably". The stabilities of completely substituted synthetic RNAs were eventually quantitated by Heidenrich et al. [54], who found inclusion of fC + fU increased the half-life of a 36-mer in cell culture supernatants from less than 2 min to greater than 17 h, and in fetal calf serum from less than 2 min to about 1 h.

The 2' positions of ribonucleotides were originally chosen for modification because of the role they play in both the enzymatic and nonenzymatic hydrolysis of RNAs. Chemical modifications of the phosphate backbone might similarly be predicted to increase polymer stability. For example, phosphorothioate substitutions have been shown to increase oligonucleotide stability while still allowing enzymatic amplification. Ueda et al. [55] incorporated combinations of Sp diastereomers of the four (A,C,G,U) 5'-O-(1-thiophosphate) nucleotides into mRNAs using T7 RNA polymerase. If only one of the four bases was included in the transcription, yields were similar to those obtained using all four natural nucleotides, while if the four phosphorothioates were simultaneously introduced, yields dropped by a factor of three. The stability of transcripts in which only one of the phosphorothioate nucleotides was included was tested in bacterial extracts: the A phosphorothioate substitution gave approximately two-fold protection in E. coli extracts at 37°C, while the G phosphorothioate substitution gave 25-fold protection in T. thermophilus extracts at 65°C.

Since natural nucleic acids display varying susceptibilities to nuclease digestion, specific secondary and tertiary structures can also potentially be exploited in trying to engineering nucleic acid pharmaceuticals. For example, Tang et al. [56] have made deoxyoligonucleotides with and without 3′ hairpin structures of various lengths. A 10 base pair hairpin was found to increase the half-life of an oligo to snake venom phosphodiesterase from 90 s to over 1,000 s, and to slow DNA polymerase I exonuclease activity by a factor of at least 40. Since one of the primary hydrolytic components of serum are exonucleases, circularization of RNAs or DNAs should also increase their stability. Been and his coworkers have found that a circular version of the HDV ribozyme has a half-life of over 14 h in either cytoplasmic or nuclear HeLa cell extracts, but lasts only 5–13 min in these extracts as a linear molecule.

Results with functional nucleic aids such as the hammerhead ribozyme suggest that the incorporation of nucleotides with modified chemical structures frequently results in an overall alteration in nucleic acid structure; similar results have also been seen for simple ribose and deoxyribose sugar substitutions [9]. Therefore, if stabilizing residues are to be included in an aptamer, they must be introduced prior to the selection, since following selection they may not be readily accommodated. The same is not necessarily true for structural modifications, such as flanking hairpins or circularization, but it is still probably best to include these during the selection itself so that the resultant aptamers will not misfold upon their later addition.

In contrast, there are also chemical modifications that can be introduced into aptamers following selection. In general, these modifications do not interfere with the selected structure because they are included only at the peripheries of an RNA or DNA molecule, and are designed primarily to reduce the processive cleavage of an aptamer by exonucleases. As an example, Heidenreich and Eckstein also demonstrated that the addition of terminal phosphorothioate residues to hammerhead ribozymes could further improve the stability of fC and fU substituted oligos. Similarly, Tang et al. [56] terminated deoxyoligonucleotides with phosphorothioate-substituted as well as natural hairpins, and found that 6, 8, or 10 base pair hairpins showed little degradation in fetal calf serum even after 16 h at 37°C. Further, the phosphorothioate substituted hairpin oligos were injected into mice and could still be detected 24 h later. Other stabilizing substitutions that should in theory severely limit cleavage by exonucleases are also possible. Sproat et al. [57] have shown that 2′-O-methyl-substituted oligos were almost completely resistant to all nucleases tested, whether RNA- or DNA- or nonspecific, except for P1 nuclease, snake venom phosphodiesterase and staphylococcal nuclease. Of these, only phosphodiesterase showed no inhibition. Similarly, Iribarren et al. [58] found that alkyl substitutions on the 2′ hydroxyl eliminated RNase H digestion of RNAs in annealed duplexes.

In considering hypothetical mechanisms for the intracellular delivery of aptamers, we will ground theory in practice by once again using Rev as an illustration. While the Rev decoys we have isolated bind Rev with high affinity in vitro, and may compete effectively with the RRE in vivo, they can only be used to treat AIDS if they can reach Rev molecules in infected cells. Therefore, it will be necessary to harness technologies for delivering Rev decoys to cells. Ideally, the delivery systems for Rev decoys should be tailored to the biology of the virus itself. HIV-1 can attack T cells and macrophages throughout the body; thus, a delivery system for Rev decoys should be able to pass freely throughout the vasculature. HIV-1 can integrate into the cellular genome, and may be difficult to entirely eliminate from an infected individual; thus, the delivery system should target both infected and uninfected cells. HIV-1 infection is persistent, and infected individuals can develop AIDS over the course of decades; thus, a delivery system for Rev decoys should be capable of being administered over a long period of time without causing side reactions or the development of viral resistance. A lemma for this consideration is that the delivery system should be reasonably economical. Finally, HIV-1 severely debilitates host defenses; thus, a delivery system should avoid challenging an already impaired immune system. While replicative gene or viral therapies may offer the best long-term option for treatment, transformed cells or live viruses may be carcinogenic and have the potential to themselves become life-threatening in immunocompromised patients. Taking these considerations into account, a nonreplicative biological agent, such as viral capsids and liposomes, should be used to shuttle Rev decoys directly into cells. Nevertheless, while this philosophy envisions the delivery of discrete "doses" of NAPs, all of the decoys and targeting systems that are described are also amenable to replicative (gene and viral) therapies. For a more complete review of these, see Ref. [59].

Viral capsid delivery is especially appealling because it exploits naturally efficient mechanisms for crossing cell membranes. Adenovirus has been widely touted as a vector for delivering nucleic acids to human cells [60]. Adenovirus is both catholic and efficient in its infection of cells, and its effects on the host organism are relatively benign. Moreover, inactivated or defective adenovirus particles [61] can be used for infection, effectively circumventing problems associated with infectious pathogenicity and integrative transformation by viral genomes. Curiel and his colleagues have proposed methods for gene delivery in which adenoviral capsid exteriors are chemically derivatized with nucleic acids [62,63]. This method has proven extremely useful in transforming cells (efficiencies approach 100% at a ratio of 10^4 virus/cell).

In order to simplify the linkage between viral carriers and "piggybacked" nucleic acids, it may be possible to isolate aptamers that can specifically recognize viral coat proteins. Although adenovirus is not normally known to bind nucleic acids on its exterior, the large protein surface should prove to be an excellent target for selection. As we have seen, other proteins not normally known to bind nucleic acids have elicited aptamers [16,64]. In addition, the fractionation of nucleic acid pools on affinity chromatography matrices frequently elicits aptamers that bind to the carbohydrate or acrylamide column matrices, probably because the large matrix surface provides many more opportunities for bonding than do the smaller ligands attached to the column. The extensive carbohydrate (or, in the case of virus, protein) surfaces can act as large molecular "hosts" to the smaller nucleic acid "guests." Finally, the capsid proteins of viruses such as picornavirus often come together to form structural indentations, a "drug binding pore," that can bind compounds that are chemically similar to nucleotides. If viral aptamers can be isolated, they could be covalently joined to Rev decoys, and used to noncovalently derivatize virus particles. Upon infection, the virus particles should carry Rev decoys into cells. This method should be simpler than the chemical derivatization of antibodies described by Curiel et al. [62], and can potentially allow much more of the viral surface to be used for delivering nucleic acids. For example, aptamer species that recognize small epitopes repeated many times on the viral surface may allow multiple decoys to be delivered by a single virus.

RNA molecules can also be shielded from degradative enzymes in serum by encapsulation in liposomes; on fusion with cell membranes, the RNAs will be released into the cytoplasm. RNAs can be easily encapsulated using procedures such as reverse evaporation and extrusion [65] or sonication [66], and standard commercial preparations are even available for this purpose (Lipofectin, Gibco BRL, Bethesda, Maryland). To the extent that a given mixture of phospholipids, cholesterol, and/or their derivatives is stable, liposomes may be able to circulate for long periods of time, and could thus provide the best option for delivering Rev decoys by perfusion.

Since liposomes are removed from circulation primarily by endocytosis, Rev decoys may be specifically targeted to cells (macrophages) that are known to harbor HIV-1 in infected individuals. Once internalized, liposomes and their contents are localized to acidic endosomes such as the lysosome [67], and may therefore partition inefficiently into other cellular compartments. In order to enhance the cytoplasmic release of nucleic acids sequestered in acidic endosomal compartments, it may also be possible to encapsulate Rev decoys in liposomes that can undergo membrane fusion upon transfer to a low pH environment. Huang and his coworkers have shown that cationic liposomes composed of dioleoylphosphatidylethanolamine (DOPE), cholesterol, and oleic acid can fuse and release their contents following endocytosis and acidification [68,69].

As a final consideration, the potential costs of treatments with nucleic acid pharmaceuticals must be taken into account. For example, when the ex vivo blood studies were performed with the thrombin aptamer, its concentration was kept at $3 \mu M$. Using a value of 5 l for the average amount of blood in a human, a one-time administration would require approximately 75 mg of DNA, at a chemical synthesis cost (not including quality control) of around $400. Both the amount of material and the potential cost are quite reasonable when compared with many other one-time pharmaceutical interventions.

Similarly, the fact that Rev sits at a critical juncture in the HIV life cycle will continue to drive efforts to commercialize these decoys. Moreover, if a critical concentration of active Rev is in fact necessary to shift from the early to late phase of the viral replication, then a modest inhibition of Rev function may be sufficient to interdict the viral life cycle and Rev decoys should be most effective at prolonging latency, and possibly reversing moderate forms of the disease state (<500 CD4+ T-cells/ml). An interpretation of the data of Pomerantz et al. [35] suggests that roughly 3,000 Rev molecules per cell is critical for viral reactivation and that a reduction of this number by only 25% may prolong latency. Thus, as few as 750 molecules of Rev per infected cell may need to be diverted to reverse disease progression. This number may be proportionately smaller if the RRE must in fact interact with multiple binding sites on the same Rev tetramer. Based on the rough estimate that 750 decoys per infected cell may be required for treatment, and on the fact that there may be ca. 1,011 leukocyte targets in the bloodstream, we can estimate that a whole body "dose" of decoys, if appropriately delivered, would require approximately 10 μg of nucleic acid at a rough cost of $0.50 per "dose" (= $50,000/g of synthetic RNA). While we obviously do not currently possess the necessary information to make a more accurate estimate, or to determine how many doses per year may be necessary, we present this figure as a benchmark to suggest that Rev decoys may be very practical anti-AIDS drugs. A unique advantage of Rev decoys compared to other forms of treatment is that, since a population of variants is selected, it will be difficult for the virus to mutate to simultaneously evade multiple different sequence combinations. In addition, the selected sequences already represent the most likely evolutionary response of the virus, mutation of the RRE.

Conclusion

In vitro selection can be used to isolate RNA and DNA molecules that can bind to proteins with extremely high affinities and specificities. Selected aptamers can successfully compete with wild-type nucleic acid ligands for binding, can inhibit the function of enzymes, and can even affect physiological processes, such as blood clotting. While nucleic acids are much larger and more complex than conventional pharmaceuticals, they may yet prove to be useful drugs. The specificity of their interactions with proteins should mean that they will have fewer metabolic side effects than more catholic compounds, such as nucleoside analogues. Aptamers composed of natural nucleotides can be readily introduced into cells by gene or viral therapies, while aptamers composed of or augmented by unnatural bases can be selected and used directly for treatment. Overall, however, these in vitro selection methods will continue to be of intense interest primarily because of the ease with which promising new drug leads can be generated.

Acknowledgement

Andrew Ellington is a Scholar of the American Foundation for AIDS Research.

References

1. Sullenger BA, Gallardo HF, Ungers GE, Gilboa E. Overexpression of TAR sequences renders cells resistant to human immunodeficiency virus replication. Cell 1990;63:601–608.
2. Hanes SD, Brent R. A genetic model for interaction of the homeodomain recognition helix with DNA. Science 1991;251:426–430.
3. Calnan BJ, Biancalana S, Hudson D, Frankel AD. Analysis of arginine-rich peptides from the HIV TAT protein reveals unusual features of RNA protein recognition. Genes Dev 1991;5:201–210.
4. Khamis MI, Casas-Finet JR, Maki AH, Murphy JB, Chase JW. Investigation of the role of individ-

212

ual tryptophan residues in the binding of *Escherichia coli* single-stranded DNA binding protein to single-stranded polynucleotides. A study by optical detection of magnetic resonance and site-selected mutagenesis. J Biol Chem 1987;262:10938–10945.

5. Shamoo Y, Sturtevant JM, Williams KR, Ghosaini LR, Konigsberg WH, Keating KM. Site-specific mutagenesis of T4 gene-32: the role of tyrosine residues in protein-nucleic acid interactions. Biochemistry 1989;28:7409–7417.

6. Major F, Tucotte D, Gautheret D, LaPalme G, Fillion E, Cedergren R. The combination of symbolic and numerical computation for three-dimensional modeling of RNA. Science 1991;253:1255–1260.

7. Jenison RD, Gill SC, Pardy A, Polisky B. High-resolution molecular discrimination by RNA. Science 1994;263:1425–1429.

8. Green R, Ellington AD, Bartel DP, Szostak JW. In vitro genetic analysis: selection and amplification of rare functional nucleic acids. Methods 1991;2:75–86.

9. Ellington AD, Szostak JW. Selection in vitro of single-stranded DNA molecules that fold into specific ligand-binding structures. Nature 1992;355:850–852.

10. Tuerk C, Gold L. Systematic evolution of ligands by exponential enrichment: RNA ligands to bacteriophage T4 DNA polymerase. Science 1990;249:505–510.

11. Giver L, Bartel DP, Zapp ML, Green MR, Ellington AD. Selection and design of high-affinity RNA ligands for HIV-1 Rev. Gene 1993;137:19–24.

12. Schuster P, Fontana W, Stadler PF, Hofacker IL. From sequences to shapes and back: a case study in RNA secondary structures. Proc R Soc London Ser B 1994;255:279–284.

13. Jellinek D, Lynott CK, Rifkin DB, Janjic N. High-affinity RNA ligands to basic fibroblast growth factor inhibit receptor binding. Proc Natl Acad Sci USA 1993;90:11227–11231.

14. Tsai DE, Kenan DJ, Keene JD. In vitro selection of an RNA epitope immunologically cross-reactive with a peptide. Proc Natl Acad Sci USA 1992;89:8864–8868.

15. Li W-X, Kaplan AV, Grant GW, Toole JJ, Leung LLK. A novel nucleotide-based thrombin inhibitor inhibits clot-bound thrombin and reduces arterial platelet thrombus formation. Blood 1994;83:677–682.

16. Bock LC, Griffin LC, Latham JA, Vermaas EH, Toole JJ. Selection of single-stranded DNA molecules that bind and inhibit human thrombin. Nature 1992;355:564–566.

17. Wang KY, Krawczyk SH, Bischofberger N, Swaminathan S, Bolton PH. The tertiary structure of a DNA aptamer which binds to and inhibits thrombin determines activity. Biochemistry 1993;32:11285–11292.

18. Wang KY, McCurdy S, Shea RG, Swaminathan S, Bolton PH. A DNA aptamer which binds to and inhibits thrombin exhibits a new structural motif for DNA. Biochemistry 1993;32:1899–1904.

19. Macaya RF, Schultze P, Smith FW, Roe JA, Feigon J. Thrombin-binding DNA aptamer forms a unimolecular quadruplex structure in solution. Proc Natl Acad Sci USA 1993;90:3745.

20. Schultze P, Macaya RF, Feigon J. Three-dimensional solution structure of the thrombin-binding DNA aptamer d(GGTTGGTGTGGTTGG). J Mol Biol 1994;235:1532–1547.

21. Padmanabhan K, Padmanabhan KP, Ferrara JD, Sadler JE, Tulinsky A. The structure of α-thrombin inhibited by a 15-mer single-stranded DNA aptamer. J Biol Chem 1993;268:17651–17654.

22. Wu Q, Tsiang M, Sadler JE. Localization of the single-stranded DNA binding site in the thrombin anion-binding exosite. J Biol Chem 1992;267:24408–24412.

23. Oliphant AR, Brandl CJ, Struhl K. Defining the sequence specificity of DNA-binding proteins by selecting binding sites from random-sequence oligonucleotides: analysis of yeast GCN4 protein. Mol Cell Biol 1989;9:2944–2949.

24. Graham IR, Chambers A. Use of a selection technique to identify the diversity of binding sites for the yeast RAP1 transcription factor. Nucleic Acids Res 1994;22:124–130.

25. Wynne J, Treisman R. SRF and MCM1 have related but distinct DNA binding specificities. Nucleic Acids Res 1992;20:3297–3303.

26. Pollock R, Treisman R. A sensitive method for the determination of protein-DNA binding specificities. Nucleic Acids Res 1990;18:6197–6204i.

27. Sun X-H, Baltimore D. An inhibitory domain of E12 transcription factor prevents DNA binding in E12 homodimers but not in E12 heterodimers. Cell 1991;64:459–470.

28. Blackwell TK, Weintraub H. Differences and similarities in DNA-binding preferences of MyoD and E2A protein complexes revealed by binding site selection. Science 1990;250:1104–1110.

29. Kenan DJ, Query CC, Keene JD. RNA recognition: towards identifying determinants of specificity. Trends Biochem Sci 1991;16:214–220.

30. Tsai DE, Harper DS, Keene JD. U1-snRNP-A protein selects a ten nucleotide consensus sequence from a degenerate RNA pool presented in various structural contexts. Nucleic Acids Res 1991;19:4931–4936.

31. Levine TD, Gao F, King PH, Andrews LG, Keene JD. Hel-N1: an autoimmune RNA-binding protein with specificity for 3' uridylate-rich untranslated regions of growth factor mRNAs. Mol Cell Biol 1993;13:3494–3504.

32. Schneider D, Tuerk C, Gold L. Selection of high affinity RNA ligands to the bacteriophage R17 coat protein. J Mol Biol 1992;228:862–869.

33. Witherell GW, Gött JM, Uhlenbeck OC. Specific interaction between RNA phage coat proteins and RNA. Prog Nucleic Acids Res Mol Biol 1991;40:185–220.

34. Pomerantz RJ, Trono D, Feinberg MB, Baltimore D. Cells nonproductively infected with HIV-1 exhibit an aberrant pattern of viral RNA expression: a molecular model for latency. Cell 1990;62:1271–1276.

35. Pomerantz RJ, Seshamma T, Trono D. Efficient replication of human immunodeficiency virus type 1 requires a threshold level of Rev: Potential implications for latency. J Virol 1992;66:1809–1813.

36. Malim MH, Cullen BR. HIV-1 structural gene expression requires the binding of multiple Rev monomers to the viral RRE: implications for HIV-1 latency. Cell 1991;65:241–248.

37. Bartel DP, Zapp ML, Green MR, Szostak JW. HIV-1 rev regulation involves recognition of non-Watson-Crick base–pairs in viral RNA. Cell 1991;67:529–536.

38. Giver L, Bartel D, Zapp M, Pawul A, Green M, Ellington AD. Selective optimization of the Rev-binding element of HIV-1. Nucleic Acids Res 1993;21:5509–5516.

39. Tuerk C, MacDougal S, Hertz G, Gold L. In vitro evolution of high-affinity RNA ligands to the HIV-1 rev protein. In: Ferre R, Mullios K, Gibbs R, Ross A (eds) The Polymerase Chain Reaction. New York: Birkhauser, Springer-Verlag, 1993 (in press).

40. Tuerk C, MacDougal-Waugh S. In vitro evolution of functional nucleic acids: high-affinity RNA ligands of HIV-1 proteins. Gene 1993;137:33–39.

41. Jensen KB, Green L, MacDougal-Waugh S, Tuerk C. Characterization of an in vitro-selected RNA ligand to the HIV-1 Rev protein. J Mol Biol 1994;235:237–247.

42. Leclerc F, Cedergren R, Ellington AD. A three-dimensional model of the rev-binding element of HIV-1 derived from analyses of aptamers. Nature Struct Biol 1994 (in press).

43. Puglisi JD, Tan R, Calnan BJ, Frankel AD, Williamson JR. Conformation of the TAR RNA-Arginine complex by NMR spectroscopy. Science 1992;257:76–80.

44. Holland SM, Chavez M, Gerstberger S, Venkatesan S. A specific sequence with a bulged guanosine residue(s) in a stem-bulge-stem structure of Rev-responsive element RNA is required for trans activation by human immunodeficiency virus type 1 Rev. J Virol 1992;66:3699–3706.

45. Gautheret D, Major F, Cedergren R. Modeling the three-dimensional structure of RNA using discrete nucleotide conformational sets. J Mol Biol 1993;229:1049–1064.

46. Tan R, Chen L, Buettner JA, Hudson D, Frankel AD. RNA recognition by an isolated a helix. Cell 1993;73:1031–1040.

47. Battiste JL, Tan R, Frankel AD, Williamson JR. Binding of an HIV Rev peptide to Rev responsive element RNA induces formation of purine-purine base pairs. Biochemistry 1994;33:2741–2747.

48. Zapp ML, Stern S, Green MR. Small molecules that selectively block RNA binding of HIV-1 Rev protein inhibit Rev function and viral production. Cell 1993;74:969–978.

214

49. Griffin LC, Tidmarsh GF, Bock LC, Toole JJ, Leung LLK. In vivo anticoagulant properties of a novel nucleotide-based thrombin inhibitor and demonstration of regional anticoagulation in extracorporeal circuits. Blood 1993;81:3271.

50. DeAnda Jr A, Coutre SE, Moon MR et al. Pilot study of the efficacy of a thrombin inhibitor for anticoagulation during cardiopulmonary bypass. Thorac Surg 1994 (in press).

51. Aurup H, Williams DM, Eckstein F. 2'-fluoro- and 2'-amino-2'-deoxynucleoside 5'-triphosphates as substrates for T7 RNA polymerase. Biochemistry 1992;31:9636–9641.

52. Pieken WA, Olsen DB, Benseler F, Aurup H, Eckstein F. Kinetic characterization of ribonuclease-resistant 2'-modified hammerhead ribozymes. Science 1991;253:314–317.

53. Heidenreich O, Eckstein F. Hammerhead ribozyme-mediated cleavage of the long terminal repeat RNA of human Immunodeficiency virus type 1. J Biol Chem 1992;267:1904–1909.

54. Heidenreich O, Benseler F, Fahrenholz A, Eckstein F. High activity and stability of hammerhead ribozymes containing 2'-modified pyrimidine nucleosides and phosphorothioates. J Biol Chem 1994;269:2131–2138.

55. Ueda T, Tohda H, Chikazumi N, Eckstein F, Watanabe K. Phosphorothioate-containing RNAs show mRNA activity in the prokaryotic translation systems in vitro. Nucleic Acids Res 1991;19:547–552.

56. Tang JY, Temsamani J, Agrawal S. Self-stabilized antisense oligodeoxynucleotide phosphorothioates: properties and anti-HIV activity. Nucleic Acids Res 1993;21:2729–2735.

57. Sproat BS, Lamond AI, Berjer B, Neuner P, Ryder U. Highly efficient chemical synthesis of 2'-O-methyloligoribonucleotides and tetrabiotinylated derivatives: novel probes that are resistant to degradation by RNA or DNA specific nucleases. Nucleic Acids Res 1989;17:3373–3386.

58. Iribarren AM, Sproat BS, Neuner P, Sulston I, Ryder U, Lamond AI. 2'-O-Alkyl oligoribonucleotides as antisense probes. Proc Natl Acad Sci USA 1990;87:7747–7751.

59. Morgan RA, Anderson WF. Human gene therapy. Annu Rev Biochem 1993;62:191–217.

60. Berkner KL. Development of adenovirus vectors for the expression of heterologous genes. Biotechniques 1988;6:616–629.

61. Cotten M, Wagner E, Zatloukal K, Phillips S, Curiel DT, Birnstiel ML. High-efficiency receptor-mediated delivery of small and large (48 kilobase) gene constructs using the endosome-disruption activity of defective or chemically inactivated adenovirus particles. Proc Natl Acad Sci USA 1992;89:6094–6098.

62. Curiel DT, Agarwal S, Wagner E, Cotten M. Adenovirus enhancement of transferrin polylysine-mediated gene delivery. Proc Natl Acad Sci USA 1991;88:8850–8854.

63. Wagner E, Zatloukal K, Cotten M et al. Coupling of adenovirus to transferrin polylysine DNA complexes greatly enhances receptor-mediated gene delivery and expression of transfected genes. Proc Natl Acad Sci USA 1992;89:6099–6103.

64. Tsai DE, Keene JD. In vitro selection of RNA epitopes using autoimmune patient serum. J Immunol 1993;150:1137–1145.

65. Szoka Jr F, Papahadjopoulos D. Procedure for preparation of liposomes with large internal aqueous space and high capture by reverse-phase evaporation. Proc Natl Acad Sci USA 1978;75:4194–4198.

66. Conner J, Yatvin MB, Huang L. pH-sensitive liposomes: acid-induced liposome fusion. Proc Natl Acad Sci USA 1984;81:1715–1718.

67. Straubinger RM, Hong K, Friend DS, Papahadjopoulos D. Endocytosis of liposomes and intracellular fate of encapsulated molecules: encounter with a low pH compartment after internalization in coated vesicles. Cell 1983;32:1069–1079.

68. Wang CY, Huang L. Highly efficient DNA delivery mediated by pH-sensitive immunoliposomes. Biochemistry 1989;28:9508–9514.

69. Maruyama K, Kennel SJ, Huang L. Lipid-composition is important for highly efficient target binding and retention of immunoliposomes. Proc Natl Acad Sci USA 1990;87:5744–5748.

Biotechnology Annual Review Volume 1
M.R. El-Gewely, editor

Towards a new concept of gene inactivation: specific RNA cleavage by endogenous ribonuclease P

Roland K. Hartmann[1], Guido Krupp[2] and Wolf-Dietrich Hardt[1]

[1]*Institut für Biochemie, Abteilung Prof. V.A. Erdmann, Freie Universität Berlin, Berlin; and* [2]*Institut für Allgemeine Mikrobiologie, Christian-Albrechts-Universität, Kiel, Germany*

Abstract. In the first part of this chapter, general concepts for gene inactivation, antisense techniques and catalytic RNAs (ribozymes) are presented. The requirements for modified oligonucleotides are discussed with their effects on the stability of base-paired hybrids and on resistance against nuclease attack. This also includes the problems in the choice of an optimal target sequence within the inactivated RNA and the options of cellular delivery systems. The second part describes the recently introduced antisense concept based on the ubiquitous cellular enzyme ribonuclease P. This system is unique, since the substrate recognition requires the proper tertiary structure of the cleaved RNA. General properties and possible advantages of this approach are discussed.

Introduction

Techniques which specifically inhibit expression of single genes are of fundamental importance for the study of gene function and open up perspectives for therapeutic applications. Potential target types include well characterized cellular genes with undesired phenotypes or foreign genes introduced by pathogens such as viruses. The various genome projects have led to an accumulation of "unidentified open reading frames", and gene inactivation is likely to be a convenient tool to assess their in vivo function.

Antisense technologies have attracted much interest because they provide a way to bind a cellular target with high specificity. In contrast to classical pharmaceuticals, where large numbers of substances have to be screened, antisense oligonucleotides can, at least in theory, be designed simply on the basis of Watson–Crick base pairing rules.

Initial experiments employing simple complementary nucleic acids [1,2] have been extended to chemically modified and catalytic oligonucleotides, as well as triple helices. Recently, the applications of nucleic acids to control gene expression have also been expanded to "sense" RNAs and DNAs which display a high specific affinity for essential proteins and inhibit gene expression by competing for these protein factors [3,4].

Address for correspondence: R.K. Hartmann, Institut für Biochemie, Medizinische Universität zu Lübeck, Ratzeburger Allee 160, 2538 Lübeck, Germany. Tel.: +49 451 500 4065; Fax: +49 451 500 4068.

For studies of gene function, the antisense concept is only rivalled by the gene disruption approach. This technique, which involves homologous recombination to destroy the gene under investigation, has only been established for a few organisms including yeast [5], *Dictyostelium* [6], plants [7,8], mice [9] and human cells [10]. Although powerful, the method is cumbersome and also unsuitable for therapeutic applications.

Besides a detailed analysis of the cellular mechanisms that underly the effects obtained with antisense-, sense- and triple helix-forming oligonucleotides, continuing efforts have been applied to the search for additional "tools" in order to improve delivery, specificity and efficiency.

A brief introduction to antisense technology and catalytic RNAs (ribozymes) is given in the first sections. In the second part, we discuss another approach for the sequence-specific inactivation of RNA which is based on a universal cellular enzyme, the structure-specific ribonuclease P (RNase P).

Targets for Antisense Strategies

Gene expression can be subdivided into several steps leading from genomic DNA to the functional protein (Fig. 1), which depend on specific interactions of nucleic acids with other nucleic acids and proteins. The main targets of antisense strategies are cellular or viral mRNAs. Antisense nucleic acids can interfere with splicing, nuclear export, mRNA stability and accessibility for components of the translational machinery.

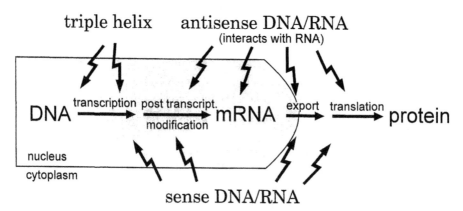

Fig. 1. Sites of interference of gene expression inside a eukaryotic cell. Transcription, processing and modification occurs in the nuclear compartment (shaded area). Subsequently, mRNAs are exported into the cytoplasm, where translation takes place. Triple helix forming oligonucleotides are thought to interfere with gene expression mainly by binding to the DNA, while sense nucleic acids act by competing for essential cellular factors. Antisense nucleic acids are designed to interfere with splicing, export into the cytoplasm, translation or RNA stability by virtue of their capability to hybridize with the target (pre-) mRNA.

Natural antisense transcripts are well known to regulate gene expression and plasmid replication in bacteria (reviewed in Ref. [11]). "Antisense" transcripts have also been found in eukaryotes, e.g. mouse RNAs complementary to herpes simplex virus transcripts [12], the myelin basic protein mRNA [13,14] or a c-*myc* pre-mRNA intron [15,16] (for reviews, see Refs. [17,18]).

Since the initial experiments by Paterson et al. [1] and Zamecnik and Stephenson [2], who had shown that artificial antisense oligonucleotides may in fact be used to interfere with gene expression, there has been a growing body of literature on antisense nucleic acids. Thus far, however, the mechanisms that underly the inhibitory effects which are frequently observed upon administration of antisense oligonucleotides, are still poorly understood. Inhibition may involve shielding of the single-stranded target sequence from essential interactions with cellular components or stimulation of cellular nucleases which are specific for double-stranded DNA–RNA or RNA–RNA hybrids. However, other mechanisms might also be involved. For example, Wassenegger et al. [19] have shown that autonomous potato spindle tuber viroid (PSTVd) RNA–RNA replication leads to complete methylation and specific inactivation of PSTVd cDNA integrated into the tobacco genome. The authors speculated that such an RNA-mediated feedback inhibition mechanism of gene expression could be operational in some of those cases where substoichiometric amounts of antisense RNA resulted in dramatic decreases of mRNA.

The lack of generally applicable rules for the design of efficient antisense oligonucleotides may reflect the diversity of possible ways to interfere with gene expression. In fact, there are only a few instances where the formation of a base-paired duplex structure between the antisense oligonucleotide and the target RNA has been documented, although this is assumed in most cases. Furthermore, high intracellular levels of antisense nucleic acid are often required to achieve appreciable effects on gene expression. Recently, dihydrofolate reductase (DHFR) expression in cultured human KB cells was found to be linearly dependent on antisense RNA levels [20]. A 600–2,800-fold excess of antisense RNA over target RNA was required to achieve a 50% reduction of DHFR mRNA. However, even excess amounts of antisense oligonucleotides may fail to cause significant inhibition. For example, a 1000-fold excess of chloramphenicol acetyltransferase (CAT) antisense RNA did not reveal any effect on CAT gene expression in monkey CV1 and tsCOS cells [21].

Antisense DNAs are generally rather short (<30 nt), chemically synthesized single-stranded oligomers complementary to the mRNA or its unspliced precursor. In vitro experiments showed that the RNA part of a DNA–mRNA hybrid can be cleaved efficiently by mammalian RNase H activities [22–24]. RNase H is thought to contribute to the observed effects of antisense DNA in vivo. This is suggested by cleavage of the target RNA in *Xenopus laevis* oocytes at the expected site [25], as well as by control experiments with modified antisense DNAs which hinder RNase H-mediated cleavage of the antisense DNA-target RNA hybrid [26,27] (see also the section on mosaic oligonucleotides). However, modified and unmodified antisense DNAs that do not render the target RNA susceptible to cleavage by RNase H may still interfere with gene expression. This has been suspected for the inhibition of in-

218

tercellular adhesion molecule 1 expression in human cell lines [26], as well as for the inhibition of splicing of herpes simplex virus [28] or in vitro translation of β-globin mRNA [29].

Antisense RNAs of considerable length have been used, sometimes extending to a few kilobases and almost covering the complete mRNA sequence. The actual cellular mechanisms underlying the observed effects on protein expression, however, can be quite diverse (Fig. 1). Interference of antisense RNA with splicing has been demonstrated in vitro [30] and in vivo [31], although others have not observed such an effect [32]. The most frequent effect in connection with antisense RNA is a simultaneous reduction in the amounts of target and antisense RNA. This is generally attributed to degradation mechanisms involving nuclear enzymes specific for double-stranded RNA, which have properties similar to those of E. coli RNase III or the "double-strand unwinding/modifying activity" [33]. Also, blockage of nuclear export of the target mRNA could be shown in cases where sense-antisense RNA hybrids were sufficiently stable to be amenable to analysis [34]. In the cytoplasmic compartment of frog oocytes, both the mRNA target as well as the microinjected (capped) antisense RNA were found to be fairly stable (half-life from hours to days

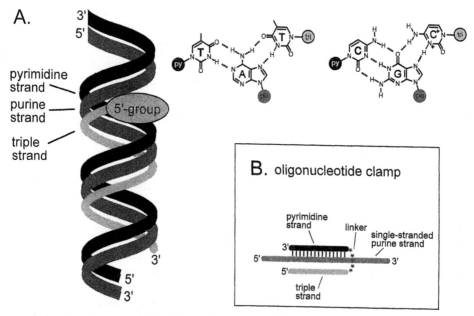

Fig. 2. Structure of a triple helix of the py*pu-py type. (A) In the case shown here, the triple strand binds in the major groove of a homopurine-homopyrimidine stretch of DNA via Hoogsteen hydrogen bonding to the purine strand; light grey, triple strand; dark grey, homopurine strand; black, homopyrimidine strand. (B) Oligonucleotide clamp as described by Giovannangeli et al. [52]. The target DNA strand containing the homopurine stretch (dark grey) is bound by the two pyrimidine stretches of the "clamp" via Watson–Crick base pairing (with the strand shown in black) and Hoogsteen hydrogen bonding (with the strand shown in light grey). A linker (asterisks), which may be of a different chemical nature, connects both parts of the clamp.

[35–37], and inhibition of gene expression was attributable to blocking of translation due to duplex formation between mRNA and antisense RNA.

Triple helix formation has been discussed as another way to interfere with gene expression. According to the initial concept, a single-stranded DNA oligonucleotide was directed against the double-stranded DNA of the target gene ("anti-gene approach", discussed in Ref. [38]; Figs. 1 and 2). The third strand generally binds in the major groove of a homopurine-homopyrimidine stretch of a Watson–Crick double helix. The various types of triple helices have distinct orientations of the third strand (parallel or antiparallel with respect to the homopurine strand of the Watson–Crick double helix) and differ in the type of hydrogen bonds (Hoogsteen or reverse Hoogsteen) which are formed between the bases of the third strand and positions N^6/O^6 and N^7 of the bases of the homopurine strand (reviewed in Ref. [39]). In the case of homopyrimidine oligonucleotides, thymine forms Hoogsteen hydrogen bonds with adenine, while adenine is simultaneously involved in (Watson–Crick) base pairing with a second thymine (Fig. 2A). Similarly, protonated cytosine can form Hoogsteen hydrogen bonds with a guanine that is engaged in a standard G–C Watson–Crick base pair [40].

Third strands containing A, G, inosine, 5-methyl-cytosine or even α-anomeric nucleotides were also shown to be capable of triple helix formation. The structures of most of these triple helices are thought to differ from that shown in Fig. 2A [39]. Recently, it has been reported that RNA–DNA hybrids and double-stranded RNA are also capable of forming triple helices [41,42]. Stability of the triple helix can be increased by tethering intercalating groups to the third strand [38,43,44] or when the third strand is made of RNA instead of DNA [41]. Attachment of reactive groups to the third strand has been applied to improve the efficiency of the triple helix approach by irreversible covalent modification of (or cleavage at) the target site [39,43]. Thus far, triple helix formation has been restricted to the presence of homopurine stretches in the target sequence.

In vitro experiments have shown that binding of the third strand can interfere with methylation of double-stranded DNA by methylases [45–47] and cleavage by restriction endonucleases or DNase I as well as with binding of a eukaryotic transcription factor [48–51]. DNA synthesis performed in vitro on a single-stranded DNA template could be inhibited by so-called oligonucleotide clamps (e.g. [52]; Fig. 2B), which are able to simultaneously form Watson–Crick and Hoogsteen base pairs with a homopurine stretch of the single-stranded target DNA. Various inhibitory effects attributed to triple helix formation have also been observed in prokaryotic and eukaryotic in vitro transcription assays [53–56]. In another study, the polypurine tract (PPT) of HIV-1, which forms an RNA–DNA hybrid during replication, was analyzed as a target site for triple helix formation ([57]; Silke Volkmann, personal communication).

There is only limited evidence that triple helix formation may be useful for the regulation of gene expression in vivo: c-*myc* mRNA levels could be specifically reduced in cultured HeLa cells when a DNA oligomer capable of forming a triple helix with the c-*myc* promoter was added to the culture medium [58]. Triple helix formation in vivo was confirmed by specific DNase I protection of the promoter sequence in extracted nuclei. Another study reported a 50% reduction of IL2α mRNA levels upon administration of a triple helix-forming DNA oligonucleotide to intact normal lymphocytes [59]. However, it has not been shown unambiguously whether the observed effects were actually due to triple helix formation in vivo. This also holds for the reduction of HIV-1 transcription observed upon addition of triple helix-forming oligonucleotides to the culture medium of acutely and chronically infected cells [60]. On the other hand, Grigoriev et al. [43] provided direct evidence that triple helix formation can occur in vivo. In their study, a psoralen-oligonucleotide conjugate was targeted to the promoter of the gene coding for the α subunit of the interleukin 2 receptor (IL-2Rα). When cells, transfected with a mixture of the target gene and the psoralen-oligonucleotide conjugate, were irradiated posttransfectionally, a significant fraction of the target gene promoter was cross-linked to the psoralen-oligonucleotide conjugate, which was paralleled by decreased target gene expression [43]. For significant inhibition, however, higher concentrations of the psoralen-oligonucleotide conjugate were required compared to experiments in which cross-linked complexes with the target gene promoter were preformed in vitro before transfection [43].

Ribozymes: Catalytic Antisense RNAs

Ribozyme strategies, which confer nucleolytic activity on the antisense RNA, have attracted much attention as an extension of the antisense concept. The ribozyme (= catalytic RNA; RNA enzyme) binds to the target RNA, catalyzes cleavage and can dissociate to enter further cycles of binding, cleavage and dissociation. In this case, RNA inactivation is irreversible and, at least in theory, one molecule of antisense (ribozyme-) RNA can catalyze the cleavage of several target molecules.

Table 1 summarizes the main features of the known RNA enzymes including their biological context, the catalyzed reactions and relevant applications. Thus far, the hammerhead motif has been the most widely used type of ribozyme in attempts to inactivate RNAs in vivo [e.g. 61–69]. Recently, the hairpin motif has extended the repertoire of tools for RNA inactivation in vivo [70,71]. The hairpin ribozyme may bear the advantage of being highly active at relatively low, i.e. physiological Mg^{2+}-concentrations [70].

Expression strategies for ribozymes have made use of efficient Pol II promoters (e.g. of the human β-actin gene [66–71]; viral promoters of the SV40 tk gene or from cytomegalovirus [69]; retroviral LTRs [62,65]) or Pol III promoters (e.g. of tRNA genes or the VAI promoter [71]). Retroviral vectors are frequently used since they are considered to be the preferred vehicles for gene delivery to human hematopoietic cells. In one study, the highest level of resistance to HIV-1 infection was obtained when ribozyme expression was controlled by the tk-promoter fused to the trans-activation-responsive (TAR) element in order to stimulate ribozyme expression in response to the presence of the HIV-1 Tat protein [69]. However, the TAR sequence may also function as a decoy to bind Tat, thus preventing viral gene expression and replication, as shown by Sullenger et al. [4]. Target sites, which have been chosen recently for ribozyme-mediated inactivation of retroviral RNAs, are within the 5'-untranslated leader region of HIV-1 [70,71] or within the regions coding for regulatory proteins, such as Rex and Tax of bovine leukaemia virus [62]. The 5'-untranslated leader sequence is a promising target for several reasons: (i) it is conserved among various HIV-1 isolates, (ii) it is present in all HIV-1 transcripts, and (iii) cleavage within this region renders the RNA capless which impairs its translation efficiency and which is likely to accelerate its degradation [70,71].

A major difficulty associated with ribozyme applications has been to prove unambiguously that ribozyme-mediated RNA destruction and not antisense activity is responsible for the observed effects, as discussed by Symons [72] and Altman [73]. This experimental difficulty is attributable to the rapid degradation of RNA fragments in mammalian cells. To our knowledge, the only case where an intact hammerhead-mediated RNA cleavage product could be isolated and verified by RNA sequencing has been reported for *Xenopus* oocytes [74]. The use of disabled control ribozymes with mutations in the catalytic core [70,71,75] also provided strong evidence that ribozyme-mediated cleavage can contribute significantly to the inhibition of RNA function in mammalian cells. Less conclusive approaches employed to analyze cleavage by ribozymes in vivo rely on target RNA quantifications by reverse transcription and PCR amplification using pairs of primers flanking the expected cleavage site [62,65,68]. In one reported case, an antisense RNA has been more effective than an antisense-ribozyme directed against the same region of the target RNA [76]. This observation underlines the value of comparative studies to assess the efficiencies of antisense versus antisense-ribozyme concepts. In addition, many in vivo experiments have been performed with cultured cells and *Xenopus* oocytes and embryos so that a final judgement of the therapeutic advantages of ribozymes over normal antisense nucleic acids would be premature.

The group I introns have been discovered in mitochondrial precursor RNAs from fungi and plants, in nuclear pre-rRNA of several lower eukaryotes, in genes of chloroplasts and bacteriophages (reviewed in [77]) and also in eubacterial tRNA genes [78]. The protein-independent self-splicing of the *Tetrahymena* intervening sequence from precursors of ribosomal RNA was the first discovery of an RNA-catalyzed reaction [79]. Group I introns share a common secondary and tertiary structure and a highly conserved catalytic core. Self-splicing involves two successive transesterification reactions, and requires a divalent metal ion (Mg^{2+} or Mn^{2+}) and a guanosine cofactor. The first step of the reaction can be designed to proceed in *trans*, mediating cleavage of a single-stranded RNA substrate ([80,81], Fig. 3). Base changes in the so-called "internal guide sequence" (IGS) allowed the *Tetrahymena* ribozyme (a shortened form of the intron lacking both splice sites) to be engineered in order to recognize and cleave a variety of substrates in vitro [80–82]. However, tertiary interactions stabilize binding of matched as well as mismatched substrates and thereby reduce the specificity of the *trans*-cleavage reaction. Discrimination against mismatched substrates can be enhanced by employing mutant versions of the *Tetrahymena* group I intron [83]. However, specific group I intron-catalyzed inactivation of

VS RNA [270]	Self-cleavage of multimeric precursors of VS RNA	*cis*-endoribonuclease	–	?
Group I intron [77]	Removal of introns during maturation of mRNAr, rRNA and mRNA primary transcripts	Two consecutive *cis* transesterifications	*trans* endoribonuclease RNA ligase	?
Group II intron [271]	Splicing of primary transcripts	Two consecutive *cis* transesterifications	–	?
RNase P (RNA) [209,262,263]	5′-maturation of primary transcripts of tRNAs	*trans*-acting endoribonuclease	*cis* endoribonuclease	?

The hammerhead, hairpin, hepatitis delta and group I ribozymes as well as eubacterial RNase P RNAs are considered true RNA enzymes, since in each case a single catalytic RNA molecule could be shown to be capable of catalyzing more than one round of reaction. The most relevant engineered properties are listed. However, other activities (e.g. for the group I introns) have been reported (reviewed in Refs. [7,271,272]). Thus far, *cis* self-cleavage and RNA self-ligation but not cleavage or ligation in trans have been documented for VS RNA, a *Neurospora* mitochondrial plasmid transcript, whose catalytic domain could be assigned to a 164-nt RNA fragment [270,273,274]. As for the VS RNA and group I intron-derived ribozymes, considerable RNA ligation activity has been observed for the hairpin ribozyme [77,102,274]. Cleavage reactions catalyzed by the hammerhead, hairpin, hepatitis delta and VS RNA ribozymes yield 2′, 3′-cyclic phosphates and 5′-OH termini. 3′-OH and 5′-phosphate termini are generated in cleavage reactions catalyzed by RNase P enzymes. Similarly, *trans*-cleavage by the *Tetrahymena* ribozyme results in 3′-OH termini [77]. Only few group II introns have been shown to self-splice in vitro. The low efficiencies and the required non-physiological conditions suggest that *trans*-acting factors support group II intron excision in vivo (reviewed in Ref. [271]). Likewise, eubacterial RNase P RNAs require elevated salt concentrations for efficient cleavage in the absence of the protein subunit [209].

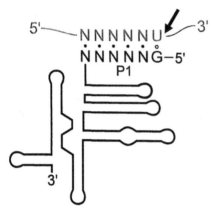

Fig. 3. Schematic outline of the secondary structure of a *trans*-acting version of the *Tetrahymena* ribozyme [77,80]. The target RNA is bound to the so-called "internal guide sequence" to form the helical element P1, here shown as a stretch of 6 consecutive base pairs including the conserved G-U base pair at the cleavage site (black arrow). The ribozyme is shown in black, the substrate in grey.

target RNAs in vivo may be disfavoured by the large size of the ribozyme and its limited target site specificity.

The hammerhead ribozyme is a small RNA motif found in a number of plant virusoids, avocado sunblotch viroid and viral satellite RNAs [84–87]. It consists of three short helices and a core harbouring about a dozen conserved bases (Fig. 4), and is thought to be responsible for self-cleavage of genomic multimers into monomeric units during the "rolling circle" replication of corresponding circular RNAs. Cleavage can also proceed in *trans* [88,89]. In such bi- (or even tri-) partite hammerhead

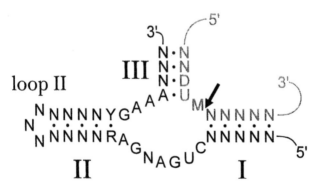

Fig. 4. General structure of a hammerhead ribozyme. The target RNA (shown in grey) is bound to the ribozyme (shown in black) via two stretches of Watson–Crick base pairs, yielding helices I and III. Cleavage occurs 3′ of a trinucleotide sequence "DUM" (GUC in most cases) at the position indicated by the black arrow. Positions where base identities are of minor importance have been indicated by N; R = purine; Y = pyrimidine; D = G, A (U); M = C, A. Conformational stability of the enzymatic domain, target specificity and kinetic performance can be influenced by varying the lengths of helices I, II and III.

constructs, the enzyme strand (= the ribozyme) has to associate with the substrate strand via base pairing to permit cleavage of the substrate strand (Fig. 4). Detailed analysis has revealed that hammerhead ribozymes of the type shown in Fig. 4 can be adapted to cleave any target RNA sequence 3' of a "DUM" sequence (D = G, A, and, with reduced efficiency, U; M = C, A and, with severely reduced activity, U [72,90]). Affinity, specificity and turnover rates mainly depend on the design of the two helices by which the ribozyme recognizes the RNA target. Extensive studies have been performed to further optimize this small ribozyme for specific inactivation of target RNAs (see also the section on In vitro synthesis of stabilized oligonucleotides). Koizumi et al. [67] have shown that a C(UUCG)G-tetraloop may stabilize helix II, thus preventing the formation of alternative conformations which reduce catalytic efficiency. Replacing the hairpin loop II (Fig. 4) by a tetra-deoxythymidine linker and providing the regions that base-pair with the target as DNA preserved wild-type levels of catalytic activity in vitro [91].

Recently, the *lacZ* gene (encoding β-galactosidase), as well as an anti-*lacZ* hammerhead, have been inserted into Moloney murine leukaemia virus (MoMLV) DNA expression vectors in order to assess the correlation between cellular co-localization and ribozyme efficiency [92]. The genomic RNA of MoMLV has two distinct fates: (i) nuclear export along certain pathways in order to become packaged into new virus particles and (ii) export into the cytoplasm for translation of viral proteins. When two MoMLV vectors, one encoding the *lacZ* message, the other coding for the anti-*lacZ* hammerhead, are brought into the nucleus of the same cell, both are transcribed as genomic MoMLV RNA. Thus, the hammerhead ribozyme may encounter its target in two different environments: (i) the packaging pathway, where both RNAs co-localize and (ii) the cytoplasm where no co-localization is expected to occur. In fact, the amount of intact MoMLV RNA encoding *lacZ* was drastically reduced in the packaging pathway where the ribozyme was co-localized with its target. Viral progeny showed a reduction of more than 90% in the proportion of *lacZ* coding genomic RNA. However, when the anti-*lacZ* hammerhead RNA was excluded from the packaging pathway due to the absence of a packaging signal sequence, its presence in the cytoplasma had no effect on *lacZ* expression. This illustrates that intracellular traffic-control patterns are likely to be key determinants for the successful down-regulation of gene expression by ribozymes and antisense nucleic acids. In addition, a distinct subcellular localization of catalytic antisense nucleic acids may also reduce the amount of non-specific RNA cleavage (see discussion in the section on Target RNA interaction), because most cellular RNAs will never be encountered. Another concept for enhanced targeting efficiency has been pursued by Larson et al. [93] who have tethered a hammerhead ribozyme to the 3'-end of human tRNA[Lys], which serves as the primer for cDNA synthesis of genomic RNA by HIV-1 reverse transcriptase.

Other experimental approaches have addressed the problem of target site accessibility. Goodchild [94] has utilized "facilitator" oligodeoxynucleotides that base-pair with the region 5' of the target site. This is intended to destroy secondary structures within the target RNA that may preclude the target site from hybridization with the ribozyme. In a related approach, the additional oligonucleotide (here termed anchor sequence) was linked directly to the ribozyme [95]. The binding site for the anchor sequence was put at some distance from the hammerhead moiety. Anchored hammerheads made it possible to cleave an aberrant chimeric mRNA (containing binding sites for both moieties), while leaving the wild-type mRNA, which contained only one binding site for either the anchor or the hammerhead moiety, unaffected. In vitro studies employing a triple hammerhead ribozyme, which contained three different ribozymes in a row, resulted in cleavage at all three target sites within an RNA fragment (367 nucleotides) of Hepatitis B virus pregenomic RNA [96]. Chen et al. [63] have designed multi-target ribozymes by linking up to nine *trans*-acting hammerhead ribozymes in series, which recognized different target sites of HIV-1 env RNA. In vitro and in vivo experiments have shown that such multi-target ribozymes were more active against HIV-1 than ribozyme monomers, and the efficacy of target RNA inactivation correlated with the number of different ribozymes in the construct. In the study by Ohkawa et al. [97], RNA inactivation efficiencies could be increased when multiple ribozymes were flanked by *cis*-acting ribozymes at both their 5'- and 3'-ends which resulted in the liberation of small monomeric *trans*-ribozymes. In some cases, hammerhead ribozymes have been reported to be almost inactive on longer RNAs [93,98]. For example, poor cleavage by hammerhead ribozymes in vitro was observed at three different target sites in RNA transcripts (approximately 1 kb in length) comprising the HIV-1 LTR region (nucleotides −525 to 386 of HIV-1), whereas cleavage of small model substrates was efficient [98]. In contrast, other target RNAs such as a 240-nt fragment of the human *mdr1* mRNA could be cleaved efficiently in vitro [66; Holm, Dietel, Krupp, unpublished results]. The *mdr1* mRNA is overexpressed in P-glycoprotein positive tumour cells, which results in resistance to multiple lipophilic cytotoxic drugs as a major impediment to cancer chemotherapy [99]. Expression of the ribozyme in a tumour cell line resulted in a drastic reduction of the drug resistance [66].

In conclusion, efficient hammerhead cleavage of long RNAs in vivo is hardly predictable based on results with small oligonucleotide substrates. This may be attributable to target RNA inaccessibility due to extensive secondary and tertiary structures, to spurious folding of ribozyme–target RNA complexes, or to mRNA–protein interactions and intracellular trafficking pathways, which may prevent any contact between the ribozyme and its target in vivo. On the other hand, structured target RNA sequences might be well suited for hybridization to antisense RNA in vivo by mechanisms involving the naturally occurring "kissing" phenomenon [100,101]. However, such potential target sites are likely to escape attention on the basis of in vitro experiments with small model substrates.

The hairpin ribozyme was identified as part of the minus strand of tobacco ring-spot virus satellite RNA [102,103] (Table 1, Fig. 5). Its catalytic core consists of a 50-nt catalytic domain and a 14-nt substrate domain, which harbours the cleavage site. Catalytic and substrate domains can reside on separate RNA molecules for cleavage to occur in *trans* [104–106] (Fig. 5). Since the 50-nt ribozyme domain binds the substrate strand via four Watson–Crick base-pairs in helix II and six or

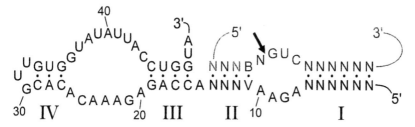

Fig. 5. Secondary structure of a hairpin ribozyme. The target RNA (shown in grey) binds to the enzymatic moiety (shown in black) via two stretches of Watson–Crick base pairing and is cleaved 5' of a GUC sequence. Helix II comprises four base pairs, while the length of helix I can be increased. The length of helix IV can also be extended without affecting activity. B = C, U, G; V = G, A, C (according to [299]).

more in helix I, any RNA can be targeted for cleavage immediately 5' of a GUC sequence. However, engineered hairpin ribozymes are often less active than the wild-type ribozyme [106,107], probably due to altered structures of helices I and II. In the case of a hairpin ribozyme targeted against HIV pol I RNA, a 20-fold reduced catalytic efficiency (k_{cat}/K_m) could be restored to wild-type levels with sequence variants selected by in vitro evolution techniques. Two bases ($A_{11} \rightarrow G$; $U_{39} \rightarrow C$; $A_{11} = V_{11}$ in Fig. 5) were exchanged in the optimized ribozyme [108]. Another hairpin ribozyme directed against the 5'-leader of HIV-1 caused a 70–90% inhibition of Tat activity and p24 expression in vivo when transcribed from a Pol II [70] or Pol III promoter [71]. A catalytically inactive control ribozyme ($A_{22}A_{23}A_{24} \rightarrow$ UGC, see Fig. 5) which caused only about 10% inhibition indicated that the observed interference with HIV-1 expression was largely attributable to the catalytic

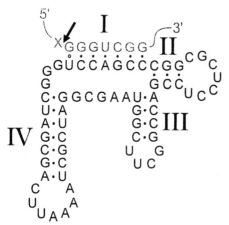

Fig. 6. Secondary structure of a trans-acting circularized version of the hepatitis delta ribozyme. The target RNA (shown in grey) is bound via a stretch of 6 consecutive base pairs and 1 G-U base pair at the cleavage site (black arrow). The ribozyme is shown in black (adopted from [114]).

properties of the ribozyme. Such an anti-HIV-1 hairpin ribozyme is scheduled for a first clinical test in 1995 [109].

The hepatitis delta ribozyme motif occurs in the genomic and antigenomic RNA of hepatitis delta virus [110]. A minimal sequence for efficient self-cleavage in vitro consists of an 85-nt RNA which cuts off its 5'-nucleotide [111,112]. As shown recently, the hepatitis delta ribozyme can be designed to act in *trans* [113, reviewed in 114]. A circularized version (Fig. 6) of the ribozyme moiety was particularly resistant to nucleolytic attack in HeLa cell extracts [115], although efficient hybridization with long target RNAs may be impeded for topological reasons. It remains to be elucidated whether the hepatitis delta ribozyme can be adapted to bind different substrate RNAs (other than the naturally occurring hepatitis delta sequences) via base pairing (helix I), and whether efficient and specific in *trans*-cleavage can be achieved in vitro and in vivo.

Delivery Systems: Entry of Antisense Nucleic Acids into Cells

Applying antisense technology to living cells or even higher organisms requires efficient strategies for the delivery of antisense nucleic acids.

Generally, two routes for the delivery of nucleic acids are employed: (i) transfection with the gene encoding the antisense RNA and (ii) in vitro synthesis of the antisense nucleic acid followed by introduction into the cell (see Fig. 7). With respect to therapeutic applications, the first approach entails the experimental and ethical problems of gene therapy, while the second approach bears problems similar to those of conventional drug applications.

Transfection allows one to introduce genes coding for antisense RNAs into eukaryotic cells. The cellular transcription machinery is then producing the antisense RNA right "in place" and high intracellular levels can be obtained (often up to several thousand times the level of target RNA). Transfection circumvents the problem of degradation due to nucleolytic attack in extracellular fluids or endocytic compartments which is normally associated with exogenous application of antisense oligonucleotides. On the other hand, this excludes the use of deoxyoligonucleotides or chemically modified antisense nucleic acids. Transfection methods for introducing foreign DNA into living cells make use of calcium precipitated vector DNA (frequently derived from viruses or retroviruses), electroporation [116], DEAE-dextran [117], cationic liposomes [118,119], polyamines [120], transferrin-poly-lysine "donuts" [121] or cross-linking of an expression plasmid to cell-targeting entities, such as specific antibodies (called antifection [122]). Coupling of the DNA to other ligands which interact with cellular receptors might be conceived.

Beyond transient transfection, stable transfection of cell lines or transgenic organisms expand the repertoire of methods for the expression of antisense nucleic acids, although the time and effort required does not make these approaches attractive for screening.

Cellular uptake mechanisms can also be exploited to introduce antisense nucleic

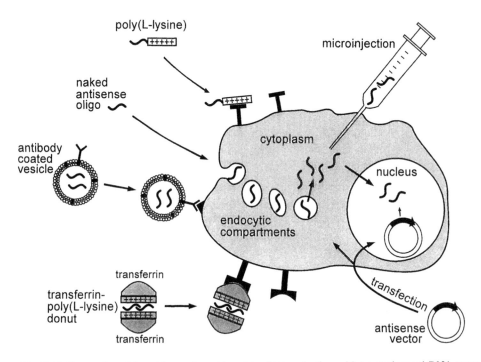

Fig. 7. Delivery of nucleic acids can be achieved by (i) transfection with an engineered DNA vector encoding the antisense RNA or by (ii) in vitro synthesis of the antisense nucleic acid followed by introduction into the cell. Nucleic acids can be microinjected into the cytoplasm or the nucleus. Exogenously applied nucleic acids are assumed to be taken up via receptor-mediated endocytosis and/or pinocytosis. A different import mechanism seems to be operational in the case of oligonucleotides coupled to peptides derived from the *Drosophila* homeoprotein Antennapedia (see text).

acids synthesized in vitro (Fig. 7). Radioactive oligonucleotides were found to bind to the cell surface and to accumulate in nuclei [123], although this may not hold for every oligonucleotide and all types of cells. For example, a fluorescently labelled DNA 20-mer was taken up inefficiently by HeLa and 3T6 cells when present in concentrations of $10\,\mu M$; cellular uptake was far more efficient in the case of phosphorothioate and methylphosphonate analogs [124]. Generally, import of polyanionic nucleic acids seems to be facilitated by specific receptor-mediated endocytosis and/or pinocytosis. Alkylating oligonucleotides were used to identify proteins of approximately 80,000 which might be involved in this process in different cell lines [125,126]. In addition, coupling of lipophilic membrane anchoring groups (such as cholesterol) to oligonucleotides may facilitate uptake into mammalian cells [127]. It is still unclear how internalized antisense oligomers can escape from the endocytic compartment. Entrapment in endocytic vesicles and degradation by lysosomal enzymes may present significant drawbacks.

The fate of oligonucleotides has also been tested in tissue drafts and whole organisms. Preliminary studies in mice have shown that a dose of 30 mg/kg (of body weight) of an oligonucleoside phosphorothioate was sufficient to make it "bio-available" in most tissues for up to 48 h [124]. A dose 100 mg/kg (daily for 14 days) showed no toxic effects. A 15% degradation was observed after 24 h in plasma, stomach, heart, spleen and intestines, while 50% degradation was found in kidney and liver. In another experiment, developing tissues from mouse embryos were cultured in the presence of $2.5 \mu M$ fluorescently labelled and phosphorothioate-stabilized DNA 19-mers [128]. Fluorescence microscopy revealed that mesenchymal cells of lung, salivary gland, kidney, ovary and testis accumulated high intracellular levels of the oligonucleotide. On the other hand, no detectable amounts of oligonucleotide were found in mature epithelial cells. This phenomenon correlated with the presence of basement membranes, as judged by laminin staining. Due to the low uptake efficiency of certain cell types, suppression of gene expression in whole organs by external administration of antisense oligonucleotides may therefore prove difficult.

Poly(L-lysine) can be linked to 3'-termini of oligonucleotides by standard methods [129]. In addition to accelerated and increased uptake by cultured cells, poly(L-lysine) seems to impart nuclease resistance to the 3'-end [130]. Poly(L-lysine)-conjugated antisense oligonucleotides directed against the nucleocapsid N protein mRNA of VSV (vesicular stomatitis virus) in infected L929 cells were 10–50 times more active than unmodified control oligomers. Concentrations below $1 \mu M$ were already effective [131], while methylphosphonate-modified oligomers inhibited VSV expression only when applied in concentrations of $50 \mu M$. Poly(L-lysine)-conjugated oligonucleotides, which can interact with negatively charged cell surface molecules, are probably taken up by non-specific, receptor-mediated endocytosis and accumulate in the acidic endocytic compartment (Fig. 7), where they might be released by proteolysis of the carrier [129]. Problems may arise from cytotoxic effects sometimes observed at high doses of poly(L-lysine). Sequence-specific antiviral activity of oligonucleotide-poly(L-lysine) conjugates can be enhanced in the presence of polyanions, such as heparin, with a concomitant reduction in cytotoxicity [129]. Conjugates of poly(L-lysine) with asialoorosomucoid were developed as a means to deliver plasmid DNA to hepatocytes [132]. Likewise, plasmid DNA complexed with transferrin-poly(L-lysine)-conjugates can be efficiently introduced into hematopoietic cells or other transferrin receptor-rich cells [121,133]. In these "Trojan Horse" approaches, plasmid DNA condensed with the positively charged L-lysine residues by electrostatic interactions is efficiently internalized by receptor-mediated endocytosis. Such methods might also prove useful for the delivery of antisense oligonucleotides. Addition of replication-deficient adenovirus as an endosome-disrupting agent has improved transferrin-poly(L-lysine)-mediated gene delivery in a number of cell lines [134]. Discovery of new natural, receptor-mediated endocytosis pathways may trigger the development of new antisense delivery systems.

Microinjection may be the method of choice for the introduction of excess amounts of in vitro synthesized antisense nucleic acids (Fig. 7). In a recent study with fluorescent synthetic 15-mers, nearly complete translocation into the nucleus was observed within a minute after microinjection [135]. Fluorescence energy transfer experiments with microinjected sense and antisense DNAs showed that hybrids can form in the cytoplasm as well as in the nuclei of cultured human 3T3 cells [136]. In this study, the 28-nt sense DNA consisted of the translation initiation region for HIV *rev* mRNA and carried a fluorescein tag (fluorescence donor) at its 5'-end. The antisense DNA was 3'-labelled with rhodamine (fluorescence acceptor) and was stabilized by phosphorothioate modifications of the phosphate backbone. Fluorescence energy transfer, which is strongly dependent on the distance between the fluorescence donor and acceptor, has been used to follow hybrid formation and destruction. Preformed sense-antisense hybrids had a half-life of 15 min, which is about 10 times less than under physiological buffer conditions in vitro. Sequentially injected single-stranded oligonucleotides formed hybrids which were transiently detected in the cytoplasm before accumulating in the nucleus, as indicated by confocal laser scanning microscopy. These experiments provide a first step towards a detailed understanding of antisense action and hybridization in living cells [136]. Similar experiments may be suitable to evaluate target accessibility and hybridization kinetics for antisense RNAs, ribozymes or triple helices.

Liposomes can be used to encapsulate and deliver nucleic acids (Fig. 7). First, they provide protection against nucleolytic degradation in the culture medium and may increase uptake efficiency. In addition, they can be targeted to specific cell types, e.g. by specific antibodies or other protein markers attached to their surface [129,137].

Phage packaging might also be a reasonable alternative. Pickett and Peabody [138] could show that *lacZ* mRNA fused with a 21-nt viral packaging signal was encapsidated in vivo by coat protein of *E. coli* phage MS2. It is still unclear, whether viral particles may prove useful for the delivery of antisense oligonucleotides. Some successful experiments employing antisense oligonucleotides entrapped in viral envelopes have been discussed by Vlassov and Yakubov [123]. A non-pathogenic form of hepatitis delta virus (HDV) may serve as a delivery vehicle, e.g. to combat hepatitis B virus, since HDV has a natural affinity for liver cells [109].

Peptide-mediated internalization of oligonucleotides extends the repertoire of cell delivery methods. It has recently been discovered that an oligopeptide of 16 amino acids (called Penetratin 1, trademark of Appligene), which corresponds to the third helix of the homeodomain of the *Drosophila* homeoprotein Antennapedia, can be efficiently internalized by numerous cell types [139; Appligene, product information]. When linked to oligonucleotides, the peptide mediates their transfer to the cytoplasm and nucleus. Several observations indicate that the route of cell entry is different from classical receptor-mediated endocytosis [139], thus bypassing the problem of oligonucleotide trapping in endocytic compartments or degradation by lysosomal nucleases.

In Vitro Synthesis of Stabilized Oligonucleotides

Synthetic oligonucleotides offer great flexibility with respect to their chemical and biological properties. Thus, each aspect relevant to the efficiency of an antisense oligonucleotide might be addressed: (i) delivery/cellular uptake; (ii) intracellular localization; (iii) hybrid stability; duplexes formed between oligonucleotide and target nucleic acid should be stable enough to allow significant complex formation in vivo; on the other hand, an increase in stability should not compromise a high specificity of interaction (the specificity issue is addressed in detail in the section on target RNA interaction); (iv) biological stability of the antisense oligonucleotide and (v) specific degradation of the target RNA.

As a result, modified oligonucleotides may be more effective at lower, non-toxic doses than their unmodified counterparts.

Modified oligonucleotides. The structures of the most frequently used chemical modification types are shown in Fig. 8.

Modified riboses. 2'-Hydroxyl groups of RNA have been substituted by 2'-H (DNA), 2'-O-methyl, 2'-O-allyl [138], 2'-fluoro or 2'-amino groups [141].

Modified phosphodiesters. In methylphosphonates, the charged oxygen is replaced by a methyl group [142,143]. In phosphorothioates [144], one non-bridging oxygen is substituted by sulphur. Due to the sulphur substitution, the phosphorus

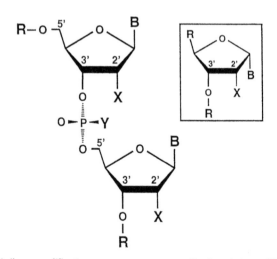

X: 2'-ribose modifications

-OH	2'-hydroxyl (in natural RNA)
-H	2'-hydrogen (in natural DNA)
-O-CH₃	2'-O-methyl-ribose
-O-CH₂CH=CH₂	2'-O-allyl-ribose
-F	2'-fluoro-2'-deoxy-ribose
-NH₂	2'-amino-2'-deoxy-ribose

Y: phosphate modifications

-O	naturally occuring phosphodiester
-S	phosphorothioate
-CH₃	methylphosphonate

R: terminal conjugates

see text for details

Fig. 8. Types of frequently used nucleic acid modifications have been indicated by letters (X, Y and R) in the dinucleotide structure. B indicates the base moiety. The framed insert shows the different stereochemical arrangement in α-anomeric nucleosides. For details, see text.

atom gains chirality giving rise to diastereomeric isomers with distinct biological properties. Only Rp-phosphorothioates can be obtained by enzymatic synthesis with DNA or RNA polymerases [145]. The Rp-isomer is shown in Fig. 8. In phosphoro-dithioates [146], both non-bridging oxygens are replaced by sulphur.

Nucleic acid conjugates. Nucleic acid conjugates include poly-L-lysine [147], lipophilic groups such as polyalkyl derivatives [148,149], cholesterol [148,150] or polyethylene glycol (PEG) [151,152]. In addition to stabilization against nucleolytic degradation, attachment of PEG with different degrees of polymerization to 3'- and 5'-ends of oligonucleotides by automated synthesis allows one to adjust the hydro-phobicity of a given oligonucleotide [151]. Since hydrophobicity of the conjugates increases with increasing PEG chain length, the properties of conjugated oligonu-cleotides may be fine-tuned by means of PEG chain length in order to optimize the desired biological effects. Conjugates with intercalating dyes and cross-linking or cleaving agents involving metal chelators have been designed to enhance interaction with or destruction of nucleic acid targets (reviewed in Refs. [153,154]).

Furthermore, α-anomeric oligonucleotides (insert of Fig. 8), where the carbon at position one of the sugar moiety has the α- instead of the common β-configuration, as well as 2'-5'-linked phosphodiesters, have been characterized [153,155] (see also below).

Polyamide derivatives (called PNA) represent the most extreme case of backbone modification. In PNA, the sugar-phosphate backbone is replaced completely by a polyamide chain, which preserves the capability of Watson–Crick base pairing with target nucleic acids [156–159]. The most evident advantage of PNA derivatives is their resistance to cellular nucleases.

Properties of modified oligonucleotides
The different modifications vary considerably in their resistance towards individual nucleases. Generally, nucleases can be functionally subdivided into two groups: Exonucleases attack the termini, whereas endonucleases cleave phosphodiester bonds at certain positions within the oligonucleotide. In combination, both classes of nucleases cause exponential degradation, since endonucleases constantly produce more 3'- and 5'-termini which exonucleases can act upon. RNA oligonucleotides are readily degraded by ribonucleases. Substitution of the 2'-hydroxyl group greatly improves resistance against RNases due to either steric hindrance or lack of the nucleophilic 2'-hydroxyl group. The absence of the 2'-hydroxyl group inhibits cleavage by those RNases which utilize this functional group in their catalytic mechanism to attack the adjacent 3',5'-phosphodiester bond. However, the stability of simple DNA oligonucleotides, which can be regarded as 2'-H RNA derivatives, is also limited, since they are degraded by DNA-specific DNases and non-specific nucleases. 2'-O-Methyl- and 2'-O-allyl-modified oligonucleotides were reported to be highly resistant to degradation by purified nucleases and in bovine serum [141,160,161]. This included resistance to endonucleases such as S1 and mung bean nuclease, as well as against DNase I and 3'-exonuclease III, and moderate resistance against micrococcal and P1 nucleases. On the other hand, the 2'-O-methyl- and 2'-

O-allyl-oligonucleotides were easily degraded by snake venom phosphodiesterase and nuclease Bal31. In addition to effects of modifications owing to their terminal or internal location within an oligonucleotide, nuclease resistance characteristics may be modulated in the context of mosaic oligonucleotides, where different types of modification are combined within a single oligonucleotide (see next section). Sequence-dependent effects on thermodynamic stability of duplexes directed against DNA have been observed with 2′-*O*-methyl substituted antisense oligonucleotides, resulting in overall stabilization or destabilization compared to the unmodified hybrid. However, hybrids of 2′-*O*-methyl-substituted antisense nucleic acids with RNA were shown to be more stable than unmodified DNA–RNA duplexes and reached the stability of RNA–RNA hybrids [162,163].

Phosphorothioate modifications have been widely used to stabilize antisense DNA and RNA against nucleolytic attack [128,164–166]. Sp-isomers are resistant to attack by snake venom phosphodiesterase, but become readily degraded by nuclease P1. In contrast, Rp-isomers (Fig. 8) are cleaved by snake venom phosphodiesterase, but are resistant to nuclease P1 [144]. Phosphorothioate backbone modifications in DNA destabilize hybridization with RNA and DNA to a similar extent (roughly 0.3 kcal/mol per modification [162]). Supposedly, this also holds for modified RNA binding to DNA or RNA targets [162]. The study by Kibler-Herzog et al. [167] indicated that the extent of destabilization depends on the sequence context of the modification. A phosphorothioate group between adenines caused less destabilization of a DNA double helix than when occurring between two thymines. Agrawal et al. [168] have addressed the toxicity problem by injecting mice with different amounts of an anti-HIV oligodeoxynucleoside phosphorothioate 20-mer. No symptoms were observed up to 40 mg/kg of body weight, while a fourfold higher dose already killed some test animals. These effects were similar to those obtained with the unmodified oligodeoxynucleotide [168]. For a comparison, HIV infection of cultured HeLa cells could already be inhibited by 50% when phosphorothioate-modified antisense oligodeoxynucleotides were added to the medium to a final concentration of 0.01–0.1 μM (cotransfectionally) or 0.1–1 μM (posttransfectionally [126]). In a study with cultured human ATH 8 cells, no cytotoxic effects were observed when up to 25 μM of a oligodeoxnucleoside phosphorothioate 14-mer were added to the medium [169]. Surprisingly, in this system, a non-specific oligodeoxycytidin phosphorothioate 14-mer was able to inhibit HTLV-III$_B$ replication at concentrations as low as 1 μM [169]. This anti-HIV activity of phosphorothioate homodeoxyoligonucleotides has also been observed by others [168]. Inhibition of HIV replication, although to a lesser extent, was also reported for unmodified homodeoxyoligonucleotides and their phosphomorpholidate analogs [168]. One possible mechanism of this apparently sequence-independent antiviral activity is competitive inhibition of binding of the primer-template substrate to HIV reverse transcriptase. A phosphorothioate oligodeoxycytidylate was shown to bind with higher affinity to the enzyme than the unmodified oligodeoxycytidylate [170]. However, when the oligonucleotide phosphorothioates were added after infection, those complementary to a specific target site were more effective than either mismatched variants or homooligomers [171].

Agrawal et al. [124] have performed a first study on the uptake of fluorescently labelled phosphorothioate-modified oligonucleotide 20-mers by cultured HeLa cells. The phosphorothioate analog seemed to enter the cells more readily than the unmodified oligonucleotide and fluorescence microscopy suggested that the phosphorothioate oligonucleotide localized to the intranuclear- and the perinuclear space.

In phosphorodithioates, both non-bridging phosphate oxygens are simultaneously replaced by sulphur. Oligonucleotides of this type were reported to be even more resistant to nucleolytic attack than phosphorothioates [146], although this may not always hold true, as suggested by Ghosh et al. [172]. Phosphorodithioate modification of one strand of a 17 base-pair DNA double helix lowered the T_m by 6°C and 17°C compared to the analogous phosphorothioate-modified double helix and the unmodified control, respectively [172].

Methylphosphonate modifications confer resistance on DNA oligonucleotides to both 3'- and 5'- exonucleases as well as serum and cellular activities [173]. Hybrid helices with one strand being DNA and the second methylphosphonate-modified DNA were less stable than the corresponding unmodified DNA duplexes, and the extent of modification correlated with the loss of hybrid stability [167]. Hybrid duplexes of methylphosphonate-modified DNA with RNA were even more unstable [174]. Furthermore, the hybrids with RNA are insusceptible to cleavage by RNase H (see next section). While standard polyanionic DNA or RNA is taken up by cells via receptor-mediated endocytosis, methylphosphono oligonucleotides with three or less phosphodiester linkages seem to enter cells by another efficient, yet rather undefined, active import mechanism [175]. Efficient cellular uptake of fluorescently labelled methylphosphono oligonucleotides has also been observed by fluorescence microscopy [124].

The properties of phosphoramidates are similar to those of methylphosphonates, although phosphoramidate modifications (as part of one strand of a DNA double helix) destabilize duplex structures more than methylphosphonate modifications [176]. According to a preliminary toxicity study, application of up to 40 mg of phosphoramidate-modified oligonucleotide per kg of body weight showed no effect on mice, while a dose of 150 mg/kg can already be lethal [168].

Resistance to nucleases has also been reported for α-anomeric oligonucleotides (Fig. 8). An α-anomeric oligodeoxythymidylate was found to be 300–500 times more resistant to exonucleolytic attack than oligo-dT [177]. In general, α-anomeric deoxyoligonucleotides do not seem to form standard antiparallel Watson–Crick hybrid duplexes with normal β-anomeric deoxyoligonucleotides; evidence for the formation of antiparallel duplexes was only found in studies of α-deoxythymidylates. However, α-anomeric deoxyoligonucleotides were shown to be capable of forming triple helices with double-helical DNA (reviewed in Ref. [38]). Duplexes of α-oligodeoxythymidylate with β-anomeric poly- or oligo-riboadenylates were found to be more stable than those with poly- or oligo-deoxyadenylates [177]. An α-anomeric dodecaribonucleotide of mixed sequence formed triple stranded structures with β-deoxyoligonucleotides, but no evidence for duplex formation with β-deoxyoligonucleotides was found [178]. In an in vitro translation system, an α-anomeric antisense DNA complementary to a region adjacent to the cap site of β-globin mRNA inhibited initiation of translation in reticulocyte lysates more effectively than the normal β-counterpart [29]. One potential disadvantage of this modification is that hybrids of α-anomeric DNA and mRNA are rendered insusceptible to cleavage by RNase H. HIV-1 infection could be inhibited in cultured cells by α-dodecaribonucleotides, although with apparent lack of sequence specificity [178].

There is only limited data available on the biological characteristics of $2',5'$-linked antisense oligonucleotides. In addition to resistance against several nucleases, $2',5'$-linked oligoribonucleotides bind complementary RNA much stronger than the complementary DNA [155]. However, $2',5'$-linkages were observed to also destabilize RNA duplexes [155,179]. A strategy involving the 2–5A-dependent RNase for the specific cleavage of RNA has been proposed recently [180]. 2–5A-dependent RNase, an endoribonuclease that mediates inhibitory effects of interferon on virus infection, is activated by $5'$-phosphorylated $2'-5'$-linked oligoadenylates. In the approach by Torrence et al. [180], a tetrameric $5'$-phosphorylated $2'-5'$-oligoadenylate was covalently tethered to an antisense oligodeoxythymidylate. This caused specific 2–5A-dependent RNase-mediated RNA cleavage at the target site of the attached antisense oligonucleotide in an extract of human lymphoblastoid Daudi cells.

Since modifications alter the hybridization properties of nucleic acids, modified antisense oligonucleotides might also be useful to improve discrimination between target and non-target sites (see section on Target RNA interaction).

Mosaic oligonucleotides

In mosaic oligonucleotides (also termed chimeric oligonucleotides), advantageous properties of different types of modifications can be combined. Their most simple versions are represented by DNA or RNA oligonucleotides which include a single type of modification confined to a subset of nucleotides. For example, oligonucleotides can be capped at both ends with short segments of phosphorothioate, methylphosphonate or phosphoramidate internucleoside linkages to confer resistance to exonucleolytic attack [181,182]. Two consecutive methylphosphonodiester linkages at each, $3'$- and $5'$-end, were able to protect a synthetic DNA from degradation by purified exonucleases [183]. Tidd [184] could show that two consecutive methylphosphonodiester linkages at the $3'$-end were already sufficient to reduce exonucleolytic attack in foetal calf serum.

Modifications of terminal phosphate groups have also been described. For example, methylphosphate $5'$-caps, which naturally occur in some small RNAs such as U6 snRNA, increased RNA stability in *Xenopus* oocytes [185]. In order to improve the efficiency of antisense oligonucleotides, termini were modified with cholesterol and phenazinium groups [127,186]. Polyethyleneglycol can also be incorporated at internal positions of oligonucleotides by automated phosphoramidite chemistry [151].

Reducing the number of unmodified nucleotides in the "core" may provide additional resistance to endonucleases. However, in practice, the number as well as the type and position of modifications will have to be optimized for each different anti-sense approach. In addition, very little is known about toxic side effects of modified nucleotides resulting from degradation of synthetic antisense nucleic acids [187].

Substrates for vertebrate RNase H
The endoribonuclease RNase H specifically cleaves the RNA strand of RNA–DNA hybrids [188]. RNase H activities have been found in prokaryotes and eukaryotes as diverse as *E. coli*, *Tetrahymena*, yeast, plants and mammals. They produce either oligomers, up to 9 nucleotides in length, or mono- and dimers with 5′-phosphate and 3′-hydroxyl termini [22]. Even though there is only limited evidence supporting a role for RNase H in oligonucleotide-mediated inhibition of gene expression in mammalian cells [e.g. 26], much attention has focused on substrate recognition by RNase H. The major isoform of human RNase H, for example, was shown to cleave double-stranded DNA containing only a single RNA residue, even when this position is not involved in base pairing [189].

2′-Fluoro- or 2′-O-methyl-, as well as methylphosphonate-, phosphoro-N-morpholidate- or phosphoro-N-butylamidate modifications of antisense oligonucleotides abolish cleavage by human RNase H [190–192]. In contrast, phosphorothioate-modified antisense oligonucleotides allow and may even enhance cleavage by RNase H from *Xenopus laevis* and HeLa nuclear extracts [165,190,191].

Several types of modification have been combined in a single mosaic antisense oligonucleotide in order to improve performance (e.g. nuclease resistance and/or cellular uptake) while retaining the ability to direct cleavage by RNase H [181,191]. For example, human RNase H was capable of cleaving a target RNA annealed to a methylphosphonate-modified antisense DNA oligonucleotide containing a stretch of 2–6 consecutive phosphodiester or phosphorothioate deoxynucleotide linkages [190]. The cellular uptake of such antisense oligonucleotides containing three or less phosphodiester bonds was more efficient than uptake of all-phosphodiester oligonucleotides [175]. A mosaic antisense oligonucleotide consisting of phosphorothioates flanked with 2′-O-methyl-phosphorothio-nucleotides was even more efficient in vivo than the all-phosphorothioate control [27]. For a set of antisense oligonucleotides of this type with different extents of 2′-O-methyl modification, cleavage by mammalian RNase H in vitro correlated with their antisense efficiency in cell culture [27].

Modified hammerhead ribozymes
Modification of catalytic antisense RNAs (ribozymes) is an even more elaborate task, since several internal functional groups of bases, sugars and phosphates are expected to affect catalytic function of the ribozyme. Most efforts have been applied toward improving nuclease resistance of the hammerhead ribozyme, and the majority of experiments were performed with bimolecular hammerhead structures of the type shown in Fig. 9. Base identities were found to be important for activity at positions 3 to 6, 8, 9, 12 to 15 and 16.1. A purine-pyrimidine (R-Y) base pair at positions 10.1 and 11.1 is required for efficient cleavage of the hammerhead type shown in Fig. 9. At position M_{17}, 5′ to the cleavage site, activity is decreased by U and abolished by

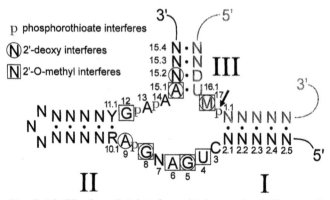

Fig. 9. Modifications that interfere with hammerhead ribozyme function: Circled nucleotides, 2'-deoxyribose modifications; boxed nucleotides, 2'-*O*-methyl-ribose modifications; p, Rp-phosphorothioates. For further details, see Fig. 4.

G (reviewed in Ref. [72]). $D_{16.2}$ can be G or A, while U reduces activity significantly. In general, results attained by mutational studies are consistent with effects of specific base modifications [140,193–198]. Yet, only minor effects on cleavage activity were observed for 7-deaza derivatives at positions A_9, A_{13}, A_{14} and $A_{15.1}$ [71], and for inosine at position 8 [193,195,197].

Inhibitory effects of 2'-deoxy-modifications were localized to the riboses of G_5, G_8, A_9, $A_{15.1}$, $N_{15.2}$ and of nucleotide M_{17} located 5' to the scissile bond (Fig. 9) [199–201]. The pattern obtained for 2'-*O*-methyl- and 2'-*O*-allyl-modifications was somewhat different, with interference at G_5, G_8 (not at A_9), but in addition at G_{12}, $A_{15.1}$ (not $N_{15.2}$) and M_{17}, and to a lesser extent at U_4 and A_6 [140] (Fig. 9).

In addition, 2'-amino- and 2'-fluoro-2'-deoxy ribose modifications have been studied [141,198]. The same interference pattern was observed with the 2'-deoxy- and 2'-amino-nucleosides, whereas a different pattern was obtained for 2'-*O*-alkyl- and 2'-fluoro-modifications. The similarity seen for 2'-deoxy and 2'-amino-nucleosides might reflect their common preference for the $C_{2'}$-endo conformation of the ribose moiety, whereas 2'-*O*-alkyl as well as 2'-fluoro nucleosides, as unmodified ribonucleosides, prefer the $C_{3'}$-endo sugar puckering mode [198,202].

Rp-phosphorothioates only interfered with the hammerhead cleavage reaction when introduced 3' of positions G_8, G_{12}, A_{13} and at the scissile phosphodiester 3' of M_{17} [203,204] (Fig. 9).

In general, positions that are sensitive towards chemical modification reside within the conserved core, while changes in the stem regions leave catalysis largely unaffected. Results of mutational and modification studies provided the basis for the construction of highly modified mosaic hammerhead oligonucleotides with appreciable catalytic performance and improved nuclease resistance. In one case, stems I, II and III were made of phosphorothioate-modified DNA. This slowed down exonucleolytic attack in human serum, but yielded only little protection against degradation in bovine serum, probably due to a high content of endonucleolytic activities. When positions C_3, U_4 and U_7 were also replaced by phosphorothio-deoxynucleosides, wild-type cleavage rates were retained, while nuclease resistance in human and bovine sera was increased at least a thousand-fold [164]. Other examples include the complete replacement of cytidines by 2′-fluoro-C and uridines by 2′-fluoro-U in concert with phosphorothioate modifications at both ends of the hammerhead ribozyme, which only caused a slight decrease in catalytic performance, but largely increased the half-life in cell culture supernatants [98]. Likewise, hammerhead ribozymes were at least 1,000 times more stable in rabbit serum than their unmodified counterparts when all U and C residues were replaced by 2′-fluoro or 2′-amino derivatives. However, the 2′-amino-C modification decreased catalytic performance by 20-fold [141]. Paolella et al. [140] used 2′-O-methyl and 2′-O-allyl modifications to stabilize the hammerhead against nucleolytic attack. A hammerhead ribozyme where all positions except nucleotides U_4, G_5, A_6, G_8, G_{12} and $A_{15.1}$ (see above) carried 2′-O-modifications showed good catalytic performance combined with a high resistance towards serum nucleases.

In conclusion, catalytic antisense RNAs are amenable to heavy modification conferring largely increased nuclease resistance in vitro, and therefore seem attractive for therapeutic applications. Of course, the use of modified nucleosides demands significant efforts on the study of potential toxic side effects.

Ribonuclease P: A Ubiquitous Cellular Enzyme

Ribonuclease P (RNase P) is an essential structure-specific endoribonuclease present in all organisms which removes 5′-flanking sequences of precursor tRNAs by a single cut to generate their mature 5′-ends. All RNase P enzymes thus far analyzed, with the exception of the activity from spinach chloroplasts [205], were found to be ribonucleoprotein particles (reviewed in Refs. [206,207]). In general, eubacterial enzymes have a higher RNA/protein ratio than eukaryotic RNase P enzymes, and archaebacterial enzymes appear to equal either their eubacterial or eukaryotic counterparts in this respect [207,208]. RNA subunits of eubacterial RNase P enzymes were shown to be catalytically active in vitro in the absence of protein components [209]; thus, eubacterial RNase P RNA subunits are true RNA catalysts or ribozymes. In vitro catalysis by RNA subunits from eukaryotic nuclear, mitochondrial, or archaebacterial RNase P enzymes in the absence of associated proteins has not been demonstrated so far. The secondary structures of the well-characterized RNase P RNAs from *Saccharomyces cerevisiae* and *E. coli* and regions protected by the protein moieties as inferred from enzymatic and chemical footprinting experiments are shown in Fig. 10 (according to Ref. [210]).

Model substrates composed of two separate RNAs, which retain certain structural features of present-day tRNAs, could be cleaved by *E. coli* RNase P [211–215]. These observations suggested the feasibility of an RNA inactivation concept involving RNase P, which is discussed in detail in the following sections.

240

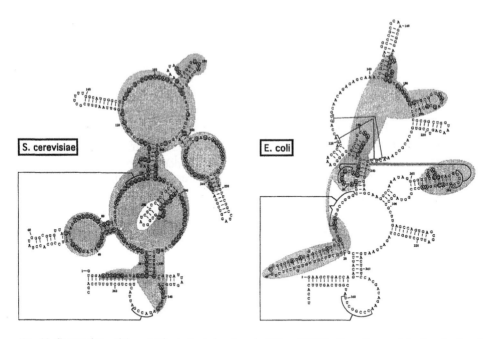

Fig. 10. Comparison of the protein-protected regions in RNase P RNAs from *S. cerevisiae* (nuclear) and
E. coli. In the *S. cerevisiae* RNA structure, circled nucleotides indicate positions of increased or new
sensitivity to enzymatic and chemical modification following deproteinization of the purified cellular
enzyme. In the *E. coli* structure, circled nucleotides identify positions protected from enzymatic and
chemical modifications in the presence of the protein subunit (C5). Shaded areas in both structures
illustrate clusters of nucleotides that are protected in the presence of the protein components (taken
from Ref. [210]; with permission of the authors and the publisher, the American Chemical Society).
Base pairing between nucleotides 120/121 and 236/237 as well as between 125/126 and 234/235 have
been included in the *E. coli* structure [267,268].

Physical and Kinetic Properties of Eukaryotic RNase P Enzymes

Catalytic rates (k_{cat}) of multiple turnover reactions catalyzed by eukaryotic RNase P
enzymes have not been reported so far. This may be attributable to: (i) unavailability
of highly purified enzyme preparations, (ii) loss of activity during purification [216],
e.g. caused by disruption of the ribonucleoprotein complex or degradation of the
RNA subunit, (iii) difficulties associated with the reconstitution of the enzyme in
vitro from isolated protein and RNA components [217], and (iv) unavailability of
sufficient amounts of (overexpressed) protein component(s). Detailed kinetic charac-
terizations of eukaryotic RNase P enzymes are expected to be performed in the near
future.

Previously, the protein subunit (105,000 Da) of mitochondrial RNase P from *Saccharomyces cerevisiae* mitochondria could be purified and its gene has been cloned
[218]. More recently, also the gene encoding a putative 100,000 Da protein compo-

nent common to *S. cerevisiae* nuclear RNase P and RNase MRP (see below) was identified [219]. Likewise, *Schizosaccharomyces pombe* RNase P (a separate mito-chondrial RNase P activity has not been identified thus far) was purified to apparent homogeneity (23,000-fold enriched), and a single 100,000 Da protein subunit was found to copurify with enzymatic activity [220]. For human RNase P, a protein com-ponent with a molecular mass of about 40,000 Da was found to be associated with the enzyme [216,221,222]. However, a 40,000 Da protein subunit cannot account for the physical properties of human RNase P which predict a high protein/RNA ratio [217]. It is conceivable that either multiple 40,000 Da subunits associate with the 340-nt H1 RNA subunit of human RNase P or additional, as yet unidentified protein subunits may be part of this ribonucleoprotein particle [217].

RNase P enzymes from *S. cerevisiae* and *S. pombe* bind precursor tRNAs with higher affinity than mature tRNA [220,223]. When purified *S. pombe* RNase P was incubated with a non-cleavable s^4-uracil-modified pre-tRNA, specific cross-linking to the 100,000 Da protein subunit was observed. In this assay, pre-tRNA inhibited cross-linking of the non-cleavable pre-tRNA more efficiently than mature tRNA [220]. Nuclease treatment of the *S. cerevisiae* enzyme did not abolish binding of (pre)-tRNA, suggesting a key role of the protein component in substrate binding [223]. The K_d for binding of the non-cleavable precursor sup3-e tRNA^{Gly} to *S. cere-visiae* RNase P, as inferred from gel retardation analysis, was in the range of 1.5 nM [223]. The K_m for the cleavage of two pre-tRNAs by *S. pombe* RNase P was reported to be 20 nM [224], and the enzyme was estimated to be present at 1,000 copies per cell [224].

Nonetheless, a more detailed knowledge of intracellular RNase P concentration and distribution, particularly for the human RNase P, will be required to better assess substrate turnover by the enzyme in vivo.

Substrate Recognition by Different RNase P Enzymes

RNase P is the only known cellular enzyme that accounts for the 5'-maturation of all precursor tRNAs in the cell. Since RNase P enzymes recognize structural features common to all tRNAs, heterologous precursor tRNAs are generally functional substrates. In contrast, aminoacyl-tRNA synthetases sense structural idiosyncrasies of their cognate tRNAs. The human mitochondrial pre-tRNA processing enzyme might be exceptional since its substrate specificity was recently found to differ from that of human nuclear RNase P (Robert Karwan, personal communication).

The design of bipartite precursor tRNA-like substrates for the specific inactivation of target RNAs requires knowledge of how RNase P enzymes recognize their natural pre-tRNA substrates. Initial studies with *E. coli* RNase P revealed that this particular enzyme can cleave small hairpin substrates mimicking co-axially stacked acceptor stem and T arm, but missing the D- and anticodon arms [226]. Even an acceptor mi-crohelix carrying the 3'-terminal CCA extension of functional tRNAs could be cleaved, although with reduced efficiency [211,212]. Unlike the *E. coli* enzyme,

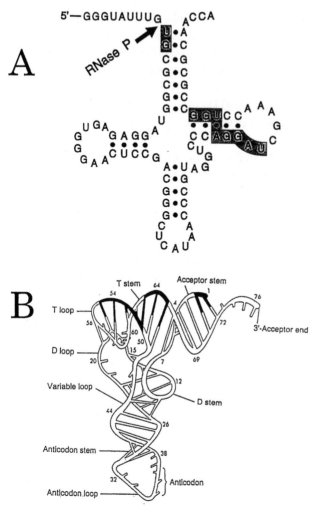

Fig. 11. Proposed RNase P contact regions in tRNAs. (A) Secondary structure of a precursor of initiator tRNA[Met] from *Schizosaccharomyces pombe* transcribed *in vitro*; boxed nucleotides indicate positions at which base modifications interfered with cleavage by *S. pombe* RNase P and *E. coli* RNase P RNA [226]. (B) Localization of the interfering positions in the tertiary structure (model of yeast tRNA[Phe]). Proposed contact regions for RNase P enzymes are marked in black (according to Ref. [226]).

RNase P from the thermophilic eubacterium *Thermus thermophilus* cleaved precursor tRNAs lacking the D-arm with substantially reduced efficiency [213]. In comparison, deletion of the anticodon arm neither affected cleavage by the *E. coli* enzyme nor by *T. thermophilus* RNase P significantly [227]. In conclusion, precursor tRNAs missing an authentic D-arm structure generally interact more weakly with RNase P enzymes [227], even though appreciable activity in vitro can be attained with some eubacterial enzymes (such as the one from *E. coli*). Likewise, eukaryotic RNase P enzymes from human HeLa cells [215,227] and the yeast *Schizosaccharo-*

myces pombe (R.K. Hartmann and G. Krupp, unpublished results) display a strict requirement for the presence of the D-arm. Several lines of evidence indicate that eubacterial RNase P enzymes [228,229], as well as their eukaryotic counterparts [228], interact mainly with the acceptor stem and the T arm, along the continuous helical segment at the top of the tRNA tertiary structure (Fig. 11), while the D-arm (and possibly the variable loop) is likely to support this interaction by indirect means [215,227,228,230]. As observed for its eubacterial counterparts, the anticodon arm is not involved in recognition by human RNase P [227,231]. In some cases, where inhibition of substrate cleavage by yeast RNase P due to alterations in the anticodon arm was observed [224,232], effects are attributable to induced changes of the overall tRNA structure and/or steric hindrance of RNase P due to spurious folding of the anticodon/intron domain [233–235].

Potential RNA Inactivation by RNase P

Specific cleavage of target RNAs by RNase P involves three components: (i) the target RNA, (ii) a second RNA, termed the external guide sequence (EGS, [211,215, 231], which forms a precursor tRNA-like hybrid structure with the target RNA, and (iii) the endoribonuclease RNase P. Although being aware of the more complex situation engendered by the involvement of three components in RNase P-based applications, some interesting advantages may be inherent in this system, e.g. the introduction of a second level of specificity provided by the interaction of the precursor tRNA-like hybrid with endogenous RNase P. In addition, we would like to note that such a three-component system is not unprecedented: concerted RNA inactivation by antisense DNA oligonucleotides and RNase H or 2–5A-dependent RNase (see section on Modified oligonucleotides) involve cellular enzymes as well; similarly, inactivation of target RNAs as part of double-stranded sense-antisense RNA/RNA hybrids may be enhanced by cellular activities such as RNase III or the unwinding/modifying activity [235,236].

With regard to an approach relying on RNase P several central questions arise: Will the EGS RNA interact specifically with its target RNA and will it form a precursor tRNA-like structure, and how far can the dissociation equilibrium be driven towards complex formation? In addition, the precursor tRNA-like structure has to be recognized by endogenous RNase P which raises the question: What are the chances of encountering an endogenous RNase P particle in the cell and will the target RNA-EGS complex be an efficient substrate for the enzyme? This will be affected by the specificity of the target RNA–EGS interaction, complex stability and its fate along cellular trafficking pathways, by the subcellular localization and abundance of RNase P particles, and by the successful competition of precursor tRNA-like substrates with natural pre-tRNAs.

Two examples of precursor tRNA-like structures have been depicted in Fig. 12. The first strategy, relying on hybrid acceptor and D-stems (Fig. 12A), was shown to permit cleavage of a target RNA by human RNase P in vitro [215,231], and has al-

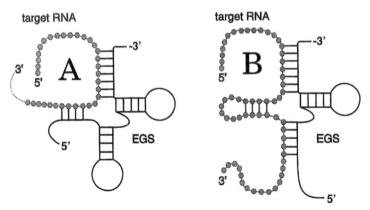

Fig. 12. Secondary structures of precursor tRNA-like hybrid RNAs. Strings of beads depict the target RNA which forms a complex with the second RNA, termed external guide sequence (EGS) [211]. (A), according to Yuan et al. [215]; (B), according to Surratt et al. [214].

ready reached the stage of being tested in cell culture [215]. The second strategy (Fig. 12B) would involve hybrid acceptor and anticodon stems [214] and may be an alternative for target sites carrying natural D-arm-like structures. In fact, a screening in the EMBL data bank for sequences equivalent to D-arms of natural tRNAs revealed that such elements occur fairly often in eukaryotic genes (unpublished observations).

Yuan et al. [215] have designed an EGS for RNase P-mediated cleavage of mRNA coding for chloramphenicol acetyltransferase (CAT). Plasmids encoding CAT and the EGS (under control of the mouse U6 snRNA gene promoter transcribed by RNA polymerase III) were cotransfected into human lung cancer cells (A549). A 60% reduction in CAT activity at 48 h posttransfection could be observed due to expression of the EGS RNA. Concomitantly, cellular levels of CAT mRNA were found to be decreased. Since expected RNase P cleavage products were not detectable, probably due to rapid degradation of mRNA fragments, further efforts are required to prove the involvement of RNase P in EGS-mediated reduction of intracellular CAT mRNA levels. A variant EGS that associates with the CAT mRNA as efficiently as the normal EGS, but that would be suboptimal for recognition by RNase P, could be a conceivable control. However, such a variant EGS may be difficult to design without affecting formation, stability and specificity of the target RNA–EGS complex (see below).

Target RNA interaction: stability and specificity of hybrid formation

Ribozymes, such as the hammerhead and hairpin versions, involve two regions of consecutive Watson–Crick-type base pairing between the target and the catalytic antisense nucleic acid (Figs. 4, 5 and 9). A minimal number of base pairs is required to form the specific complex. However, expanding base pairing beyond a certain

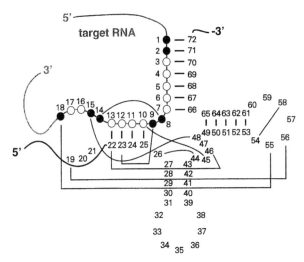

Fig. 13. Secondary structure of a precursor tRNA-like hybrid RNA based on the structure of yeast tRNA[Phe]. Circles indicate nucleotides provided by the target RNA, and nucleotides of the EGS RNA are solely marked by numbers, according to the numbering system for yeast tRNA[Phe]. Filled circles indicate bases that are involved in tertiary interactions in yeast RNA[Phe], and which are likely to increase stability and support authentic folding of precursor tRNA-like structures. Lines connect bases involved in tertiary interactions in yeast tRNA[Phe] (according to [269]). Bases corresponding to nucleotides 44–48 and 57–60 were randomized by Yuan and Altman [231] for in vitro selection in the context of another tRNA construct. Base pairing in the anticodon stem has been omitted since the most efficient EGS selected by Yuan and Altman [231] had a disrupted anticodon stem.

length (also depending on the overall stability of the hybrid as defined by the A/U content) may decrease the specificity of the interaction, since mismatches do not destabilize complex formation enough so as to be discriminated against [238]. If the dissociation of ribozyme–substrate complexes containing mismatches is sufficiently slow, cleavage will occur whenever a complex is formed. Furthermore, substrate turnover catalyzed by the ribozyme nucleic acid may be accelerated at non-specific sites due to better product dissociation of mismatched hybrids [239]. Based on theoretical considerations regarding the specificity of antisense oligonucleotides, 11–15 nucleotides are likely to represent the shortest sequence which is unique to an RNA pool of a typical higher eukaryotic cell, assuming the occurrence of 10^4 different mRNA species at an average length of 2 kilobases corresponding to a sequence complexity of 2×10^7 bases [25,153]. A DNA decamer was shown to be sufficient to elicit RNase H cleavage at the RNA target site in *Xenopus* oocytes [25]. However, a 10-mer was estimated to have 19 complementary sites in the RNA pool just mentioned [25]. Accordingly, a 13-mer, which provides four overlapping windows of 10 nucleotides, would perfectly match 76 complementary stretches of 10 nucleotides in the RNA pool. Although a 10-bp interaction may not always represent the minimum length for concerted target site inactivation by antisense oligonucleotides and endogenous RNase H (see below), such considerations illustrate that nonspecific RNA inactivation can become a serious drawback of DNA antisense approaches.

Formation of precursor tRNA-like hybrids also involves short regions of regular Watson–Crick base pairing (Figs. 12 and 13). This raises similar questions as discussed above with respect to the specificity of such interactions. However, tRNA molecules involve a set of tertiary interactions as inferred from X-ray analyses and supported by numerous studies on tRNA mutants (reviewed in Refs. [202,240]). Single-stranded nucleotides provided by the target RNA (Fig. 13), such as nucleotides 8 and 9 connecting acceptor stem and D stem, as well as one or more nucleotides in the D-loop, are likely to increase the sequence-specific information contributed by the target site. On the one hand, such tertiary interactions in addition to base pairing will stabilize the precursor tRNA-like structure. Furthermore, the specificity of target RNA inactivation by endogenous RNase P will be increased, since the enzyme will discriminate between a properly folded pre-tRNA-like structure at the target site and suboptimal hybrid structures formed at non-target sites. For example, efficient processing of a bacterial precursor tRNA^{Gly} by human RNase P was dependent on the presence of a genuine D-arm [227]. The 5′- and 3′-end maturation of plant tRNA^{Ser} transcripts in HeLa cell nuclear extracts was found to be inhibited by two concomitant mismatches, one occurring in the acceptor stem and the other in the D-stem [241]. Disruption of the first or second acceptor stem base-pair in the context of a *Schizosaccharomyces pombe* sup3-e tRNA^{Ser} also prevented cleavage by RNase P enzymes from *S. pombe* [224] and *Saccharomyces cerevisiae* [223]. Cleavage was also reduced for sup3-e tRNA^{Ser} variants carrying a mismatch at the third acceptor stem base-pair or by mutations in the D-stem and D-loop [233]. Elimination of the first acceptor stem base-pair in a precursor tRNA^{Leu} from yeast caused a 50% reduction of the initial cleavage rate by *Xenopus laevis* RNase P, and simultaneous disruption of the 5′-terminal three acceptor stem base-pairs almost completely abolished cleavage by the enzyme [230]. An exchange of G19 to C in the D-loop of a yeast suppressor pre-tRNA^{Leu} (an amber allele of the tRNA^{Leu} used in [230]) abolished cleavage by RNase P from *S. cerevisiae* nuclei [232]. Yuan et al. [215] have analyzed some structural requirements for cleavage by human RNase P of a precursor tRNA-like hybrid of the type shown in Fig. 12A and Fig. 13. Changes in the number of base pairs in the acceptor and D-stems as well as switches of base identities at positions 8 and 9 (Fig. 13) yielded inactive substrates. Anticodon and variable stems and loops, however, were found to have marginal effects on cleavage by human RNase P [215], as was observed for the 3′-terminal CCA sequence (W.-D. Hardt and R.K. Hartmann, unpublished data). In accordance, a tRNA, in which the anticodon arm had been replaced by a tetra-nucleotide linker, was cleaved with similar efficiency by human RNase P as the wild-type tRNA [227], supporting the notion that the anticodon arm per se is not recognized by the enzyme. Base pairings and base identities which may be critical for the formation of precursor tRNA-like structures involving hybrid acceptor and D-stems are summarized in Fig. 13. Note that detailed effects on RNase P processing owing to mismatches or disruption of potential tertiary interactions by switches of base identities at positions 8 to 10 or in the D-loop

(Fig. 13) are likely to depend on the particular tRNA context, and may also vary for different eukaryotic RNase P enzymes.

In conclusion, it is reasonable to assume that most disruptions of hybrid stem regions and tRNA-like tertiary interactions will easily abolish cleavage of precursor tRNA-like hybrid structures by endogenous RNase P which senses an intricate three-dimensional RNA structure. Accordingly, complexes formed by an EGS RNA at non-target sites, which deviate from the target RNA sequence in one or more positions, will have a low probability of being cleaved by the enzyme, thus increasing specificity. It may turn out that mismatches in RNA–RNA and DNA–RNA hybrids are more easily tolerated in ribozyme- or RNase H-catalyzed reactions. Woolf et al. [25] have provided evidence that partial destruction of non-target RNAs by RNase H due to imperfectly matched antisense oligonucleotides is likely to occur, as discussed above. This may impair the specificity of antisense effects, and may give rise to cyto-toxic effects [56], although the question remains whether results obtained with *Xenopus* oocytes (non-dividing immature eggs) can be extrapolated to other cell types [25,165]. Furthermore, frog embryos grow at lower temperatures than mammalian cells, thus lowering the stringency of sense-antisense nucleic acid hybridization conditions. In addition, RNA pools of many cells turn over rapidly, which may alleviate effects of non-specific RNA degradation. Finally, intracellular traffic-control patterns may preclude some mRNAs from being inactivated by antisense oligonucleotides.

In general, one expects RNA binding proteins to mask RNA targets and interfere with antisense applications. However, the findings of Tsuchihashi et al. [246] have demonstrated that the p7 nucleocapsid protein (NCp7) of human immunodeficiency virus-type 1 can increase the activity and specificity of hammerhead ribozymes. Similar properties were also reported for the mammalian heterogeneous nuclear ribonucleoprotei*n* A1 (hnRN*P* A1) [242,243]. By virtue of binding preferentially to single-stranded nucleic acids, these proteins can enhance multiple turnover reactions of the ribozyme by increasing the rate of annealing between substrate and ribozyme, and by facilitating dissociation of the cleavage products [242–244]. Such RNA binding proteins of broad specificity seem to act as RNA chaperones, preventing the trapping of RNA in inactive structures or enhancing the resolution of misfolded RNAs. For example, ribosomal protein S12 of *E. coli* was reported to facilitate adoption of a catalytically active *trans*-splicing group I intron as well as the action of a hammerhead ribozyme, and splicing enhancement was still observed when S12 was removed by proteinase K treatment before initiating the splicing reaction by the addition of GTP [245]. Activities of RNA binding proteins were largely retained in short peptide derivatives, containing two basic amino acid clusters of the amino terminus of NCp7 [244] or the carboxy-terminal domain of hnRNP A1 [243]. However, higher concentrations of the proteins inhibit ribozyme activity. Thus, improvements of ribozyme efficiency may be limited when such proteins are overexpressed in the target cell, or when active peptide subdomains are covalently linked to synthetic ribozymes in approaches relying on external ribozyme application. Nonetheless, based on the observations that NCp7 not only accelerates multiple turnover reactions but also enhances the discrimination between matched and mismatched hammerhead substrates in vitro [243,244], one could speculate that RNA chaperones may alleviate the specificity problem in vivo or might open up the perspective to utilize extended base pairing between substrate and enzyme RNAs without compromising specificity.

Optimization of RNase P substrates

The cellular function of tRNAs and their interaction with a variety of cellular components have defined the constraints for the evolution of present-day tRNA structures. If processing of precursor tRNAs by RNase P were the only selective constraint, introducing novel intramolecular interactions or disrupting others may modulate the tRNA structure into becoming a more efficient substrate for this particular enzymatic activity, e.g. by increasing k_{cat} and lowering the K_m of the substrate. In case of pre-tRNA-like hybrid structures, association rates of the two RNAs may become rate-limiting for cleavage by RNase P. Since optimizing bipartite pre-tRNA-like structures for processing by RNase P may require simultaneous changes of several base identities, a rationale for the design of mutations is lacking. Thus, in vitro evolution of fast-cleaving substrate variants appears to be the most promising approach. Yuan and Altman [231] have applied an in vitro selection protocol to pick out pre-tRNAs efficiently cleaved by human RNase P. The substrate was a monomolecular chimeric pre-tRNA in which nucleotides 1–18 were derived from the target RNA, while the remainder of the tRNA corresponded to the EGS. In this particular approach, nucleotides in the T-loop and variable loop were randomized (Fig. 13), and fast-cleaving pre-tRNA variants were selected by eight cycles of selection and amplification. Subsequently, hybrids of target RNA and EGS, designed on the basis of selected pre-tRNA variants, were analyzed for processing by human RNase P. The most efficient EGS (EGS 9 [231]) promoted cleavage of the target RNA by human RNase P at a rate only 1.5 times slower than observed with the selected monomolecular pre-tRNA. It is amazing that the EGS 9 directed specific target mRNA cleavage by human RNase P in vitro with a catalytic efficiency (v_{max}/K_m) equal to that achieved with a wild-type pre-tRNA^{Tyr}. Conspicuously, the EGS 9 had a disrupted anticodon stem, and nucleotides in the anticodon loop region became accessible to the double-strand specific cobra venom nuclease, suggesting that novel tertiary interactions occurred in the complex of target RNA and EGS RNA.

Cellular localization of human RNase P

Once a precursor tRNA-like complex is formed between the target RNA and the EGS, an RNase P molecule has to be encountered for specific inactivation of the target RNA. RNase P is thought to be located in the nucleus [217]. Since significant amounts of RNase P activity leak out of nuclei during the isolation procedure [217], the actual intracellular distribution of the enzyme in vivo has been unclear. Fluorescently-labelled H1 RNA (the RNA subunit of human RNase P) moves very rapidly to nucleolar sites when injected into nuclei of either normal rat kidney (NRK) cells or HeLa cells [247]. In contrast, microinjected human pre-tRNA was observed throughout the nucleoplasm [247]. H1 RNA appears to co-localize with fibrillarin, a protein component of snRNPs involved in pre-rRNA processing, as inferred from immunocytochemical studies [248]. On the other hand, preliminary in situ hybridization data suggest that although the majority of endogenous H1 RNA in rat fibroblasts is localized in the nucleoplasm and cytoplasm, a small percentage of signal is observed in nucleoli (M.R. Jacobson and T. Pederson, pers. commun.). This is consistent with biochemical analyses, based on fractionation of HeLa cells into nucleolar, nucleoplasmic and cytoplasmic fractions. The presence of H1 RNA in total RNA from each fraction was examined by primer extension. Again, a small amount of H1 RNA was observed in the nucleolar fraction, while the majority was detected in the nucleoplasmic and cytoplasmic fractions (M.R. Jacobson and T. Pederson, pers. commun.). Furthermore, hybrid selection of RNA isolated from nucleolar, nucleoplasmic and cytoplasmic fractions of [³H]uridine pulsed HeLa cells reveals that H1 RNA is transiently present in the nucleolar fraction and subsequently accumulates in the nucleoplasmic and cytoplasmic fractions (M.R. Jacobson and T. Pederson, pers. commun.). The data suggest that RNase P may either have some unknown function in the nucleolus, and/or assembly of the enzyme from RNA and protein components occurs at this location.

In the study by Jacobson et al. [247], MRP RNA, the RNA subunit of RNase MRP, another structure-specific ribonucleoprotein endoribonuclease, showed a rapid nucleolar localization similar to H1 RNA after microinjection into HeLa nuclei, and appears to co-localize with fibrillarin as well [248]. RNase MRP RNA mutated in the binding site for the 40,000 Da To-antigen (see below) did not localize to the nucleolus upon microinjection into the nucleus of normal rat kidney (NRK) cells [249]. RNase MRP is thought, as one of its biological functions, to be involved in processing of primer RNA for mitochondrial DNA replication [250–253]. In HeLa cells, most if not all of the RNase MRP is found in the nucleus [254], localized to the granular compartment of the nucleolus [255] suggesting a role in ribosome biogenesis. In fact, RNase MRP depletion in Saccharomyces cerevisiae led to altered 5.8S rRNA maturation [256]. Recent data suggest the possibility that RNase MRP is responsible for cleavage at site A3 in the internal transcribed spacer 1 (ITS1) of pre-rRNA, which is a prerequisite for the generation of the major short form of 5.8S rRNA in S. cerevisiae [257]. It is interesting to note that RNase P and RNase MRP share several common features: (i) both endoribonucleases generate 5'-phosphates and 3'-OH termini in a reaction dependent on divalent metal ions [251], (ii) human anti-Th/To antibodies immunoprecipitate RNase P and RNase MRP [222,258], and the 40,000 Da To-antigen interacts with human MRP RNA and RNase P RNA [258]. Very recently, a mutation in the pop1 gene (encoding a 100,000 Da protein, POP1) was shown to affect 5.8S rRNA processing as well as pre-tRNA processing in S. cerevisiae, and immunoprecipitation revealed that the 100,000 Da POP1 protein associates with both RNase P and RNase MRP RNA [219]. This suggests that the

40,000 Da protein (To-antigen) and POP1 are shared by both ribonucleoprotein particles; (iii) RNase P RNA subunits from bacteria, eukaryotic nuclei and mitochondria as well as MRP RNAs (also including 7–2 RNA, the plant equivalent to mammalian MRP RNA) can be folded into a similar, cage-shaped secondary structure involving a highly conserved pseudoknot [259–261]. It remains to be shown whether the striking similarities between RNase P and RNase MRP may also indicate a common functional context of the two ribonucleoprotein enzymes.

In conclusion, a detailed picture with respect to the subcellular distribution of mammalian RNase P is still not at hand. However, the existing experimental evidence discussed above does not argue against the notion that the enzyme is homogeneously dispersed in the nucleoplasm and the cytoplasm. This may be taken as a promising indication since the availability of RNase P in vivo will be a key determinant with respect to the enzyme's ability to gain access to precursor tRNA-like hybrid structures formed on RNA polymerase II transcripts. Analysis of the in vivo fate of reporter gene transcripts synthesized by RNA polymerase II which carry an insertion of a monomolecular tRNA may be a first step in assessing whether intracellular traffic-control patterns permit RNase P to act on RNA polymerase II transcripts. A reporter gene construct harbouring an altered tRNA structure, which is cleaved with low efficiency by RNase P, may serve as a control.

RNase P RNA–EGS RNA conjugates

Recently, self-cleaving RNA constructs were described in which a complete pre-tRNA moiety was attached to one end of *E. coli* RNase P RNA. When the tRNA structure was linked to the 3'-end, intramolecular site-specific cleavage was observed [262]. Similarly, conjugates of bacterial RNase P RNAs and EGS RNAs were proposed as a method for efficient target RNA inactivation. In the approach by Liu and Altman (Fenyong Liu and Sidney Altman, pers. commun.), an EGS RNA tethered to the 3'-end of *E. coli* RNase P RNA was reported to be active under conditions of low monovalent salt and 10 mM Mg^{2+}. Frank et al. [263] coupled the EGS-equivalent RNA (here called internal guide sequence (IGS)) to internal nucleotides of circularly permuted RNase P RNA. This was intended to position the hybrid tRNA adjacent to the substrate binding region of RNase P RNA in order to facilitate cleavage of the target RNA. Cleavage was most efficient in vitro at high concentrations of monovalent cations (2.5–3.0 M ammonium acetate).

Successful application of such conjugates as specific tools for RNA inactivation will depend on the stability and proper folding of bacterial RNase P RNA moieties in the eukaryotic environment, as well as on efficient cleavage in the ionic milieu of mammalian cells.

Exogenous application of EGS RNAs

As outlined earlier, exogenous application of antisense oligonucleotides offers the potential to synthesize modified, nuclease-resistant RNAs (by enzymatic or chemical methods). In addition, the modified and thus more stable RNAs may bear the advantage of being more effective tools at lower doses than their unmodified counterparts. As mentioned before, a potential drawback might arise from the possibility that modified mononucleotides generated by RNA degradation enter the cellular nucleotide pool, which may cause toxic side effects owing to their in-corporation by cellular polymerases or to inhibition of enzymes involved in nucleo-tide metabolism.

The most commonly used modifications include phosphorothioate and ribose 2'-hydroxyl modifications. Utilization of such modifications for the synthesis of modi-fied EGS RNAs requires identifying non-bridging phosphate oxygens (in the case of phosphorothioate modifications) and ribose 2'-hydroxyls in the EGS RNA that do not interfere with the formation of the target RNA-EGS complex and/or with rec-ognition by RNase P. As a first step, modification interference studies with natural precursor tRNAs may be performed. For this purpose, pre-tRNAs partially modified with Rp-phosphorothioates (obtained by in vitro transcription employing bacterio-phage polymerases) are subjected to cleavage by RNase P [264]. Substrates readily cleaved by the enzyme can be separated by denaturing polyacrylamide gel electro-phoresis from those showing retarded cleavage (due to modifications at particular positions). Subsequent iodine treatments of the unprocessed and mature tRNA frac-

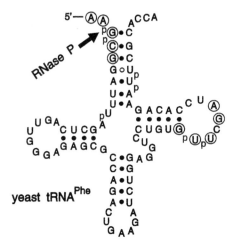

p : Rp-phosphorothioate interferes

Ⓝ : 2'-deoxy interferes

Fig. 14. Sites of Rp-phosphorothioate and 2'-deoxyribose-Rp-phosphorothioate modifications in tRNA[Phe] that interfere with cleavage by *E. coli* RNase P RNA. Analysis was performed with in vitro transcribed tRNA[Phe] carrying a dinucleotide 5'-flank. The RNase P cleavage site is marked by an ar-row. The letter p identifies Rp-phosphorothioate-modified phosphodiesters and encircled nucleotides indicate 2'-deoxyribose-Rp-phosphorothioate modifications that interfered with cleavage by *E. coli* RNase P RNA [264; F. Conrad, A. Hanne and G. Krupp, unpublished results].

tion leads to backbone breakages at phosphorothioate-modified positions. Modifications that interfere with processing by RNase P will accumulate in the unprocessed substrate fraction, thus yielding a stronger signal in the iodine cleavage pattern. This approach has recently been extended by incorporating dNTPαS or even 2'-O-methyl-NTPαS nucleotides into precursor tRNAs [264,265; G. Krupp, unpublished data], which provides a means to identify functionally important 2'-hydroxyls at positions where an Rp-phosphorothioate-modification by itself has no effect on processing by RNase P. The potential of modification interference studies is illustrated by results obtained with *E. coli* RNase P RNA (Fig. 14), although eukaryotic RNase P enzymes may respond differently to site-specific modifications. In a next step, phosphorothioate- and ribose 2'-modifications having only marginal effects on RNase P cleavage as part of monomolecular tRNA substrates may be introduced into a corresponding EGS RNA (Fig. 13). This allows one to evaluate whether such modifications have the same effects on cleavage in the context of bipartite pre-tRNA-like hybrid substrates.

Prospects

Since pre-tRNA-like hybrid structures have been identified [231] which are cleaved as efficiently as natural pre-tRNA transcripts by human RNase P in vitro, an important prerequisite for specific RNA inactivation by this enzyme seems to have been accomplished. Successful in vivo applications will depend on efficient and specific complex formation of target RNA and EGS and concomitant availability of RNase P. This may also include coupling of EGS RNAs to cellular localization signals. Clearly, approaches of specific RNA inactivation involving RNase P have now to surmount the hurdle of successful and conclusive testing in cell culture.

Acknowledgements

We are grateful to Marty R. Jacobson and Thoru Pederson (Shrewsbury, MA, USA), Norman R. Pace (Bloomington, IN, USA), Sidney Altman (New Haven, CT, USA), Lisa Hegg (Worcester, MA, USA), Karin Mölling (Zürich, Switzerland), David Tollervey (Heidelberg, Germany), Robert Karwan (Vienna, Austria) and E. Gerhart H. Wagner (Uppsala, Sweden) and their coworkers for providing unpublished results, David Engelke (Ann Arbor, MI, USA) and the American Chemical Society for their agreement to use Fig. 10, and to Abraham Shevack (Berlin), Silke Volkmann (Berlin), Leif Kirsebom (Uppsala, Sweden), Jens Peter Fürste (Berlin), Anita Marchfelder (Berlin), Matthias Marget (Kiel), Judith Schlegl (Berlin) and Volker A. Erdmann (Berlin) for helpful comments and critical reading of the manuscript. Financial support from the Deutsche Forschungsgemeinschaft to R.K.H. (SFB344/C2, Ha 1672/4-1) and to G.K. (Kr 817/3-2) is acknowledged. W.-D.H. is a stipend of the Boehringer Ingelheim Fonds.

References

1. Paterson BM, Roberts BE, Kuff EL. Structural gene identification and mapping by DNA-mRNA hybrid-arrested cell-free translation. Proc Natl Acad Sci USA 1977;74:4370–4374.
2. Zamecnik PC, Stephenson ML. Inhibition of Rous sarcoma virus replication and cell transformation by a specific oligodeoxynucleotide. Proc Natl Acad Sci USA 1978;75:280–284.
3. Giver L, Bartel D, Zapp M, Pawul A, Green M, Ellington AD. Selective optimization of the Rev-binding element of HIV-1. Nucleic Acids Res 1993;21:5509–5516.
4. Sullenger BA, Gallardo HF, Ungers GE, Gilboa E. Overexpression of TAR sequences renders cells resistant to human immunodeficiency virus replication. Cell 1990;63:601–608.
5. Winston F, Chumley F, Fink GR. Eviction and transplacement of mutant genes in yeast. Methods Enzymol 1983;101:211–228.
6. De-Lozanne A, Spudich JA. Disruption of the *Dictyostelium* myosin heavy chain gene by homologous recombination. Science 1987;236:1086–1091.
7. Lee KY, Lund P, Lowe K, Dunsmuir P. Homologous recombination in plant cells after *Agrobacterium*-mediated transformation. Plant Cell 1990;2:415–425.
8. Paszowski J, Baur M, Bogucki A, Potrykus I. Gene targeting in plants. EMBO J 1988;7:4021–4026.
9. Schwartzberg PL, Goff SP, Robertson EJ. Germ-line transmission of a c-*abl* mutation produced by targeted gene disruption in ES cells. Science 1989;246:799–803.
10. Song KY, Schwartz F, Maeda N, Smithies O, Kucherlapati R. Accurate modification of a chromosomal plasmid by homologous recombination in human cells. Proc Natl Acad Sci USA 1987;84:6820–6824.
11. Thomas CM. Regulation of gene expression and function by antisense RNA in bacteria. In: Murray JAH (ed) Antisense RNA and DNA. New York: Wiley-Liss 1992;51–76.
12. Stevens JG, Wagner EK, Devi-Rao GB, Cook ML, Feldman LT. RNA complementary to a herpesvirus α gene mRNA is prominent in latently infected neurons. Science 1987;235:1056–1059.
13. Okano H, Ikenaka K, Mikoshiba K. Recombination within the upstream gene of duplicated myelin basic protein genes of myelin deficient shi^{mld} mouse results in the production of antisense RNA. EMBO J 1988;7:3407–3412.
14. Tosic M, Roach A, de Rivaz J-C, Dolivo M, Matthieu J-M. Post-transcriptional events are responsible for low expression of myelin basic protein in myelin deficient mice: role of natural antisense RNA. EMBO J 1990;9:401–406.
15. Nepveu A, Marcu KB. Intragenic pausing and antisense transcription within the murine c-myc locus. EMBO J 1986;5:2859–2865.
16. Piechaczyk M, Blanchard J-M, Bonnieu A, Fort P, Mechti N, Rech J, Cuny M, Marty L, Ferre F, Lebleu B, Jeanteur P. Role of RNA structures in c-myc and c-fos gene regulations. Gene 1988;72:287–295.
17. Kimelman D. Regulation of eukaryotic gene expression by natural antisense transcripts. In: Erickson RP, Izant JG (eds) Gene Regulation: Biology of antisense RNA and DNA. New York: Raven Press, 1992;1–10.
18. Krystal GW. Regulation of eukaryotic gene expression by naturally occurring antisense RNA. In: Erickson RP, Izant JG (eds) Gene Regulation: Biology of Antisense RNA and DNA. New York: Raven Press, 1992;11–20.
19. Wassenegger M, Heimes S, Riedel L, Sänger HL. RNA-directed de novo methylation of genomic sequences in plants. Cell 1994;76:567–576.
20. Wang S, Dolnick BJ. Quantitative evaluation of intracellular sense: antisense RNA duplexes. Nucleic Acids Res 1993;21:4383–4391.
21. Kerr SM, Stark GR, Kerr IM. Excess antisense RNA from infectious recombinant SV40 fails to inhibit expression of a transfected, interferon-inducible gene. Eur J Biochem 1988;175:65–73.

254

22. Crouch RJ, Dirksen M-L. Ribonucleases H. In: Linn SM, Roberts RJ (eds) Nucleases. Cold Spring Harbor, NY: Cold Spring Harbor Laboratory 1982;211–241.

23. Eder PS, Walder JA. Ribonuclease H from K562 human erythroleukemia cells. J Biol Chem 1991;266:6472–6479.

24. Walder RY, Walder JA. Role of RNase H in hybrid-arrested translation by antisense oligonucleotides. Proc Natl Acad Sci USA 1988;85:5011–5015.

25. Woolf TM, Melton DA, Jennings CGB. Specificity of antisense oligonucleotides *in vivo*. Proc Natl Acad Sci USA 1992;89:7305–7309.

26. Chiang M-Y, Chan H, Zounes MA, Freier SM, Lima WF, Bennett CF. Antisense oligonucleotides inhibit intracellular adhesion molecule 1 expression by two distinct mechanisms. J Biol Chem 1991;266:18162–18171.

27. Monia BP, Lesnik EA, Gonzalez C, Lima WF, McGee D, Guinosso CJ, Kawasaki AM, Cook PD, Freier SM. Evaluation of 2′-modified oligonucleotides containing 2′-deoxy gaps as antisense inhibitors of gene expression. J Biol Chem 1993;268:14514–14522.

28. Kulka M, Smith CC, Aurelian L, Fishelevich R, Meade K, Miller P, Ts'o POP. Site specificity of the inhibitory effects of oligo(nucleoside methyl-phosphonate)s complementary to the acceptor splice junction of herpes simplex virus type 1 immediate early mRNA 4. Proc Natl Acad Sci USA 1989;86:6868–6872.

29. Boiziau C, Kurfurst R, Cazenave C, Roig V, Thuong NT, Toulmé J-J. Inhibition of translation initiation by antisense oligonucleotides via an RNase H independent mechanism. Nucleic Acids Res 1991;19:1113–1119.

30. Munroe SH. Antisense RNA inhibits splicing of pre-mRNA *in vitro*. EMBO J 1988;7:2523–2532.

31. Stout JT, Caskey CT. Antisense RNA inhibition of HPRT synthesis. Somat Cell Mol Genet 1990; 16:369–382.

32. Ch'ng JLC, Mulligan RC, Schimmel P, Holmes EW. Antisense RNA complementary to 3′ coding and noncoding sequences of creatine kinase is a potent inhibitor of translation *in vivo*. Proc Natl Acad Sci USA 1989;86:10006–10010.

33. Bass BL. The double-stranded RNA unwinding/modifying activity. In: Murray JAH (ed) Antisense RNA and DNA. New York: Wiley-Liss 1992;159–174.

34. Kim SK, Wold BJ. Stable reduction of thymidine kinase activity in cells expressing high levels of anti-sense RNA. Cell 1985;42:129–138

35. Harland R, Weintraub H. Translation of mRNA injected into *Xenopus* oocytes is specifically inhibited by antisense RNA. J. Cell Biol 1985;101:1094–1099.

36. Melton DA. Injected anti-sense RNAs specifically block messenger RNA translation *in vivo*. Proc Natl Acad Sci USA 1985;82:144–148.

37. Wormington WM. Stable repression of ribosomal protein L1 synthesis in *Xenopus* oocytes by microinjection of antisense RNA. Proc Natl Acad Sci USA 1986;83:8639–8643.

38. Hélène C. Control of gene expression by antisense and antigene oligonucleotide-intercalator conjugates. In: Erickson RP, Izant JG (eds) Gene Regulation: Biology of Antisense RNA and DNA. New York: Raven Press 1992;109–118.

39. Thuong NT, Hélène C. Sequence-specific recognition and modification of double helix-DNA by oligonucleotides. Angew Chem Int Ed Engl 1993;32:666–692.

40. Fossella JA, Kim YJ, Shih H, Richards EG, Fresco JR. Relative specificities in binding of Watson-Crick base pairs by third strand residues in a DNA pyrimidine triplex motif. Nucleic Acids Res 1993;21:4511–4515.

41. Escude C, Francois J-C, Sun J-S, Ott G, Sprinzl M, Garestier T, Hélène C. Stability of triple helices containing RNA and DNA strands:experimental and molecular modeling studies. Nucleic Acids Res 1993;21:5547–5553.

42. Han H, Dervan PB. Sequence-specific recognition of double helical RNA and RNA·DNA by triple helix formation. Proc Natl Acad Sci USA 1993;90:3806–3810.

43. Grigoriev M, Praseuth D, Guieysse AL, Robin P, Thuong NT, Hélène C, Harel-Bellan A. Inhibi-

tion of gene expression by triple helix-directed DNA cross-linking at specific sites. Proc Natl Acad Sci USA 1993;90:3501–3505.

44. Grigoriev M, Praseuth D, Robin P, Hemar A, Saison-Behmoaras T, Dautry-Varsat A, Thuong NT, Hélène C, Harel-Bellan A. A triple helix-forming oligonucleotide-intercalator conjugate acts as a transcriptional repressor via inhibition of NFκB binding to interleukin-2 receptor α-regulatory sequence. J Biol Chem 1992;267:3389–3395.

45. Hanvey JC, Shimizu M, Wells RD. Site-specific inhibition of EcoRI restriction/modification enzymes by a DNA triple helix. Nucleic Acids Res 1990;18:157–161.

46. Strobel SA, Dervan PB. Single-site enzymatic cleavage of yeast genomic DNA mediated by triple helix formation. Nature 1991;350:172–174.

47. Strobel SA, Doucette-Stamm LA, Riba L, Housman DE, Dervan PB. Site-specific cleavage of chromosome 4 mediated by triple-helix formation. Science 1991;254:1639–1642.

48. Blume SW, Gee JE, Shrestha K, Miller DM. Triple helix formation by purine-rich oligonucleotides targeted to the human dihydrofolate reductase promoter. Nucleic Acids Res 1992;20:1777–1784.

49. François J-C, Saison-Behmoaras T, Thuong NT, Hélène C. Inhibition of restriction endonuclease cleavage via triple helix formation by homopyrimidine oligonucleotides. Biochemistry 1989;28:9617–9619.

50. Gee JE, Blume S, Snyder RC, Ray R, Miller DM. Triplex formation prevents Sp1 binding to the dihydrofolate reductase promoter. J Biol Chem 1992;267:11163–11167.

51. Maher III LJ, Wold B, Dervan PB. Inhibition of DNA binding proteins by oligonucleotide-directed triple helix formation. Science 1989;245:725–730.

52. Giovannangeli C, Thuong NT, Hélène C. Oligonucleotide clamps arrest DNA synthesis on a single-stranded DNA target. Proc Natl Acad Sci USA 1993;90:10013–10017.

53. Cooney M, Czernuszewicz G, Postel EH, Flint SJ, Hogan ME. Site-specific oligonucleotide binding represses transcription of the human c-myc gene in vitro. Science 1988;241:456–459.

54. Duval-Valentin G, Thuong NT, Hélène C. Specific inhibition of transcription by triple helix-forming oligonucleotides. Proc Natl Acad Sci USA 1992;89:504–508.

55. Maher III LJ, Dervan PB, Wold B. Analysis of promoter-specific repression by triple-helical DNA complexes in a eukaryotic cell-free transcription system. Biochemistry 1992;31:70–81.

56. Young SL, Krawczyk SH, Matteucci MD, Toole JJ. Triple helix formation inhibits transcription elongation in vitro. Proc Natl Acad Sci USA 1991;88:10023–10026.

57. Volkmann S, Dannull J, Mölling K. The polypurine tract, PPT, of HIV as target for antisense and triple-helix-forming oligonucleotides. Biochimie 1993;75:71–78.

58. Postel EH, Flint SJ, Kessler DJ, Hogan ME. Evidence that a triplex-forming oligodeoxyribonucleotide binds to the c-myc promoter in HeLa cells, thereby reducing c-myc mRNA levels. Proc Natl Acad Sci USA 1991;88:8227–8231.

59. Orson FM, Thomas DW, McShan WM, Kessler D.J, Hogan ME. Oligonucleotide inhibition of IL2Rα mRNA transcription by promoter region collinear triplex formation in lymphocytes. Nucleic Acids Res 1991;19:3435–3441.

60. McShan WM, Rossen RD, Laughter AH, Trial JA, Kessler DJ, Zendegui JG, Hogan ME, Orson FM. Inhibition of transcription of HIV-1 in infected human cells by oligodeoxynucleotides designed to form DNA triple helices. J Biol Chem 1992;267:5712–5721.

61. Cameron FH, Jennings PA. Specific gene suppression by engineered ribozymes in monkey cells. Proc Natl Acad Sci USA 1989;86:9139–9143.

62. Cantor GH, McElwain TF, Birkebak TA, Palmer GH, Ribozyme cleaves rex/tax mRNA and inhibits bovine leukemia virus expression. Proc Natl Acad Sci USA 1993;90:10932–10936.

63. Chen C-J, Banerjea AC, Harmison GG, Haglund K, Schubert M. Multitarget-ribozyme directed to cleave at up to nine highly conserved HIV-1 env RNA regions inhibits HIV-1 replication - potential effectiveness against most presently sequenced HIV-1 isolates. Nucleic Acids Res 1992;20:4581–4589.

64. Cotten M, Birnstiel ML. Ribozyme mediated destruction of RNA *in vivo*. EMBO J 1989;8:3861–3866.

65. Dropulic B, Lin NH, Martin MA, Jeang K-T. Functional characterization of a U5 ribozyme: intracellular suppression of human immunodeficiency virus type 1 expression. J Virol 1992;66:1432–1441.

66. Holm PS, Scanlon KJ, Dietel M. Reversion of multidrug resistance in the P-glycoprotein positive human pancreatic cell line (EPP85–181RDB) by introduction of a hammerhead ribozyme. Br J Cancer 1994 (in press).

67. Koizumi M, Kamiya H, Ohtsuka E. Inhibition of c-Ha-*ras* gene expression by hammerhead ribozymes containing a stable C(UUCG)G hairpin loop. Biol Pharm Bull 1993;16:879–883.

68. Sarver N, Cantin EM, Chang PS, Zaia JA, Ladne PA, Stephens DA, Rossi JJ. Ribozymes as potential anti-HIV-1 therapeutic agents. Science 1990;247:1222–1225.

69. Weerasinghe M, Liem SE, Asad S, Read SE, Joshi S. Resistance to human immunodeficiency virus type 1 (HIV-1) infection in human CD4+ lymphocyte-derived cell lines conferred by using retroviral vectors expressing an HIV-1 RNA-specific ribozyme. J Virol 1991;65:5531–5534.

70. Ojwang JO, Hampel A, Looney DJ, Wong-Staal F, Rappaport J. Inhibition of human immunodeficiency virus type 1 expression by a hairpin ribozyme. Proc Natl Acad Sci USA 1992;89:10802–10806.

71. Yu M, Ojwang J, Yamada O, Hampel A, Rappaport J, Looney D, Wong-Staal F. A hairpin ribozyme inhibits expression of diverse strains of human immunodeficiency virus type 1. Proc Natl Acad Sci USA 1993;90:6340–6344.

72. Symons RH. Small catalytic RNAs. Annu Rev Biochem 1992;61:641–671.

73. Altman S. RNA enzyme-directed gene therapy. Proc Natl Acad Sci USA 1993;90:10898–10900.

74. Saxena SK, Ackerman EJ. Ribozymes correctly cleave a model substrate and endogenous RNA *in vivo*. J Biol Chem 1990;265:17106–17109.

75. Scanlon KJ, Jiao L, Funato T, Wang W, Tone T, Rossi JJ, Kashani-Sabet M. Ribozyme-mediated cleavage of c-fos mRNA reduces gene expression of DNA synthesis enzymes and metallothionein. Proc Natl Acad Sci USA 1991;88:10591–10595.

76. Lo KMS, Biasolo MA, Dehni G, Palú G, Haseltine WA. Inhibition of replication of HIV-1 by retroviral vectors expressing *tat*-antisense and anti-*tat* ribozyme RNA. Virology 1992;190;176–183.

77. Cech TR. Self-splicing of group I introns. Annu Rev Biochem 1990;59:543–568.

78. Reinhold-Hurek B, Shub DA. Self-splicing introns in tRNA genes of widely divergent bacteria. Nature 1992;357:173–176.

79. Kruger K, Grabowski PJ, Zaug AJ, Sands J, Gottschling DE, Cech TR. Self-splicing RNA: autoexcision and autocyclization of the ribosomal RNA intervening sequence of *Tetrahymena*. Cell 1982;31:147–157.

80. Murphy FL, Cech TR. Alteration of substrate specificity for the endoribonucleolytic cleavage of RNA by the *Tetrahymena* ribozyme. Proc Natl Acad Sci USA 1989;86:9218–9222.

81. Zaug AJ, Been MD, Cech TR. The *Tetrahymena* ribozyme acts like an RNA restriction endonuclease. Nature 1986;324:429–433.

82. Doudna JA, Cormack BP, Szostak JW. RNA structure, not sequence, determines the 5′ splice-site specificity of a group I intron. Proc Natl Acad Sci USA 1989;86:7402–7406.

83. Young B, Herschlag D, Cech TR. Mutations in a nonconserved sequence of the *Tetrahymena* ribozyme increase activity and specificity. Cell 1991;67:1007–1019.

84. Epstein LM, Gall JG. Self-cleaving transcripts of satellite DNA from the newt. Cell 1987;48:535–543.

85. Forster AC, Symons RH. Self-cleavage of plus and minus RNAs of a virusoid and a structural model for the active sites. Cell 1987;49:211–220.

86. Hutchins CJ, Rathjen PD, Forster AC, Symons RH. Self-cleavage of plus and minus RNA transcripts of avocado sunblotch viroid. Nucleic Acids Res 1986;14:3627–3640.

87. Prody GA, Bakos JT, Buzayan JM, Schneider IR, Bruening G. Autolytic processing of dimeric plant virus satellite RNA. Science 1986;231:1577–1580.
88. Haseloff J, Gerlach WL. Simple RNA enzymes with new and highly specific endoribonuclease activities. Nature 1988;334:585–591.
89. Uhlenbeck OC. A small catalytic oligoribonucleotide. Nature 1987;328:596–600.
90. Ruffner DE, Stormo GD, Uhlenbeck OC. Sequence requirements of the hammerhead RNA self-cleavage reaction. Biochemistry 1990;29:10695–10702.
91. McCall MJ, Hendry P, Jennings PA. Minimal sequence requirements for ribozyme activity. Proc Natl Acad Sci USA 1992;89:5710–5714.
92. Sullenger BA, Cech TR. Tethering ribozymes to a retroviral packaging signal for destruction of viral RNA. Science 1993;262:1566–1569.
93. Larson GP, Bertrand E, Rossi JJ. Designing and testing of ribozymes as therapeutic agents. Methods 1993;5:19–27.
94. Goodchild J. Enhancement of ribozyme catalytic activity by a contiguous oligodeoxynucleotide (facilitator) and by 2'-O-methylation. Nucleic Acids Res 1992;20:4607–4612.
95. Pachuk CJ, Yoon K, Moelling K, Coney LR. Selective cleavage of bcr-abl chimeric RNAs by a ribozyme targeted to non-contiguous sequences. Nucleic Acids Res 1994;22:301–307.
96. Weizsäcker F, von Blum HE, Wands JR. Cleavage of hepatitis B virus RNA by three ribozymes transcribed from a single DNA template. Biochem Biophys Res Commun 1992;189:743–748.
97. Ohkawa J, Yuyama N, Takebe Y, Nishikawa S, Taira K. Importance of independence in ribozyme reactions: Kinetic behavior of trimmed and of simply connected multiple ribozymes with potential activity against human immunodeficiency virus. Proc Natl Acad Sci USA 1993;90:11302–11306.
98. Heidenreich O, Eckstein F. Hammerhead ribozyme-mediated cleavage of the long terminal repeat RNA of human immunodeficiency virus type 1. J Biol Chem 1992;267:1904–1909.
99. Endicott J, Ling V. The biochemistry of P-glycoprotein mediated multidrug resistance. Annu Rev Biochem 1989;58:137–171.
100. Hjalt T, Wagner EGH. The effect of loop size in antisense and target RNAs on the efficiency of antisense RNA control. Nucleic Acids Res 1992;20:6723–6732.
101. Wagner EGH, Simons RW. Antisense RNA control in bacteria, phage and plasmids. Annu Rev Microbiol 1994 (in press).
102. Buzayan JM, Gerlach WL, Bruening G. Non-enzymatic cleavage and ligation of RNAs complementary to a plant virus satellite RNA. Nature 1986;323:349–353.
103. Haseloff J, Gerlach WL. Sequences required for self-catalysed cleavage of the satellite RNA of tobacco ringspot virus. Gene 1989;82:43–52.
104. Feldstein PA, Buzayan JM, Bruening G. Two sequences participating in the autolytic processing of satellite tobacco ringspot virus complementary RNA. Gene 1989;82:53–61.
105. Feldstein PA, Buzayan JM, van Tol H, DeBaer J, Gough GR, Gilham PT, Bruening G. Specific association between an endoribonucleolytic sequence from a satellite RNA and a substrate analogue containing a 2'-5' phosphodiester. Proc Natl Acad Sci USA 1990;87:2623–2627.
106. Hampel A, Tritz R, Hicks M, Cruz P. Hairpin catalytic RNA model: evidence for helices and sequence requirement for substrate RNA. Nucleic Acids Res 1990;18:299–304.
107. Kikuchi Y, Sasaki N. Site-specific cleavage of natural mRNA sequences by newly designed hairpin catalytic RNAs. Nucleic Acids Res 1991;19:6751–6755.
108. Joseph S, Burke JM. Optimization of an anti-HIV hairpin ribozyme by in vitro selection. J Biol Chem 1993;268:24515–24518.
109. Barinaga M. Ribozymes: killing the messenger. Science 1993;262:1512–1514.
110. Sharmeen L, Kuo MYP, Dinter-Gottlieb G, Taylor J. Antigenomic RNA of human hepatitis delta virus can undergo self-cleavage. J Virol 1988;62:2674–2679.
111. Perrotta AT, Been MD. The self-cleaving domain from the genomic RNA of hepatitis delta virus:sequence requirements and the effects of denaturant. Nucleic Acids Res 1990;18:6821–6827.

258

112. Perrotta AT, Been MD. A pseudoknot-like structure required for efficient self-cleavage of hepatitis delta virus RNA. Nature 1991;350:434–436.

113. Been MD, Perrotta AT, Rosenstein SP. Secondary structure of the self-cleaving RNA of hepatitis delta virus: applications to catalytic RNA design. Biochemistry 1992;31:11843–11852.

114. Been MD. *Cis*- and *trans*-acting ribozymes from a human pathogen, hepatitis delta virus. Trends Biochem Sci 1994;19:251–256.

115. Puttaraju M, Perrotta AT, Been MD. A circular trans-acting hepatitis delta virus ribozyme. Nucleic Acids Res 1993;21:4253–4258.

116. Neumann E, Schaefer-Ridder M, Wang Y, Hofschneider PH. Gene transfer into mouse lyoma cells by electroporation in high electric fields. EMBO J 1982;1:841–845.

117. Sambrook J, Fritsch EF, Maniatis T. Molecular Cloning. Cold Spring Harbor, NY: Cold Spring Harbor Laboratory Press, 1989.

118. Dwarki VJ, Malone RW, Verma IM. Cationic liposome-mediated RNA transfection. Methods Enzymol 1993;217:644–654.

119. Felgner PL, Gadek TR, Holm M, Roman R, Chan HW, Wenz M, Northrop JP, Ringold GM, Danielsen M. Lipofection:A highly efficient, lipid-mediated DNA-transfection procedure. Proc Natl Acad Sci USA 1987;84:7413–7417.

120. Loeffler J-P, Behr J-P. Gene transfer into primary and established mammalian cell lines with lipopolyamine-coated DNA. Methods Enzymol 1993;217:599–618.

121. Cotten M, Wagner E, Birnstiel ML. Receptor-mediated transport of DNA into eukaryotic cells. Methods Enzymol 1993;217:618–643.

122. Hirsch F, Poncet P, Freeman S, Gress RE, Sachs DH, Druet P, Hirsch R. Antifection: a new method for targeted gene transfection. Transpl Proc 1993;25:138–139.

123. Vlassov VV, Yakubov LA. Oligonucleotides in cells and in organisms. Pharmacological considerations. In: Wickstrom E (ed) Prospects for Antisense Nucleic Acid Therapy of Cancer and AIDS. New York: Wiley-Liss, 1991;243–266.

124. Agrawal S, Sarin PS, Zamecnik M, Zamecnik PC. Cellular uptake and anti-HIV activity of oligonucleotides and their analogs. In: Erickson RP, Izant JG (eds) Gene Regulation:Biology of Antisense RNA and DNA. New York: Raven Press, 1992;273–283.

125. Loke SL, Stein CA, Zhang XH, Mori K, Nakanishi M, Subasinghe C, Cohen JS, Neckers ML. Characterization of oligonucleotide transport into living cells. Proc Natl Acad Sci USA 1989;86:3474–3478.

126. Yakubov LA, Deeva EA, Zarytova VF, Ivanova EM, Ryte AS, Yurchenko LV, Vlassov VV. Mechanism of oligonucleotide uptake by cells:Involvement of specific receptors? Proc Natl Acad Sci USA 1989;86:6454–6458.

127. Boutorin AS, Gus'kova LV, Ivanova EM, Kobetz ND, Zarytova VF, Ryte AS, Yurchenko LV, Vlassov VV. Synthesis of alkylating oligonucleotide derivatives containing cholesterol or phenazinium residues at their 3'-terminus and their interaction with DNA within mammalian cells. FEBS Lett 1989;254:129–132.

128. Rothenpieler UW, Dressler GR. Differential distribution of oligodeoxynucleotides in developing organs with epithelial-mesenchymal interactions. Nucleic Acids Res 1993;21:4961–4966.

129. Leonetti JP, Degols G, Clarenc JP, Mechti N, Lebleu B. Cell delivery and mechanisms of action of antisense oligonucleotides. In: Cohn WE, Moldave K (eds) Progress in Nucleic Acid Research and Molecular Biology 44. New York: Academic Press, 1993;143–166.

130. Degols G, Leonetti J-P, Mechti N, Lebleu B. Antiproliferative effects of antisense oligonucleotides directed to the RNA of c-*myc* oncogene. Nucleic Acids Res 1991;19:945–948.

132. Wu GY, Wu CH. Receptor-mediated *in vitro* gene transformation by a soluble DNA carrier system. J Biol Chem 1987;262:4429–4432.

131. Lemaitre M, Bayard B, Lebleu B. Specific antiviral activity of a poly(L-lysine)-conjugated oligodeoxyribonucleotide sequence complementary to vesicular stomatitis virus N protein mRNA initiation site. Proc Natl Acad Sci USA 1987;84:648–652.

133. Zenke M, Steinlein P, Wagner E, Cotten M, Beug H, Birnstiel ML. Receptor-mediated endocytosis of transferrin-polycation conjugates: an efficient way to introduce DNA into hematopoietic cells. Proc Natl Acad Sci USA 1990;87:3655–3659.

132. Curiel DT, Agarwal S, Wagner E, Cotten M. Adenovirus enhancement of transferrin-polylysine-mediated gene delivery. Proc Natl Acad Sci USA 1991;88:8850–8854.

135. Leonetti JP, Mechti N, Degols G, Gagnor C, Lebleu B. Intracellular distribution of microinjected antisense oligonucleotides. Proc Natl Acad Sci USA 1991;88:2702–2706.

136. Sixou S, Szoka Jr FC, Green GA, Giusti B, Zon G, Chin DJ. Intracellular oligonucleotide hybridization detected by fluorescence resonance energy transfer (FRET). Nucleic Acids Res 1994;22: 662–668.

137. Leonetti JP, Machy P, Degols G, Lebleu B, Leserman L. Antibody targeted liposomes containing oligodeoxyribonucleotides complementary to viral RNA selectively inhibit viral replication. Proc Natl Acad Sci USA 1990;87:2448–2451.

138. Pickett GG, Peabody DS. Encapsidation of heterologous RNAs by bacteriophage MS2 coat protein. Nucleic Acids Res 1993;21:4621–4626.

139. Derossi D, Joliot AH, Chassaing G, Prochiantz A. The third helix of the Antennapedia homeodomain translocates through biological membranes. J Biol Chem 1994;269:10444–10450.

140. Paolella G, Sproat BS, Lamond AI. Nuclease resistant ribozymes with high catalytic activity. EMBO J 1992;11:1913–1919.

141. Pieken WA, Olsen DB, Benseler F, Aurup H, Eckstein F. Kinetic characterization of ribonuclease-resistant 2′-modified hammerhead ribozymes. Science 1991;253:314–317.

142. Miller PS. Antisense oligonucleoside methylphosphonates. In: Murray JAH (ed) Antisense RNA and DNA. New York: Wiley-Liss, 1992;241–253.

143. Miller PS, Cushman CD, Levis JT. Synthesis of oligo-2′-deoxyribonucleoside methylphosphonates. In: Eckstein F (ed) Oligonucleotides and analogues. A practical approach. Oxford: IRL Press 1991;137–154.

144. Eckstein F. Nucleoside phosphorothioates. Annu Rev Biochem 1985;54:367–402.

145. Griffiths AD, Potter BVL, Eperon IC. Stereospecificity of nucleases towards phosphorothioate-substituted RNA:stereochemistry of transcription by T7 RNA polymerase. Nucleic Acids Res 1987;15:4145–4162.

146. Beaton G, Dellinger D, Marshall WS, Caruthers MH. Synthesis of oligonucleotide phosphorodithioates. In: Eckstein F (ed) Oligonucleotides and Analogues. A Practical Approach. Oxford: IRL Press, 1991;109–135.

147. Leonetti JP, Rayner B, Lemaitre M, Gagnor C, Milhaud PG, Imbach JL, Lebleu B. Antiviral activity of conjugates between poly-(L-lysine) and synthetic oligodeoxyribonucleotides. Gene 1988; 72:323–332.

148. MacKellar C, Graham D, Will DW, Burgess S, Brown J. Synthesis and physical properties of anti-HIV antisense oligonucleotides bearing terminal lipophilic groups. Nucleic Acids Res 1992; 20:3411–3417.

149. Shea RG, Marsters JC, Bischofberger N. Synthesis, hybridization properties and antiviral activity of lipid-oligodeoxynucleotide conjugates. Nucleic Acids Res 1990;18:3777–3783.

150. Letsinger RL, Zhang G, Sun DK, Ikeuchi T, Sarin PS. Cholesteryl-conjugated oligonucleotides: Synthesis, properties, and activity as inhibitors of replication of human immunodeficiency virus in cell culture. Proc Natl Acad Sci USA 1989;86:6553–6556.

151. Jäschke A, Fürste JP, Cech D, Erdmann VA. Automated incorporation of polyethylene glycol into synthetic oligonucleotides. Tetrahedron Lett 1993;34:301–304.

152. Jäschke A, Fürste JP, Erdmann VA, Cech D. Hybridization-based affinity partitioning of nucleic acids using PEG-coupled oligonucleotides. Nucleic Acids Res 1994;22:1880–1884.

153. Hélène C, Toulmé J-J. Specific regulation of gene expression by antisense, sense and antigene nucleic acids. Biochim Biophys Acta 1990;1049:99–125.

154. Knorre DG, Zarytova VF. Novel antisense derivatives:antisense DNA intercalators, cleavers, and

alkylators. In: Wickstrom E (ed) Prospects for antisense nucleic acid therapy of cancer and AIDS. New York: Wiley-Liss, 1991;195–218.

155. Giannaris PA, Damha MJ. Oligoribonucleotides containing 2',5'-phosphodiester linkages exhibit binding selectivity for 3',5'-RNA over 3',5'-ssDNA. Nucleic Acids Res 1993;21:4742–4749.

156. Egholm M, Buchardt O, Christensen L, Behrens C, Freier SM, Driver DA, Berg RH, Kim SK, Norden B, Nielsen PE. PNA hybridizes to complementary oligonucleotides obeying the Watson-Crick hydrogen-bonding rules. Nature 1993;365:566–568.

157. Hanvey JC, Peffer NJ, Bisi JE, Thomson SA, Cadilla R, Josey JA, Ricca DJ, Hassman CF, Bonham MA, Au KG, Carter SG, Bruckenstein DA, Boyd AL, Noble SA, Babiss LE. Antisense and antigene properties of peptide nucleic acids. Science 1992;258:1481–1485.

158. Nielsen PE, Egholm M, Berg RH, Buchardt O. Sequence-selective recognition of DNA by strand displacement with a thymine-substituted polyamide. Science 1991;254:1497–1500.

159. Nielsen PE, Egholm M, Berg RH, Buchardt O. Sequence specific inhibition of DNA restriction enzyme cleavage by PNA. Nucleic Acids Res 1993;21:197–200.

160. Iribarren AM, Sproat BS, Neuner P, Sulston I, Ryder U, Lamond AI. 2'-O-alkyl oligoribonucleotides as antisense probes. Proc Natl Acad Sci USA 1990;87:7747–7751.

161. Sproat BS, Lamond AI, Beijer B, Neuner P, Ryder U. Highly efficient chemical synthesis of 2'-O-methyloligoribonucleotides and tetrabiotinylated derivates;novel probes that are resistant to degradation by RNA or DNA specific nucleases. Nucleic Acids Res 1989;17:3373–3386.

162. Freier SM, Lima WF, Sanghvi YS, Vickers T, Zounes M, Cook PD, Ecker DJ. Thermodynamics of antisense oligonucleotide hybridization. In: Erickson RP, Izant JG (eds) Gene regulation: biology of antisense RNA and DNA. New York: Raven Press, 1992;95–107.

163. Inoue H, Hayase Y, Imura A, Iwai S, Miura K, Ohtsuka E. Synthesis and hybridization studies on two complementary nona(2'-O-methyl)ribonucleotides. Nucleic Acids Res 1987;15:6131–6148.

164. Shimayama T, Nishikawa F, Nishikawa S, Taira K. Nuclease-resistant chimeric ribozymes containing deoxyribonucleotides and phosphorothioate linkages. Nucleic Acids Res 1993;21:2605–2611.

165. Woolf TM, Jennings CGB, Rebagliati M, Melton DA. The stability, toxicity and effectiveness of unmodified and phosphorothioate antisense oligodeoxynucleotides in *Xenopus* oocytes and embryos. Nucleic Acids Res 1990;18:1763–1769.

166. Zamecnik PC, Goodchild J, Taguchi Y, Sarin PS. Inhibition of replication and expression of human T-cell lymphotropic virus type III in cultured cells by exogenous synthetic oligonucleotides complementary to viral RNA. Proc Natl Acad Sci USA 1986;83:4143–4146.

167. Kibler-Herzog L, Zon G, Uznanski B, Whittier G, Wilson WD. Duplex stabilities of phosphorothioate, methylphosphonate and RNA analogs of two DNA 14-mers. Nucleic Acids Res 1991;19:2979–2986.

168. Agrawal S, Goodchild J, Civeira MP, Thornton AH, Sarin PS, Zamecnik PC. Oligodeoxynucleoside phosphoramidates and phosphorothioates as inhibitors of human immunodeficiency virus. Proc Natl Acad Sci USA 1988;85:7079–7083.

169. Matsukura M, Shinozuka K, Zon G, Mitsuya H, Reitz M, Cohen JS, Broder S. Phosphorothioate analogs of oligodeoxynucleotides: inhibitors of replication and cytopathic effects of human immunodeficiency virus. Proc Natl Acad Sci USA 1987;84:7706–7710.

170. Majumdar C, Stein CA, Cohen JS, Broder S, Wilson SH. Stepwise mechanism of HIV reverse transcriptase: primer function of phosphorothioate oligodeoxynucleotide. Biochemistry 1989;28:1340–1346.

171. Agrawal S, Ikeuchi T, Sun D, Sarin PS, Konopka A, Maizel J, Zamecnik PC. Inhibition of human immunodeficiency virus in early infected and chronically infected cells by antisense oligodeoxynucleotides and their phosphorothioate analogues. Proc Natl Acad Sci USA 1989;86:7790–7794.

172. Ghosh MK, Ghosh K, Dahl O, Cohen JS. Evaluation of some properties of a phosphorodithioate oligodeoxyribonucleotide for antisense application. Nucleic Acids Res 1993;21:5761–5766.

173. Quartin RS, Brakel CL, Wetmur JG. Number and distribution of methylphosphonate linkages in oligodeoxynucleotides affect exo- and endonuclease sensitivity and ability to form RNase H substrates. Nucleic Acids Res 1989;17:7253–7262.

174. Giles RV, Tidd DM. Enhanced RNase H activity with methylphosphonodiester/phosphodiester chimeric antisense oligodeoxynucleotides. Anticancer Drug Des 1992;7:37–48.

175. Giles RV, Spiller DG, Tidd DM. Chimeric oligodeoxynucleotide analogues:enhanced cell uptake of structures which direct ribonuclease H with high specificity. Anticancer Drug Des 1993;8:33–51.

176. Froehler B, Ng P, Matteucci M. Phosphoramidate analogues of DNA:synthesis and thermal stability of heteroduplexes. Nucleic Acids Res 1988;16:4831–4839.

177. Thuong NT, Asseline U, Roig V, Takasugi M, Hélène C. Oligo(α-deoxynucleotide)s covalently linked to intercalating agents: differential binding to ribo- and deoxyribopolynucleotides and stability towards nuclease digestion. Proc Natl Acad Sci USA 1987;84:5129–5133.

178. Debart F, Rayner B, Degols G, Imbach J-L. Synthesis and base-pairing properties of the nuclease-resistant α-anomeric dodecaribonucleotide α-[r(UCUUAACCCACA)]. Nucleic Acids Res 1992; 20:1193–1200.

179. Kierzek R, He L, Turner DH. Association of $2'$-$5'$ oligoribonucleotides. Nucleic Acids Res 1992; 20:1685–1690.

180. Torrence PF, Maitra RK, Lesiak K, Khamnei S, Zhou A, Silverman RH. Targeting RNA for degradation with a $(2'$-$5')$oligoadenylate-antisense chimera. Proc Natl Acad Sci USA 1993;90:1300–1304.

181. Dagle JM, Walder JA, Weeks DL. Targeted degradation of mRNA in *Xenopus* oocytes and embryos directed by modified oligonucleotides:studies of An2 and cyclin in embryogenesis. Nucleic Acids Res 1990;18:4751–4757.

182. Hoke GD, Draper K, Freier SM, Gonzalez C, Driver VB, Zounes MC, Ecker D.J. Effects of phosphorothioate capping on antisense oligonucleotide stability, hybridization and antiviral efficacy versus herpes simplex virus infection. Nucleic Acids Res 1991;19:5743–5748.

183. Agrawal S, Goodchild J. Oligodeoxynucleoside methylphosphonates: synthesis and enzymatic degradation. Tetrahedron Lett 1987;28:3539–3542.

184. Tidd DM. Anticancer drug design using modified antisense oligonucleotides. In: Murray JAH (ed) Antisense RNA and DNA. New York: Wiley-Liss, 1992;227–240.

185. Shumyatsky G, Wright D, Reddy R. Methylphosphate cap structure increases the stability of 7SK, B2 and U6 small RNAs in *Xenopus* oocytes. Nucleic Acids Res 1993;21:4756–4761.

186. Krieg AM, Tonkinson J, Matson S, Zhao Q, Saxon M, Zhang L-M, Bhanja U, Yakubov L, Stein CA. Modification of antisense phosphodiester oligodeoxynucleotides by a $5'$ cholesteryl moiety increases cellular association and improves efficacy. Proc Natl Acad Sci USA 1993;90:1048–1052.

187. Agrawal S. Antisense oligonucleotides:A possible approach for chemotherapy of AIDS. In:Wickstrom E (ed) Prospects for Antisense Nucleic Acid Therapy of Cancer and AIDS. New York:Wiley-Liss, 1991;143–158.

188. Hausen P, Stein H. Factors influencing the activity of mammalian RNA polymerase. Cold Spring Harbor Symp Quant Biol 1970;35:709.

189. Eder PS, Walder RY, Walder JA. Substrate specificity of human RNase H1 and its role in excision repair of ribose residues misincorporated in DNA. Biochimie 1993;75:123–126.

190. Agrawal S, Mayrand SH, Zamecnik PC, Pederson T. Site-specific excision from RNA by RNase H and mixed-phosphate-backbone oligodeoxynucleotides. Proc Natl Acad Sci USA 1990;87:1401–1405.

191. Furdon PJ, Dominski Z, Kole R. RNase H cleavage of RNA hybridized to oligonucleotides containing methylphosphonate, phosphorothioate and phophodiester bonds. Nucleic Acids Res 1989; 17:9193–9204.

192. Kawasaki AM, Casper MD, Freier SM, Lesnik EA, Zounes MC, Cummins LL, Gonzalez C, Cook

PD. Uniformly modified 2′-deoxy-2′-fluoro phosphorothioate oligonucleotides are nuclease-resistant compounds with high affinity and specificity for RNA targets. J Med Chem 1993;36: 831–841.

193. Fu D-J, McLaughlin LW. Importance of specific purine amino and hydroxyl groups for efficient cleavage by a hammerhead ribozyme. Proc Natl Acad Sci USA 1992;89:3985–3989.

194. Fu D-J, McLaughlin LW. Importance of specific adenosine N^7-nitrogens for efficient cleavage by a hammerhead ribozyme. A model for magnesium binding. Biochemistry 1992;31:10941–10949.

195. Fu D-J, Rajur SB, McLaughlin LW. Importance of specific guanosine N^7-nitrogens and purine amino groups for efficient cleavage by a hammerhead ribozyme. Biochemistry 1993;32:10629–10637.

196. Grasby JA, Butler PJG, Gait MJ. The synthesis of oligoribonucleotide containing O^6-methylguanosine:the role of conserved guanosine residues in hammerhead ribozyme cleavage. Nucleic Acids Res 1993;21:4444–4450.

197. Tuschl T, Ng MMP, Pieken W, Benseler F, Eckstein F. Importance of exocyclic base functional groups of central core guanosines for hammerhead ribozyme activity. Biochemistry 1993;32: 11658–11668.

198. Williams DM, Pieken WA, Eckstein F. Function of specific 2′-hydroxyl groups of guanosines in a hammerhead ribozyme probed by 2′-modifications. Proc Natl Acad Sci USA 1992;89:918–921.

199. Olsen DB, Benseler F, Aurup H, Pieken WA, Eckstein F. Study of a hammerhead ribozyme containing 2′-modified adenosine residues. Biochemistry 1991;30:9735–9741.

200. Perreault J-P, Wu T, Cousineau B, Ogilvie KK, Cedergren R. Mixed deoxyribo- and ribooligonucleotides with catalytic activity. Nature 1990;344:565–567.

201. Yang J, Usman N, Chartrand P, Cedergren R. Minimum ribonucleotide requirement for catalysis by the RNA hammerhead domain. Biochemistry 1992;31:5005–5009.

202. Saenger W. Principles of nucleic acid structure. New York-Berlin-Heidelberg-Tokyo: Springer, 1984.

203. Buzayan JM, Van Tol H, Feldstein PA, Bruening G. Identification of non-junction phosphodiester that influences an autolytic processing reaction of RNA. Nucleic Acids Res 1990;18:4447–4451.

204. Ruffner DE, Uhlenbeck OC. Thiophosphate interference experiments locate phosphates important for the hammerhead RNA self-cleavage reaction. Nucleic Acids Res 1990;18:6025–6029.

205. Wang MJ, Davis NW, Gegenheimer P. Novel mechanisms for maturation of chloroplast transfer RNA precursors. EMBO J 1988;7:1567–1574.

206. Altman S, Kirsebom L, Talbot S. Recent studies of ribonuclease P. FASEB J 1993;7:7–14.

207. Darr SC, Brown JW, Pace NR. The varieties of ribonuclease P. Trends Biochem Sci 1992;17: 178–182.

208. LaGrandeur TE, Darr SC, Haas ES, Pace NR. Characterization of the RNase P RNA of *Sulfolobus acidocaldarius*. J Bacteriol 1993;175:5043–5048.

209. Guerrier-Takada C, Gardiner K, Marsh T, Pace N, Altman S. The RNA moiety of ribonuclease P is the catalytic subunit of the enzyme. Cell 1983;35:849–857.

210. Tranguch AJ, Kindelberger DW, Rohlman CE, Lee J-Y, Engelke DR. Structure-sensitive RNA footprinting of yeast nuclear ribonuclease P. Biochemistry 1994;33:1778–1787.

211. Forster AC, Altman S. External guide sequences for an RNA enzyme. Science 1990;249:783–785.

212. Li Y, Guerrier-Takada C, Altman S. Targeted cleavage of mRNA *in vitro* by RNase P from *Escherichia coli*. Proc Natl Acad Sci USA 1992;89:3185–3189.

213. Schlegl J, Fürste JP, Bald R, Erdmann VA, Hartmann RK. Cleavage efficiencies of model substrates for ribonuclease P from *Escherichia coli* and *Thermus thermophilus*. Nucleic Acids Res 1992;20:5963–5970.

214. Surratt CK, Lesnikowski Z, Schifman AL, Schmidt FJ, Hecht SM. Construction and processing of transfer RNA precursor models. J Biol Chem 1990;265:22506–22512.

215. Yuan Y, Hwang E-S, Altman S. Targeted cleavage of mRNA by human RNase P. Proc Natl Acad Sci USA 1992;89:8006–8010.
216. Gold HA, Craft J, Hardin JA, Bartkiewicz M, Altman S. Antibodies in human serum that precipitate ribonuclease P. Proc Natl Acad Sci USA 1988;85:5483–5487.
217. Bartkiewicz M, Gold H, Altman S. Identification and characterization of an RNA molecule that copurifies with RNase P activity from HeLa cells. Genes Dev 1989;3:488–499.
218. Morales MJ, Dang YL, Lou YC, Sulo P, Martin NC. A 105-kDa protein is required for yeast mitochondrial RNase P activity. Proc Natl Acad Sci USA 1992;89:9875–9879.
219. Lygerou Z, Mitchell P, Petfalski E, Séraphin B, Tollervey D. The *POP1* gene encodes a protein component common to the RNase MRP and RNase P ribonucleoproteins. Genes Dev 1994;8: 1423–1433.
220. Zimmerly S, Drainas D, Sylvers LA, Söll D. Identification of a 100-kDa protein associated with nuclear ribonuclease P activity in *Schizosaccharomyces pombe*. Eur J Biochem 1993;217:501–507.
221. Mamula MJ, Baer M, Craft J, Altman S. An immunological determinant of RNase P protein is conserved between *Escherichia coli* and humans. Proc Natl Acad Sci USA 1989;86:8717–8721.
222. Yuan Y, Tan E, Reddy R. The 40-kilodalton To autoantigen associates with nucleotides 21 to 64 of human mitochondrial RNA processing/7–2 RNA *in vitro*. Mol Cell Biol 1991;11:5266–5274.
223. Nichols M, Söll D, Willis I. Yeast RNase P: catalytic activity and substrate binding are separate functions. Proc Natl Acad Sci USA 1988;85:1379–1383.
224. Drainas D, Zimmerly S, Willis I, Söll D. Substrate structural requirements of *Schizosaccharomyces pombe* RNase P. FEBS Lett 1989;251:84–88.
225. Krupp G, Cherayil B, Frendewey D, Nishikawa S, Söll D. Two RNA species co-purify with RNase P from the fission yeast *Schizosaccharomyces pombe*. EMBO J 1986;5:1697–1703.
226. McClain WH, Guerrier-Takada C, Altman S. Model substrates for an RNA enzyme. Science 1987;238:527–530.
227. Hardt W-D, Schlegl J, Erdmann VA, Hartmann RK. Role of the D arm and the anticodon arm in tRNA recognition by eubacterial and eukaryotic RNase P enzymes. Biochemistry 1993;32: 13046–13053.
228. Kahle D, Wehmeyer U, Krupp G. Substrate recognition by RNase P and by the catalytic M1 RNA:identification of possible contact points in pre-tRNAs. EMBO J 1990;9:1929–1937.
229. Thurlow DL, Shilowski D, Marsh TL. Nucleotides in precursor tRNAs that are required intact for catalysis by RNase P RNAs. Nucleic Acids Res 1991;19:885–891.
230. Carrara G, Calandra P, Fruscoloni P, Doria M. Tocchini-Valentini GP. Site selection by *Xenopus laevis* RNAase P. Cell 1989;58:37–45.
231. Yuan Y, Altman S. Selection of guide sequences that direct efficient cleavage of mRNA by human ribonuclease P. Science 1994;263:1269–1273.
232. Leontis N, DaLio A, Strobel M, Engelke D. Effects of tRNA-intron structure on cleavage of precursor tRNAs by RNase P from *Saccharomyces cerevisiae*. Nucleic Acids Res 1988;16:2537–2552.
233. Pearson D, Willis I, Hottinger H, Bell J, Kumar A, Leupold U, Söll D. Mutations preventing expression of *sup3* tRNA[Ser] nonsense suppressors of *Schizosaccharomyces pombe*. Mol Cell Biol 1985;5:808–815.
234. Willis I, Frendewey D, Nichols M, Hottinger-Werlen A, Schaack J, Söll D. A single base change in the intron of a serine tRNA affects the rate of RNase P cleavage *in vitro* and suppressor activity *in vivo* in *Saccharomyces cerevisiae*. J Biol Chem 1986;261:5878–5885.
235. Bass BL, Weintraub H. An unwinding activity that covalently modifies its double-stranded RNA substrate. Cell 1988;55:1089–1098.
236. Wagner RW, Smith JE, Cooperman BS, Nishikura K. A double-stranded RNA unwinding activity introduces structural alterations by means of adenosine to inosine conversions in mammalian cells and *Xenopus* eggs. Proc Natl Acad Sci USA 1989;86:2647–2651.

264

237. Willis I, Nichols M, Chisholm V, Söll D, Heyer W-D, Szankasi P, Amstutz H, Munz P, Kohli J. Functional complementation between mutations in a yeast suppressor tRNA gene reveals potential for evolution of tRNA sequences. Proc Natl Acad Sci USA 1986;83:7860–7864.

238. Herschlag D. Implications of ribozyme kinetics for targeting the cleavage of specific RNA molecules *in vivo*: more isn't always better. Proc Natl Acad Sci USA 1991;88:6921–6925.

239. Herschlag D, Cech TR. Catalysis of RNA cleavage by the *Tetrahymena thermophila* ribozyme. 2. Kinetic description of the reaction of an RNA substrate that forms a mismatch at the active site. Biochemistry 1990;29:10172–10180.

240. Cantor CR, Schimmel PR. Biophysical Chemistry, Part I: The Conformation of Biological Macromolecules. San Francisco, CA: WH Freeman, 1980.

241. Beier D, Beier H. Expression of variant nuclear *Arabidopsis* tRNA^{Ser} genes and pre-tRNA maturation differ in HeLa, yeast and wheat germ extracts. Mol Gen Genet 1992;233:201–208.

242. Bertrand E, Rossi JJ. Facilitation of hammerhead ribozyme catalysis by the nucleocapsid protein of HIV-1 and the heterogeneous nuclear ribonucleoprotein A1. EMBO J 1994;13:2904–2912.

243. Herschlag D, Khosla M, Tsuchihashi Z, Karpel RL. An RNA chaperone activity of non-specific RNA binding proteins in hammerhead ribozyme catalysis. EMBO J 1994;13:2913–2924.

244. Müller G, Strack B, Dannull J, Sproat BS, Surovoy A, Jung G, Moelling K. Amino acid requirements of the nucleocapsid protein of HIV-1 for increasing catalytic activity of a *Ki-ras* ribozyme *in vitro*. J Mol Biol 1994 (in press).

245. Coetzee T, Herschlag D, Belfort M. *Escherichia coli* proteins, including ribosomal protein S12, facilitate in vitro splicing of phage T4 introns by acting as RNA chaperones. Genes Dev 1994;8: 1575–1588.

246. Tsuchihashi Z, Khosla M, Herschlag D. Protein enhancement of hammerhead ribozyme catalysis. Science 1993;262:99–102.

247. Jacobson MR, Cao L-G, Wang Y-L, Pederson T. Rapid nucleolar localization of RNase P RNA microinjected into the nucleus of living cells. Presented at the RNA Processing Meeting. Cold Spring Harbor Laboratory, USA, 1993;Abstract 135 (pers. commun.).

248. Jacobson MR, Cao L-G, Wang Y-L, Pederson T. The RNA subunit of ribonuclease P rapidly localizes in the nucleolus after microinjection into the nucleus of living cells. J Cell Biochem 1994;18C(Suppl):115.

249. Jacobson MR, Cao L-G, Wang Y-L, Pederson T. Nucleolar localization of the RNA subunit of RNase MRP microinjected into the interphase nucleus. Presented at the RNA Processing Meeting. Madison, Wisconsin, USA, 1994;Abstract 194 (pers. commun.).

250. Chang DD, Clayton DA. A mammalian mitochondrial RNA processing activity contains nucleus-encoded RNA. Science 1987;235:1178–1184.

251. Chang DD, Clayton DA. A novel endoribonuclease cleaves at a priming site of mouse mitochondrial DNA replication. EMBO J 1987;6:409–417.

252. Stohl LL, Clayton DA. *Saccharomyces cerevisiae* contains an RNase MRP that cleaves at a conserved mitochondrial RNA sequence implicated in replication priming. Mol Cell Biol 1992;12: 2561–2569.

253. Topper JN, Bennett JL, Clayton DA. A role for RNAase MRP in mitochondrial RNA processing. Cell 1992;70:16–20.

254. Kiss T, Filipowicz W. Evidence against a mitochondrial location of the 7–2/MRP RNA in mammalian cells. Cell 1992;70:11–16.

255. Reimer G, Raska I, Scheer U, Tan EM. Immunolocalization of 7–2-ribonucleoprotein in the granular component of the nucleolus. Exp Cell Res 1988;176:117–128.

256. Schmitt ME, Clayton DA. Nuclear RNase MRP is required for correct processing of pre-5.8S rRNA in *Saccharomyces cerevisiae*. Mol Cell Biol 1993;13:7935–7941.

257. Henry Y, Wood H, Morissey JP, Petfalski E, Kearsey S, Tollervey D. The 5′ end of yeast 5.8S rRNA is generated by exonucleases from an upstream cleavage site. EMBO J 1994;13:2452–2463.

258. Gold HA, Topper JN, Clayton DA, Craft J. The RNA processing enzyme RNase MRP is identical to the Th RNP and related to RNase P. Science 1989;245:1377–1380.

259. Forster AC, Altman S. Similar cage-shaped structures for the RNA components of all ribonuclease P and ribonuclease MRP enzymes. Cell 1990;62:407–409.

260. Kiss T, Marshallsay C, Filipowicz W. 7–2/MRP RNAs in plant and mammalian cells: association with higher order structures in the nucleolus. EMBO J 1992;11:3737–3746.

261. Schmitt ME, Bennett JL, Dairaghi DJ, Clayton DA. Secondary structure of RNase MRP RNA as predicted by phylogenetic comparison. FASEB J 1993;7:208–213.

262. Kikuchi Y, Sasaki-Tozawa N, Suzuki K. Artificial self-cleaving molecules consisting of a tRNA precursor and the catalytic RNA of RNase P. Nucleic Acids Res 1993;21:4685–4689.

263. Frank DN, Harris ME, Pace NR. Remodeling active-site structure in a ribozyme: rational design of self-cleaving pre-tRNA-ribonuclease P conjugates. Biochemistry 1994 (in press).

264. Gaur RK, Krupp G. Modification interference approach to detect ribose moieties important for the optimal activity of a ribozyme. Nucleic Acids Res 1993;21:21–26.

265. Gaur RK, Krupp G. Enzymatic RNA synthesis with deoxynucleoside 5'-O-(1-thiotriphosphates). FEBS Lett 1993;315:56–60.

266. Anderson P, Monforte J, Tritz R, Nesbitt S, Hearst J, Hampel A. Mutagenesis of the hairpin ribozyme. Nucleic Acids Res 1994;22:1096–1100.

267. Haas ES, Brown JW, Pitulle C, Pace NR. Further perspective on the catalytic core and secondary structure of ribonuclease P RNA. Proc Natl Acad Sci USA 1994;91:2527–2531.

268. Tallsjö A, Svärd SG, Kufel J, Kirsebom LA. A novel tertiary interaction in M1 RNA, the catalytic subunit of *Escherichia coli* RNase P. Nucleic Acids Res 1993;21:3927–3933.

269. Quigley GJ, Rich A. Structural domains of transfer RNA molecules. Science 1976;194:796–806.

270. Saville BJ, Collins RA. A site-specific self-cleavage reaction performed by a novel RNA in *Neurospora* mitochondria. Cell 1990;61:685–696.

271. Saldanha R, Mohr G, Belfort M, Lambowitz AM. Group I and group II introns. FASEB J 1993;7; 15–24.

272. Wittop Koning TH, Schümperli D. RNAs and ribonucleoproteins in recognition and catalysis. Eur J Biochem 1994;219:25–42.

273. Collins RA, Olive JE. Reaction conditions and kinetics of self-cleavage of a ribozyme derived from Neurospora VS RNA. Biochemistry 1993;32:2795–2799.

274. Saville BJ, Collins RA. RNA-mediated ligation of self-cleavage products of a *Neurospora* mitochondrial plasmid transcript. Proc Natl Acad Sci USA 1991;88:8826–8830.

© 1995 Elsevier Science B.V. All rights reserved
Biotechnology Annual Review Volume 1
M.R. El-Gewely, editor

Artificial cells with emphasis on bioencapsulation in biotechnology

Thomas Ming Swi Chang

Artificial Cells and Organs Research Centre, Medicine and Biomedical Engineering, Faculty of Medicine, McGill University, Montreal, Quebec, Canada H3G 1Y6

Abstract. The most common use of artificial cells is for bioencapsulation of biologically active materials. Each artificial cell can contain combinations of materials. The permeability, composition and shape of an artificial cell membrane can be varied using different types of synthetic or biological materials. These possible variations in contents and membranes allow for large variations in the properties and functions of artificial cells. Artificial cells containing adsorbents have been a routine form of treatment in hemoperfusion for patients. This includes acute poisoning, high blood aluminum and iron, and supplement to dialysis in kidney failure. Artificial red blood cell substitutes based on modified hemoglobin are already in Phase I and Phase II clinical trials in patients. Artificial cell encapsulated cell cultures are being studied for the treatment of diabetes, liver failure, gene therapy and other conditions. Research on artificial cells containing enzymes includes their use for treatment in hereditary enzyme deficiency diseases and other diseases. Recent demonstration of extensive enterorecirculation of amino acids in the intestine has allowed oral administration to deplete specific amino acids. One example is phenylketonuria, an inborn error of metabolism resulting in high systemic phenylalanine levels. Preliminary clinical studies in patients using bioencapsulation of cells or enzymes have started. Artificial cells containing complex enzyme systems convert wastes like urea and ammonia into essential amino acids. Artificial cells are being used for the production of monoclonal antibodies, interferon and other biotechnological products. Other areas of biotechnological uses include drug delivery, and other areas of biotechnology, chemical engineering and medicine.

Introduction

"Artificial Cell is not a specific physical entity. It is an idea involving the preparation of artificial structures of cellular dimensions for possible replacement or supplement of deficient cell functions. It is clear that different approaches can be used to demonstrate this idea." This description comes from this author's 1972 monograph on "Artificial Cells" [1].

Figure 1 summarizes the basic principle of artificial cells. Like biological cells, artificial cells contain biologically active materials. However, the content of artificial cells can be more varied than biological cells. These include both biological and synthetic materials. Figure 1 shows some possible contents. Any one artificial cell can contain one or a combination of the materials shown in Fig. 1. The membranes of artificial cells can also be extensively varied using many different types of syn-

Address for correspondence: Thomas Ming Swi Chang Artificial Cells and Organs Research Centre, Faculty of Medicine, McGill University, Montreal, Quebec, Canada H3G 1Y6

268

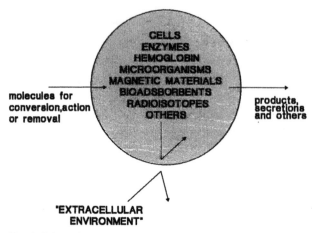

Fig. 1. Schematic representation of artificial cells.

thetic or biological materials. These possible variations in contents and membrane materials allow for unlimited variations in the properties of artificial cells.

The permeability can be controlled over a wide range. For example, ultrathin membranes can retain macromolecules like proteins and enzymes. Simultaneously it allows the rapid diffusion of smaller molecules. This way, each artificial cell retains the enclosed materials and separates them from undesirable external impermeant materials, while the large surface area and the ultrathin membrane allow permeant substrates and products to diffuse rapidly (Fig. 1). Ten milliliters of 20 μm diameter artificial cells have a total surface area of about 20,000 cm². The membrane thickness is 200 Å. Thus, mass transfer of permeant molecules across 10 ml of artificial cells can be 100 times higher than that for a standard hemodialysis machine. Liposomes, which are an extension of artificial cells, do not follow this principle.

Variations in dimensions are also possible. Dimensions depend on the type of use and contents (Table 1).

Historical Background of Artificial Cells

The first reports on artificial cells were published by this author as early as 1957 [2] and 1964 [3]. Shortly after this, our research showed the feasibility of artificial cells

Table 1. Artificial cells: terminology and dimension

Terminology	Dimensions	Examples of contents
Encapsulation	Millimeters	Cell cultures, microorganisms, tissues, others
Microencapsulation	Micrometers	Enzymes, protein, peptides, microorganisms, organelles, cells, etc.
Nanoencapsulation	Nanometers	Enzymes, protein, peptides, antibiotics, hemoglobin, etc.
Crosslinkage	Angstroms	Polyhemoglobin red blood cell substitutes

in biotechnology. This includes blood substitutes, enzyme technology, bio-encapsulation of cell cultures, drug delivery, biosorbents and other applications [1–7]. However, at that time there was not much international research interest in this type of biotechnology. Others were more interested in our research on artificial cells in hemoperfusion [4,8]. There was particular interest in our clinical use of this for patients with acute poisoning, kidney failure and liver failure [1,9]. Extensive international research and routine clinical application of artificial cells in hemo-perfusion followed very shortly after.

Interest in our research on artificial cells in biotechnology only started in the early 1980s. This followed the explosive international interests in all areas of biotechnology around that time [10–15]. The major impetus was in the second half of the 1980s. This was a time of concern regarding potential immunodeficiency disease (AIDS) due to HIV in donor blood. This has resulted in extensive research and developments on the biotechnological approach of modified hemoglobin. This is based on our original idea of microencapsulated and crosslinked hemoglobin for red blood cell substitutes [1–7]. The intense international effort [16] has led to rapid progress. Thus, within 7 years of this, Phase I and Phase II clinical trials using modified hemoglobin in patients are already in progress. Research and applications of artificial cells related to biotechnology and medicine now cover many areas (Table 2). This review can only cover a very general overview of these areas. The emphasis is on the bioencapsulation of cells, microorganisms and enzymes. These are promising areas in biotechnology that still need more extensive research. Only brief summaries of

Table 2. Artificial cells: research and applications

Acute poisoning	Routine clinical treatment
Aluminum and iron overload	Routine clinical treatment
End-stage kidney failure	Routine clinical treatment as supplement to dialysis
Liver failure	Routine clinical application for limited types of acute liver failure
Red blood cell substitutes for transfusion	Phase II clinical trial
Enzyme defects in inborn errors of metabolism	Clinical trial
Monoclonal antibodies production from hybridomas	Experimental and production
Diabetic mellitus (bioencapsulated islets)	Phase I clinical trial
Bioartificial liver (bioencapsulated hepatocytes)	Experimental
Gene therapy using bioencapsulated cells or microorganisms	Experimental
Drug delivery systems	Experimental
Blood group antibodies removal	Clinical trial
Clinical laboratory analysis	Experimental
Conversion of cholesterol into carbon dioxide	Experimental
Bilirubin removal	Experimental
Production of fine biochemical	Industrial application
Aquatic culture	Industrial application
Conversion of wastes (e.g. urea and ammonium) into useful products (e.g. essential amino acids)	Industrial application
Other biotechnological and medical applications in progress	

other important areas such as modified hemoglobin blood substitutes, drug delivery and biosorbents are given. Later reviews will deal in more detail with these and other specific topics.

Preparation, Materials and Contents

Methods of preparation

Many methods are now available for the preparation of artificial cells. This review is not the place to describe these methods in detail. The most commonly used approaches are based on the following:

Microencapsulation and nanoencapsulation
This is usually prepared by modifications and variations of the basic procedures [1–7] (Fig. 2). The contents intended for artificial cells are prepared as fine emulsions. Membranes are then formed on each microdroplet. There are many ways to form fine emulsions. However, bioencapsulation of sensitive biological materials requires gentler methods. Recent descriptions containing examples of updated methods for preparing these artificial cells [14] and other approaches [15] are available.

Encapsulation
Drop techniques are generally used to prepare larger artificial cells, especially for encapsulation of cells or tissues like islets. The original drop technique [1,2,4–6] forms the basis of these methods. This and more recent methods are discussed later under cell encapsulation.

Crosslinkage
There are many approaches for preparing hemoglobin artificial cells. One way is to use bifunctional agents to crosslink hemoglobin molecules on the surface of a microdroplet to form crosslinked hemoglobin membrane [1,3,5–7]. If we keep

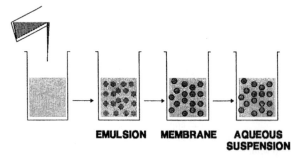

EMULSION   MEMBRANE   AQUEOUS
                        SUSPENSION

Fig. 2. Preparation of small artificial cells by the emulsification method [1–7,14]. Figure reproduced from Ref. [11] with permission from the publisher.

decreasing the size of the microdroplet of hemoglobin, the bifunctional agents would eventually crosslink all the hemoglobin molecules to form "polyhemoglobin" [1,3,5–7]. Crosslinking together less than 10 hemoglobin molecules, resulted in a soluble polyhemoglobin [16].

Artificial cells containing adsorbents, biosorbents or immunosorbents
This makes use of ultrathin membrane coating of sorbent granules [1,8].

Variations in membrane materials and configurations

Synthetic polymers

Table III. Artificial cells containing biologically active materials: variations in membranes, configurations and contents (first reference cited)

1957, 1964 Chang	Ultrathin (200 Å) synthetic polymer membranes artificial cells containing enzyme, multienzymes, proteins
1964 Chang	Cross-linked protein membrane artificial cells containing single enzyme, multienzymes, protein
1964 Chang	Cross-linked protein-enzyme conjugates and polyhemoglobin
1965 Chang	Artificial cells containing smaller artificial cells for intracellular multicompartmental systems
1965 Chang	Artificial cells containing intact biological cells
1965 Chang	Solid polymer microspheres containing enzymes and biological materials
1966 Mosbach	Solid polymer microspheres containing microorganisms
1966 Chang	Artificial cells containing protein and magnetic material
1966 Chang	Artificial cells containing both enzyme and adsorbent for product of enzyme reaction
1967 Chang	Heparin-complexed membrane artificial cells
1968 Mueller et al.	Bilayer lipid membrane artificial cells containing hemoglobin
1969 Chang	Ultrathin polymer membrane coated adsorbent particles
1969 Chang	Bilayer lipid-protein or bilayer lipid-polymer membrane artificial cells containing hemolysate
1970 Sessa and Weissman	Multilamellar lipid liposome containing enzymes
1972 Gregoriadis and Ryman	Multilamellar lipid liposomes as enzymes carrier
1972 May and Li	Liquid hydrocarbon emulsion containing enzyme microdroplets
1973 Ihler	Use of red blood cells as carrier to entrap biologically active materials
1976 Chang	Biodegradable synthetic polymer membrane (e.g. polylactic acid) artificial cells containing hormones (e.g. insulin) and other biologically active materials
1980 Johnston and Chapman	Polymerized liposomes as carrier
1980 Lim and Sun	Alginate-polylysine artificial cells containing islets
1980 Chang	Ultrathin blood compatible polymer coated immunosorbent for removal of antibodies from blood in animal study
1989 Garofalo and Chang	Macroporous Agar artificial cells containing microorganisms, permeable to macromolecules like lipoprotein to convert cholesterol into carbon dioxide
1993 Prakash and Chang	Artificial cells containing genetically engineered *E. coli* to remove urea

Overview of cell encapsulation: see Table 4

Reference [9] contains the detail journal sources of the above reference up to 1980.

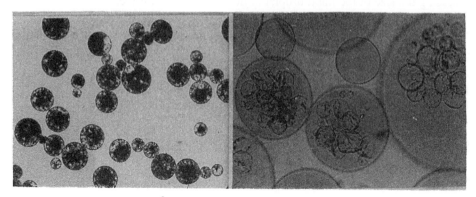

Fig. 3. Artificial cells with 200 Å ultrathin polymer membrane containing proteins and enzymes. The other side shows larger artificial cells containing smaller cells to form intracellular compartments. From Ref. [1] with permission of the copyright holder.

Different types of synthetic polymers can be used (Table 3). Variations in configuration are also possible (Table 3). A single ultrathin polymer membrane is the most common one (Fig. 3). The unlimited type of polymers used allow for possible variations in permeability, biocompatibility and other characteristics [1–13]. Artificial cells can contain smaller "intracellular compartments" [1,4–6] (Fig. 3). Others can be prepared to form solid polymer microspheres containing microdroplets of biologically active materials [5]. Liquid hydrocarbons microdroplets containing biologically active materials are useful in biotechnology and other applications [17].

Biodegradable or biological materials
These types of artificial cells do not stay in the body after they have completed their functions. There are several biodegradable materials (Table 3). Protein membrane artificial cells and polyhemoglobin are two examples [1,3,5,6]. The use of lipid is another common approach. This includes the use of lipid-protein membrane [1], concentric lipid membranes and submicrometer lipid vesicles [18]. Another approach is the use of biodegradable synthetic polymers. This author first used polylactide [19]. Many types of polylactides and polyglycolic acids are now available for artificial cells [20]. Investigators have also used other types of synthetic biodegradable polymers. Polyanhydride is one example [21]. Research on biodegradable artificial cells is now a very active field.

Bioencapsulation of cells

General

"… Microencapsulation of intact cells or tissue fragments … the enclosed material might be protected from destruction and from participation in immunological processes, while the enclosing membrane would be permeable to small molecules of specific cellular product which could then enter the general extracellular compartment of the recipient. For instance, encapsulated endocrine cells might survive and maintain an effect supply of hormone ... . The situation would then be comparable to that of a graft placed in an immunologically favorable site." "There would be the further advantage that implantation could be accomplished by a simple injection procedure rather than by a surgical operation."

"Microencapsulation of intact cells ... . The erythrocytes were suspended in hemolysate rather than in the diamine solution; and a silicone oil [Dow Corning 200 fluid] was substituted for the stock organic liquid. The microencapsulation was then carried out ... by the principle of interfacial polymerization for membranes of cross-linked proteins ... . A large number of human erythrocytes suspended in hemolysate within a microcapsule of about 500 μm diameter was prepared by the syringe [drop] method" (Fig. 4).

The above excerpts came from a 1965 publication by the author [5]. Excerpts of this have also appeared in other earlier publications [1,6]. However, there was then no interest in this type of biotechnology. It had to wait until the early 1980s before other groups started to investigate this approach. For instance, we had earlier encouraged other groups including Connaught Laboratory to explore this approach, especially in encapsulation of islets. In 1980, Sun and his collaborators at Connaught Laboratory reported their results with this approach using a modified technique based on polylysine-alginate [22]. Since then, many groups have investigated the

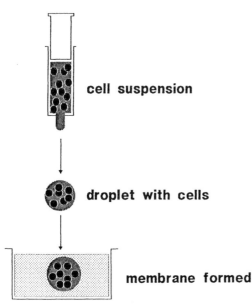

cell suspension

droplet with cells

membrane formed

Fig. 4. The author's original method for encapsulation of cells [5]. Drop methods continue to be the basis for different modified methods for cell encapsulation. Figure reproduced from Ref. [11] with permission of the publisher.

274

Table IV. Artificial cell microencapsulated cells and microorganisms (first reference cited)

1965–1972 Chang	Cross-linked protein membrane artificial cells containing intact cells; preparation and basic research. Proposed use in implantations of islets for diabetes mellitus and hepatocytes for liver
1980 Lim and Sun	Alginate-polylysine-PEI artificial cells containing islets for implantation in diabetes mellitus in rats
1981 Damon Co	Artificial cells containing hybridoma and fibroblasts for production of monoclonal antibodies and interferon
1982 Sefton et al	Endragit RL and Alginate-Endragit RL membrane artificial cells
1984 Sun et al	Artificial cells containing islets can maintain diabetic rats normoglycemic for up to 1 year after implantation
1985 Tice and Meyers	Cross-linked albumin membrane artificial cells containing hybridomas
1985 Goosen and Sun	Alginate-polylysine-polyethylene artificial cells containing islets
1986 Wong and Chang	Implantation of artificial cells containing liver cells increases survival time of acute liver failure rats
1986 Sun et al	Polyacrylate membrane artificial cells containing hybridomas
1986 Shiotami	Chitosan-CM-Cellulose artificial cells containing hybridomas
1989 Garofalo and Chang	Macroporous Agar artificial cells containing microorganisms to convert lipoprotein bound cholesterol in serum into carbon dioxide
1989 Wong and Chang	Implantation of rat liver cells in artificial cells continued to survive in mice
1989 Aghazaman-Kashani and Chang	Hepatotrophic factor secreted by encapsulated hepatocytes is retained in artificial cells
1989 Bruni and Chang	Implantation of artificial cells containing liver cells lowered hyperbilirubinemia in Gunn rats
1991 Wong and Chang	Two-step method to prevent cells extrusion and rejection on surface of artificial cells
1993 Prakash and Chang	Artificial cells containing genetically engineered *E. coli* to remove urea and ammonium

bioencapsulation of islets, other endocrine cells, hepatocytes and genetically engineered microorganisms [23] (Table 4). Alginate-polylysine-alginate membrane and other approaches are being used.

Bioencapsulated islets

Many groups have maintained normal blood glucose levels in diabetic mice, rats, dogs and monkey after one implantation of islet artificial cells [22–30]. More recently one group has maintained a diabetic patient normoglycemic after implantation [31]. This is a very active area and will be the subject of a separate review.

With the present interest in gene therapy and the ability to produce genetically engineered microorganisms, this review concentrates on the following two areas: encapsulation of hepatocytes and encapsulation of microorganisms.

Bioencapsulated hepatocytes as a model for xenograft transplantation and gene therapy

Introduction

In some types of gene therapy, induction can be carried out in hepatocytes. At present, instead of direct injection of vectors into humans, the following is carried out. Small amounts of liver tissue are obtained by surgery. These are cultured in vitro. Induction is then carried out in vitro. A second surgical procedure is then required to reintroduce the cells back into the patients. There are potential problems related to this. One is the need for two surgical procedures for each treatment, especially in small infants, e.g. phenylketonuria patients. The other problem is that the reintroduced cells may not survive infinitely. Thus, one needs to repeat the procedure. Obtaining other sources of liver cells will solve these problems. Unfortunately, nonautologous hepatocytes would be immunorejected, and the patients requires immunosuppression. The other possibility is to use encapsulation. This way, hepatocytes can be obtained from a human or nonhuman donor. If needed, this can include induction by vectors in vitro to further increase specific enzyme activity. Artificial cells would protect them from rejection after implantation. Furthermore, administration would be by injection rather than by surgery. We have carried out studies in model systems using hepatocytes to analyze the feasibility of this method. The same principles apply to xenografts of cells or tissues.

Implantation of encapsulated rat hepatocytes increases the survival of fulminant hepatic failure rats

We implanted encapsulated rat hepatocytes into rats with galactosamine induced acute liver failure [32]. This significantly increased their survival time when compared to the control group. In these studies, hepatocytes from 125–135 g young Wistar rats were isolated. They were enclosed within alginate-polylysine-alginate microcapsules with mean diameter of 300-μm. Each artificial cell contained 120 ± 20SD hepatocytes. Implantations involved a simple injection procedure using syringes with 20-gauge needles. Wistar rats (275–285 g) with galactosamine-induced fulminant hepatic failure were used. Each rat in the control group received one peritoneal injection of 4 ml of artificial cells containing no hepatocytes. Each rat in the treated group received one peritoneal injection of 4 ml of 62,000 artificial cells containing a total of 7.4×10^6 hepatocytes. Two other groups have supported this finding of increase survival after implantation [33,34].

Implantation of encapsulated rat hepatocytes lowers the plasma bilirubin level in Gunn rats with hyperbilirubinemia

The homozygous Gunn rat is an animal model for human nonhemolytic hyperbilirubinemia (Crigler-Naijar type I). Gunn rats are mutants of the Wistar rat strain. They have a defect in the enzyme bilirubin uridine diphosphate glucuronyltransferase. These rats cannot conjugate bilirubin. As a result they have a very high systemic bilirubin level. Our studies showed that a single intraperitoneal injection of artificial cells containing hepatocytes significantly lowered the bilirubin levels [35–38].

In the first experiment, 3.5 months old Gunn rats weighing 258 ± 12 g were used. During the 16-day control period, the serum bilirubin increased at a rate of 0.32 ± 0.07 mg/100 ml per day. This reached 14 ± 1 mg/100 ml at the end of the control

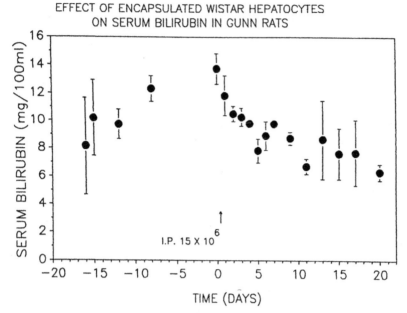

Fig. 5. Changes in systemic bilirubin levels in 3.5 months old Gunn rats. The first 20 days are control periods. This was followed by an intraperitoneal injection of 1.1 ml of artificial cells into each rat. Each 1.1 ml of artificial cells contained 1.5×10^6 viable Wistar rat hepatocytes. Each point represents the mean and standard deviation. From Bruni and Chang [37]. Reproduced with permission of the publisher.

period. On day 16, each animal received an intraperitoneal injection of 1.1 ml of artificial cells containing 15×10^6 viable Wistar rat hepatocytes. Twenty days after implantation, the serum bilirubin decreased to a level of 6 ± 1 mg/100 ml (Fig. 5). The level remained low 90 days after implantation. Microcapsules recovered from the peritoneal cavity remained intact.

In the second experiment, we used four groups of Gunn rats. The bilirubin levels did not decrease in the control group and the group which received control microcapsules containing no hepatocytes. The rate of decrease of indirect bilirubin was the same for the group receiving encapsulated hepatocytes and the group receiving free hepatocytes. This shows that the hepatocytes in the microcapsules have the same ability as free hepatocytes to conjugate bilirubin. The alginate-polylysine-alginate microcapsules therefore do not form a barrier for the transport of bilirubin.

Another group has carried out similar studies in the Gunn rats and supported our findings [39–41]. They have in addition showed that it is possible to carry out repeated implantation for long-term treatment [40,41].

Immunoisolation of encapsulated rat hepatocytes when implanted into mice
Thus, encapsulated rat hepatocytes are effective in rats with acute liver failure and in Gunn rats. The next step is to see if encapsulation protects the hepatocytes from

immunological rejection. We therefore studied whether artificial cells protect rat hepatocytes from rejection when implanted into mice [42].

Rat liver cells were isolated from 125–150 g male Wistar rats following the standard method. Exclusion of trypan blue dye is used to measure the viability of hepatocytes. After isolation, the viability was 80%. We then encapsulation the hepatocyes within alginate-polylysine-alginate microcapsules [42] The percent of viable cells decreased to 63.4% after the microencapsulation procedure.

Free or encapsulated rat hepatocytes were implanted intraperitoneally into 20–22 g male CD-1 Swiss mice. At set intervals mice were sacrificed. The peritoneal cavity of each mouse was washed twice with 2.0 ml of iced Hank's balanced salt solution. There were varying degrees of aggregation and clumping of microcapsules. When this happens, the encapsulated hepatocytes are no longer viable. A subsequent section deals with clumping and aggregation including a novel technique to prevent this. The present study is a basic study to look only at microcapsules that remain free-floating. We want to first see whether under optimal conditions, rat hepatocytes can remain viable when implanted into mice.

There was a significant increase ($P < 0.001$) in the percentage of viable hepatocytes in the free-floating microcapsules. The percentage of viable cells increased with time. Thus 29 days after implantation, the viability increased from the original 62% to nearly 100%. There was no significant change in the total number of hepatocytes in the microcapsules (Fig. 6). On the other hand, mice rapidly rejected free rat

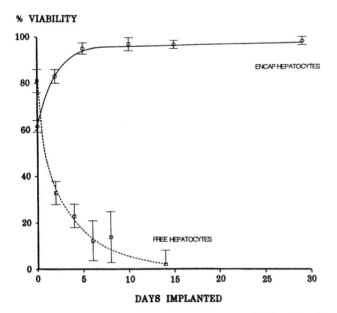

RAT XENOGRAFT IN NORMAL MICE

Fig. 6. Artificial cells encapsulated rat hepatocytes implanted peritoneally into mice. Viability of rat hepatocytes inside free-floating artificial cells. From Wong and Chang [42]. Reproduced with permission of the publisher.

hepatocytes implanted without encapsulation. By the 14th day, there were no intact free hepatocytes detected in the mice.

Thus rat hepatocytes inside free-floating microcapsules are immunoisolated. As a result, they are not rejected after implantation into mice. Instead of rejection, the percentage of viable hepatocytes increases.

Accumulation of hepatic stimulatory factor released by hepatocytes inside artificial cells

After implantation into mice, the increase in viability of microencapsulated rat hepatocytes was surprising and unexpected [42]. There are hepatic stimulatory factors (HSF) which can stimulate liver regeneration. Thus the purpose of our next study [43,44] was to see if HSF are released from encapsulated hepatocytes. Furthermore, are they retained inside the artificial cells after they are released ?

Hepatocytes were isolated from 70–90 g Wistar-Lewis rats. Free or microencapsulated hepatocytes were cultured separately. HSF were analyzed by its ability to increase survival time and stimulate incorporation of [³H]thymidine in liver of rats with galactosamine-induced fulminant hepatic failure. Supernatant from free hepatocyte culture medium contained significant HSF activity. Supernatant from hepatocyte microcapsules culture medium did not have significant HSF activity. In the next study, the microcapsules containing hepatocytes were ruptured. The supernatant of the microcapsule contents was found to have significant HSF activity. This suggests that hepatocytes in the microcapsules secrete HSF that can stimulate liver regeneration. This factor is retained inside the microcapsules after secretion [43]. Sephacryl gel chromatography showed that this factor has a molecular weight of over 110,000 Da [44]. This factor loses its hepatic stimulatory effect after heat treatment or trypsin treatment. Its elution profile on Sephacryl gel does not change after such treatments.

The peritoneal cavity is an excellent cell culture medium. Encapsulated rat hepatocytes are not rejected after implantation into mice. Furthermore, rat hepatocytes secrete HSFs which are retained in the microcapsules with increasing concentration. All the above factors may help to increase the observed viability and recovery of microencapsulated hepatocytes. We used trypan blue dye exclusion to measure viability. The increase in viability most likely reflects recovery of membrane integrity.

New Approaches in Cell Encapsulation

Methods to improve biocompatibility and immunoisolation

One problem with microencapsulated cells is the variation in biocompatibility from batch to batch. This is especially a problem in xenografts. There has been much research to investigate different types of membrane materials. The most commonly used material is alginate-polylysine-alginate. Some groups have investigated the use of other materials [52–57]. For instance, Sefton's group have studied the use of

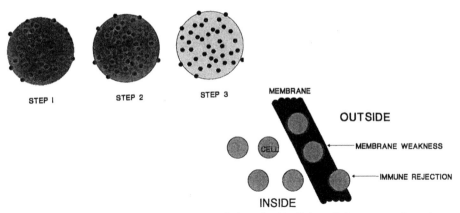

Fig. 7. Above: standard method for bioencapsulation of cells. Below: Enlargement of membrane. In encapsulation of hepatocytes or microorganisms, some cells may become entrapped in the membrane matrix. This can result in membrane weakness or perforation. Figure reproduced from [11] with permission of the publisher.

polyacrylates and other polymeric materials for cell encapsulation [52,53]. Another group has used polyelectrolyte complexes [57]. For islets, improvement in purity of the material alone seems to help in solving the problem. Thus, Soon-Shiong's group used purified alginate with a high glucuronic acid content to form alginate-polylysine-alginate encapsulated islets [32]. They have improved the biocompatibility of bioencapsulation of islets to such an extent that they no longer observe a fibrous reaction after implantation even in large animals like dogs. They have also obtained successful results in their preliminary clinical trial in a diabetic patient [33]. Islets are large aggregates of cells. Furthermore, each artificial cell only needs to contain one or more islets. The situation is quite different in the encapsulation of high concentrations of dispersed small cells like hepatocytes and microorganisms. Here, other factors are also important. We have therefore carried out basic studies in animals [45]. Based on these findings, we have devised a novel two-step procedure [46,47] to solve the specific problem of encapsulation of high concentrations of smaller cells like hepatocytes and microorganisms.

Problems with standard methods of cell bioencapsulation
In bioencapsulating hepatocytes, there were occasionally hepatocytes incorporated into the membrane matrix [45]. When this happened the capsular membrane over these sites appeared thin and poorly formed [45]. Cells embedded in the membrane matrix could perforate this imperfection.

After implantation into mice, rat hepatocytes exposed on the surface attracted an acute cell mediated host response [45]. Even if the membrane embedded cells did not perforate the membrane, macrophages and lymphocytes could perforate these imperfections. The activation of macrophages and lymphocytes could be followed by cytokine release and fibrous deposition. Cellular entrapment into the capsular mem-

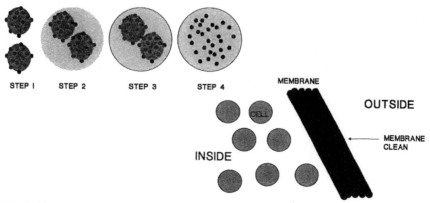

Fig. 8. Above: new two-step method for bioencapsulation of a large number of cells. Below: enlargement of membrane. This helps to prevent entrapment of cells in the membrane matrix. Figure reproduced from [11] with permission of the publisher.

brane would statistically increase with increasing concentration of cells encapsulated. Severe problems usually start at hepatocyte concentrations of 10×10^6 to 20×10^6.

A novel approach to solve the above problems
We devised a two-step method to prevent the above problems [46,47]. Figure 8 is a summary of the steps.

The steps are as follows. (1) Small calcium alginate gel microspheres containing entrapped cells were first formed. (2) The small microspheres were entrapped within larger calcium alginate gel microspheres. This is an important additional step to lessen the chance of cells protruding on the surface of the gel. (3) The alginic acid on the surface of the larger microsphere was reacted with poly-l-lysine to form a microcapsule membrane. By doing this, cells are much less likely to be embedded in the membrane matrix. (4) The entire content of the microcapsule was then liquefied by citrate to remove calcium. This also liquefied the smaller calcium alginate gel microspheres inside the microcapsule. This released the hepatocytes to float freely in the microcapsule.

Microscopic studies showed that there was no cell embedded within the walls of the microcapsular membrane [46]. The cells were in free suspension inside the microcapsules. After 7 and 11 days of implantation into mice, there was significantly more free-floating microcapsules compared to those prepared by the standard procedure. When seen in cross-section, the capsular membrane appeared uniform with no encapsulated cells seen embedded within the membrane matrix. There were also no lymphocytes or macrophages adhering to the microcapsule surface.

Improvements in mass transfer, culture methods and preservation
Further studies include developing this approach for possible applications. We are

using dextran to analyze mass transfer of large molecules across these microcapsules [48]. This is important because hepatocytes synthesize several proteins and peptides required by the body. Dextran preparation has a wide molecular size distribution. By using chromatography, we can study the mass transfer of many molecular sizes simultaneously. Hepatocyte conformation in the microcapsules is another area of study. We have studied the formation of multicellular hepatocyte spheroids in microcapsules [49]. Another group studied hepatocytes microencapsulated with a three-dimensional collagen matrix [50]. In another study they showed that they could cryopreserve microencapsulated hepatocytes [51]. When the cyropreseved preparations were reconstituted and implanted into Gunn rats, they continued to carry out their functions. This approach of cryopreservation allows the long-term storage of the preparation until required.

Other studies on cell bioencapulsation

Cell bioencapsulation is being actively explored for biotechnological applications. There are many other approaches [54–64]. Sun's group have studied bioencapsulated parathyroid tissues. We have also bioencapsulated rat kidney cells which continued to secrete erythropoietin in response to the proper stimulus [64].

Bioencapsulation of Microorganisms

General

With genetic engineering, there is much potential for using different types of microorganisms to carry out useful biotechnological functions. However, for many of the biotechnological and medical applications, the microorganisms may need to be encapsulated.

Encapsulation of microorganism for converting lipoprotein bound cholesterol into carbon dioxide

We have studied the use of a bacteria immobilized in artificial cells to convert cholesterol into carbon dioxide [65,66]. *Pseudomonas pictorum* [ATCC #23328] was used as a model system. The standard method of cell encapsulation does not allow for free permeation of cholesterol bound to lipoprotein. We therefore devised another approach based on high porosity agar. These high porosity agar beads stored at 4°C did not show any sign of deterioration. The beads retained their activity even after 9 months of storage. There was no evidence of leakage of the enclosed bacteria. We optimized the permeability to allow the diffusion of lipoprotein-cholesterol. Detailed kinetic studies were carried out. Open pore agar beads containing microorganisms were incubated in serum. Their ability to remove cholesterol was compared to control capsules and free bacteria. There were no significant differences between the bioencapsulated and free bacteria. Bacteria available at present have a low enzyme activity for cholesterol conversion. Indeed, bacterial reaction was the limiting step in the overall reaction of the bioencapsulated bacteria. For practical applications, we shall require genetically engineered microorganisms with high enzyme activities. In the meantime, this basic finding of bioencapsulation may be applicable to future use of microorganisms in this and other areas. This would be especially applicable in the conversion of macromolecular substrates or substrate bound to macromolecules.

Artificial cells encapsulated genetically engineered E. coli *containing* K. aerogenes *gene to remove urea or ammonium*

In kidney failure, the complicated dialysis machine is still in common use. Artificial cells containing absorbent in hemoperfusion can remove all organic uremic metabolites except urea. Extensive research has not resulted in an effective urea removal system. Earlier studies by another group using soil bacteria were not successful because it did not have a high urea utilization rate. Genetic engineering can now enhance the urea utilization capacity of *E. coli*. We are therefore studying the bioencapsulation of this genetically engineered *E. coli* that contains *K. aerogenes* gene [67,68]. Bacteria *E. coli* DH5 strain were encapsulated in alginate-polylysine-

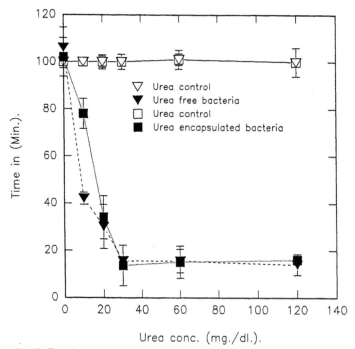

Fig. 9. Genetically engineered *E. coli* containing *K. aerogenes* gene. Free bacteria compared to bioencapsulated bacteria in removing urea. From Prakash and Chang [68] with permission of the publisher.

alginate microcapsules [67,68]. This preparation could efficiently remove urea without any increase in ammonia [67,68]. A 100 mg alginate encapsulated bacteria rapidly reduces urea in a 100 ml solution. The original urea concentration 100.00 ± 1.00 mg/dl fell to 1.55 ± 0.13 mg/dl in 30 min (Fig. 9). There was no increase in the ammonia in the reaction medium.

The original ammonia concentration in the medium, was $758.00 \pm 70.00 \mu M$. Within 20 min this was lowered to $90.42 \pm 38.05 \mu M$. The encapsulation process did not affect the urea depletion efficiency of the bacteria. Using a single pool model, we calculated urea removal for a total body water volume of 40 l. This shows that 40 g of microencapsulated bacterial cells can reduce the urea in the 40 l (100.00 mg/dl) to 6.00 mg/dl in 30 min. Furthermore, the encapsulated bacteria can remove 98.50% of urea from this 40 l in 30 min without producing ammonia. This is 10 times more efficient than oxystarch gel and 30 times more efficient than microencapsulated zirconium phosphate-urease.

Optimization in the methods and mass transfer of bioencapsulated microorganisms

A number of groups are optimizing the bioencapsulation of microorganisms [69–71]. For instance we have optimized the preparation and mass transfer of bioencapsulated microorganisms [69].

Artificial Cells Containing Enzymes for Inborn Errors of Metabolism and Other Metabolic Diseases

Artificial cells containing enzymes for inborn errors of metabolism and others

Our earlier basic studies of implantation of artificial cells containing enzymes showed their effectiveness [3–4,72–76]. A typical example is the implantation of catalase artificial cells into acatalesemic mice [72,73]. These studies show that enzymes in artificial cells effectively replace the deficient catalase after implantation. Furthermore, enzyme in artificial cells does not leak out and the artificial cells remained intact. As a result, this did not cause hypersensitivity or immunological reactions even with repeated injection [73]. This approach is different from the use of liposomes to deliver enzymes [18]. In the case of liposomes, the enzymes have to be released to carry out their functions [18]. In all cases of parental administration, enzymes are not stable at 37°C; we need to repeat the injections weekly. Treatment of enzyme deficiency is a long-term treatment. This results in the accumulation of injected materials. The use of enzyme artificial cells in hemoperfusion [4,77,78] avoids the need for repeated injections. However, this requires extracorporeal circulation of blood.

Oral administration of phenylalanine-ammonia-lyase artificial cells for phenylketonuria

We have investigated the oral administration approach [79,80] in a phenylketonuria

PKU rat model using artificial cells containing phenylalanine ammonia-lyase (PAL). Oral ingestion lowered systemic phenylalanine in PKU rats to normal levels in 7 days ($P < 0.010$). Unlike the control PKU group, these treated PKU rats were symptom-free with steady increase in body weight of $1.84 \pm 0.06\%$ per day. This is dose related. The thinking then was that, in the intestine, the artificial cells removed phenylalanine derived from digested dietary proteins.

Oral administration of xanthine-oxidase artificial cells for Lesch–Nyhan disease
In another study we treated a patient with Lesch–Nyhan disease, using oral artificial cells containing xanthine oxidase to lower systemic hypoxanthine [81,82]. This is an inborn error of metabolism due to hypoxanthine phosphoribosyltransferase deficiency. Hypoxanthine being lipid-soluble diffuses rapidly from the blood into the intestinal tract. Within 1 week there was a fall in plasma hypoxanthine level. There was also a reduction of CSF hypoxanthine and inosine after 2 weeks. There were no adverse effects.

Present status
At the beginning these findings with PKU and Lesch–Nyhan disease did not stimulate too much interest. Since there were not many metabolites from the body that diffuse readily from blood across the intestinal wall. Only very small water soluble molecules like urea and lipid-soluble molecules like hypoxanthine belong to this group. In the case of PKU, it was thought that artificial cells functioned by removing phenylalanine from ingested protein. This being the case, it was thought that this would not be different from using a low phenylalanine diet. However, two recent findings have renewed interest in this approach. One of these is the recent report of the UK MRC report related to problems encountered with low phenyl-alanine diet. The other is our recent new finding of an extensive enterorecirculation of endogenous amino acids.

Low phenylalanine diets for patients with phenylketonuria
The UK MRC Study Group on Phenylketonuria published their recommendations in the British Journal of Medicine in 1993 [83]. It mentions the difficulties and expense of the PKU diet. It involves rigorous restriction of natural protein intake and ingestion of unpalatable substitutes for protein. It is also difficult to maintain smooth phenylalanine control during periods of minor fevers or low caloric intake. It also points out that there is a high risk of fetal damage in offspring of PKU mothers. It emphasizes the need for an alternative to PKU diet.

Enzyme therapy based on the recent new finding of enterorecirculation of amino acids

There is a somewhat analogous situation being used clinically for cholesterol removal. For cholesterol, much of the bile acid secreted in the bile is reabsorbed as it passes down the intestine. The body uses the reabsorbed bile acid to form choles-

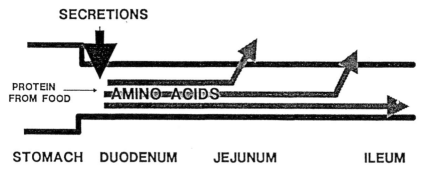

SECRETIONS

PROTEIN ——→
FROM FOOD   AMINO ACIDS

STOMACH   DUODENUM   JEJUNUM   ILEUM

Fig. 10. Enterorecirculation of amino acids. Large amount of endogenous sources of amino acids that enters the intestine are reabsorbed. This endogenous source of amino acids is much high than that from ingested proteins in food (Chang and Lister [84–86]). Figure reproduced and modified from Ref. [11] and published here with permission of the publisher.

terol. Oral adsorbents remove bile acid and prevent its reabsorption, thereby lowering systemic cholesterol levels. Our research has recently showed that amino acids behave in a similar although not identical manner [84–86]. Very unexpectedly it was found that the major source of amino acids in the intestine did not come from dietary protein. Very surprisingly and unexpectedly, we found that the major source of amino acids in the intestine comes from gastric, pancreatic, intestinal and other secretions (Fig. 10). These secretions contain high concentrations of proteins, enzymes, polypeptides and peptides. Tryptic enzymes in the intestine break these down into amino acids. As these amino acids pass down the intestine, they are reabsorbed into the body. There is therefore a large recirculation of amino acids between the body and intestine (Fig. 10) [84,85]. The dietary source of amino acids is negligible when compared to the endogenous source (Fig. 10) [84,85].

Preliminary studies show that orally administered artificial cells containing an enzyme can deplete the corresponding amino acid from the intestinal tract [84–85]. The specific amino acid is therefore removed and not reabsorbed into the body. This selective removal of specific amino acid allows for the removal of undesirable amino acids from the body. This can form the basis for the treatment of inborn errors of amino acid metabolism. Our earlier observation of the effectiveness of oral enzyme artificial cells in PKU rats [79,80,84–86] is most like due to their removing phenylalanine from this enterorecirculation. Other conditions include tyrosinase for tyrosinemia [84,85] and histidinase for histidinemia [87]. Asparaginase artificial cells to remove asparagine for some types of leukemia is another example [84,85].

Artificial cells containing cell culture in enzyme therapy

Oral uses of artificial cells discussed above are ready for clinical use. Since artificial cells pass through the intestine and are excreted once they have carried out their functions, it is easier to ensure their safety in patients. Furthermore, earlier clinical results of oral administration have already shown their safety. However, implantation

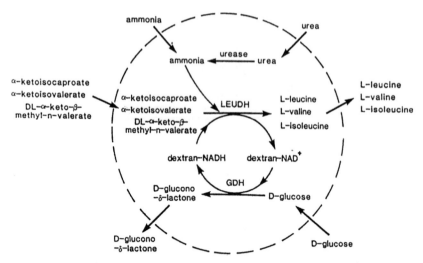

Fig. 11. Artificial cells containing a multienzyme system for the conversion of urea or ammonia into essential amino acids (leucine, isoleucine and valine). From Gu and Chang [94–96].

of artificial cells containing cells is a different matter. This requires the implantation of artificial cells into the body and therefore requires much more research. On the other hand, implantation of cells in artificial cells could be a more effective treatment. It only requires one implantation, rather than repeated oral ingestion.

Artificial cells containing multienzyme systems

We have studied the use of artificial cells containing multienzyme systems [14,88–93]. One example for possible application is for the conversion of urea or ammonia into essential amino acids (leucine, valine and isoleucine) [94–96]. In liver failure, one needs to remove the elevated ammonium. There is also a marked decrease in the ratio of branched chain amino acids (valine, leucine and isovaline). In kidney failure there is marked elevation of urea. We prepared artificial cells containing glucose dehydrogenase (EC 1.1.1.47), leucine dehydrogenase (EC 1.4.1.9), urease (EC 3.5.1.5) and dextran-NAD [94–96]. These artificial cells can convert ammonia or urea into L-leucine, L-valine and L-isoleucine. Glucose dehydrogenase effectively regenerates the dextran-NADH required using glucose at blood concentration. Not only is ammonium removed, they are also converted into the required essential amino acids. We now wait for genetic engineering to produce these enzymes with higher specific activities for actual applications.

Bioencapsulation of Biosorbents and Immunosorbents

This is the simplest of artificial cells for bioencapsulation of biologically active material. This has been used in routine clinical applications for some time [9,104–

106]. Sorbents like activated charcoal, resins and immunosorbents cannot be used in direct blood perfusion. This is because biosorbents release fine particles into the blood. The biosorbent itself can also damage blood cells. This author found that sorbents like activated charcoal inside artificial cells did not release particles or damage blood cells [1,4,8]. We therefore developed and used this successfully in patients [1,8,9,100].

Preparation

Each artificial cell is formed by applying an ultrathin coating of ultrathin membrane on each $100\,\mu$m diameter activated charcoal microspheres. Seventy grams of these artificial cells are placed in a cylindrical container similar to a chromatography column. Screens on either side of the column retain the artificial cells. The openings in the screens are about $40\,\mu$m. This retains the artificial cells in the device but allows blood to perfuse freely through the column. Each hemoperfusion device being used in patients contains 70 g of artificial cells. The mass transfer for this small device is much higher than that for a standard dialysis machine. This is because of the large surface area and ultrathin membrane of the artificial cells. Extensive clinical data are now available from this and other groups. Since this is a very large area, only a brief summary and a few review articles on clinical uses are included here [9,13,100–102].

Acute poisoning

We first used this effectively in patients with acute poisoning in 1973 [9,100]. Other groups including those of Shreiner, Winchester, Agishi, Odaka, Ota, Bonominni and others have supported this finding [13,101,102]. This is now a routine treatment for acute poisoning in adult and pediatric patients [101]. There are certain requirements. (1) Drug or toxin can be adsorbed by activated charcoal in the artificial cells. (2) Volume distribution of drug should allow sufficient level in circulating blood for removal. (3) Treatment must be started before irreversible damage.

Chronic renal failure

Our study between 1971 and 1980 [1,9,99] showed that it is more effective than hemodialysis in removing organic metabolites in uremic patients. This is being used in two ways. (1) In series with dialysis, we showed that it shortened dialysis time and improved dialysis resistant symptoms [9,100]. This has been supported by other groups including those of Ota, Beber, Qules, Winchester, Martin, Trznadel, Asaba, Otsubo, Agishi, Stefoni, Bonimini, Splendiani [13,99–101]. (2) In series with a small ultrafiltrator it can replace the dialysis machine. Here we used oral adsorbents to control potassium and phosphates. An urea removal system is being developed to complete the hemoperfusion-ultrafiltrator approach.

Liver failure

In 1972 the author found that the detoxifying functions of hemoperfusion resulted in temporary recovery from coma in grade IV hepatic coma patients [9,13,99]. Williams, Odaka, Winchester, Agishi, Blume, Bartels, Yamazaki, Amano, Gelfand, Gimson, Takahashi, Keneedy and others have supported this observation [9,13,100]. Our studies in fulminant hepatic failure (FHF) rats and Williams' studies in FHF patients showed that hemoperfusion increased the survival rates if started in earlier coma. Hemoperfusion is more effective than the liver in detoxication. On the other hand, it does not fulfill the other complex hepatic functions. Progress in bioencapsulation of hepatocyes would lead to a combine use with hemoperfusion to supplement its functions.

Aluminum and iron overload

Chang and Barre [13,100,101] used hemoperfusion with desferroxaimine in patients with high aluminum levels. Nephrologists now use this as a routine treatment for removing aluminum or iron in patients [13,100].

Immunosorbents and other adsorbents

The author used artificial cells for other sorbents [4,7] and immunosorbents [13]. This is another exciting area under laboratory and clinical investigation by a number of groups.

Modified Hemoglobin as Red Blood Cell Substitute

This is a very extensive area. It will be covered in a separate review. The present review contains only a very brief summary. Since the mid-1980s there has been ongoing concern regarding potential immunodeficiency disease (AIDS) due to HIV in donor blood. This has resulted in extensive research and developments on the biotechnological approach of our earlier idea of modified hemoglobin [1–5]. Until this time, there was little interest in this. Since 1987, this has developed into a very large area [102–106]. The large number of papers in this area is reflected in the need to add "Blood Substitutes" to the name of a journal in this rapidly advancing field [106].

Artificial red blood cells from encapsulated hemoglobin

The author prepared the first modified hemoglobin as artificial red cells microencapsulating hemoglobin [2]. The enclosed hemoglobin continued to reversibly carry oxygen. We showed that these artificial red blood cells had no blood group antigens [1,5]. The circulating time after infusion was short [1,3,5,6]. We modified their surface properties and diameter. This increased the circulating time. In 1980, Djordjevich and Miller prepared lipid membrane artificial red cells of submicrometer sizes [102,103]. This resulted in further increase in circulating time. Many other groups have since made further important progress, Hunt, Farmer, Tsuchida, Schmidt, Rudolph, Beissinger, Chang and many others [102–106]. Research included variations in surface properties, new techniques in preparation and improvements in membrane strength. This resulted in increase in circulation time useful for short-term blood replacement in animal studies.

Crosslinked hemoglobin

In 1964 [3] and later [1,5,6,] the author used a bifunctional agent (a diacid) to prepare modified hemoglobin blood substitutes: (1) crosslinking hemoglobin on the surface to form artificial cell membrane; (2) crosslinking all the hemoglobin to form polyhemoglobin aggregates. The reaction is as follows:

$$Cl-CO-(CH_2)_8-CO-Cl + HB-NH_2 = HB-NH-CO-(CH_2)_8-CO-NH-HB$$

Diacid   Hemoglobin   Crosslinked hemoglobin

In 1968, Bund and Jandl prepared soluble polyhemoglobin by decreasing the number of hemoglobin crosslinked to less than 10 to increase circulating time [120]. In 1975, Benesch pyridoxalated hemoglobin to improve the P50 of polyhemoglobin [99–100]. Further studies by many groups include new crosslinking agents, intramolecular crosslinkage, conjugated hemoglobin, bovine hemoglobin, physiopathology and clinical trials [102–106]. These groups include Abuchowski, Agishi, Bakker, Biro, Bucci, Chang, DeVenuto, Estep, Faivre, Feola, Greenburg, Hedlund, Hori, Hsia, Iwashita, Jesch, Messmer, Moss, Nose, Pristoupil, Sekiguchi, Sideman, Valeri, Winslow, Wong and many others [102–106]. Shoemaker's group [105] uses recombinant techniques to produce human hemoglobin in microorganisms. Most modified hemoglobins are effective as short-term blood substitutes [102–106].

    Crosslinked hemoglobin is the simplest, and therefore the first to be ready for human use. Initially, animal safety results were not completely applicable to humans [103]. Chang and Lister [103] designed a test, based on adding modified hemoglobin to human plasma in a test tube. We then analyze for complement activation by modified hemoglobin. This test helped to bridge the gap between animal studies and clinical trials in humans [103–104]. The groups of Estep, Gould Przybelski, and Shoemaker are carrying out clinical trials in humans [105]. Gould's group is now starting Phase II clinical trials on efficacy using the following crosslink method:

$$H-CO-(CH_2)_3-CO-H + HB-NH_2 = HB-NH-CO-(CH_2)_3-CO-NH-HB$$

Dialdehydes   Hemoglobin   Crosslinked hemoglobin

Future perspectives

Artificial cells can contain many types of biologically active materials. There are therefore many other areas of applications and research. For example, artificial cells prepared from crosslinked protein, biodegradable polymer membrane or lipid membrane, are being investigated for use as drug carriers. This is such a large area that it requires a separate review. In other examples, the author has enclosed magnetic and biological materials together inside artificial cells [4]. This allows for localization with external magnetic fields [4] either in a bioreactor or in the body. Others have used artificial cells in laboratory analysis of free and protein-bound hormones in patients [10,11,13]. We have studied its use for 1-shot vaccine [10,11] and for removing large lipohyllic molecules from small hydrophyllic molecules. Still others have used artificial cells for aquatic culture for shrimps and lobsters [10,11].

The author wrote in his 1972 book on Artificial Cells [1]: "Artificial cell is a concept; the examples described ... are but physical examples for demonstrating this idea. In addition to extending and modifying the present physical examples, completely different systems could be made available to further demonstrate the clinical implications of the idea of "artificial cells." An entirely new horizon is waiting impatiently to be explored."

Acknowledgements

The author acknowledges the support of the Medical Research Council of Canada in the operating grants and the MRC career investigatorship. He also acknowledges the Quebec Ministry of Education, Science and Technology for awarding him the Virage Award of Centre of Excellence in Biotechnology in 1985 and its permanent integration into this research center in 1990.

References

1. Chang TMS. Artificial Cells. Springfield: CC Thomas, 1972.
2. Chang TMS. Hemoglobin corpuscles. Research Report for Honours Physiology, Medical Library, McGill University 1957. Also reprinted in "30th Anniversary in Artificial Red Blood Cell Research" Biomater, Artif Cells Artif Org 1988;16:1–9.
3. Chang TMS. Semipermeable microcapsules. Science 1964;146:524–525.
4. Chang TMS. Semipermeable aqueous microcapsules ["artificial cells"]: with emphasis on experiments in an extracorporeal shunt system. Trans Am Soc Artif Intern Org 1966;12:13–19.
5. Chang TMS. Semipermeable aqueous microcapsules. PhD Thesis, McGill University, 1965.
6. Chang TMS, MacIntosh FC, Mason SG. Semipermeable aqueous microcapsules: I. Preparation and properties. Can J Physiol Pharmacol 1966;44:115–128.
7. Chang TMS, MacIntosh FC, Mason SG. Encapsulated hydrophilic compositions and methods of making them. Canadian Patent, 873, 815, 1971.
8. Chang TMS. Removal of endogenous and exogenous toxins by a microencapsulated absorbent. Can J Physiol Pharmacol 1969;47(12):1043–1045. (Also in Chang TMS. Nonthrombogenic mi-

crocapsules. US Patent, 3, 522, 346, 1970. Chang TMS. Blood compatible microcapsules containing detoxicants. US Patent, 3, 725, 113, 1973).

9. Chang TMS. Microencapsulated adsorbent hemoperfusion for uremia, intoxication and hepatic failure. Kidney Int 1975;7:S387-S392.

10. Chang TMS. Biotechnol of artificial cells including application to artificial organs. In: Moo-Young M (ed) Comprehensive Biotechnology: The Principles, Applications and Regulations of Biotechnol. In: Industry Agriculture and Medicine. New York: Pergamon Press, 1985:53–72.

11. Chang TMS. Artificial cells in immobilization biotechnology. Biomater, Artif Cells Immobil Biotechnol 1992;20:1121–1143.

12. Chang TMS. Artificial cells. In: Dulbecco R (ed) Encyclopedia of Human Biology. San Diego, CA: Academic Press, 1991;1:377–383.

13. Chang TMS. Recent advances in artificial cells with emphasis on biotechnological and medical approaches based on microencapsulation. In: Donbrow M (ed) Microcapsules and Nanoparticles in Medicine and Pharmacology. Boca Raton, FL: CRC Press, 1992;323–339.

14. Chang TMS. Recycling of NAD[P] by multienzyme systems immobilized by microencapsulation in artificial cells. Methods Enzymol 1987;136:67–82.

15. Donbrow M (ed). Microcapsules and Nanoparticles in Medicine and Pharmacology. Boca Raton, FL: CRC Press, 1992.

16. Chang TMS (ed). Blood Substitutes and Oxygen Carriers (Special Volume). Biomater, Artif Cells Immobil Biotechnol 1992;20:154–941.

17. May SW, Li NN. The circulation of urease using liquid-surfactant membranes. Biochem Biophys Res Commun 1972;47:1179.

18. Gregoriadis F. Liposomes as Drug Carriers: Recent Trends and Progress. New York: Wiley, 1989.

19. Chang TMS. Biodegradable semipermeable microcapsules containing enzymes, hormones, vaccines, and other biological. J Bioeng 1976;1:25–32.

20. Jalil R, Nixon JR. Biodegradable poly(lactic acid) and poly(lactide-co-glycolide) microspheres. J Microencaps 1990;7:297.

21. Mathiowitz E, Langer R. Polyanhydride microspheres as drug delivery systems. In: Donbrow M (ed) Microcapsules and Nanoparticles in Medicine and Pharmacology. Boca Raton, FL: CRC Press, 1992;99–123.

22. Lim F, Sun AM. Microencapsulated islets as bioartificial endocrine pancreas. Science 1980;210: 908–909.

23. Chang TMS, Prakash S. Artificial cells for biomicroencapsulation of cells and genetically engineered E. coli for cell therapy, gene therapy and removal of urea and ammonia. In: Tuan RS (ed) Expression and Detection of Recombinant Genes. Methods in Molecular Biology Series. New York: Humana Press, in press.

24. Fan MY, Lum ZP, Fu XW, Levesque L, Lai IT, Sun AM. Reversal of diabetes in BB rats by transplantation of encapsulated pancreatic islets. Diabetes 1990;39:519–522.

25. Lum ZP, Tai IT, Krestow M, Norton J, Vacek I, Sun AM. Prolonged reversal of diabetic states in NOD mice by xenografts of microencapsulated rat islets. Diabetes 1991;40:1511–1516.

26. Brunetti P, Basta G, Faloerni A, Calcinaro F, Pietrapaolo M, Calafiore R. Immunoprotection of pancreatic islet grafts within artificial microcapsules. Int J Artif Org 1991;14:789–791.

27. Sun AM, Vecek I, Sun YL, Ma X, Zhou D. In vitro and in vivo evaluation of microencapsulated porcine islets. ASAIO J, 1992;38j:125–127.

28. Calafiore R. Transplantation of microencapsulated pancreatic human islets for therapy of diabetes mellitus, ASAIO J 1992;38j:34–37.

29. Goosen MFA, O'Shea GM, Gharapetian HM, Chou S, Sun AM. Optimization of microencapsulation parameters: Semipermeable microcapsules as a bioartificial pancreas. Biotechnol Bioeng 1985;27:146–150.

30. Soon-Shiong P, Otterlie M, Skjak-Braek G, Smidsrod O, Heintz R, Lanza RP, Espevik T. An

immunologic basis for the fibrotic reaction to implanted microcapsules. Transpl Proc 1991;23:758–759.

31. Soon-Shiong P, Heintz RE, Merideth N, Yao QX, Yao Z, Zheng T, Murphy M, Moloney MK, Schmehll M, Harris M, Mendez R, Mendez R, Sandford PA. Insulin independence in a type 1 diabetic patient after encapsulated islet transplantation. Lancet 1994;343:950–951.

32. Wong H, Chang TMS. Bioartificial liver: implanted artificial cells microencapsulated living hepatocytes increases survival of liver failure rats. Int J Artif Org 1986;9:335–336.

33. Cai ZH, Shi ZQ, Sherman M, Sun AM. Development and evaluation of a system of microencapsulation of primary rat hepatocytes. Hepatology 1989;10:885–860.

34. Dixit V, Gordon VP, Pappas SC, Fisher MM. Increased survival in galactosamine induced fulminant hepatic failure in rats following intraperitoneal transplantation of isolated encapsulated hepatocytes. In: Baquey C, Dupuy B (eds) Hybrid Artificial Organs, vol 177. Paris, France: Colloque ISERM, 1989;257–264.

35. Bruni S, Chang TMS. Hepatocytes immobilized by microencapsulation in artificial cells: effects on hyperbilirubinemia in Gunn rats. J Biomater Artif Cells Artif Org 1989;17:403–412.

36. Bruni S, Chang TMS. Encapsulated hepatocytes for controlling hyperbilirubinemia in Gunn rats. Int J Artif Org 1991;14:239–241.

37. Bruni S, Chang TMS. Kinetic analysis of UDP-glucoronosyl-transferase in bilirubin conjugation by encapsulated hepatocytes for transplantation into rats. Artif Org 1995;19, in press.

38. Bruni S, Chang TMS. Effects of donor strains and age of the recipient in the use of microencapsulated hepatocytes to control hyperbilirubinemia in the Gunn rat. Int J Artif Org 1995;18, in press.

39. Dixit V, Darvasi R, Arthur M, Brezina M, Lewin K, Gitnick G. Restoration of liver function in Gunn rats without immunosuppression using transplanted microencapsulated hepatocytes. Hepatology 1990;12:1342–1349.

40. Dixit V, Arthur M, Gitnick G. Repeated transplantation of microencapsulated hepatocytes for sustained correction of hyperbilirubinemia in Gunn rats. Cell Transpl 1992;1:275–279.

41. Dixit V, Arthur M, Gitnick G. A morphological and functional evaluation of transplanted isolated encapsulated hepatocytes following long term transplantation in Gunn rats. Biomater Artif Cells Immobil Biotechnol 1993;21:119–133.

42. Wong H, Chang TMS. The viability and regeneration of artificial cell microencapsulated rat hepatocyte xenograft transplants in mice. Biomater Artif Cells Artif Org 1988;16:731–740.

43. Kashani S, Chang TMS. Physical chemical characteristics of hepatic stimulatory factor prepared from cell free supernatant of hepatocyte cultures. Biomater Artif Cells Immobil Biotechnol 1991; 19:565–578.

44. Kashani S, Chang TMS. Effects of hepatic stimulatory factor released from free or microencapsulated hepatocytes on galactosamine induced fulminant hepatic failure animal model. Biomater Artif Cells Immobil Biotechnol 1991;19:579–598.

45. Wong H, Chang TMS. Microencapsulation of cells within alginate poly-L-lysine microcapsules prepared with standard single step drop technique: Histologically identified membrane imperfections and the associated graft rejection. J Biomater Artif Cells Immobil Biotechnol 1991;182:675–686.

46. Wong H, Chang TMS. A novel two step procedure for immobilizing living cells in microcapsules for improving xenograft survival. Biomater Artif Cells Immobil Biotechnol 1991;19:687–698.

47. Chang TMS, Wong H. A novel method for cell encapsulation in artificial cells. USA Patent No 5, 084, 350, 1992.

48. Coromili V, Chang TMS. Polydisperse dextran as a diffusing test solute to study the membrane permeability of alginate polylysine microcapsules. Biomater Artif Cells Immobil Biotechnol 1993; 21:323–335.

49. Ito Y, Chang TMS. In-vitro study of multicellular hepatocytes spheroid formed in microcapsules. J Artif Org 1992;16:422–426.

50. Dixit V, Darvasi R, Arthur M, Lewin K, Gitnick G. Improved function of microencapsulated hepatocytes in a hybrid bioartificial liver support system. Artif Org 1992;16: 336–341.

51. Dixit, V, Darvasi R, Arthur M, Lewin K, Gitnick G. Cryopreserved microencapsulated hepatocytes: transplantation studies in Gunn rats. Transplantation 1993;55:616–622.

52. Sugamori ME and Sefton MV. Microencapsulation of pancreatic islets in a water insoluble polyacrylate. Trans ASAIO 1989;35:791–799.

53. Uluday H, Sefton MV. Metabolic activity of CHO fibroblasts in HEMA-MMA microcapsules. Biotechnol Bioeng 1992;39:672–678.

54. Goosen MFA, King GA, McKnight CA, Marcotte N. Animal cell culture engineering using alginate polycation microcapsules of controlled membrane molecular weight cut-off. J Membrane Sci 1989;40:233–243.

55. Pernot JM, Brun H, Pouyet B. Main parameters involved in microencapsulation by in situ polycondensation. Biomater Artif Cells Immobil Biotechnol 1993;21:415–420.

56. Kersulec A, Bazinet C, Corbineau F, Come D, Barbotin JN, Hervagault JF, Thomas D. Physiological behaviour of encapsulated somatic embryos Biomater Artif Cells Immobil Biotechnol 1993; 21:375–382.

57. Dautzenberg, Holzapfel G, Lukanoff B. Methods for a comprehensive characterization of microcapsules based on polyelectrolyte complexes Biomater Artif Cells Immobil Biotechnol 1993;21: 399–406.

58. Spirin AS. Cell free protein synthesis bioreactor. In: Todd P, Sikdur SK, Bier M (eds) Frontiers in Bioprocessing II. 1991.

59. Stevenson W, Sefton MV. Development of Polyacrylate Microcapsules. In: Goosen MFA (ed) Fundamentals of Animal Cell Encapsulation and Immobilization. Boca Raton, FL: CRC Press, 1993;143–182.

60. Chang TMS. Living cells and microorganisms immobilized by microencapsulation inside artificial cells. In: Goosen MFA (ed) Fundamentals of Animal Cell Encapsulation and Immobilization. Boca Raton, FL: CRC Press, 1993;143–182;183–196.

61. Bugarski B, Jovanovic G, Vunjak-Novakovi G. Bioreactor systems based on micro-encapsulated animal cell cultures. In: Goosen MFA (ed) Fundamentals of Animal Cell Encapsulation and Immobilization. Boca Raton, FL: CRC Press, 1993;143–182,267–296.

62. Goosen MFA (ed). Fundamentals of Animal Cell Encapsulation and Immobilization. Boca Raton, FL: CRC Press, 1993;326.

63. Poncelet D, Poncelet B, Beaulieu C, Neufeld RJ. Scale-up of gel bead and microcapsule production in cell immobilization. In: Goosen MFA (ed) Fundamentals of Animal Cell Encapsulation and Immobilization. Boca Raton, FL: CRC Press, 1993;113–142;143–182.

64. Koo J, Chang TMS. Secretion of erythropoietin from microencapsulated rat kidney cells: preliminary results. Int J Artif Org 1993;16:557–560.

65. Garofalo F, Chang TMS. Immobilization of P. pictorum in open pore agar, alginate polylysine-alginate microcapsules for serum cholesterol depletion. Biomater Artif Cells Artif Org 1989;17: 271–290.

66. Gàrofalo F, Chang TMS. Effects of mass transfer and reaction kinetics on serum cholesterol depletion rates of free and immobilized Pseudomonas pictorum. Appl Biochem Biotechnol 1991;27: 75–91.

67. Prakash S, Chang TMS. Genetically engineered E. coli cells containing K. aerogenes gene, microencapsulated in artificial cells for urea and ammonia removal. Biomater Artif Cells Immobil Biotechnol 1993;21:629–636.

68. Prakash S, Chang TMS. Preparation and in vitro analysis of microencapsulated genetically engineered E. coli DH5 cells for urea and ammonia removal. Biotech Bioeng 1995;47, in press.

69. Lloyd-George I, Chang TMS. Free and microencapsulated erwinia herbicola for the production tyrosine. J Biomater Artif Cells Immobil Biotechnol 1993;21:323–335.

294

70. Digat B. A new bioencapsulation technology for microbial inoculants. Biomater Artif Cells Immobil Biotechnol 1993;21:299–306.

71. Tamponnet C, Binot R, Lasseur C. Bioencapsulation: a biotechnological tool for biological life support for manned missions by the European Space Agency. Biomater Artif Cells Immobil Biotechnol 1993;21:307–316.

72. Chang TMS, Poznansky MJ. Semipermeable microcapsules containing catalase for enzyme replacement in acatalasemic mice. Nature 1968;218(5138):242–245.

73. Poznansky MJ, Chang TMS. Comparison of the enzyme kinetics and immunological properties of catalase immobilized by microencapsulation and catalase in free solution for enzyme replacement. Biochim Biophys Acta 1974;334:103–115.

74. Chang TMS. The in vivo effects of semipermeable microcapsules containing L-asparaginase on 6C3HED lymphosarcoma. Nature 1971;229(528):117–118.

75. Chang TMS. L-Asparaginase immobilized within semipermeable microcapsules: in vitro and in vivo stability. Enzyme 1973;14(2):95–104.

76. Chong ES, Chang TMS. In vivo effects of intraperitoneally injected L-asparaginase solution and L-asparaginase immobilized within semipermeable nylon microcapsules with emphasis on blood L-asparagine, 'body' L-asparagine, and plasma L-asparagine levels. Enzyme 1974;18:218–239.

77. Shu CD, Chang TMS. Tyrosinase immobilized within artificial cells for detoxification in liver failure. I. Preparation and in vitro studies. Int J Artif Org 1980;3(5):287–291.

78. Shu CD, Chang TMS. Tyrosinase immobilized within artificial cells for detoxification in liver failure. II. In vivo studies in fulminant hepatic failure rats. Int J Artif Org 1981;4:82–84.

79. Bourget L, Chang TMS. Phenylalanine ammonia-lyase immobilized in semipermeable microcapsules for enzyme replacement in phenylketonuria. FEBS Lett 1985;180:5–8.

80. Bourget L, Chang TMS. Phenylalanine ammonia-lyase immobilized in microcapsules for the depletion of phenylalanine in plasma in phenylketonuria rat model. Biochim Biophys Acta 1986; 883:432–438.

81. Chang TMS. Preparation and characterization of xanthine oxidase immobilized by microencapsulation in artificial cells for the removal of hypoxanthine. Biomater Artif Cells Artif Org 1989;17: 611–616.

82. Palmour RM, Goodyer P, Reade T, Chang TMS. Microencapsulated xanthine oxidase as experimental therapy in Lesch-Nyhan disease. Lancet 1989;2(8664):687–688.

83. Medical Research Council Working Party on Phenylketonuria. Br Med J 1933;396(9th January): 115–119.

84. Chang TMS, Lister C. Plasma/intestinal concentration patterns suggestive of entero-portal recirculation of amino acids: effects of oral administration of asparaginase, glutaminase and tyrosinase immobilized by microencapsulation in artificial cells. Biomater Artif Cells Artif Org 1988/89;16: 915–926.

85. Chang TMS, Bourget L, Lister C. US patent granted, 1992.

86. Bourget L, Chang TMS. Effects of oral administration of artificial cells immobilized phenylalanine ammonia-lyase on intestinal amino acids of phenylketonuria rats. Biomater Artif Cells Artif Org 1989;17:161–182.

87. Khanna R, Chang TMS. Characterization of L-histidine ammonia-lyase immobilized by microencapsulation in artificial cells: Preparation kinetics, stability, and in vitro depletion of histidine. Int J Artif Org 1990;13:189–195.

88. Campbell J, Chang TMS. Enzymatic recycling of coenzymes by a multi-enzyme system immobilized within semipermeable collodion microcapsules. Biochim Biophys Acta 1975;397:101–109.

89. Cousineau J, Chang TMS. Formation of amino acid from urea and ammonia by sequential enzyme reaction using a microencapsulated multienzyme system. Biochem Biophys Res Commun 1977; 79(1):24–31.

90. Grunwald J, Chang TMS. Continuous recycling of NAD using an immobilized system of collo-

dion microcapsules containing dextran-NAD, alcohol dehydrogenase, and malic dehydrogenase. J Appl Biochem 1979;1:104–114.

91. Chang TMS, Malouf C. Effects of glucose dehydrogenase in converting urea and ammonia into amino acid using artificial cells. Artif Org 1979;3(1):38–41.

92. Wahl HP, Chang TMS. Recycling of NAD+ cross-linked to albumin or hemoglobin immobilized with multienzyme systems in artificial cells. J Mol Catal 1986;39:147–154.

93. Yu YT, Chang TMS Ultrathin lipid-polymer membrane microcapsules containing multienzymes, cofactors and substrates for multistep enzyme reactions. FEBS Lett 1981;125(1):94–96.

94. Gu KF, Chang TMS. Conversion of ammonia or urea into L-leucine, L-valine, and L-isoleucine using artificial cell immobilizing multienzyme system and dextran-NADH+. I. Glucose dehydrogenase for cofactor recycling. ASAIO - Official J Am Soc Artif Intern Org 1988;11:24–28.

95. Gu KF, Chang TMS. Conversion of ketoglutarate into L-glutamic acid with urea as ammonium source using multienzyme system and dextran-NAD+ immobilized by microencapsulation with artificial cells in a bioreactor. J Bioeng Biotechnol 1988;32:363–368.

96. Gu KF, Chang TMS. Production of essential L-branched-chained amino acids, in bioreactors containing artificial cells immobilized multienzyme systems and dextran-NAD. Appl Biochem Biotechnol 1990;26:263–269.

97. Daka JN, Chang TMS. Bilirubin removal by the pseudoperoxidase activity of free and immobilized hemoglobin and hemoglobin co-immobilized with glucose oxidase. Biomater Artif Cells Artif Org 1989;17:553–562.

98. Chang TMS, Daka JN. Removal of bilirubin by the pseudoperoxidase activity of immobilized hemoglobin. US Patent No 4820416, 1989.

99. Chang TMS. Hemoperfusion alone and in series with ultrafiltration or dialysis for uremia, poisoning and liver failure. Kidney Int 1976;S305-S311.

100. Winchester JF. Hemoperfusion. In: Maher J (ed) Replacement of Renal Function by Dialysis, 3rd edn. Boston, MA: Kluwer, 1988;439–460.

101. Bonomini V (ed). Biotechnol in Renal Replacement Therapy. London: Karger, 1989.

102. Chang TMS. In: Geyer R (ed) Blood Substitutes, New York: Marcel Decker 1988. (also simultaneously published in Biomater Artif Cells Artif Org 1988;1–3;1–704.)

103. Chang TMS (ed). Blood Substitutes and Oxygen Carriers. New York: Marcel Decker, 1992. (also simultaneously published in Biomater Artif Cells Immobil Biotechnol 1992;2–4:155–1120.

104. Chang TMS, Reiss J, Winslow R (eds). Blood Substitutes-General. Artificial Cells, Blood Substitutes Immobilization Biotechnology 1994;2:123–360.

105. Abstracts from the Vth International Symposium on Blood Substitutes, San Diego, CA, USA.

106. Chang TMS, Weinstok S (eds). Artificial cells, blood substitutes immobilization biotechnology. In: Blood Substitutes 1995;23:257–459.

Biotechnology Annual Review Volume 1
M.R. El-Gewely, editor

The production of recombinant human erythropoietin

Noboru Inoue[1], Makoto Takeuchi[1], Hideya Ohashi[2] and Takamoto Suzuki[2]

[1]Central Laboratories for Key Technology and [2]Pharmaceuticals Division, Kirin Brewery Co. Ltd., Yokohama Kanagawa, Japan

Abstract. Erythropoietin (EPO) is the glycoprotein hormone that promotes differentiation of erythroid progenitor cells in bone marrow. The normal kidney produces EPO to maintain erythrocyte for oxygen supply. This hormone activity was found in the serum of anemic animals in the 1890s. Renal failure results in severe anemia because of reduced EPO production, therefore anemia patients expected EPO treatment for long time. However, this was difficult due to the limited amount of EPO. Many researchers have tried to isolate EPO since the 1950s. Finally Miyake and Goldwasser purified highly active EPO from the urine of aplastic anemia patients. Since then, the characteristics and structural information from the purified material accelerated the cloning of the EPO gene. Mammalian cells were essential to produce EPO, because EPO contains 40% carbohydrate that plays some important roles in its activity, stability and biosynthesis. In 1984, two groups succeeded in cloning the EPO gene and expressing this gene in mammalian cells. Recombinant human EPO is currently available for anemia treatment. In this paper, we review production in mammalian cells, molecular characterization, especially carbohydrate moieties, and clinical applications of recombinant EPO.

Introduction

Recently, many factors that stimulate the proliferation and differentiation of cells have been identified and become commercially available. Among them, hematopoietic factors are expected to improve anemia treatments and cancer therapy by increasing the peripheral blood cell numbers.

As shown in Fig. 1, all of the peripheral blood cells are derived from pluripotent stem cells, which are selfreproducible and differentiate into any type of blood cells corresponding to the specific stimulation. In the clinical application of differentiation factors, it is very important to control such specific stimulation. Erythrocytes play a major role in supplying oxygen to peripheral tissues. In erythroid lineage, pluripotent stem cells initially differentiate into proerythroblasts via burst-forming unit erythroid (BFU-E) and colony-forming unit erythroid (CFU-E) cells. These progenitor cells then differentiate into erythroblasts, which synthesize hemoglobin. Followed by erythroblast denucleation and differentiation into reticulocytes, they are finally released from the marrow into the blood stream as mature erythrocytes. The differentiation process from erythroblastic precursor cells to erythroblasts acts as a rate-limiting step for erythrocyte regeneration. Erythropoietin (EPO) stimulates the matu-

Address for correspondence: N. Inoue, Central Laboratories for Key Technology, Kirin Brewery Co. Ltd., 1-13-5 Fukuura Kanazawa-ku, Yokohama Kanagawa 236, Japan.

298

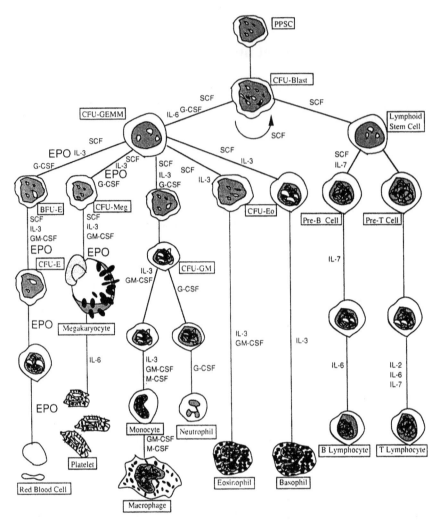

Fig. 1. Hemopoietic lineages. Differentiation lineages into various cells and abbreviations are shown. PPSC, pluripotent stem cell; CFU-Blast, colony forming unit blast, CFU-GEMM, colony forming unit granulocyte erythroid macrophage megakaryocyte; BFU-E, burst forming unit erythroid; CFU-Meg, colony forming unit megakaryocyte; CFU-Eo, colony forming unit eosinophil; CFU-E, colony forming unit erythroid; CFU-GM, colony forming unit granulocyte macrophage.

ration of CFU-E cells into proerythroblasts and regulates this step to maintain the number of circulating erythrocytes. EPO is a glycoprotein produced in the kidney in adults and in the fetal liver. Varying partial oxygen pressure in the blood and nutritional state regulate the production of this hemopoietic factor. Unlike other hematopoietic factors, such as granulocyte colony-stimulation factor (G-CSF), granulocyte and macrophage colony-stimulation factor (GM-CSF) and macrophage colony-stimulation factor (M-CSF) which are all produced close to their target cells,

EPO is produced in the kidney, and moves to the bone marrow via the blood stream. EPO then binds to specific receptors on target cells and signal transduction promotes the differentiation of erythroid progenitor cells. The concentration of EPO in normal human serum varies from 15 to 30 milliunits/ml. Supplemental EPO is the desirable treatment in cases of renal failure with decreased EPO secretion. EPO has become a familiar drug that can specifically control the erythrocyte numbers. As a result, new EPO therapies have been developed for the control of not only renal anemia but also other types of anemia.

Carnot and Deflendre [1] predicted the presence of EPO as a hormonal factor for the first time in 1906. Erslev et al. [2] injected serum obtained from exsanguinated anemic rabbits into normal rabbits and found that the number of peripheral reticulocytes increased markedly. From this result, they suggested the presence of EPO as a hormonal factor in 1953. In 1957, Jacobson et al. [3] discovered that the kidney is the main producer of EPO. However, the identification and purification of EPO was extremely difficult and was not accomplished until 1977 when Miyake et al. [4] purified human EPO from urine. Thereafter, Anagnostou [5] showed that the administration of crude EPO improves the condition of anemic, partially nephrectomized rats. These findings suggested that a decline in the production and/or release of EPO may cause severe anemia with chronic renal failure. We are able to produce recombinant human EPO(rHuEPO) by cell culture using recombinant DNA technology and subsequently achieve high levels of purification. rHuEPO has been demonstrated to be applicable to autologous blood donation and to yield sufficient blood levels as required for orthopedic and cardiac surgery.

Manufacturing Methods

Cloning of EPO gene

Although the existence of EPO has been known for the past 100 years, the EPO molecule was identified only 17 years ago. In 1977, Miyake et al. purified EPO from the urine of patients with severe aplastic anemia. The purified EPO has a specific activity of 70,400 U/mg with a molecular weight of 34 k. This sample provided enough information about amino acid sequence to clone the human EPO gene.

In 1985, Lin et al. [6] and Jacobs et al. [7] finally succeeded in cloning the human EPO gene independently. Figure 2 shows the process of Lin from gene cloning of the EPO to its expression. Two peptides, which had relatively low codon degeneracy, were selected from internal tryptic fragments. Two mixed oligonucleotide pools, corresponding to these amino acid sequences, were then synthesized and used as nonoverlapping probes. One probe mixture contained 20-nucleotide-long oligonucleotides containing all possible coding sequences for an internal heptapeptide (EpQ); the second mixture contained 17-nucleotide-long oligonucleotides directed against the coding sequence for a hexapeptide (EpV). There was the possibility of 128 combinations of each of these oligonucleotides (Fig. 3). These two set of radio-

300

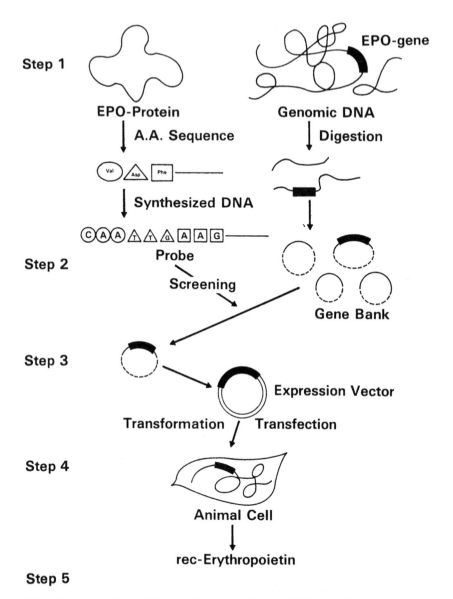

Step 1

EPO-Protein

A.A. Sequence

Val Asp Phe ————

Synthesized DNA

C A A A A A A A G ————

Probe

EPO-gene

Genomic DNA

Digestion

Step 2

Screening

Gene Bank

Step 3

Expression Vector

Transformation / Transfection

Step 4

Animal Cell

rec-Erythropoietin

Step 5

Fig. 2. The process from EPO gene cloning to expression. EPO gene cloning and expression are illustrated in Steps 1–5. Step 1: the oligonucleotide pools corresponding to the amino acid sequence of EPO tryptic fragments were synthesized and used as cloning probes. Step 2: EPO gene was cloned from human fetal liver genomic library using synthesized oligonucleotides. Step 3: expression vector was constructed with the full-length EPO gene. Step 4: EPO expression vector was transfected to CHO cells. Step 5: the working cell for EPO production was selected from EPO expressing CHO cells.

labeled nucleotide probes were used for the sequential hybridization of a Charon 4A phage human fetal liver genomic library. Gene clones which hybridized with both probe mixtures were selected to eliminate false positive clones.

Probe mixture
EpV = Va l- Asn- Phe- Tyr- A l a- T r p- Lys
    3' CAA TTG AAG ATG CGA ACC TT 5'
         G   A   A   A   T
         C   G
                             C

Probe mixture
EpQ = G l n- P r o- T r p- G l u- P r o- L e u
    3' GTT CGA ACC CTT GGA GA 5'
        C   T   C   T A
             G   G
             C   C

Fig. 3. Oligonucleotide probes corresponding to the EPO peptide fragment. The mixed oligonucleotide pools were synthesized corresponding to the amino acid sequence of a hexapeptide and a heptapeptide derived from internal tryptic fragments of EPO. Each probe mixture contain a pool of 128 oligonucleotides.

Southern blot analysis revealed that one of these clones coded for the full-length EPO gene. The two approaches verified the EPO coding clone identified by sequential hybridization using oligonucleotide probes. One of them was the comparison of the amino acid sequence predicted from the clone and the partial amino acid sequence of urinary human EPO. The other was based on the immunological and biological EPO activity of a glycoprotein expressed in mammalian cells. Human EPO gene clones have subsequently been determined using a genomic gene library of human fetal liver. To determine the cDNA of monkey EPO, an oligonucleotide mixture of human EPO was used to screen the cDNA library constructed from an anemic monkey kidney induced by phenylhydrazine treatment. Murine EPO gene clones were also isolated using human and monkey EPO gene sequences as probes [8].

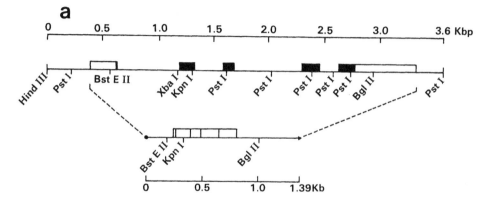

Fig. 4a. See following page for figure legend.

b

```
AAGCTTCTGGGCTTCCAGACCCAGCTACTTTGCGGAACTCAGCAACCCAGGCATCTCTGAGTCTCCGCCCAAGACCGGGATGCCCCCCAGGGGAGGTGTCCGGGAGCCCAGCCTTTCCCA    120

GATAGCACGCTCCGCCAGTCCCAAGGGTGCGCAACCGGCTGCACTCCCCTCCCGCGACCCAGGGGCCCGGGAGCAGCCCCCATGACCCACACGCACGTCTGCAGCAGCCCCGCTCACGCCC    240

CGGCGAGCCTCAACCCAGGCGGTCCTGCCCCTGCTCTGACCCCGGGTGGCCCCTACCCCTGGCGACCCCTCACGCACACAGCCTCTCCCCCACCCCCCACCCGCGGCACGCACACATGCAGAT    360

AACAGCCCCGACCCCCGGCCAGAGCCGCAGAGTCCCTGGGCCACCCGGCCGCTGCTGCGCTGCGCCGCACCGCGCTGTCCTCCGGGAGCCGGACCGGGGCCACCGCGCCCGCTCTGCT    480

CCGACACCGCGCCCCCTGGACAGCCGCCCTCTCCTCTAGGCCCGTGGGGCTGGCCCTGCACCGCCGAGCTTCCCGGGATGAGGGCCCCGGTGTGGTCACCCGGCGCGCCCCAGGTCGCT    600
```

```
                -27    -24
                MetGlyValHisG
CACCCACCCCGGCCAGGCGCGGAGATGGGGGTGCACGGTGAGTACTCGCGGGCTGGGCGCTCCCGGCGCCCGGGTCCCTGTTTGAGCGGGGATTTAGCGCCCCGGCTATTGGCCAGGAGG    720

TGGCTGGGTTCAAGGACCGGCGACTTGTCAAGGACCCCGGAAGGGGGAGGGGGTGGGGCAGCCTCCACGTGCCAGCGGGGACTTGGGGGAGTCCTTGGGGATGGCAAAAACCTGACCTG    840

TGAAGGGGACACAGTTTGGGGGTTGAGGGGAAGAAGGTTTGGGGGTTCTGCTGTGCCAGTGGAGAGGAAGCTGATAAGCTGATAACCTGGGCGCTGGAGCCACCACTTATCTGCCAGAGG    960

GGAAGGCCTCGTCACACCAGGATTGAAGTTTGGCCGGAGAAGTGGATGCTGGTAGCTGGGGGTGGGGTGTGCACACGGCAGCAGGATTGAATGAAGGCCAGGGAGGCAGCACCTGAGTGC    1080

TTGCATGGTTGGGGACAGGAAGGACGAGCTGGGGCAGAGACGTGGGGATGAAGGAAGCTGTCCTTCCACAGCCACCCTTCTCCCTCCCCCCCTGACTCTCAGCCTGGCTATCTGTTCTAG    1200
```

```
        -22                           -10                        1                          10
        luCysProAlaTrpLeuTrpLeuLeuLeuSerJeuLeuSerLeuProLeuGlyLeuProValleuGlyAlaProproArgLeulleCysAspSerArgValLeuGluArgTyrLeuLeuG
        AATGTCCTGCCTGGCTGTGGCTTCTCCTGTCCCTGCTGTCGCTCCCTCTGGGCCTCCCAGTCCTGGGCGCCCCACCACGCCTCATCTGTGACAGCCGAGTCCTGGAGAGGTACCTCTTGG    1320
```

```
        20                          26
        luAlaLysGluAlaGluAsnlleThr
        AGGCCAAGGAGGCCGAGAATATCACGGTGAGACCCCTTCCCCAGCACATTCCACAGAACTCACGCTCAGGGCTTCAGGGAACTCCTGCCAGATCCAGGAACCTGGCACTTGGTTTGGGGT    1440

        GGAGTTGGGAAGCTAGACACTGCCCCCCTACATAAGAATAAGTCTGGTGGCCCCAAACCATACCTGGAAACTAGGCAAGGAGCAAAGCCAGCAGATCCTACGGCCTGTGGGCCAGGGCCA    1560
```

```
                                27    30           35           40           45          50
                                ThrGlyCysAlaGluHisCysSerLeuAsnGluAsnlleThrValProAspThrLysValAsrPheTyrAlaTrpL
        GAGCCTTCAGGGACCCTTGACTCCCCGGGCTGTGTGCATTTCAGACGGGCTGTGCTGAACACTGCAGCTTGAATGAGAATATCACTGTCCCAGACACCAAAGTTAATTTCTATGCCTGGA    1680
```

```
        55
        ysArgMetGlu
        AGAGGATGGAGGTGAGTTCCTTTTTTTTTTTTTTTTTCCTTTCTTTTGGAGAATCTCATTTGCGAGCCTGATTTTGGATGAAAGGGAGAATGATCGGGGGAAAGGTAAAATGGAGCAGCAGA    1800

        GATGAGGCTGCCTGGGCGCAGAGGCTCACGTCTATAATCCCAGGCTGAGATGGCCGAGATGGGAGAATTGCTTGAGCCCTGGAGTTTCAGACCAACCTAGGCAGCATAGTGAGATCCCCC    1920

        ATCTCTACAAACATTTAAAAAAATTAGTCAGGTGAAGTGGTGCATAGTGGTAGTCCCAGATATTTGGAAGGCTGAGGCGGGAGGATCGCTTGAGCCCAGGAATTTGAGGCTGCAGTGACC    2040

        TGTGATCACACCACTGCACTCCAGCCTCAGTGACAGAGTGAGGCCCTGTCTCAAAAAAGAAAAGAAAAAGAAAAATAATGAGGGCTGTATGGAATACATTCATTATTCATTCACTCACT    2160

        CACTCACTCATTCATTCATTCATTCATTCAACAAGTCTTATTGCATACCTTCTGTTTGCTCAGCTTGGTGCTTGGGGCTGCTGAGGGGCAGGAGGGGAGAGGGTGACATGGGTCAGCTGAC    2280
```

```
                        56    60             65            70          75         80
                        ValGlyGlnGlnAlnAlaGluValTrpGlnGlyLeuAlaLeuLeuSerGluAlaValLeuArgGlyGlnAlaLeuLeuValAsnSerSerGlnProTr
        TCCCAGAGTCCACTCCCTGTAGGTCGGGCAGCAGGCCGTAGAAGTCTGGCAGGGCCTGGCCCTGCTGTCGGAAGCTGCCTGCGGGGCCAGGCCCTGTTGGTCAACTCTTCCCAGCCGTG    2400
```

```
        90                         100          105         110          115
        pGluProLeuGlnLeuHisValAspLysAlaValSerGlyLeuArgSerLeuThrThrLeuLeuArgAlaLeuGlyAlaGln
        GGAGCCCCTGCAGCTGCATGTGGATAAAGCCGTCAGTGGCCTTCGCAGCCTCACCACTCTGCTTCGGGCTCTGGGAGCCCAGGTGAGTAGGAGCGGACACTTCTGCTTGCCCTTTCTGTA    2520
```

```
                                                                            116          120
                                                                            LysGluAlalleSerProProAsp
        AGAAGGGGAGAAGGGTCTTGCTAAGGAGTACAGGAACTGTCCGTATTCCTTCCCTTTCTGTGGCACTGCAGCGACCTCCTGTTTTCTCCTTGGCAGAAGGAAGCCATCTCCCCTCCAGAT    2640
```

```
        130                        140          145         150          155         160
        AlaAlaSerAlaAlaProLeuArgThrlleThrAlaAspThrPheArgLysLeuPheArgValTyrSerAsnPheLeuArgGlyLysLeuLysLeuTyrThrGlyGluAlaCysArgThr
        GCGGGCCTCAGCTGCTCCGAACAATCACTGCTGACACTTTCCGCAAACTCTTCCGAGTCTACTCCAATTTCCTCCGGGGAAAGCTGAAGCTGTACACAGGGGAGGCCTGCAGGACA    2760
```

```
        166
        GlyAspArgEnd
        GGGACAGATGACCAGGTGTGTCCACCTGGGCATATCCACCACCTCCCTCACCAACATTGCTTGTGCCACACCTCCCCGCCACTCCTGAACCCCGTCGAGGGGCTCTCAGCTCAGCGC    2880

        CAGCCTGTCCCATGGACACTCCAGTGCCAGCAATGACATCTCAGGGGCCAGAGGAACTGTCCAGAGAGCAACTCTGAGATCTAAGGATGTCACAGGGCCAACTTGAGGGCCCAGAGCAGG    3000

        AAGCATTCAGAGAGCAGCTTTAAACTCAGGGACAGAGCCATGCTGGGAAGACGCCTGAGCTCACTCGGCACCCTGCAAAATTTGATGCCAGGACACGCTTTGGAGGCGATTTACCTGTTT    3120

        TCGCACCTACCATCAGGGACAGGATGACCTGGAGAACTTAGGTGGCAAGCTGTGACTTCTCCAGGTCTCACGGGCATGGGCACTCCCTTGGTGGCAAGAGCCCCCTTGACACCGGGGTGG    3240

        TGGGAACCATGAAGACAGGATGGGGGCTGGCCTCTGGCTCTCATGGGGTCCAAGTTTTGTGTATTCTTCAACCTCATTGACGAGAACTGAAACCACCAATATGACTCTTGGCTTTTCTGT    3360

        TTTCTGGGAACCTCCAAATCCCCTGGCTCTGTCCCAACTCCTGGCAGCAGTGCAGCAGGTCCAGGTCCGGGAAATGAGGGGTGGAGGGGGCTGGGCCCTACGTGCTGTCTCACACAGCCTG    3480

        TCTGACCTCTCGACCTACCGGCCTAGGCCACAAGCTCTGCCTACGCTGGTCAATAAGGTGTCTCCATTCAAGGCCTCACCGCAGTCAAGGCAGCTGCCACCCTGCCCAGGGCAAGGCTGCAG    3602
```

Fig. 4. (a) Map of the human EPO gene. Exons are indicated by boxes; the solid areas indicate the approximate position of the coding protein of the gene. (b) Nucleotide sequence of the human EPO gene. The sequence of the 193-amino acid primary translation product is shown. The N-terminus of the mature protein begins at amino acid 1.

Figure 4a shows the schematic diagrams of human EPO gene and mRNA transcript and the nucleotide sequence of the human genomic gene. Figure 4b shows the amino acid sequence of the primary translation product. Restriction fragment analysis of human lymphocyte DNA, using human EPO cDNA as a probe, showed that a single band appeared after restriction enzyme digestion. The size of the hybridized bands is similar to that of the originally isolated EPO genomic clone, and hybridization at lower stringency did not show any additional bands. These results indicated that there is only a single copy of the human EPO gene and there are no other genes closely related to the EPO gene or pseudogene. Several transcriptional initiation sites have been mapped to nucleotides 105, 118, 123, 135, 141 and 151, and the polyadenylation site has been located at nucleotides 3341–2. The human EPO gene has about 2,800 bp and comprises 5 exons and 4 introns which are transcribed and spliced to form a 1,200 bp mRNA. This mRNA codes for a protein of 193 amino acids, of which the first 27 amino acids are predominantly hydrophobic, consistent with a signal sequence. The mature protein of 166 amino acids was predicted from the nucleotide sequence. Gene analysis using a hybridoma with chromosome translocated human cells showed that the EPO gene is located at q11-q22 of the long arm of chromosome 7 [9]. To determine homology with the human EPO gene, counterpart genes from monkey and mouse were cloned using the human EPO gene. Like the human EPO gene, these genes were found to have 5 exons and 4 introns and to be very similar to the human EPO gene, with the amino acid and DNA sequences of the monkey EPO showing 92% and 94% homology with the human, respectively, and showing 80% amino acid sequence homology with the mouse EPO [10,11].

Expression of recombinant EPO in animal cells

Since the sugar chains are essential for the biological activity of EPO, we chose animal cells as host for the EPO production. The construction of a high-expression vector and selection of host cell line are very important factors for the large-scale production of EPO in animal cells. Figure 5 illustrates the construction of the EPO expression vector.

This vector contains:
1. an SV40-derived late promoter as a EPO gene expression promoter, the enhancer sequence and origin of replication;
2. an SV40-derived late polyadenylation sequence as a terminator [12];
3. the mouse-derived dihydrofolate reductase (DHFR) mini-gene for gene amplification;
4. pBR322 plasmid-derived replication origin and ampicillin-resistant genes (Amp) so that replication in *E. coli* is possible. (however, sequences which inhibit replication in animal cells are deficient [13]);
5. a BamHI restriction site for insertion of the gene.

The DHFR mini-gene is important in the expression of the rHuEPO gene because DHFR is an essential enzyme for the *de novo* synthesis of nucleic acids. Methotrexate (MTX), an analogue of folic acid which inhibits dihydrofolate reductase,

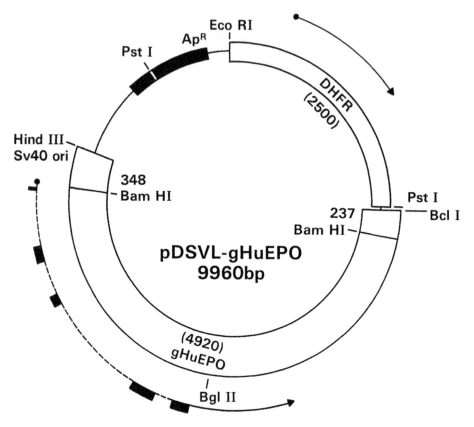

Fig. 5. EPO expression plasmids. EPO expression plasmid contains a DHFR minigene, simian virus 40 (SV40) origin of replication and early/late promoters, and genomic human EPO gene. The SV40 late promoter is used to drive the expression of the EPO gene. Arrows indicate the orientation of transcription.

which presents in the medium, amplifies the DHFR gene and the EPO gene simultaneously in cells [14]. Since these two genes are connected, cells containing the amplified EPO gene can produce a large amount of EPO in culture medium. We have used a mutant cell line, CHO DUK XB 11 as the host cell to make it easier to select cells containing the amplified DHFR-gene. The CHO DUK XB11 cell [15] is a subline of the Chinese hamster ovary cell, CHO K1, which was isolated as a DHFR deficient strain by the tritium suicide method after exposing CHO K1 cells to γ-rays. EPO-producing transformants were actually obtained using the above method in which the EPO expression vector was introduced into the host cell (CHO DUK XB11) using the calcium phosphate method. These transformants were cultured in the presence of MTX, whose concentration was increased in a stepwise manner to amplify DHFR genes connected to the EPO gene. These EPO-producing candidate cells were obtained for the preparation of master working cells, which is used in the production plant and which secrete the highest amount of EPO in culture medium. Cells with high productivity were selected using this method and were repeatedly cloned. The master cells were then prepared as production seeds from such cells. The master working cells were stored in polycarbonate tubes in liquid nitrogen and each tube was used to initiate the production of EPO.

Mass production of rHuEPO

Prior to clinical application of rHuEPO, various tests have to be done to assure the homogeneity and safety of the product. For example, the growth profiles and genetic characteristics of the master working cells were verified, the genetic stability of the vector was determined, and the absence of tumorgenicity and viruses, fungi, bacteria and mycoplasma in the master working cells was clearly demonstrated. In addition, to guarantee product quality such as molecular homogeneity and absence of contamination, various quality control tests at every stage of EPO production were carried out. EPO production was carefully carried out under the guidelines of Good Industrial Large-scale Practice (GILSP), for drug production using recombinant DNA technology. In the first step of EPO production in culture medium, frozen master

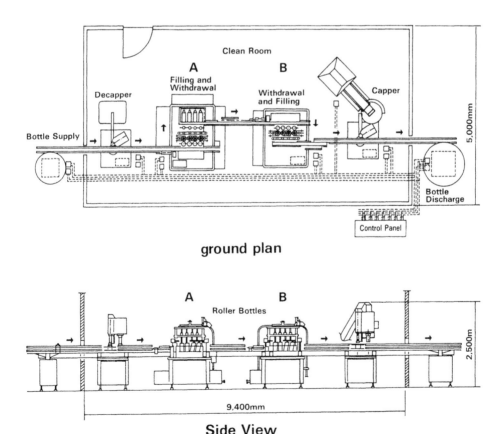

Fig. 6. Automatic roller bottle handling system. The handling system was shown in a gland plane (upper) and a side view (lower).

working cells are subjected to a pre-cultivation process in a T flask and spinner flask. In this step cells are grown in both stationary culture and suspension culture. Cells producing rHuEPO are cultured in roller bottles and the product harvested from the supernatant. Roller bottles are standard containers often used in mass culture of adhesive cells. Roller bottle culture is a very simple method: after the inoculation of precultured cells, cultivation is controlled only by temperature at 37°C and the rotation speed of the bottles. In addition, removal of serum from the medium is a very important step in EPO purification because of the similarity between serum albumin and EPO in terms of molecular weight and molecular electron charge. In the case of adhesive cells in roller bottles, serum medium can easily be exchanged for serum-free medium. In our production facility, an automatic roller bottle handling system has been developed and can easily be scaled up for large scale production. This system performs the following functions; an automatic culture process including inoculation of preculture cells, putting the bottles on the rotation stand and removing them, and changing and harvesting the medium (Fig. 6). Recently, a number of companies have manufactured bottles with an undulate surface to double the internal surface area, and these bottles can increase production efficiency per bottle. In our experience, culture using these roller bottles has numerous advantages over other culture methods using fermenters. For example, an operating factor such as good reproducibility, high yield, easy control of culture scale, easy monitoring of cell growth and reduced risk of contamination are all benefits of this system. Therefore, this seems to be an effective method for manufacturing profitable and useful proteins. The purification process is conducted in a clean room designed to be able to efficiently prevent intrusion of any kind of contamination such as xenogenic proteins, pyrogens and bacteria. After concentrating the harvested medium using ultrafiltration, the final EPO is purified from culture supernatant by a series of chromatography steps, anion exchange chromatography, reversed-phase chromatography and gel filtration chromatography. Finally, purified EPO is membrane filtered to sterilize it. This product (Fig. 7) is then subjected to various quality checks, e.g. for biological activity, immunoreactivity, molecular size using SDS-PAGE, analysis of amino acid composition and sugar content. These data are compared with those of the standard sample as a reference.

Characteristics of EPO

EPO peptide molecule

As mentioned in the previous section, human EPO cDNA codes 193 amino acids. The N-terminal 27 amino acid residues constitute a signal peptide, and the mature EPO molecule is predicted to be composed of the remaining 166 amino acid residue, which is secreted into the medium. However, the size of the mature protein has been determined to be 165 amino acids, one residue shorter than the predicted length. Amino acid sequence analysis of recombinant EPO expressed by CHO cells and

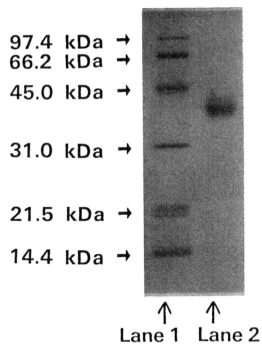

97.4 kDa →
66.2 kDa →

45.0 kDa →

31.0 kDa →

21.5 kDa →

14.4 kDa →

↑ ↑
Lane 1   Lane 2

Fig. 7. SDS-PAGE analysis of purified rHuEPO. rHuEPO was purified to single band on SDS-PAGE (lane 2). The sample was incubated in the presence of dithiothreitol and SDS prior to electrophoresis. Proteins on the gel were detected by CBB staining. Lane 1 consists of a molecular weight marker.

human EPO purified from patient urine has revealed that the carboxyl terminal arginine residue is missing from the isolated mature protein [16]. Therefore, the molecular weight of the EPO peptide moiety calculated from the amino acid sequence is 18,242. The mature EPO expressed by BHK cells contains 166 amino acids. It was also confirmed that recombinant EPO molecule contain four cysteine residues (Cys) and that there are two disulfide bonds between positions 29 and 33 and between positions 7 and 161. Similar findings regarding the processing of amino acids at the carboxyl terminal and disulfide bonds were obtained with urine-derived human EPO and monkey EPO. Circular dichroism spectral analysis suggested that the protein moiety of recombinant EPO contains an α-helix as a major secondary structure and that the ratio of the α-helix moiety is about 50%.

EPO contains three Asn-X-Ser/Thr sequences. Asn-linked sugar chains attach to all of these potential N-glycosylation sites at positions 24, 38 and 83. A mucin-type sugar chain also attaches to the serine residue at position 126 [17]. The molecular weight of recombinant EPO was 38,000–42,000 on SDS-PAGE and 29,000 by HPLC.

These findings are in good accordance with those of human urinary EPO. Isoelectric focusing also showed similar profiles for human urinary EPO and rHu EPO, with isoelectric points of pI 3.5–4.5. The specific biological activity of recombinant

EPO determined using hypoxic hypervolemic mice is not less than 1.5×10^5 IU/mg. Very similar dose-response curves of human urinary EPO and recombinant EPO using this method indicate that the two types of EPO have similar biological characteristics.

Sugar chain structures and the relationship between sugar chain and biological function in EPO

EPO contains three Asn-linked sugar chains and one mucin-type sugar chain. The sugar chains comprise 40% of EPO's total molecular weight. The sugar chains, especially sialic acids and branching structures, in EPO have been shown to play an important role in the expression of its biological activity in vitro [18] as well as in vivo [19]. The relationship between the sugar chain structures and their biological function has been studied extensively [20] and several functions of these sugar chains have been clarified. The author compared sugar chain structures of human urinary EPO and recombinant human EPO (derived from CHO cells). Of course the sugar chain structure of recombinant EPO is not exactly the same as that of human urinary EPO, because the two host cells (human kidney and Chinese hamster ovary cell) are different. The mucin-type sugar chains are clearly different, the urinary human EPO has ±NeuAc α2–6GalNAc (Isialyl Tn-antigen structure), while recombinant EPO has ±NeuAc α2–3Gal β1–4(±NeuAc) α2–6GalNAc [21]. It is widely known that the latter structure is the more common in the human body than the Tn-antigen structure. The Asn-linked sugar chains of urinary human EPO have greater variety in structure than those of the recombinant form; the former has α2–6 linked sialic acids (40%) as well as the major α2–3 linked one, while the latter have only α2–3 linkage [22]. There are also quantitative differences between the two. CHO cells may have a tendency to produce poly-N-acetyllactosamine repeating sugar chain structure, which requires a tetraantennary structure for its base construction. The ratio of biantennary to tetraantennary sugar chains is different.

Other criteria for comparison of the sugar chains are similar between the two. Most importantly, there is no sugar chain structure in the recombinant human EPO that does not exist in urinary human EPO. The major sugar chains of EPO were fucosylated tetra-antennary complex-type with or without N-acetyllactosamine (Gal β1–4GlcNAc) repeating units at their outer chain moieties. In addition, 2,4- and 2,6-branched triantennary and biantennary structures were detected (Fig. 8) [22,23]. Vliegenthart et al. [24] has reported that recombinant EPO has a trace amount of N-glycolyl neuraminic acid (HD antigen). So far, there is no report of patients who have raised antibody against HD antigen introduced by treatment with recombinant EPO. A Gal β1–3GlcNAc structure in recombinant human EPO was produced in BHK cells [25]. In recombinant human EPO, there is the tendency for sugar chains at each glycosylation site to share the same variety of antennary structures. The ratio of larger sugar chains increased in the order of Asn24 < Asn38 < Asn83 [26].

The sugar moieties of EPO have a number of actual roles in the molecule's biological functions. Our previous review discusses these points in detail [27]. EPO has

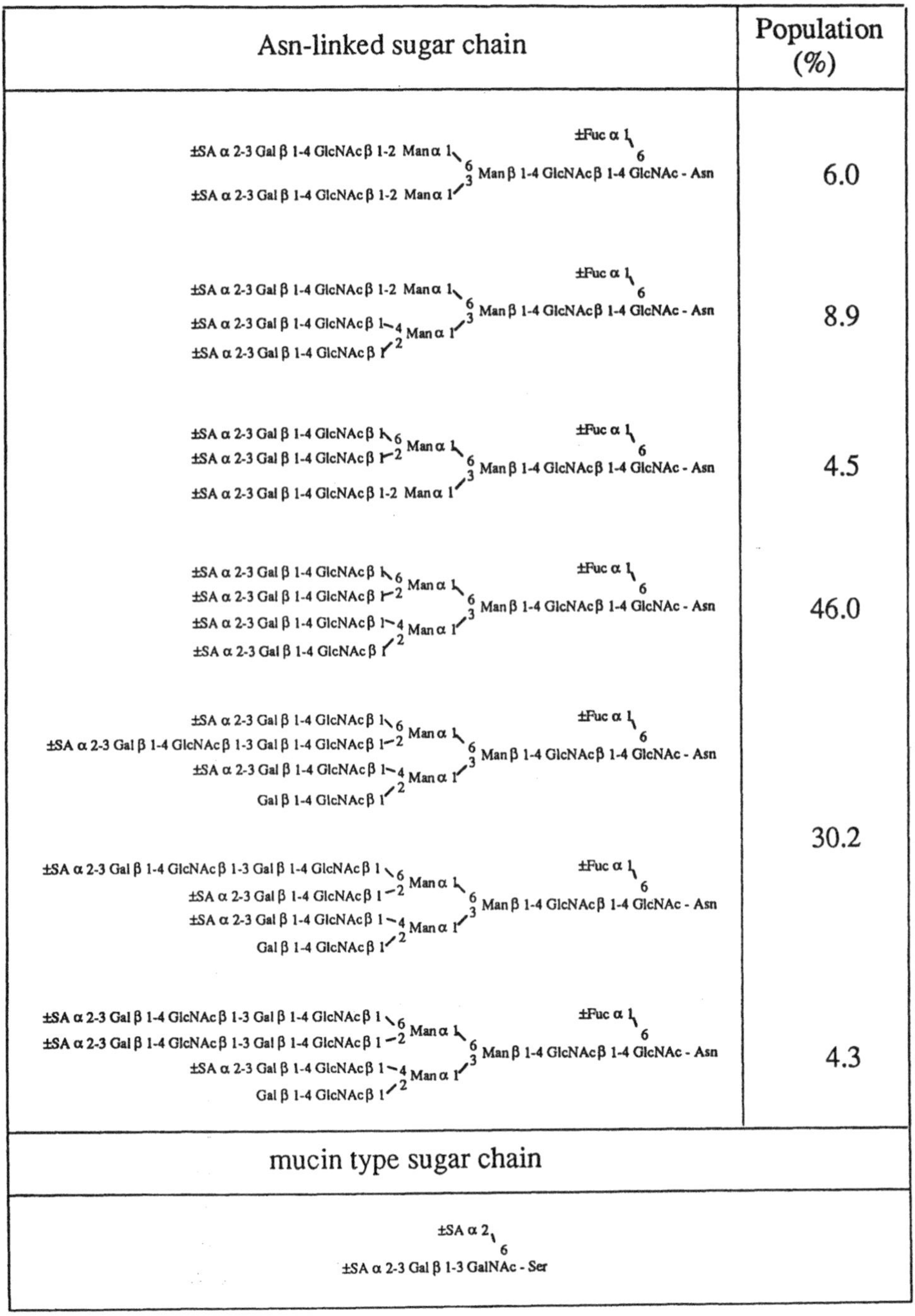

Asn-linked sugar chain	Population (%)

±SA α 2-3 Gal β 1-4 GlcNAc β 1-2 Man α 1
     ⁶
     ₃ Man β 1-4 GlcNAc β 1-4 GlcNAc - Asn   (±Fuc α 1→6)
±SA α 2-3 Gal β 1-4 GlcNAc β 1-2 Man α 1
	6.0

±SA α 2-3 Gal β 1-4 GlcNAc β 1-2 Man α 1 ±SA α 2-3 Gal β 1-4 GlcNAc β 1-4 Man α 1 ±SA α 2-3 Gal β 1-4 GlcNAc β 1-2 Man β 1-4 GlcNAc β 1-4 GlcNAc - Asn (±Fuc α 1→6)	8.9
±SA α 2-3 Gal β 1-4 GlcNAc β 1-6 Man α 1 ±SA α 2-3 Gal β 1-4 GlcNAc β 1-2 ±SA α 2-3 Gal β 1-4 GlcNAc β 1-2 Man α 1 Man β 1-4 GlcNAc β 1-4 GlcNAc - Asn (±Fuc α 1→6)	4.5
±SA α 2-3 Gal β 1-4 GlcNAc β 1-6 Man α 1 ±SA α 2-3 Gal β 1-4 GlcNAc β 1-2 ±SA α 2-3 Gal β 1-4 GlcNAc β 1-4 Man α 1 ±SA α 2-3 Gal β 1-4 GlcNAc β 1-2 Man β 1-4 GlcNAc β 1-4 GlcNAc - Asn (±Fuc α 1→6)	46.0
±SA α 2-3 Gal β 1-4 GlcNAc β 1-6 Man α 1 ±SA α 2-3 Gal β 1-4 GlcNAc β 1-3 Gal β 1-4 GlcNAc β 1-2 ±SA α 2-3 Gal β 1-4 GlcNAc β 1-4 Man α 1 Gal β 1-4 GlcNAc β 1-2 Man β 1-4 GlcNAc β 1-4 GlcNAc - Asn (±Fuc α 1→6)	
±SA α 2-3 Gal β 1-4 GlcNAc β 1-3 Gal β 1-4 GlcNAc β 1-6 Man α 1 ±SA α 2-3 Gal β 1-4 GlcNAc β 1-2 ±SA α 2-3 Gal β 1-4 GlcNAc β 1-4 Man α 1 Gal β 1-4 GlcNAc β 1-2 Man β 1-4 GlcNAc β 1-4 GlcNAc - Asn (±Fuc α 1→6)	30.2
±SA α 2-3 Gal β 1-4 GlcNAc β 1-3 Gal β 1-4 GlcNAc β 1-6 Man α 1 ±SA α 2-3 Gal β 1-4 GlcNAc β 1-3 Gal β 1-4 GlcNAc β 1-2 ±SA α 2-3 Gal β 1-4 GlcNAc β 1-4 Man α 1 Gal β 1-4 GlcNAc β 1-2 Man β 1-4 GlcNAc β 1-4 GlcNAc - Asn (±Fuc α 1→6)	4.3

mucin type sugar chain
±SA α 2→6 ±SA α 2-3 Gal β 1-3 GalNAc - Ser

Fig. 8. Structure of sugar chain from rHuEPO. Asn-Linked sugar chains and a mucin type sugar chain were analyzed by gel filtration with sequential glycosidases digestion and methylation analysis.

two different types of sugar chains; Asn-linked and mucin-type sugar chains. Each type of sugar chain has a different structure with a variety of functions. It is convenient to separate the Asn-linked sugar chains into three portions; terminal, branch and core sugar chain [27]. The terminal sugar chain contains sialic acids, repeating ploy-*N*-acetyllactosamine and galactose. These saccharides are located at the most outward region of the molecule which includes the polypeptide portion, suggesting that this region could have roles in interaction with other molecules. Sialic acids have a negative charge which electrostatically repels other molecules. Removal of sialic acids, which causes exposure of the penultimate galactose residues, increases the affinity for EPO receptors and hepatic asialoglycoprotein binding lectin [28,29]. This trapping system in the blood circulation results in decrease of in vivo activity (measured with animals) in spite of the enhancement of in vitro biological activity (measured with cultured erythroid progenitor cells). Fukuda et al. reported that EPO rich in *N*-acetyllactosamine triantennary structure, binds with hepatic lectin regardless of the presence of sialic acids [30].

The branching sugar chain portion contains GlcNAc branches attached to the core sugar chain. So far, mono- to pentaantennary structures are generally reported. It has been proved that the ratio of tetraantennary to biantennary correlates positively with the in vivo biological activity of EPO [20]. Because the number of ligands and the distance between them are critical to recognition by lectin, the branching sugar chain portion supports the function of the terminal sugar chain portion. The branching portion also forms the bulky structure which should prevent protease attack and filtration from blood veins. In other words, this sugar chain portion increases the stability of EPO in circulation.

The core sugar chain portion contains Manα1–3(Manα1–6)Manβ1–4GlcNAcβ1–4GlcNAc structure. All of the Asn-linked sugar chains have this structure as a common "core". This portion is located nearest to the polypeptide portion suggesting that it may be involved in maintaining active conformation of the polypeptide portion.

Because the mucin-type sugar chain contributes no more than 3.3% of the total mass of EPO, the functional roles of the mucin-type sugar chain cannot be as significant as those of Asn-linked sugar chains. The removal of this sugar chain and the prevention of *O*-*N*-acetylgalactosaminylation caused a slight loss of in vitro biological activity of EPO [31]. Site-directed conversion of Ser^{126}(*O*-*N*-acetylgalactosaminylation site) to Ala^{126} results in no secretion [32].

The number of Asn-linked sugar chains also affects the activity of EPO. Site directed conversion of the *N*-glycosylation site strongly inhibited the secretion of EPO [32]. Partial removal of Asn-linked sugar chains increases the in vitro biological activity two- to three-fold. Complete removal of these sugar chains from EPO results in variable activities. The in vitro activity of nonglycosylated EPO appears to differ from 0 to 200% of the intact EPO depending on the method of deglycosylation [31]. When EPO is produced in *E. coli*, nonglycosylated EPO forms insoluble aggregates. Only trace amount of EPO produced in *E. coli* stay in solution and have approximately 50% in vitro biological activity of the intact form.

The authors suppose that the Asn-linked sugar chains of EPO could be important for folding of the polypeptide portion during the biosynthetic pathway. Support for this hypothesis comes from the finding of a molecule that assists the folding of newly synthesized glycoproteins through interaction with Asn-linked sugar chains. Recently Hammond et al. has found a protein named calnexin, which appears to be a candidate for such a molecule [32]. They indicate that calnexin acts as a chaperon by binding glycoproteins, via a precursor Asn-linked sugar chain. Although much more evidence is needed to prove this hypothesis, it is worthwhile to prove the notion that Asn-linked sugar chains have a role in the folding of newly synthesizing polypeptides.

Clinical Application

Human recombinant EPO is widely used throughout the world to treat anemia patients with chronic renal failure, cancer and HIV infection. Eschbach et al. [33] studied the benefit of rHuEPO for anemia associated with chronic renal failure in 1987. Recently, Goodnough [34] reported the application of rHuEPO in surgery and oncology in 1991. Additionally, there are many reports about new applications to anemia in prematurity [35], multiple myeloma [36] and aplastic anemia [37]. At the beginning of its clinical development, it was assumed that rHuEPO might not be so effective when given by subcutaneous administration because the rHuEPO molecule is a highly glycosylated protein with high molecular weight, and only intravenous administration was expected to be suitable for clinical application. However, clinical efficacy by subcutaneous administration was better than predicted and this form of administration can lead to much longer retention of rHuEPO in the blood than same-dose intravenous administration. We have developed a new high-concentration formulation for subcutaneous injection, and this has made it possible to reduce the frequency of administration and to expand the clinical applications. Many clinical applications of rHuEPO are still under trial in various form of anemia, such as anemia associated with cancer treatment and anemia accompanying BMT.

Conclusion

The large-scale production of EPO has been achieved through utilization of "new biotechnology" such as recombinant DNA methods, large-scale cell culture and structure analysis of carbohydrates. EPO is now clinically available as a drug to treat renal anemia and other types of anemia associated with various diseases. rHuEPO is a good example to demonstrate the successful application of new biotechnology. We expect that many new physiologically active substances, such as EPO, will be developed using biotechnology and these will provide many benefits in the future.

312

References

1. Carnot P, Deflandre N. Sur l'activite hemopoietique du serum au cours de la regeneration du sang. C R Acad Sci Paris 1906;143:384–386.
2. Erslev AJ, Lavietes PH, Van Wagener G. Erythropoietin stimulation induced by anemic serum. Proc Soc Exp Biol Med 1953;83:548–550.
3. Jacobson LO, Goldwasser E, Fried W, Plzak LF. The role of the kidney in erythropoiesis. Nature 1957;179:633–634.
4. Miyake T, Kung CK, Goldwasser E. Purification of human erythropoietin. J Biol Chem 1977;252: 5558–5564.
5. Anagnostou A, Barone J, Fried W, Kedo A. Effect of erythropoetin therapy on the red cell volume of uraemic and non-uraemic rats. Br J Hematol 1977;37:85–91.
6. Lin FK, Suggs S, Lin CH, Brown JK, Smalling R, Egrie JC, Chen KK, Fox GM, Martin F, Stabinsky Z, Badrawi SM, Lai PH, Goldwasser E. Cloning and expression of the human erythropoietin gene. Proc Natl Acad Sci USA 1985;82:7580–7584.
7. Jacobs K, Shoemaker C, Rudersdorf R, Neill SD, Kaufman RJ, Mufson A, Seehra J, Jones SS, Hewick R, Fritsch EF, Kawakita M, Shimizu T, Miyake T. Isolation land characterization of genomic and cDNA clones of human erythropoietin. Nature 1985;313:806–809.
8. Goldwasser E, Lin FK, McDonald JD. Cloning, sequence and evolutionary analysis of the mouse erythropoietin. Mol Cell Biol. 1986;6:842–848.
9. Law ML, Cai GY, Lin FK, Wei Q, Huang S-Z, Hartz JH, Morse H, Lin C-H, Jones C, Kao FT. Chromosomal assignment of the human erythropoietin gene and its DNA polymorphism. Proc Natl Acad Sci USA 1986;83:6920–6924.
10. McDonald JD, Lin F-K, Goldwasser E. Cloning, sequencing, and evolutionary analysis of the mouse erythropoietin gene. Mol Cell Biol 1986;6:842–848.
11. Shoemaker CB, Mitosock LD. Murine erythropoietin gene: cloning, expression, and human gene homology. Mol Cell Biol 1986;6:849–858.
12. Cole CN, Santangelo GM. Analysis in cos-1 cells of processing and polyadenylation signals by using derivatives of the herpes simplex virus type 1 thymidine kinase gene. Mol Cell Biol 1983;3:267–279.
13. Lusky M, Botchan M. Inhibition of SV40 replication in simian cells by specific pBR322 DNA sequence. Nature 1981;293:79–81.
14. Shimke RT (ed). Gene Amplication. Cold Spring Harbor Laboratory, Cold Spring Harbor, NY, 1982.
15. Urlaub G, Chasin LA. Isolation of Chinese hamster cell mutants deficient in dihydrofolate reductase activity. Proc Natl Acad Sci USA 1980;77:4216–4220.
16. Recny MA, Scoble HA, Kim Y. Structural characterization of natural human urinary and recombinant DNA-derived erythropoietin. J Biol Chem 1987;262:17156–17163.
17. Lai PH, Everett R, Wang FF, Arakawa T, Goldwasser E. Structural characterization of human erythropoietin. J Biol Chem 1986;261:3116–3121.
18. Takeuchi M, Takasaki S, Masae Shimada, Kobata A. Role of sugar chains in the in vitro biological activity of human erythropoietin produced in recombinant Chinese hamster ovary cells. J Biol Chem 1990;265:12127–12130.
19. Goldwasser E, Kung CK, Eliason J. On the mechanism of erythropoietin-induced differentiation. The role of sialic acid in erythropoietin action. J Biol Chem 1974;249:17516–17521.
20. Takeuchi M, Inoue N, Strickland TW, Kubota M, Wada M, Shimizu R, Hoshi S, Kozutsumi H, Takasaki S, Kobata A. Relationship between sugar structure and biological activity of recombinant human erythropoietin produced in Chinese hamster ovary cells. Proc Natl Acad Sci USA 1989;86:7819–7822.
21. Inoue N, Takeuchi M, Asano K, Shimizu R, Takasaki S, Kobata A. Structure of mucin-type sugar

chains on human erythropoietins purified from urine and the culture medium of recombinant Chinese hamster ovary cells. Arch Biochim Biophys 1993;301:375–378.

22. Takeuchi M, Takasaki S, Miyazaki H, Kato T, Hoshi S, Kochibe N, Kobata A. Comparative study of the asparagine-linked sugar chains of human erythropoietins purified from urine and the culture medium of recombinant Chinese hamster ovary cells. J Biol Chem 1988;263:3657–3663.

23. Sasaki H, Bothner B, Dell A, Fukuda M. Carbohydrate structure of erythropoietin expressed in Chinese hamster ovary cells by a human erythropoietin cDNA. J Biol Chem 1987;262:12059–12076.

24. Hokke CH, Bergwerff AA, Dedem GWK, Oostrum J, Kamerling JP, Vliegenthart JFG. Sialylated carbohydrate chain of recombinant human glycoproteins expressed in Chinese hamster ovary cells contain traces of N-glycolylneuraminic acid. FEBS Lett 1990;275:9–14.

25. Tsuda E, Goto M, Murakami A, Akai K, Ueda M, Kawanishi G, Takahashi N, Sasaki R, Chiba H, Ishihara H, Mori M, Tejima S, Endo S, Arata Y. Comparative structural study of N-linked oligosaccharides of urinary and recombinant erythropoietin. Biochemistry 1988;27:5646–5654.

26. Sasaki H, Ochi N, Dell A, Fukuda M. Site-specific glycosilation of human recombinant erythropoietin: analysis of gycopeptides or peptides at each glycosylation site by fast atom bombardment mass spectrometry. Biotechnology 1988;27:8618–8626.

27. Takeuchi M, Kobata A. Structures and functional roles of the sugar chains of human erythropoietin. Glycobiology 1991;1:337–346.

28. Kawasaki T, Ashwell G. Chemical and physical properties of an hepatic membrane protein that specifically binds asialoglycoproteins. J Biol Chem 1976;251:1296–1302.

29. Morell AG, Gregoriadis G, Scheinberg IH, Hickman J, Ashwell G. The role of sialic acid in determining the survival of glycoproteins in the circulation. J Biol Chem 1971;246:1461–1467.

30. Fukuda MN, Sasaki H, Lopez L, Fukuda N. Survival of recombinant erythropoietin in the circulation: the role of carbohydrates. Blood 1989;73:84–89.

31. Takeuchi M, Takasaki S, Shimada M, Kobata A. Role of sugar chains in the in vitro biological activity of human erythropoietin produced in recombinant Chinese hamster ovary cells. J Biol Chem 1990;265:12127–12130.

32. Dube S, Fisher JW, Powell JS. Glycosylation at specific sites of erythropoietin is essential for biosynthesis, secretion, and biological function. J Biol Chem 1988;263:17516–17521.

33. Hammond C, Braakman I, Helenius A. Role of N-linked oligosaccharide recognition, glucose trimming, and calnexin in glycoprotein folding and quality control. Proc Natl Acad Sci USA 1994; 91:913–917.

34. Eshbach JW, Egrie JC Downing NR, Brown JK, Adamson JW. Correction of the anemia of end-stage renal disease with recombinant human erythropoietin. N Engl J Med 1987;316:73–76.

35. Goodnough LT, Rudnick S, Price TH, Ballas SK, Collins ML, Crowley JP, Kosmin M, Kruskall MS, Lenes BA, Menitove JE, Silverstein LE, Smith KJ, Wallas CH, Abels RI, Von Tress M. Increased preparative collection of autologous blood with recombinant human erythropoietin therapy. N Engl J Med 1989;321:1163–1168.

36. Halperin DS, Wacker P, Lacourt G, Felix M, Babel JF, Aapro M, Wyss M. Effects of recombinant human erythropoietin in infants with the anemia of prematurity: a pilot study. J Pediatr 1990;116: 779–786.

37. Ludwig H, Fritz E, Kotzmann H, Hocker P, Gisslinger H, Barnas U. Erythropoietin treatment of anemia associated with multiple myeloma. N Engl J Med 1990;322:1693–1699.

38. Stein RS, Abels RI, Krantz SB. Pharmacologic doses of recombinant human erythropoietin in the treatment of myelodysplastic syndromes. Blood 1991;78:1658–1663.

© 1995 Elsevier Science B.V. All rights reserved
Biotechnology Annual Review
M.R. El-Gewely, editor

Lipases and esterases: a review of their sequences, structure and evolution

Henrik W. Anthonsen[1], António Baptista[1], Finn Drabløs[1], Paulo Martel[1], Steffen B. Petersen[1,*], Maria Sebastião[2] and Louis Vaz[3]

[1]*MR-Center, SINTEF UNIMED, Trondheim, Norway;* [2]*Laboratório de Engenharia Bioquímica, Instituto Superior Técnico, Lisboa, Portugal; and* [3]*Limetree Rd 19, Liverpool, New York, USA*

Abstract. This chapter aims to provide a brief review on the enzyme family of lipases and esterases. The sequences, 3D structures and pH dependent electrostatic signatures are presented and analyzed. Since the family comprises more than 100 sequences, we have tried to focus on the most interesting features from our perspective, which translates into finding similarities and differences between members of this family, in particular in and around the active sites, and to identify residues that are partially or totally conserved. Such residues we believe are either important for maintaining the structural scaffold of the protein or to maintain activity or specificity. The structure function relationship for these proteins is therefore of central interest. Can we uniquely identify a protein from this large family of sequences – and if so, what is the identifier? The protein family displays some highly complex features: many of the proteins are interfacially activated, i.e. they need to be in physical contact with the aggregated substrate. Access to the active site is blocked with either a loop fragment or an α-helical fragment in the absence of interfacial contact. Although the number of known, relevant protein 3D structures is growing steadily, we are nevertheless faced with a virtual explosion in the number of known or deduced amino acid sequences. It is therefore unrealistic to expect that all protein sequences within the foreseeable future will have their 3D structure determined by X-ray diffractional analysis or through other methods. When feasible the gene and/or the amino acid sequences will be analyzed from an evolutionary perspective. As the 3D folds are often remarkably similar, both among the triglyceride lipases as well as among the esterases, the functional diversities (e.g. specificity) must originate in differences in surface residue utilization, in particular of charged residues. The pH variations in the isopotential surfaces of some of the most interesting lipases are presented and a qualitative interpretation proposed. Finally we illustrate that NMR has potential for becoming an important tool in the study of lipases, esterases and their kinetics.

Lipase and Esterase Catalytic Function

Although the focus of the present paper is on triacyl glyceride lipases and esterases, it is important to point out some similarities to other enzymes that have similar substrates, such as the phospholipases. In Fig. 1 is shown a model substrate, a phospholipid, where X is typically choline. Thus, the substrate for phospholipases share the basic skeleton with the triacylglyceride lipases. Not surprisingly, several lipases and esterases display activity towards phospholipids in addition to tri-

Address for correspondence: Professor Steffen B. Petersen, MR Center, SINTEF-UNIMED, N-7034, Trondheim, Norway. Tel.: +47 73997700; Fax: +47 73997708; E-mail: sbp@marvin.mr.sintef.no.

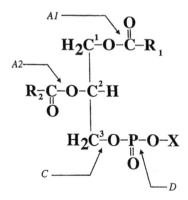

Fig. 1. Classification of phospholipases. A schematic representation of the point of hydrolysis for different phospholipases. The acyl ester bond at position 1 of 3-*sn*-phosphomonoglycerides is attacked by phospholipase A1, whereas the acyl ester bond at position 2 is attacked by phospholipase A2. Phospholipase B displays both A1 and A2 activity, in addition to activity towards lysophospholipids. Two phospholipases attack the phosphodiester bond. The action of phospholipase C leads to the release of diacylglycerol, whereas phospholipase D results in the release of diacyl glycerophosphate and free polar head group (adapted from [106]).

acylglycerides, e.g. the guinea-pig pancreatic lipase. The active site in lipases and esterases is very similar to that found in serine proteases; it consists of a Ser, a His and an Asp acid. As is seen in Fig. 2A, the active sites of bovine trypsin (PDB:4PTP) and the *Rhizomucor miehei* (PDB:4TGL) overlap extremely well. In Fig. 2B the active site residues are shown as they appear when entering the active site. It is interesting to note that, although the geometrical distances between the residues are maintained, the mutual location of the serine and aspartate with respect to the central

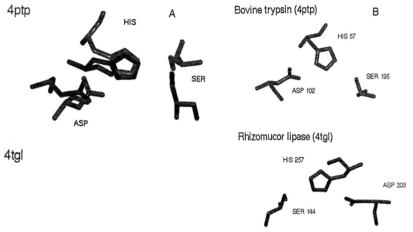

Fig. 2. Superposition of active sites from lipases and trypsin. The active site of *Rhizomucor miehei* lipase superimposed upon the active site of trypsin. The key active site residues are identical in the two proteins and, as is shown in the figure, the geometrical configuration is almost identical in the two proteins.

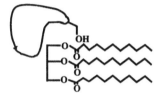

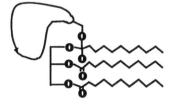

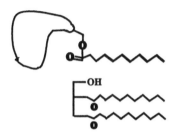

Fig. 3. The catalytic mechanism in lipases. Triacylglyceride hydrolysis. The active site serine is interacting with the 1 or the 3 glycerol ester bond and a tetrahedral intermediate is formed. The diacylglycerol is released and the remaining acyl chain is attached to the active site serine, from which it is subsequently hydrolyzed and then released as a free fatty acid (adapted from [107]).

His has interchanged. The catalytic mechanism assumed for triacyl glyceride lipases and esterases is centered on the active site Ser (Fig. 3). The nucleophile oxygen of the active site Ser forms a tetrahedral hemiacetal intermediate with the triacyl glyceride. The ester bond of the hemiacetal is hydrolyzed and the diacyl glyceride is released. The active site serine acyl ester is then reacted with an activated water molecule, and the acyl enzyme is subsequently cleaved and the fatty acid is dissociated. At this stage of the catalytic process, product release from the active site crevase is of special importance. If the fatty acid is too tightly bound to the active site it will inhibit the enzyme. It is therefore interesting that our electrostatic calculations for *Candida antarctica B* and human pancreatic lipase predict a negative electrostatic potential in the active site crevase in the pH range where the enzymes display high activity (vide infra). Thus an ionized free fatty acid will be ejected from the active site due to electrostic repulsion between a negatively charged carboxyl group and the negative charges giving rise to the negative electrostatic potential in the active site.

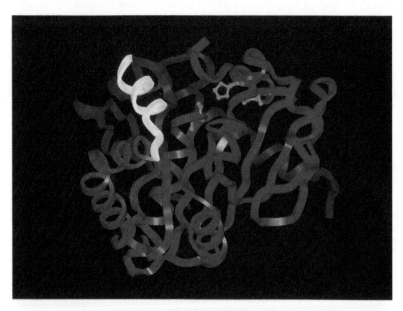

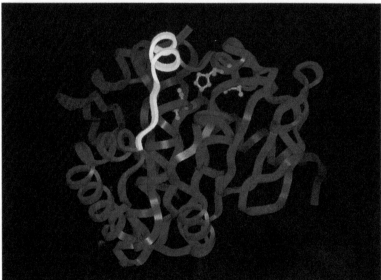

Fig. 4. The opening of the lid of *Rhizomucor miehei*. The figure shows the 3D fold of *Rhizomucor miehei* lipase with the lid open and closed.

Inhibitors of Lipases

Several organophosphorous compounds can inhibit both lipases as well as serine proteases (see Fig. 5) [1,2]. The mechanism of action is based upon an irreversible phosphorylation of the active site serine. Phospholipase A2s are not inhibited, as one would expect, based upon the differences in catalytic mechanism. Other compounds

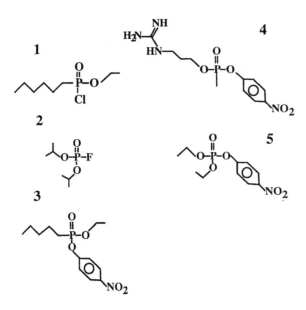

Fig. 5. Examples of lipase inhibitors (adapted from [2]).

may also interfere with the lipase catalytic function. Surfactants can activate the lipase at low concentrations and inhibit at high. Quarternary ammonium salts have increased the catalytic rate of the *Rhizomucor miehei* lipase. A correlation was found between the effect of the surfactant and absorption of lipase to the lipid interface. Similarly the non-ionic surfactant alcohol, ethoxylate monoalkyl deca-(oxyethylene) ether, enhanced the hydrolysis rate of the lipase at a surfactant concentration of 50% of its critical micelle concentration (CMC). Higher concentrations of this surfactant inhibited the *Rhizomucor miehei* lipase [1]. Bile salts also exhibit inhibitory effects towards mammalian lipases. At high concentrations, the bile salt inhibits the pancreatic lipase; however this effect can be reversed by colipase, which prevents the interaction with bile salt. Presumably the pancreatic lipase binding sites for the colipase and the bile salt are the same.

Protein based inhibitors of serine hydrolases are well documented [3]. Lipase inhibiting proteins have been reported, but detailed studies are lacking, and the concept of a mechanistic understanding of such inhibition is still lacking – to a large extent again due to the complex interfacial activation associated with lipase function. A lipoprotein lipase inhibiting protein was found in a melanoma cell line [4].

Since the protein surface is preferentially populated with charged and hydrophilic amino acid residues, metal ions may easily find a binding site consisting of, e.g. aspartic acids and carbonyl groups. Whether or not such binding will interfere with catalytic function depends on the mutual location of the metal ion binding site and the active site. With respect to lipase action, metal ions may actually stimulate the activity, since they can bind to free fatty acids, and thereby remove products. Certain iron salts have been reported to inhibit lipases from *Aspergillus niger* [5] and lipases from fungi of the genus *Geotrichum* [6].

320

Interfacial Activation

In water, lipases have only a marginal but measurable activity towards dissolved substrates. In the case of porcine pancreatic lipase acting on tripropionin, the reaction velocity towards soluble substrate is only 0.3% of the value observed for hydrolysis of emulsified substrate. Interestingly, the reaction rate with monomeric tripropionin was increased 10–20 times by addition of dioxane, acetonitrile, formamide and tert-butanol [7]. In another experiment, porcine pancreatic lipase was added to siliconized glass beads. A reversible, diffusion controlled binding was observed, and concurrently a 1000-fold increase in specific activity [8]. It is reasonable to assume that the association to the siliconized glass beads facilitates the opening of the lid covering the active site.

A fundamental difference between lipases and esterases is their ability to act on solublized substrates. In Fig. 6 shows schematically how an esterase responds to an increase in substrate. On the horizontal axis, the value 1 denotes saturation, i.e. beyond this concentration increasing amounts of emulsified substrate are present. The esterase displays normal Michaelis–Menten activity, the lipase is activated at the substrate concentration, where aggregates start to form (saturation = 1). In a simple but ingenious setup, monolayer techniques have been developed for studying lipase kinetics at an interface [9,10]. The lipid substrate is spread on an aqueous surface and the surface tension of the lipid is kept constant by mechanical means. The enzyme is then injected into the aqueous phase and the change in surface area of the lipid necessary to maintain surface tension is measured. A wide range of physical parameters can be measured this way using both very little lipid and enzyme: enzyme kinetics at an interface, surface pressure, electric potential.

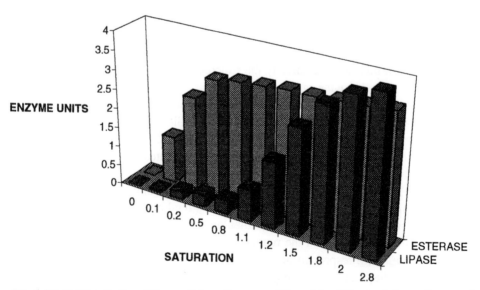

Fig. 6. Interfacial activation of lipases. Schematic representation of the difference between lipases and esterase enzymatic activity.

Reversed micelles

Reversed micelles represent a unique physical state of a two-phase system with a very large and stable interfacial area. Lipases encapsulated in reverse micelles catalyze hydrolysis, synthesis and glycerolysis of fats and other esters [11]. Sometimes the lipase affinity for the reversed micellar phase is high enough to be an aid in purification of lipases [12]. The physical presence of this large surface area may maintain the encapsulated lipase in a mostly activated state. Unfortunately, it can be difficult to monitor lipase kinetics in reversed micelles, and the physical interpretation of the role of the surfactant for substrate interaction with the lipase is not clear at present. Also, the surfactant may restrict the diffusion of the substrate and may even interact directly with the active site of the lipase. Recently, we studied how the protein extraction yield in reversed micellar systems could be improved using surface charge engineering [13].

Specificity

The lipases can be grouped into some major groups: 1,3 specific, 2-specific, partial glyceride specific, fatty acid specific and non-specific lipases [14]. In addition most of the lipases exhibit stereospecificity and some display clear phospholipase activity in addition to their activity towards acylglycerides. The latter has been observed for the guinea pig lipase [15], where this particular feature has been associated with the absence of a structural domain, called the lid, that is presumed to block access to the active site of the normal lipases.

Most triglyceride lipases display 1,3 specificity towards triacylglycerides. Thus the pancreatic, gastric as well as lipoprotein lipases are all 1,3 specific, whereas the bile salt activated lipase and the pancreatic carboxyl ester lipase are non-specific. Among the microbial lipases the *Aspergillus niger, Chromobacterium viscosum, Humicola lanuginosa, Rhizomucor miehei, Pseudomonas fluorescens, Rhizopus arrhizus* and *Rhizopus delemar* are 1,3-specific. The *Candida antarctica A* and *B, Geotrichum candidum, Penicillum cyclopium* and the *Staphylococcus hyiscus* are non-specific. The *S. hyiscus* also displays phospholipase activity [14]. Lipases with a specificity towards the 2-position are rare. Recently, it has been reported that the *Geotrichum candidum* produces one major and four minor forms of lipase, of which two of the minor forms showed a clear preference for the 2-position when the substrate was triolein. Interestingly, the four lipases showed quite different activities towards simple triglycerides. High activities were observed towards triglycerides with a fatty acid chain length of 8–10 (the major and one of the minor forms) as well as 20 (for all four forms), but activities towards other intermediate and shorter chain lengths were very small [16]. Whether such observations can be explained, at least in part, by the physical state of the substrate is not clear. One company claims that *C. cylindracea* lipase does have activity towards the 2-position as well. The partial glyceride lipases displays substantially higher activity towards mono- and diglyc-

Table 1. Amino acid sequences of lipases and esterases: tabulation of lipases, esterases and related sequences

Id	Len	Description	Source	EMBL	PDB	Pos	Ser	Codon
ACES_BOVIN	583	Acetylcholinesterase (EC 3.1.1.7)	*Bos taurus*			203	DPTSVTLFGESAGAASVGMHLLSPP	
ACES_DROME	649	Acetylcholinesterase precursor (EC 3.1.1.7)	*Drosophila melanogaster*	DMACHE		276	NPEWMTLFGESAGSSSVNAQLMSPV	TCG
ACES_HUMAN	614	Acetylcholinesterase precursor (EC 3.1.1.7)	*Homo sapiens*	HSACHE		234	DPTSVTLFGESAGAASVGMHLLSPP	AGC
ACES_MOUSE	614	Acetylcholinesterase precursor (EC 3.1.1.7)	*Mus musculus*	MMACHE		234	DPMSVTLFGESAGAASVGMHILSLP	AGT
ACES_TORCA	596	Acetylcholinesterase precursor (EC 3.1.1.7)	*Torpedo californica*	TCACER	1ACE	221 226	DPKTVTIFGESAGGASVGMHILSPG TIFGESAGGASVGMHILSPGSRDLF	AGT TCT
ACES_TORMA	599	Acetylcholinesterase precursor (EC 3.1.1.7)	*Torpedo marmorata*	TMACHE		224	DPKTVTLFGESAGRASVGMHILSPG	AGT
AOAH_HUMAN	575	Acyloxyacyl hydrolase precursor	*Homo sapiens*	HSACYLHY		263	QPRGIILLGDSAGAHFHISPEWITA	TCA
BAL_BOVIN	>597	Bile-salt-activated lipase precursor (EC 3.1.1.3) (EC 3.1.1.13)	*Bos taurus*			9 132 212	LGASRLGPSPGCLAVASAAKLGS IYGGAFLMGASQGANFLSNYLYDGE DPDNITLFGESAGGASVSLQTLSPY	
BAL_HUMAN	742	Bile-salt-activated lipase precursor (EC 3.1.1.3) (EC 3.1.1.13)	*Homo sapiens*	HSBSSL HSCHE HSLIPBSA HSCEL		214	DPNNITLFGESAGGASVSLQTLSPY	TCT
BAL_RAT	612	Bile-salt-activated lipase precursor (EC 3.1.1.3) (EC 3.1.1.13)	*Rattus norvegicus*	RNCHLEST RNLPL		214	DPDNITIFGESAGAASVSLQTLSPY	TCT
CHEB_SALTY	349	Protein-glutamate methyl-esterase (EC 3.1.1.61)	*Salmonella typhimurium*			164	SSEKLIAIGASTGGTEAIRHVLQPL	
CHLE_BOVIN	>141	Cholinesterase (EC 3.1.1.8)	*Bos taurus*	BBBCHEAA		131	NPKSVTLFGESAGAASVSLHL	AGT
CHLE_CANFA	>141	Cholinesterase (EC 3.1.1.8)	*Canis familiaris*	CFBCHEAA		131	NPKSVTLFGESAGAGSVGLHL	AGT
CHLE_HUMAN	602	Cholinesterase precursor (EC 3.1.1.8)	*Homo sapiens*	HSCHEBG1 HSCHEBG2 HSCHEBG3 HSCHEBG4		226	NPKSVTLFGESAGAASVSLHLLSPG	AGT

Code	Length	Description	Species	Accession	Position	Sequence	Codon
CHLE_MACMU	>141	Cholinesterase (EC 3.1.1.8)	Macaca mulatta	HSCHEB / HSCHEF / MMBCHEAA	131	NPKSVTLFGESAGAASVSLHL	AGT
CHLE_MOUSE	603	Cholinesterase precursor (EC 3.1.1.8)	Mus musculus	MMACCHES	227	NPKSITIFGESAGAASVSLHLLCPQ	AGT
CHLE_PIG	>141	Cholinesterase (EC 3.1.1.8)	Sus scrofa	SSBCHEAA	131	NPKSVTLFGESAGAVSVSLHL	AGT
CHLE_RABIT	581	Cholinesterase precursor (EC 3.1.1.8)	Oryctolagus cuniculus	OCBCHEX1 / OCBCHEX2 / OCBCHEX3 / OCBCHEAA	205	NPKSVTLFGESAGAASVSLHLLSPR	AGT
CHLE_SHEEP	>141	Cholinesterase (EC 3.1.1.8)	Ovis aries	OOBCHEAA	131	NPKSVTLFGESAGAASVSLHL	AGT
CLCD_PSEPU	236	Carboxymethylene-butenolidase (EC 3.1.1.45)	Pseudomonas putida	ECCLC / PSCLCD	113	AIRYARHQPYSNGKVGLVGYCLGGA	AGC
CRYS_DICDI	550	Crystal protein precursor	Dictyostelium discoideum	DDCPCP	123	SNGKVGLVGYCLGGALAFLVAAKGY	TGC
					215	DKNQVTIYGESAGAFSVAAHLSSEK	TCT
CUTI_ASCRA	223	Cutinase precursor (EC 3.1.1.-)	Ascochyta rabiei	ARCUT	66	GSTEIGNMGVSAGPAVASALEAYGA	TCC
					135	PSTPIVAGGYSQGTAVMAGAIPKLD	AGC
CUTI_COLCA	228	Cutinase precursor (EC 3.1.1.-)	Colletotrichum capsici	CCCUTB	69	ASTEPGNMGISAGPIVADALESRYG	AGC
					140	PNSAVVAGGYSQGTAVMASSISELS	AGC
CUTI_COLGL	224	Cutinase precursor (EC 3.1.1.-)	Colletotrichum gloeosporioides	CGCUTA	66	ASTEPGNMGISAGPIVADALERIYG	AGC
					136	PNAAIVSGGYSQGTAVMAGSISGLS	AGC
CUTI_FUSSO	230	Cutinase precursor (EC 3.1.1.-)	Fusarium solani	FSCUTR / FSCUTA	136	PDATLLAGGYSQGAALAAASIEDLD	AGC
CUTI_MAGGR	>162	Cutinase (EC 3.1.1.-)	Magnaporthe grisea	MGCUT1	3	GLSAGTNVASRLEREFR	AGC
					72	PNAAVVAGGYSQGTAVMFNAVSEMP	AGC
D2_DICDI	535	cAMP-regulated D2 protein precursor	Dictyostelium discoideum	DDD2 / DDD21 / DDD22 / DDD23 / DDD24	213	NKEMITIWGESAGAFSVSAHLTFTY	TCT
DMPD_PSEPU	283	2-Hydroxymuconic semialdehyde hydrolase (EC 3.1.1.-)	Pseudomonas putida	PPDMPCD	107	EIEQADLVGNSFGGGIALALAIRHP	TCC
EST1_CAEBR	562	Gut esterase precursor (EC 3.1.1.1)	Caenorhabditis briggsae	CBGES1A	199	DPDDITIWGYSAGAASVSQLTMSPY	AGT

Table 1 (continued)

Id	Len	Description	Source	EMBL	PDB	Pos	Ser	Codon
EST1_CAEEL	562	Gut esterase precursor (EC 3.1.1.1)	Caenorhabditis elegans	CEGES1B		198	DPNQITIWGYSAGAASVSQLTMSPY	AGT
EST1_CULPI	540	Esterase B1 precursor (EC 3.1.1.1)	Culex pipiens	CPESTB1		191	DPKRVTLAGHSAGAASVQYHLISDA	AGC
EST1_HUMAN	567	Liver carboxylesterase precursor (EC 3.1.1.1)	Homo sapiens	HSCARAA HSCARBOX		221	NPGSVTIFGESAGGESVSVLVLSPL	TCA
EST1_MOUSE	>156	Esterase 1 (EC 3.1.1.1)	Mus musculus	MMEST				
EST1_RABIT	539	Liver 60 kDa carboxyl-esterase 1 (EC 3.1.1.1)	Oryctolagus cuniculus			195	DPGSVTIFGESAGGQSVSILLLSPL	
EST1_RAT	549	Liver 60 kDa carboxyl-esterase 1 precursor (EC 3.1.1.1)	Rattus norvegicus	RNCARA RNCEEI1B		221	NPDSVTIFGESAGGVSVSALVLSPL	TCA
EST2_CULPI	>215	Esterase B2 (EC 3.1.1.1)	Culex pipiens					
EST2_RABIT	532	Liver 60 kDa carboxyl-esterase 2 (EC 3.1.1.1)	Oryctolagus cuniculus			201	NPGRVTIFGESAGGTSVSSHVLSPM	
EST4_DROMO	>40	Esterase-4 (EC 3.1.1.1)	Drosophila mojavensis					
EST5_DROMO	>38	Esterase-5 (EC 3.1.1.1)	Drosophila mojavensis					
EST6_DROME	544	Esterase-6 precursor (EC 3.1.1.1)	Drosophila melanogaster	DMEST6P DMEST6A DMEST6		209	EPQNVLLVGHSAGGASVHLQMLRED	TCC
ESTA_CANFA	260	Arginine esterase precursor (EC 3.4.21.35)	Canis familiaris	CFAER CFARGES		212	EGKKDTCKGDSGGPLICDGELVGIT	TCA
ESTA_DROPS	544	Esterase-5A precursor (EC 3.1.1.1)	Drosophila pseudoobscura	DPEST5A		209	EPENILVVGHSAGGASVHLQMLRED	TCC
ESTA_STRSC	345	Esterase precursor (EC 3.1.1.-)	Streptomyces scabies	SSESTA				
ESTB_DROPS	545	Esterase-5B precursor (EC 3.1.1.1)	Drosophila pseudoobscura	DPEST5		207	EPENILVIGHSAGGGSVHLQVLRED	TCT
ESTC_DROPS	545	Esterase-5C precursor (EC 3.1.1.1)	Drosophila pseudoobscura	DPEST5		207 / 231	EPENIIVVGHSAGGASVHLQMLRED / DFAQVAKAGISFGGNAMDPWVIHQS	TCC / TCC

ESTD_HUMAN	>297	Esterase D (EC 3.1.1.1)	*Homo sapiens*	HSETRD		164	DPQRMSIFGHSMGGHGALICALKNP	TCC
ESTJ_HELVI	564	Juvenile hormone esterase precursor (EC 3.1.1.59)	*Heliothis virescens*	HVESTJ		220	DPSDITIAGQSAGASAAHLLTLSKA	AGC
ESTJ_MANSE	>15	Juvenile hormone esterase (EC 3.1.1.59)	*Manduca sexta*					
ESTP_DROME	544	Esterase-P precursor (EC 3.1.1.1)	*Drosophila melanogaster*	DMEST6P		206	MPDNIVLIGHSAGGASAHLQLLHED	TCT
ESTP_RAT	565	Pi 6.1 esterase precursor (EC 3.1.1.1)	*Rattus norvegicus*	RNPI61E		221	NPGSVTIFGESAGGFSVSALVLSPL	TCT
EST_ACICA	290	Esterase (EC 3.1.1.-)	*Acinetobacter calcoaceticus*	ACESTERAS		131	KPKDIIISGDSCGANLHLALSLRLK	TCA
EST_MOUSE	554	Carboxylesterase precursor (EC 3.1.1.1)	*Mus musculus*	MMCXE		221	NPDSVTIFGESSGGISVSVLVLSPL	TCA
LCAT_HUMAN	440	Phosphatidylcholine-sterol acyltransferase precursor (EC 2.3.1.43)	*Homo sapiens*	HSLCAT, HSLCAT1, HSLCATG, HSLCATGA		205	YGKPVFLIGHSLGCLHLLYFLLRQP	AGC
LCAT_MOUSE	438	Phosphatidylcholine-sterol acyltransferase precursor (EC 2.3.1.43)	*Mus musculus*	MMLCATX		205	YGKPVFLIGHSLGCLHVLHFLLRQP	AGC
LCAT_PIG	>188	Phosphatidylcholine-sterol acyltransferase precursor (EC 2.3.1.43)	*Sus scrofa*					
LCAT_RAT	440	Phosphatidylcholine-sterol acyltransferase precursor (EC 2.3.1.43)	*Rattus norvegicus*	RNLCATMR		205	YGKPVFLIGHSLGCLHVLHFLLRQP	AGC
LIP1_CANRU	549	Lipase 1 precursor (EC 3.1.1.3)	*Candida rugosa*	CCLIP1, CCLIPASE	1CRL	224	DPTKVTIFGESAGSMSVMCHILWND	CTG
LIP1_GEOCN	563	Lipase 1 precursor (EC 3.1.1.3)	*Geotrichum candidum*		1TRH	236	DPDKVMIFGESAGAMSVAHQLVAYG	
LIP1_MORSP	319	Lipase 1 (EC 3.1.1.3)	*Moraxella sp*	MSLIPASE		189	DPKRLGAIGWSMGGGGALKLATERS	TCA
LIP1_PSYIM	317	Lipase 1 precursor (EC 3.1.1.3)	*Psychrobacter immobilis*	PILIPAA		142	LASNIHVGGNSMGGAISVAYAAKYP	TCG
LIP2_CANRU	548	Lipase 2 precursor (EC 3.1.1.3)	*Candida rugosa*	CCLIP2		223	DPSKVTIYGESAGSMSTFVHLVWND	CTG

Table 1 (continued)

Id	Len	Description	Source	EMBL	PDB	Pos	Ser	Codon
LIP2_GEOCN	>557	Lipase 2 precursor (EC 3.1.1.3)	Geotrichum candidum	GCLIP2	1THG	230	DPDKVMIFGESAGAMSVAHQLIAYG	TCC
LIP2_MORSP	433	Lipase 2 (EC 3.1.1.3)	Moraxella sp	MSLIP2		239	SPSRIVLSGDSAGGCLAALVAQQVI	AGT
LIP3_CANRU	549	Lipase 3 precursor (EC 3.1.1.3)	Candida rugosa	CCLIP3		224	DPSKVTIFGESAGSMSVLCHLIWND	CTG
LIP3_MORSP	315	Lipase 3 precursor (EC 3.1.1.3)	Moraxella sp	MSLIP3		142	LASNTHVGGNSMGGAISVAYAAKYP	TCG
LIP4_CANRU	549	Lipase 4 precursor (EC 3.1.1.3)	Candida rugosa	CCLIP4		224	DPSKVTIFGESAGSMSVMCQLLWND	CTG
LIP5_CANRU	549	Lipase 5 precursor (EC 3.1.1.3)	Candida rugosa	CCLIP5		224	DPSKVTIFGESAGSMSVLCHLLWNG	CTG
LIPG_CANFA	>40	Triacylglycerol lipase (EC 3.1.1.3) (gastric)	Canis familiaris					
LIPG_HUMAN	398	Triacylglycerol lipase precursor (EC 3.1.1.3) (gastric)	Homo sapiens	HSGLR		118	DAGYDVWLGNSRGNTWARRNLYYSP	AGC
						172	GQKQLHYVGHSQGTTIGFIAFSTNP	TCC
LIPG_RAT	395	Triacylglycerol lipase precursor (EC 3.1.1.3) (lingual)	Rattus norvegicus	RNLIP		117	DAGYDVWLGNSRGNTWSRKNVYYSP	AGT
						171	GQEKIHYVGHSQGTTIGFIAFSTNP	TCT
LIPH_HUMAN	499	Triacylglycerol lipase precursor (EC 3.1.1.3) (hepatic)	Homo sapiens	HSTGLH HSHTGL HSLIPH		168	SRSHVHLIGYSLGAHVSGFAGSSIG	AGC
						180	GAHVSGFAGSSIGGTHKIGRITGLD	TCC
						231	HTFTREHMGLSVGIKQPIGHYDFYP	AGC
LIPH_MOUSE	510	Triacylglycerol lipase precursor (EC 3.1.1.3) (hepatic)	Mus musculus	MMHTL		34	QGVGTEPFGRSLGATEASKPLKKPE	AGC
						169	SRSKVHLIGYSLGAHVSGFAGSSMD	AGC
						232	HTFTREHMGLSVGIKQPIAHYDFYP	AGT
LIPH_RAT	494	Triacylglycerol lipase precursor (EC 3.1.1.3) (hepatic)	Rattus norvegicus	RNHL RNHLP		169	SRSKVHLIGYSLGAHVSGFAGSSMG	AGC
						181	GAHVSGFAGSSMGGKRKIGRITGLD	TCC
						232	HTFTREHMGLSVGIKQPIAHYDFYP	AGT
LIPL_BOVIN	>465	Lipoprotein lipase precursor (EC 3.1.1.34)	Bos taurus	BTLPL		149	PLGNVHLLGYSLGAHAAGIAGSLTN	AGC
						210	HTFTRGSPGRSIGIQKPVGHVDIYP	AGT
LIPL_CAVPO	465	Lipoprotein lipase precursor (EC 3.1.1.34)	Cavia porcellus	CPLPPL CPGPLPL1		149	SVDNVHLLGYSLGAHAAGVAGSRTN	AGC
						210	HTFTRGSPGRSIGIQKPVGHVDIYP	AGT

ID	Length	Description	Species	Database codes	Position	Sequence	Code
LIPL_CHICK	490	Lipoprotein lipase precursor (EC 3.1.1.34)	*Gallus gallus*	CPGPLPL2 CPGPLPL3 CPGPLPL4 CPGPLPL5 GGLPL	159	PLNNVHLLGYSLGAHAAGIAGSLTK	AGT
LIPL_HUMAN	475	Lipoprotein lipase precursor (EC 3.1.1.34)	*Homo sapiens*	HSLPL HSLPLR HSLIPAS HSLPLFI	159	PLDNVHLLGYSLGAHAAGIAGSLTN	AGC
					220	HTFTRGSPGRSIGIQKPVGHVDIYP	AGC
LIPL_MOUSE	474	Lipoprotein lipase precursor (EC 3.1.1.34)	*Mus musculus*	MMLPLA MMLPL	159	PLDNVHLLGYSLGAHAAGVAGSLTN	AGC
					220	HTFTRGSPGRSIGIQKPVGHVDIYP	AGT
LIPL_RAT	474	Lipoprotein lipase precursor (EC 3.1.1.34)	*Rattus norvegicus*		159	PLDNVHLLGYSLGAHAAGVAGSLTN	
					220	HTFTRGSPGRSIGIQKPVGHVDIYP	
LIPP_CANFA	467	Triacylglycerol lipase precursor (EC 3.1.1.3) (pancreatic)	*Canis familiaris*	CFPLIP	171	SPSQVQLIGHSLGAHVAGEAGSRTP	AGC
LIPP_HORSE	>461	Triacylglycerol lipase precursor (EC 3.1.1.3) (pancreatic)	*Equus caballus*	1HPL	165	SPSNVHIIGHSLGSHAAGEAGRRTN	
LIPP_HUMAN	465	Triacylglycerol lipase precursor (EC 3.1.1.3) (pancreatic)	*Homo sapiens*	HSTGLIP	169	SPSNVHVIGHSLGAHAAGEAGRRTN	AGC
LIPP_MOUSE	468	Triacylglycerol lipase precursor (EC 3.1.1.3) (pancreatic)	*Mus musculus*	MMCTLL	170	SPENVHLIPHSLGSHVAGEAGRRLE	AGC
LIPP_PIG	449	Triacylglycerol lipase precursor (EC 3.1.1.3) (pancreatic)	*Sus scrofa*		152	SPSNVHVIGHSLGSHAAGEAGRRTN	
LIPP_RABIT	465	Triacylglycerol lipase precursor (EC 3.1.1.3) (pancreatic)	*Oryctolagus cuniculus*	OCTRIL	170	SPSNIHVIGHSLGAHAAGEVGRRTN	AGC
LIPP_RAT	465	Triacylglycerol lipase precursor (EC 3.1.1.3) (pancreatic)	*Rattus norvegicus*	RNPANLI	169	PPDNVHLIGHSLGSHVAGEAGKRTF	AGC

Table 1 (continued)

Id	Len	Description	Source	EMBL	PDB	Pos	Ser	Codon
A43357	467	Pancreatic lipase related protein 1	*Homo sapiens*			171	PPSKVHLIGHSLGAHVAGEAGSKTP	
B43357	469	Pancreatic lipase related protein 2	*Homo sapiens*			171	SLEDVHVIGHSLGAHTAAEAGRRLG	
LIPS_HUMAN	786	Hormone sensitive lipase (EC 3.1.1.-)	*Homo sapiens*	HSHSLA		423	TGERICLAGDSAGGNLCFTVALRAA	AGT
LIPS_BOVIN	>27	Hormone sensitive lipase (EC 3.1.1.-)	*Bos taurus*					
LIPS_RAT	757	Hormone sensitive lipase (EC 3.1.1.-)	*Rattus norvegicus*	RATLIP		423	TGERICLAGDSAGGNLCITVSLRAA	AGC
LIPT_BURCE	>56	Lipase, thermostable (EC 3.1.1.3)	*Burkholderia cepacia*					
LIP_BURCE	364	Lipase precursor (EC 3.1.1.3)	*Burkholderia cepacia*	PCLIPAA		131	GATKVNLVGHSQGGLSSRYVAAVAP	AGC
LIP_PSEAE	311	Lactonizing lipase precursor (EC 3.1.1.3)	*Pseudomonas aeruginosa*	PALIPAG PALIPAB		108	GQPKVNLIGHSHGGPTIRYVAAVRP	AGC
LIP_PSEFL	449	Lipase precursor (EC 3.1.1.3) (triacylglycerol lipase)	*Pseudomonas fluorescens*	PFTAL		206	SGKDVLVSGHSLGGLAVNSMADLST	AGC
LIP_PSEFR	277	Lipase precursor (EC 3.1.1.3)	*Pseudomonas fragi*	PFLIP PFLIPG		83	GAQRVNLIGHSQGALTARYVAAIAP	AGC
LIP_PSES5	364	Lipase precursor (EC 3.1.1.3)	*Pseudomonas* sp.			131	GATKVNLVGHSQGGLTSRYVAAVAP	
LIP_PSESP	311	Lactonizing lipase precursor (EC 3.1.1.3)	*Pseudomonas* sp.	PSLIPA PSLIPASE PSLIPL		108	GQPKVNLIGHSHGGPTIRYVAAVRP	AGC
LIP_RHIDL	392	Lipase precursor (EC 3.1.1.3)	*Rhizopus delemar*	RDLIP		268	PTYKVIVTGHSLGGAQALLAGMDLY	TCA
LIP_RHIMI	363	Lipase precursor (EC 3.1.1.3)	*Rhizomucor miehei*		1TGL 3TGL 4TGL 5TGL	238	PSYKVAVTGHSLGGATALLCALDLY	

LIP_STAAU	690	Lipase precursor (EC 3.1.1.3)	*Staphylococcus aureus*	SAGEH	412	PGKKVHLVGHSMGGQTIRLMEEFLR	AGT
LIP_STAHY	641	Lipase precursor (EC 3.1.1.3)	*Staphylococcus hyicus*	SHLIP	369	PGHPVHFIGHSMGGQTIRLLEHYLR	AGT
MDLA_PENCA	305	Mono- and diacylglycerol lipase precursor (EC 3.1.1.-)	*Penicillium camembertii*	PCMDGL	171	PNYELVVVGHSLGAAVATLAATDLR	AGC
MSAS_PENPA	1774	6-Methylsalicylic acid synthase (EC 2.3.1.-)	*Penicillium patulum*	PPMSAS	653	GITPQAVIGHSVGEIAASVVAGALS	TCC
PCD_ARTOX	493	Phenmedipham hydrolase (EC 3.1.1.-)	*Arthrobacter oxidans*	AOPMPH	188	DPNRITLVGQSGGAYSIAALAQHPV	TCA
PHAB_PSEOL	283	Poly(3-hydroxyalkanoate) depolymerase (EC 3.1.1.-)	*Pseudomonas oleovorans*	POPHABC	102	DYGQVNVIGVSWGGALAQQFAHDYP	TCT
TCBE_PSESP	238	Carboxymethylene-butenolidase (EC 3.1.1.45)	*Pseudomonas* sp.	PSTCBCDE	123	CDGGVAVIGYCLGGALAYEVAAEGF	TGC
TFDE_ALCEU	234	Carboxymethylene-butenolidase (EC 3.1.1.45)	*Alcaligenes eutrophus*	PPTFDCD	113 123	AIEYARALPFSNGRVAVVGYCLGGA SNGRVAVVGYCLGGALAFDVAARSL	TCG TGC
TODF_PSEPU	276	2-Hydroxy-6-oxo-2,4-heptadienoate hydrolase (EC 3.1.1.-)	*Pseudomonas putida*	PPF1HYOX PPTODC1C	105	ELDRVDLVGNSFGGALSLAFAIRFP	TCG
TPES_PSEPU	272	Tropinesterase (EC 3.1.1.10)	*Pseudomonas putida*		110	GLHNTTVIGHSMGSMTAGVLASIHP	TCG
VIT1_CERCA VIT1_DROME	437 439	Vitellogenin I precursor Vitellogenin I precursor	*Ceratitis capitata* *Drosophila melanogaster*	CCVG1G DMYOLK	176	NGNKDYDYGSSQGNQGATSSEEDYS	AGC
VIT2_CERCA VIT2_DROME	422 442	Vitellogenin II precursor Vitellogenin II precursor	*Ceratitis capitata* *Drosophila melanogaster*	CCVG2G DMYP2	334	ANVDFFPNGPSTGVPGADNVVEATM	TCG
VIT3_DROME	420	Vitellogenin III precursor	*Drosophila melanogaster*	DMYP3 DMYP3G	289 323	SKRPQILGGLSRGDADFVDAIHTST GDVDFYPNGPSTGVPGSENVIEAVA	TCC TCC
XYLF_PSEPU	281	2-Hydroxymuconic semialdehyde hydrolase (EC 3.1.1.-)	*Pseudomonas putida*	PPXYL	106	GIQQGDIVGNSFGGGLALALAIRHP	TCG

Table 1 (continued)

Id	Len	Description	Source	EMBL	PDB	Pos	Ser	Codon
XYNC_CALSA	266	Acetyl esterase (EC 3.1.--)	Caldocellum saccharolyticum	CSXYNAB		125	KREKTFIGGLSMGGYGALRNGLKYN	TCA
YHLB_VIBCH	171	Hypothetical 18.3 kDa protein	Vibrio cholerae	VCHLYB		137	GAKKVNLIGHSHGGPTIRYVASVRP	AGT
CHEB_ECOLI	349	Protein-glutamate methyl-esterase (EC 3.1.1.61)	Escherichia coli	ECCHE3		164	SSEKLIAIGASTGGTEAIRHVLQPL	TCA
FRZG_MYXXA	334	Protein-glutamate methyl-esterase (EC 3.1.1.61)	Myxococcus xanthus	MXFRZGF		260	CDMLLTSAGESFGPRCIGVILTGMG	TCG
SAST_ANAPL	251	S-Acyl fatty acid synthase thioesterase (EC 3.1.2.14)	Anas platyrhynchos	APSFAST APFASA		90	QEKPFALFGHSFGSFVSYALAVHLK	AGT
SAST_RAT	263	S-Acyl fatty acid synthase thioesterase (EC 3.1.2.14)	Rattus norvegicus	RNMCH RNTER		101	QDKAFAFFGHSFGSYIALITALLLK	AGT
SAST_VIBAN	252	Probable S-acyl fatty acid synthase thioesterase (EC 3.1.2.14)	Vibrio anguillarum	VAANGRA VAANGRG		92	RIEDTIIVGHSMGAQVAYEASKKLV	AGT
BAH_STRHY	299	Acetyl-hydrolase (EC 3.1.--)	Streptomyces hygroscopicus	SHBAHBRP		143	PPGRVTLAGDSAGAGLAVAALQALR	TCG
ERY3_SACER	3170	Erythronolide synthase, modules 5 and 6 (EC 2.3.1.94)	Saccharopolyspora erythraea	SEERYAB		299	RGSAVNQDGASNGLTAPSGPAQQRV	AGC
						642	GVEPAAVVGHSQGEIAAAHVAGALT	TCG
						1757	AGSAINQDGASNGLAAPSGVAQQRV	AGC
						1814	ASALLATYGKSRGSSGPVLLGSVKS	TCG
						2107	GVSPSAVIGHSQGEIAAAVVAGVLS	TCG
						3029	GDKPFVVAGHSAGALMAYALATELL	TCC
GRST_BACBR	256	Gramicidin S biosynthesis GRST protein	Bacillus brevis	BBGRSTAB BBGRSAB		95	INIPFAFLGHSMGALISFELARTIR	AGC
TGL1_YEAST	548	Triglyceride lipase-cholesterol esterase (EC 3.1.1.-)	Saccharomyces cerevisiae	SC82KBXIA SCYKL140W		201	KVDKVICIGFSQGSAQMFAAFSLSE	TCT

Id	Len	Description	Source	EMBL	PDB	Pos	Ser	Codon
YEIG_ECOLI	278	Hypothetical 31.3 kDa protein	Escherichia coli	ECU00007		145	VSDRCAISGHSMGGHGALIMALKNP	TCA
S37291	358	Triacylglycerol lipase (EC 3.1.1.3)	Pseudomonas glumae	PGLIP	1TAH	126	GATKVNLIGHSQGGLTSRYVAAVAP	AGC
PDB1TCA	317	Candida antarctica B lipase	Candida antarctica		1TCA 1TCB 1TCC	105	GNNKLPVLTWSQGGLVAQWGLTFFP	

Most data are taken directly from the SwissProt data files. Id, identification code in SwissProt; Len, length of sequence; > is used to indicate fragments of sequences; Description, a short description of the protein; Source, source organism for the protein; EMBL, entry codes for corresponding DNA sequences; PDB, entry codes for corresponding structure entries; Pos, sequence position of Ser in active site type motif; Ser, Ser type active site motif (GXSXG), or related motifs; Codon, codon of corresponding Ser, all motifs with an active site like motif are shown, including carboxymethylene butenolidase sequences with an GXCXG motif. Included in the motif is 10 residues before and 14 residues after the Ser; in the available 3D structures this corresponds approximately to a loop–strand–loop–helix–loop construct, where the active Ser is located in the second loop. This structure seems to be very well conserved in all esterases and lipases.

erides than triglycerides. The lipase from *P. cyclopium* belongs to this class whereas the lipase from *P. expansum* is non-specific [17].

Using the enantiomeric substrates, *R*- and *S*-*p*-nitrophenyl 2-methyl decanoate dissolved in water, it has been shown that *Rhizomucor miehei* and *Candida cylindracea* have different enantioselectivities. Whereas *R. miehei* and *C. cylindracea* display very similar substrate binding constants, the observed reaction velocities differed markedly, thus resulting in a specificity for the *R* for the *R. miehei* and for *S* for the *C. cylindracea* [18].

The Known Lipases and Esterases

The 130 lipases, esterases or related proteins for which the amino acid sequence have been determined or deduced from the gene sequence are given in Table 1. The sequence length, description and source are also given. In addition when available, the corresponding gene sequence is given in the EMBL column. If a 3D structure has been determined, the PDB code is also given. Since most of the proteins dealt with in the present paper are believed to be serine hydroxylases, the local context around the active site SER is given together with the codon usage for the active site SER. In some cases, it may be interesting to relate the chromosomal location of the various lipases and esterases to specific chromosomes. A tabulation of the chromosomal location of some proteins is given in Table 2.

Distinct Families of Lipases and Esterases

With the large number of lipase and esterase sequences available (see Table 1), it

Table 2. Chromosomal location: selected mammalian lipases, esterases and colipases and their chromosomal location

Protein	Organism	Reference	Chromosome
Pancreatic lipase	Mouse	[124]	19
Pancreatic lipase	Human	[125]	10
Colipase	Mouse	[124]	17
Colipase	Human	[126]	6
Colipase	Human	[125]	6
Hepatic lipase	Mouse	[124]	9
Apolipoproptein A-I	Mouse	[124]	9
Lipoprotein lipase	Mouse	[124]	8
Hormone-sensitive lipase	Mouse	[124]	7
Apolipoprotein C-II	Mouse	[124]	7
Apolipoprotein E	Mouse	[124]	7
Acetylcholinesterase	Human	[128]	7
Buturylcholinesterase	Human	[128]	3
Carboxyl ester lipase	Mouse	[124]	2
Hormone sensitive lipase	Human	[129]	?

Fig. 7. Sequence alignment of the lipoprotein lipase family. Sequence alignment of the lipoprotein lipase family, showing the active site Ser at position 183. For a description of how the alignment was generated, see Fig. 8. All sequences are given by their SWISSPROT entry code. Sequence numbers follow LIPH_HUMAN, human hepatic lipase.

becomes feasible to identify conserved sequence features and also to relate these to a 3D structure of one of the proteins. In Fig. 7 the lipoprotein lipase family has been aligned in a region including the active site SER (SER183 in human hepatic lipase; includes precursor sequence). In particular in the vicinity of the active site SER all 16 proteins display a large area of sequence identity. It is also apparent in this alignment that the lipoprotein lipases are more closely related to the hepatic lipases than to the pancreatic. The 7 pancreatic lipases all have a His preceding the active site Ser, whereas both the hepatic and lipoprotein lipases have a Tyr at this position. The sequence for the pancreatic lipase from mouse appears unique. In several locations it differs in residues which otherwise are totally conserved among the 15 other lipases given in Fig. 7. It has Pro instead of a Gly in the active site Ser context at position 181, it has a Leu in position 155, where all others have valine and it has an Ile at position 159. All sequence positions relate to LIPH_HUMAN.

In Fig. 8 the local sequence context around the active site serine in 42 esterases has been depicted. The majority of these sequences display the characteristic GESAG motif around the active site serine. However, in several of the carboxyl esterases the glutamic acid has been substituted with another large hydrophilic residue, e.g. histidine, glutamine or tyrosine. Interestingly, if one searches the sequence data base for exact matches with the consensus motif, W***N***FGG*P, neurotacin from *Drosophila* (NRT_DROME) also shows this motif. By inspection, it also displays the two glycines around the active site serine, but the serine itself has been substituted with arginine. In the SWISSPROT entry, it is noted that homology to thyroglobulin exists.

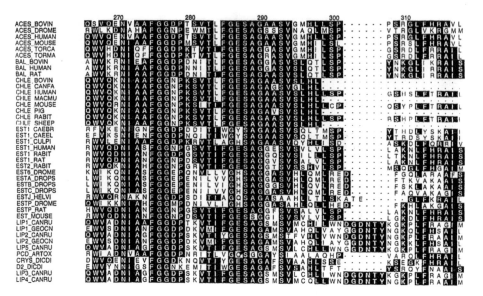

Fig. 8. Sequence alignment of the esterase family. Alignment of esterases, showing the active site Ser at position 288 in the alignment. A well conserved FGGXP motif seems to be important for formation of the initial loop before the β-strand. The alignment was generated with Clustal [108] and printed with Alscript [109].

Since the 3D structure is known for some of the esterases, an interesting analysis is possible: we can align the sequence family of esterases and compare the number of accepted amino acid substitutions at a given sequence position with the solvent accessibility of the corresponding residue in the 3D structure. In Fig. 9 it can be seen that the solvent accessibility correlates strongly with the number of accepted amino acid substitution at any given residue. Also, a very distinct insertion and deletion (INDEL) pattern is observable, as shown in the bottom trace of Fig. 9. The INDELs seem to correlate well with the surface accessible regions. This type of analysis is of course very dependent upon a proper sequence alignment, since any mis-alignment will give rise to erroneous mutation counts at a given sequence location.

From Fig. 9 it is also apparent that few residues are totally conserved in the esterase family. The accepted mutations (in green) show that less than 50 residues are partially conserved and only 28 totally conserved. In Fig. 10 we have mapped the most conserved residues onto the 3D structure of acetylcholinesterase from *Torpedo california*. The spatial location of many of these residues are predominantly in the loops connecting to the central β-sheet, and a preference for glycine and proline is noted. These residues are typically found in loop regions, but it is surprising that they seem to be so well conserved.

In order to investigate the general sequence similarities that exist between the various lipases and esterases, we have made all possible pairwise comparisons and

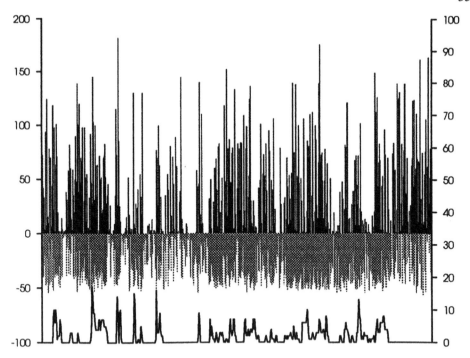

Fig. 9. Mutation and deletion analysis of the esterase family. Correlation between accessibility, variability and insertion frequency for acetylcholinesterase (1ACE). The data were taken from the corresponding HSSP file [110]. The upper blue curve is the solvent accessibility, computed from the 1ACE structure. The middle green curve is the sequence variability, computed from a multiple sequence alignment. The lower black curve is the number of insertions and deletions at a specific position, computed from the same alignment. A clear correspondence between solvent exposure, high variability and high probability of insertions can be seen.

have scored them with respect to the number of amino acid identities resulting from automatic alignment. The result is shown graphically in Fig. 11. The picture highlights the closest families, such as the lipoprotein lipase family, and the esterase family. It also depicts clearly the large number of sequences, for which the relationship to the other proteins are less clear, although several regions with identities above 20% have been found. It is possible that more exhaustive analysis based upon similarity and not identity, as well as the use of randomized sequence comparison in order to determine the noise background could improve the sensitivity of the analysis. However, the necessary CPU time for accomplishing such a task is at the moment prohibitive for us.

Colipases and apolipoproteins

Both the pancreatic and the lipoprotein lipases require the presence of another protein based molecule in order to acquire full enzymatic activity. In the case of the human pancreatic lipase, the colipase interacts with the small domain of the lipase, and is involved in the major conformational changes around the lid associated with

336

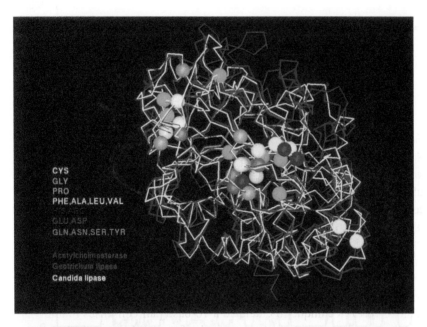

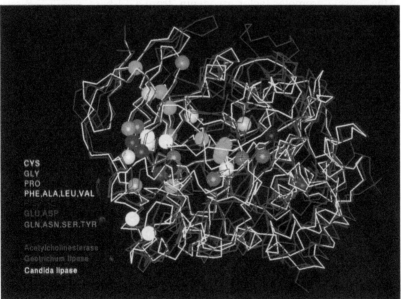

Fig. 10. Mapping of conserved residues onto the 3D structure of acetylcholinesterase. The top picture shows aligned C$_\alpha$ traces of esterase-like structures. In the central region of the molecules we are looking into the active site, where the active His can be seen. Residues that are highly conserved in a multiple alignment of several esterase sequences (Fig. 9) are shown as colored spheres. In the bottom part of the picture the structures have been rotated 90° around the vertical axis, so that the entrance to the active site is from the right-hand side of the picture. It can be seen that the conserved residues are located in two regions, at each end of the structurally conserved central β-sheet. A number of conserved Gly and Pro residues are probably important for stabilizing the protein fold in this region.

interfacial activation of the lipase. The known colipases are given in Table 3, together with their sequence database entry code.

In Fig. 12 we have shown a sequence comparison matrix for a large set of colipases and apolipoproteins. Similar to the analysis for Fig. 11, here we display the

a

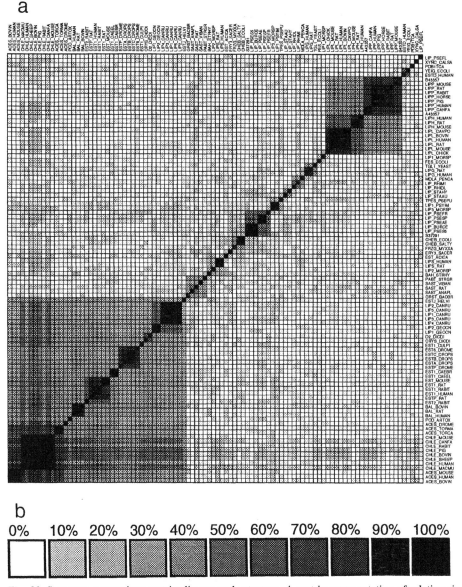

b

0%	10%	20%	30%	40%	50%	60%	70%	80%	90%	100%

Fig. 11. Sequence comparison matrix: lipases and esterases. A matrix representation of relative similarities between lipases, esterases and related sequences. The similarities are based on pairwise alignments generated with the Align program of the Fasta package [111,112], and are computed as the percentage of identical residues based on the length of the shortest of the two sequences. In particular the esterases and several lipases can be seen as well-defined groups.

338

Table 3. Colipases: the known colipase amino acid sequences including their SWISSPROT codes as well as the length of the determined sequence

Protein	Code	Length	Organism	Reference
Procolipase II	COL2_PIG	96	Pig	[130]
Procolipase A	COLA_HORSE	97	Horse	[131]
Procolipase B	COLB_HORSE	97	Horse	[132]
Colipase precursor	COL_CANFA	113	Dog	[133]
Colipase	COL_CHICK	35 (fragment)	Chicken	[134]
Colipase precursor	COL_HUMAN	113	Human	[130]
Colipase precursor	COL_RAT	113	Rat	[136]
Colipase	COL_SQUAC	40 (fragment)	Spiny dogfish	[135]

similarity between the different proteins measured by the number of amino acid identities seen in the pairwise alignments. The matrix clearly identifies the apolipo-proteins as one family and the colipases as another. Inside the apolipoprotein family, the different sub-families are well defined and their homologies between such sub-families are given by the dark gray background.

Although the approximate size of the colipase and apolipoprotein C-II (APO-C-II) is similar, no obvious sequence similarity has been reported previously. It is interest-ing to note that in the case of the lipoprotein lipases, the apolipoprotein C-II belongs to a larger family of apolipoproteins. The human colipase can be aligned using DOTMAT (Petersen, unpublished) with human APO-C-II as shown in Fig. 13 using a windowsize of 30 and a cut off of 7.0. There is a 30 AA region of homology in the central part of the sequences:

```
XLHU     GIIINLENGELCMNSAQCKSNCCQHSSALGLARCTSMASENSECSVKTLYGIYYKCPCE
LPHUC2   MGTRLLPALFLVLLVLGFEVQGTQQPQQDEMPSPTRLTQVKESLSSYWESAKTAAQNLYKI---TCE
           ...  .  *.   .    . . .... ...*.  *... ....**. .   **

XLHU     RGLTYLPAVDEKLRDLYSKSTAAMSTYTGIFTDQVLSVLKGEE*
LPHUC2   GDKTIVGSITNTNFGICHDAG*
           *.. .. .    . ....
```

The positions labeled with a '.' are regarded as similar, whereas those labeled with a '*' are identical. The number of identities is very small, but the length of the stretch of uninterrupted homologous residues may indicate that the two proteins share a common ancestor. However, the data presented here are far from conclusive. Curiously, in the case of APO-C-II it has been reported that the protein by itself acts as an esterase. Synthetic fragments of the APO-C-II also displayed such an activity [19].

Sequence repeats in apolipoproteins

An interesting feature is present in some of the apolipoproteins, in particular in the apolipoprotein 4 from human (APA4_HUMAN). The protein is 396 AA long, and a DOTMAT analysis revealed extensive self-similarity in the sequence, as judged by

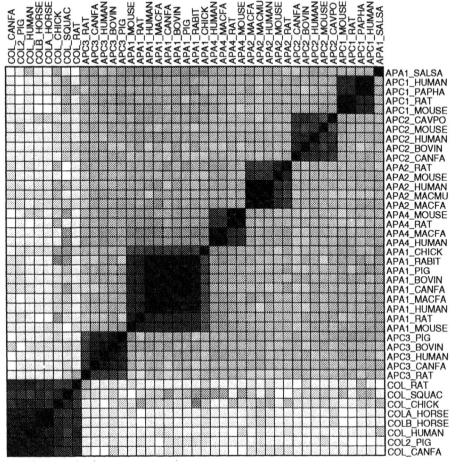

Fig. 12. Sequence comparison matrix: colipases and apolipoproteins. A matrix representation of relative similarities between colipases and apolipoproteins. For a description of the generation of this matrix, see Fig. 11. The apolipoproteins show a global similarity as well as several groups with high internal similarity.

the massive occurrence of lines parallel to the diagonal of the plot in Fig. 14A. In order to document the apparent size of the repeat, a Fourier analysis of the data underlying the graphical presentation in the dot matrix was performed. This is shown in Fig. 14B. Here the distinct peak at channel 281 corresponds to an average repeat size of 3.7 amino acids. This number is obviously not an integer, and this may either be interpreted as a result of repeats of non-uniform length, or alternatively that an extensive amphipathic α-helical structure forms a large part of the structural scaffold of this protein. The latter can be deduced from the fact that α-helices have 3.6 residues per turn, and if they are located at the interface between the protein interior and the solvent, they will display an alternating pattern of hydrophobic and hydrophilic residues.

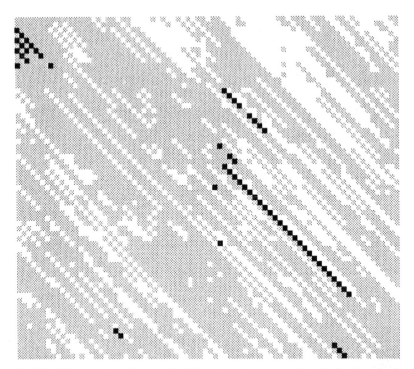

Fig. 13. Colipase and apolipoprotein C-II are homologous proteins. In the figure is shown a DOTMAT (Petersen S.B., unpublished method) alignment of human pancreatic colipase with apolipoprotein CII, also from human. The averaging window used for this comparison was 30 AA and a low threshold (7.0) was set to highlight specifically weak homologies. A clearly defined homologous region can be seen from position 21 in the human colipase (XLHU) corresponding to position 30 in human apolipoprotein CII (LPHUC2). Due to the averaging used in the DOTMAT approach, the C-terminal parts are truncated at sequence length – windowsize, thus the matrix shown in the figure has the dimensions (101–30) × (86–30); see also text for further details.

If indeed the first interpretation is true, then the fact that such extensive repeats are visible can be taken as an indication that the evolutionary history of this protein is relatively brief. One would expect that such repeats would be modified in nature's continuous effort to optimize structure and function. In particular one would expect the folding and the overall stability of such a protein to be poorly optimized in this protein.

Evolution

In order to illustrate the various relationships that exists at the sequence level between the different lipases, esterases as well as related proteins, we have constructed a non-rooted tree based on 116 such sub-sequences, all selected around the active site serine. The overall structure of the tree is given in Fig. 15A and detailed information is shown in Fig. 15B. The schematic overview in Fig. 15A highlights the distinction into the two classes, lipases and esterases, and indicates that other proteins, although having different function, are related to these two classes, e.g. protein glutamate methyl esterase and sterol acyl transferases. The figure also attempts to relate the protein groups inside a class: in the esterase class the mammalian acetylcholinesterases are closely related to the cholinesterases as well as the bile-salt stimulated lipases. Similarly in the lipase class, the lipoprotein lipases are most closely related to the hepatic lipases, and less closely to the pancreatic lipases. These three protein groups together make the lipoprotein lipase family. The gastric lipases are distinct from the other mammalian lipases. Cutinases form a very homologous group, whereas the microbial lipases show considerable spread, although for clarity they have been put in one category in Fig. 15A.

In Fig. 15B the full picture is given. In addition to the entry code from either

a

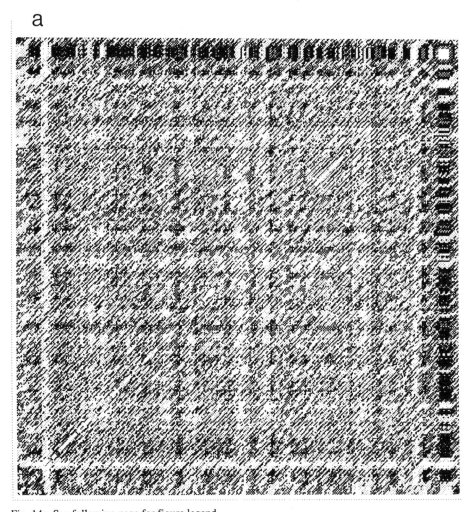

Fig. 14a. See following page for figure legend.

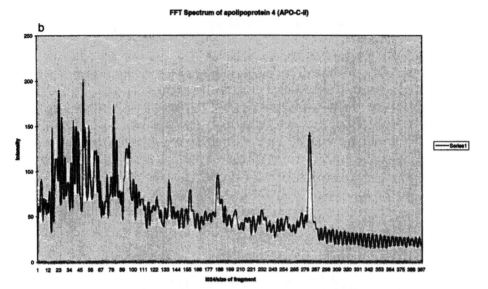

Fig. 14. Sequence repeats in apolipoprotein 4. In the figure is shown the remarkable tandem repeat features found in APO-C-II. In (A) the DOTMAT representation is given, clearly showing the repeat features as bands parallel to the diagonal. In (B) the Fourier transform spectrum is given, confirming the graphical representation in (A); an intense and distinct peak is observed slightly below 4.0 AA fragment size corresponding to a frequency of 1024/280.

SWISSPROT or PIR, the different entries have been color coded with respect to their codon usage for the active site Ser. In previous work, we have proposed that a specific pattern exists for the active site Ser codon usage with a specific lipase and esterase codon usage. The present compilation of data does not warrant any clear differentiation into such groups. Rather its seems that most, but not all of the mammalian lipases and esterases use the AGY type of codon for serine, whereas many, but again not all of the microbial lipases use the TCN type of codon.

Experimentally Determined 3D Structures of Lipases and Esterases

In 1989 no 3D structures were known for either triglyceride lipases or esterases. In part due to a sudden commercial interest in these enzymes, our knowledge has shown a remarkable growth. In 1994 we have public access to more than 10 3D structures (see Table 4), all of them solved by X-ray crystallography. In the case of the Rhizomucor lipase, four structures have been deposited in the protein data bank (PDB), of which two have inhibitors bound to the active site [20–22]. Two mammalian pancreatic lipases, from horse and human, have also been solved [23–25]. In Fig. 4, the open and closed lid structures of the *Rhizomucor miehei* are shown; these were obtained by co-crystallizing the lipase with a phosphate based inhibitor.

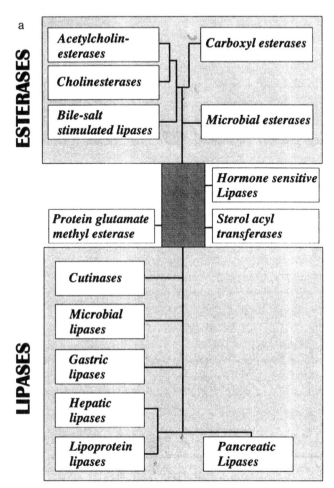

Fig. 15. Evolutionary analysis. An evolutionary analysis of a 13 AA wide context around the active site serine. In (A) the general layout of the relationship between the 113 sequences are given and in (B) the full details are shown. Each sequence is coded with respect to the codon usage for the active site serine. Note: the species *Candida rugosa* uses a different codon for serine. The figure displays a possible relationship between several lipase and esterase sequences, based on the active site subsequences shown in Table 1. The sequences believed to represent the active site region were aligned (without gaps), and a tree structure was generated using the PROTPARS and DRAWTREE programs of the PHYLIP package (version 3.5) [113]. The entry codes are color-coded according to codon usage for the active Ser, red is AG type codon, green is TC type codon, light blue other or unknown codon type. This data set should probably not be analyzed in terms of strict evolutionary relationships, as a large set of very disperse proteins are compared based on a short subsequence. However, it can be seen that the data set can be grouped into very much the expected subclasses; esterases, lipases and cutinases. It can also be seen that although codon usage seems to be well conserved within closely related sequences, there is no clear selective codon usage within all lipases or all esterases.

b

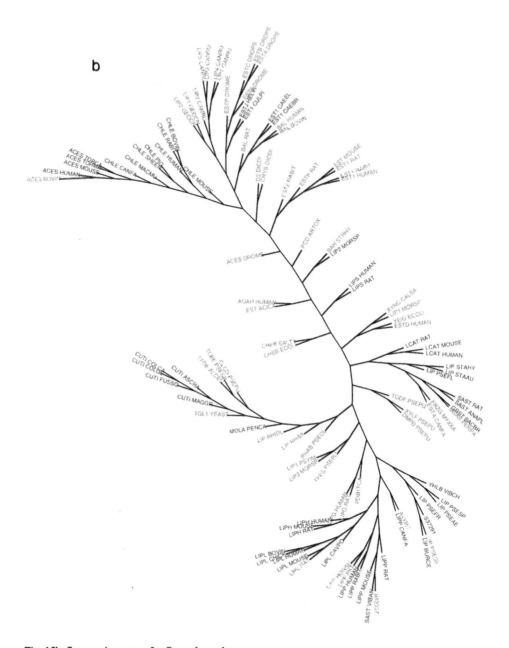

Fig. 15b. See previous page for figure legend.

Common Features Between Lipases and Esterases

In order to investigate to what extent the lipases and esterase share structural features, we have compared seven 3D structures (see Table 5): *Candida antarctica B* lipase, *Pseudomonas glumae* lipase, *Rhizomucor miehei* lipase, *Fusarium solani*

Table 4. Crystal structures of lipases and esterase

Protein	Entry	PDB	Resolution	Reference
Acetylcholinesterase				
w/acetylcholine	ACES_TORCA	1ACE	2.8	[114]
Candida rugosa lipase	LIP1_CANRU	1CRL	2.06	[115]
Horse pancreatic lipase	LIPP_HORSE	1HPL	2.3	[23]
Pseudomonas glumae	S37291	1TAH	3.0	[116]
Candida antarctica B		1TCA	1.55	[117]
		1TCB	2.1	[117]
		1TCC	2.5	[117]
Rhizomucor miehei lipase	LIP_RHIMI	1TGL[a]	1.9	[20]
Rhizomucor miehei lipase	LIP_RHIMI	3TGL	1.9	[22]
Rhizomucor miehei lipase				
w/diethylphosphate	LIP_RHIMI	4TGL	2.6	[118]
Rhizomucor miehei lipase				
w/*N*-hexylphosphonate				
ethyl ester	LIP_RHIMI	5TGL	3.0	[21]
Geotricum candidum lipase	LIP2_GEOCN	1THG	1.8	[119]
Candida rugosa lipase	LIP1_CANRU	1TRH	2.1	[120]
Human pancreatic lipase	LIPP_HUMAN	?	2.9	[121]
Human pancreatic lipase	LIPP_HUMAN	?	3.04	[25]
	fCOL2_PIG			
Human pancreatic lipase	LIPP_HUMAN	?	3.0	[24]
w/colipase and	COL2_PIG			
phosphatidylcholine				
Pseudomonas sp B 13				
Dienelactone hydrolase	CLCD_PSEPU	?	1.8	[122]
Fusarium solani cutinase	CUTI_FUSSO	?	0.9	[123]

The table lists the published crystal structures of lipases and esterases. The majority have been deposited in PDB, and their entry codes are given. The sequence database code for the protein(s) in a structure is also given in column two.
[a]C_α-coordinates only.

cutinase, human pancreatic lipase, acetylcholinesterase and *Geotricum candidum* lipase. *Candida antarctica B* lipase was used as a reference. Table 5 shows the RMS values obtained when superimposing the protein backbone of the shared structural

Table 5. Comparison of the 3D structures of 7 lipases and esterases

Protein	Active site	Residues aligned	RMS for range	Length of protein
Candida B	TWSQG	–	–	317
Human pancreatic lipase	GHSLG	125	2.70	590
Pseudomonas glumae	GHSQG	135	2.79	318
Rhizomucor miehei	GHSLG	92	2.95	269
Acetylcholinesterase	GESAG	130	2.64	534
Geotricum candidum	GESAG	134	2.86	544
Fusarium solani cutinase	GYSQG	115	2.19	213

A tabulation of a structural comparison between various lipases and esterases for which the 3D structures are known. The RMS value indicates how well two 3D structures could be overlapped. The number of residues included in the comparison is also given; typically more than 100 residues could be used for the comparison.

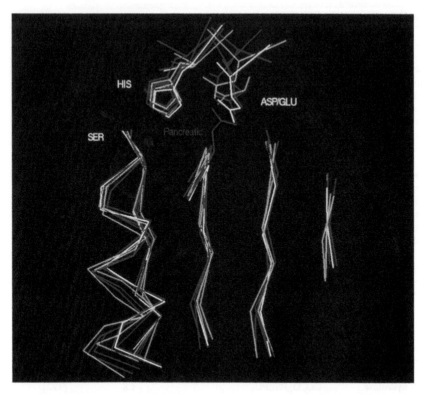

Fig. 16. Superposition of the 3D structures of 7 homologous lipases and esterases. Selected backbone fragments have been superimposed using the active site Ser local context as a guide.

features. The number of aligned residues varies since the key message of the table is to show how extensive the structural homology is. In Fig. 16 the α-carbon trace of seven lipases have been superimposed graphically. It is interesting to see how closely the Ser and the His of the active sites match, whereas the Asp (or Glu) does not. The active site Asp migrates to a totally different position in the human pancreatic lipase [26].

Electrostatics

The molecular structure and composition is the basis for all aspects of protein function. The relationship between structure and function may be subtle: although a global structure compatible with a given function is necessary, the detailed nature of the latter is often predominantly caused by a few sites, directly involved in the recognition and/or catalytic roles. That is, structure is the support of function only because it supports the atomic framework responsible for the forces that drive all functional processes. Therefore, a full understanding of protein function is only possible if the role of such forces can be understood in each particular case.

All protein functional processes share the common feature that they depend on the encounter of two or more molecules: substrate-enzyme, ligand-receptor, etc. Electrostatic forces are the strongest long-range forces acting at molecular level and they can play an important role in this diffusional encounter, either by steering or orienting the approaching molecule (see [27,28] and references therein). An additional requirement in the case of enzymatic function is that the protein has to stabilize the reaction intermediate (otherwise it would not act as a catalyst, in this case an enzyme). This stabilization also seems to be mainly of electrostatic nature [29]. Therefore, electrostatic interactions seem to play a particularly important role in protein function, and their proper modeling is crucial to understanding that function and also to designing alternative engineered forms.

This section gives a brief overview on protein electrostatics, for which several good reviews exist [27,30–35]. The existing methods to model electrostatic interactions in proteins are briefly described, together with the closely related problem of calculating pH-dependent properties. Finally, a study on the pH-dependent electrostatics of several lipases is presented.

Molecular dynamics

The most familiar general method in the simulation of biomolecular systems is Molecular Mechanics, either as Molecular Dynamics (MD) or energy minimization (for a recent review on MD and related simulation methods in chemical and biochemical systems, see [36]). In this type of model, electrostatic interactions are usually represented using a simple coulombic form (q_1q_2/Dr_{12}), which is just one of the several possible types of interaction terms usually considered (bond, angle, torsion, van der Waals, etc.), from which the energy and forces acting on the system can be computed. In principle, a MD simulation of a fully represented protein system, including water molecules and solvent ions, should give a reasonably complete description of the system. This would include not only average quantities, as the protein mean conformation, but also more detailed information, such as the different conformations, the magnitude of the observed fluctuations, the time order of global displacements, etc.

Unfortunately, the simulation time required to equilibrate the system and properly sample the possible conformations makes MD a computationally demanding technique. This leads to the search for time-saving techniques, including reduced accuracy in the modeling of the system. A typical approximation is to neglect non-bonded interactions over a given cut-off distance. However, although valid for the short-ranged van der Waals forces, this procedure can lead to serious miscalculations when applied to long-range electrostatic forces [37].

Another way of simplifying the system is by excluding the solvent from the simulation. Here again the effect on electrostatic interactions is particularly serious, since, in the absence of water molecules, the polar and charged groups on the protein surface can only be stabilized by turning into the protein itself during the simulation, thus distorting the protein surface. A less drastic solution is to simulate the screening effect of water by using a dielectric constant greater than 1, usually proportional to the distance between the interacting atoms. Several other methods have been proposed [38–41] which, although considering the screening effect, can never describe the hydrogen bond network characteristic of the protein in solution (basically also of electrostatic nature), that may be essential to its functional role.

Although the above simplifications may lead to a miscalculation of electrostatic interactions (with the consequences thereof), a sensible choice of methodology can give a qualitatively good description of the simulated protein. However, an aspect that is hardly included in MD simulations is the fact that titratable residues in a protein are continuously being protonated and deprotonated in solution, at rates dependent on the pH of the solution. Actually, the temporal order of such a process is still clearly beyond those achieved by MD and, its simulation would have to include the reaction process itself, which is still not possible. The only obvious way to include pH effects is by choosing charges that are intuitively consistent with the pH considered; e.g. at pH 7 that would mean using the charged form of all titratable residues except maybe for histidine and N-terminal ones, whose typical pK_a values are close to that value. Unfortunately, this is not necessarily the predominant form existing at the pH considered, since it is known that residue pK_a values in proteins may be shifted by several units from typical values. An alternative procedure suggested by Gilson [42] is discussed (vide infra).

In conclusion, MD simulations can give valuable information about the structural and dynamical properties of a protein molecule, at the expense of heavy or moderated computer power, depending on the detail and accuracy of the modeling. However, the usual time-saving model simplifications are particularly severe on electrostatic interactions. Moreover, the inclusion of pH effects is approximate at best. Keeping in mind that there is no 'best' modeling method, we now look at other methods, whose underlying models greatly simplify the atomic description and focus on other aspects.

Electrostatic continuum models

With electrostatic continuum models, a clear departure is made from the MD approach. The atomic detail is reduced as much as necessary and all the system is treated as a static entity. Instead, the laws of classical electrostatics are assumed to be valid at the molecular level, with protein and solvent treated as dielectric materials where point charges exist. The idea is to implicitly consider all electrostatically relevant mobility through the use of an appropriate value for the dielectric constant of the solvent and protein regions. The dielectric constant reflects the reorientation of dipoles caused by an electric field. These dipoles are usually of two types: permanent and induced. Permanent dipoles exist when the charge is not equally distributed among nearby atoms, as in the peptide bond or the water molecule. Induced dipoles occur due to the deformability of electron clouds by an electric field (electronic polarizability). The dielectric constant of water is mainly due to the reorientation of the permanent dipoles, each of them associated with a relatively free molecule. In the case of a protein, however, most atoms have relatively fixed positions and so it is believed that the dielectric constant is mostly or exclusively due to induced dipoles. The typical values used by electrostatic continuum models are the macroscopic measured value for water (~80) and 2–4 for the protein [43]. Eventually, ionic strength effects can be included by considering that the distribution of solution counterions around the protein is dictated by the electrostatic environment created by the protein (this region with counterions only starts at a given distance from the protein surface, since the ions cannot approach more than allowed by their radii), like in the Debye-Huckel theory of electrolytes. The general picture of the protein in solution emerging from this type of model is the one shown in Fig. 17. It consists on a completely formulated electrostatic 'scenario', which in principle allows for the calculation of the electrostatic potential and forces everywhere in the system (the problem can be reduced to the solution of the Poisson or, if counterions are considered, Poisson–Boltzman equations [44]). However, the actual computation was not a trivial matter before the advent of computers, which led to simplifying approximations of this general model.

The most radical approximation is to assume that $D_p = D_s$ and ignore the counterions, which actually reduces the problem to a trivial calculation of coulombic interactions. The problem is that the use of a low value for the single dielectric constant overestimates the interaction of charges near the protein surface, and the use of a high one underestimates the interaction of buried ones. Thus, even though the use of an intermediate value may provide same qualitative results, the use of this severe approximation should be avoided if other methods are available.

Another approximation that greatly simplifies the calculations is to assume a spherical shape to the protein, as originally done by Tanford and Kirkwood [45], which yields an analytical solution for the problem. Some methods were proposed to fit real, irregularly shaped proteins in this model [46,47]. Unfortunately, the Tanford-Kirkwood model was developed before protein crystallography by assuming titratable residues to be located at the surface and, although perfectly consistent with its assumptions, there are conceptual limitations that preclude its application to proteins

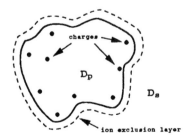

Fig. 17. A simplified representation of the electrostatic environment of a protein. A protein has a number of charges distributed, predominantly on the solvent accessible surface. The interior of the protein is assumed to behave as if it had one dielectric constant, D_p that is different from the solvent dielectric constant D_s. In the calculations all ions contained in the solvent are assumed to be excluded from direct interaction with the protein surface charges.

with buried charges. Despite these problems, the modified method of Shire et al. [46] has been successfully used in several studies (for a review see [31]).

The increase of computer power made possible the application of several numerical techniques to solve the Poisson or Poisson–Boltzman equations. These numerical techniques, being general, do not require any spherical or other symmetry conditions, and thus can solve the problem for any protein shape. In protein applications the most common in use is the technique of finite differences [48,49]. It corresponds to dividing the space using a grid and assuming that derivatives can be locally approximated by differences between the grid point values. The smaller the spacing between grid points, the better the approximation will be. The method permits a complete treatment of buried charges [50]. (For other numerical techniques applied to solve this problem in proteins, see e.g. [27].)

Among other quantities, electrostatic continuum methods allow us to compute total electrostatic energies. The difference of this electrostatic energy between two possible states of a protein molecule can be interpreted as the electrostatic contribution for the free energy change of the corresponding process. The two states may differ in charge, conformation, etc. In this way, one can estimate the electrostatic contribution for binding processes, conformational changes, alterations of pK_a values or redox potentials caused by mutations, etc.

An interesting application of continuum methods involves the use of the electrostatic potential around the protein, necessarily computed in some of the methods (e.g. finite differences). The electrostatic attraction or repulsion energy experienced by a charge is proportional to this potential, which then can be seen as steering the approach of any charged molecule. One can actually simulate this diffusional process by considering also the thermal and frictional forces due to the solvent, a methodology called Brownian Dynamics simulation. In this way rate constants for the encounter process can be computed [27,28]. A case that clearly shows the potentiality of this approach is its use in the successful design of a faster mutant of superoxide dismutase [51].

This electrostatic potential field can be visualized in several ways. The most common are by displaying surfaces where the potential is constant (equipotential surfaces) or lines corresponding to ideal charge trajectories under the action of such a field (field lines). A good example of the later is a recent study of acetylcholinesterase, where the visualization of field lines provides valuable insight on the catalytic role of a transient channel to the active site, revealed by MD simulation [52]. Equipotential surfaces correspond also to regions where a charge has a given electrostatic energy (which is simply the value of the charge times the electrostatic potential) and they are usually displayed in kT/e units (k being the Boltzman constant, T the temperature and e the unitary atomic, i.e. proton, charge). Since we know that a charged molecule in solution has a translational kinetic energy of $3kT/2$, we can roughly interpret a $3kT/2e$ surface as the one 'below' which a unitary charged molecule has not enough kinetic energy to escape (or to stay, if the charge and potential have the same sign). The use of equipotential surfaces is illustrated with several lipases in Figs. 20 and 21.

We finally refer to an alternative approach to continuum models, due to Warshel and Levitt [53]. The protein is modeled by including, in addition to partial charges (the ones responsible for permanent dipoles), an inducible dipole in each atom to represent its electronic polarizability. An iterative process is then used to obtain a consistent set of partial charges and dipoles. Solvent is usually represented as a grid of dipoles [30], which represents the water permanent dipoles. Although strictly discrete, this model is closely related to continuum ones, since the use of point dipoles is roughly equivalent to considering that the corresponding atoms are placed in regions of different dielectric constant. This induced dipole approach provided some insight on the catalytic mechanism of serine proteases [29].

pH-Dependent electrostatics

The continuum models we have described so far do not consider by themselves the effect of pH. They simply solve the electrostatic problem for a given set of charges. What those charges are and how they should depend on pH is not considered by the continuum approach itself. A protein with N titratable groups (C- and N-termini included) has 2^N possible charge sets. The properties of the protein in solution at a given pH depend on the relative amount of each of these forms. This would be simple to solve if we knew the pK_a values of each titratable group. However, as noted above, these can be shifted from typical values by several units, due to the protein environment. Even worse, the 'effective' pK_a of each titratable group is also affected by the particular charge in the others, so that it will change with pH.

As discussed above, if we consider two different protein states and compute their electrostatic energy difference, we can interpret it as the electrostatic contribution to the free energy of the process between the two. If the two states consist of different charge sets, the differences between them would be mainly electrostatic (considering the protein conformation to be unaltered) and we can use the electrostatic energy as the true free energy. By relating this with the protonation or deprotonation reaction between the two, one can compute their relative amounts at a given pH. The usual procedure is to choose a charge set (e.g. all groups neutral) and compute the relative amounts of all other forms [45,50]. However, although the result is formally simple, to treat all the 2^N cases by an electrostatic continuum method is not a trivial calculation, specially if one wants to perform the calculation at several pH values. The simplest approximation is to consider as independent the titration of the several groups [54,55], which yields a very fast calculation. Actually, most applications of the original or modified Tanford–Kirkwood method (vide supra) are used together with this approximation. More accurate methods have been proposed [42,55–57] and in principle one can obtain any desired precision, provided the necessary computer power is available.

The fractions of the possible charge sets thus obtained can now be used as weights to compute the average of any property related to the titration of the groups. The total charge of the protein, for example, can be computed at several pH values, which yields the titration curve of the protein. One can instead compute the individual mean

352

charge of each titratable group. In some cases virtually a single charge set will occur at a given pH, and it is reasonable to speak of 'the' charge set corresponding to that pH. Otherwise, if there is not a clearly predominant charge set, the best electrostatic

TITRA

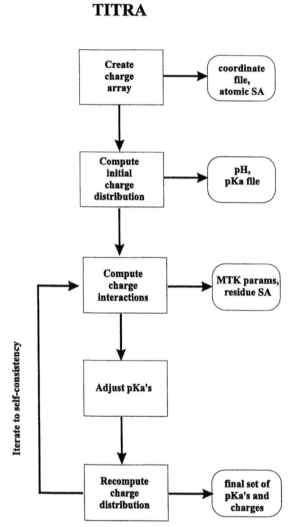

Fig. 18. pH calculations using TITRA: a flowchart. TITRA is a program that as input takes the atomic coordinates for the protein, as well as the solvent accessibility (SA) for the individual atoms. The titratable groups should be identified, an initial charge distribution is calculated using the given parameters of pH and intrinsic pK_a values. Using the SAs as well as implementing the modified Tanford–Kirkwood model (MTK) the charge–charge interactions are calculated. With these assumptions the electrostatic potential is then known at the locations of every titratable group on the protein. This allows us to modify the pK_a values, taking into account the effect of the electrostatic potential. The change in pK_a values results in a change in charge distribution and the new charge distribution is now used as input for a new calculation of the charge–charge interactions. This iterative process proceeds until the difference between one set of pK values and the previous one is below a threshold value.

picture one can have of the protein at a given pH is probably by using the mean charge of each titratable group. We can even use this mean charge set to do an additional continuum electrostatic calculation, which yields an average electrostatic potential field characteristic of the considered pH. Figs. 20 and 21 illustrate the use of this approach in several lipases.

We can now have a second look at MD simulations, in the light of this treatment of pH effects. If at a given pH there is a highly predominant charge set for our protein, we could choose that charge set to run a MD simulation characteristic of that pH value, as suggested by Gilson [42]. That is, instead of using the typical pK_a to decide the charges for the titratable groups, one can implicitly use the right ones by choosing the predominant state determined by the above procedure. But since we are going to perform a MD simulation, the protein conformation may change to some extent and one may ask if the original predominant charge set (calculated using a particular conformation) would still be valid. This is closely related to another problem, namely the fact that the usually available protein conformations are experimental ones obtained at specific pH conditions and its use in calculations corresponding to arbitrary pH values may be questionable. In fact, we know that conformation is dependent on pH, as acidic and basic denaturation illustrate. To overcome this problem we have recently developed a method that simultaneously considers the conformational freedom of the protein and the effect of pH (in preparation), which should help in future studies dealing with pH-dependent functional aspects.

Table 6. Comparison of measured and computed pK_a values in hew lysozyme

Residue	Measured	Calculated	Delta	Solvent accessibility
Asp 18	2.90	3.52	0.62	0.47
Asp 48	4.30	3.67	−0.63	0.67
Asp 52	3.60	3.76	0.16	0.20
Asp 66	2.00	3.04	1.04	0.00
Asp 87	3.62	3.40	−0.22	0.66
Asp 101	4.12	3.22	−0.90	0.42
Asp 119	2.50	3.78	1.28	0.74
Glu 7	2.60	4.11	1.51	0.76
Glu 35	6.10	6.38	0.28	0.11
His 15	5.80	6.15	0.35	0.27
Lys 1	10.80	10.42	−0.38	0.57
Lys 13	10.50	10.92	0.42	0.43
Lys 33	10.60	10.23	−0.37	0.21
Lys 96	10.80	10.62	−0.18	0.28
Lys 116	10.40	10.32	0.40	0.20
Tyr 20	10.30	10.54	0.24	0.20
Tyr 23	9.80	10.68	0.88	0.25
Tyr 53	12.10	11.14	−0.96	0.04
Ntr 1	7.90	8.30	0.4	0.20
Ctr 129	2.75	3.08	0.33	0.40

The measured values are from NMR experiments [55]. The delta column is the difference between the measured and calculated pK values.

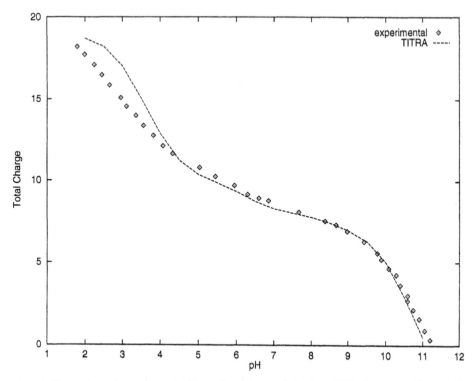

Fig. 19. Comparison of experimental with predicted titration behavior of HEW lysozyme. Using TITRA and the known crystal structure for hen eggwhite lysozyme, the computed and predicted titration behavior can be compared.

Electrostatic isopotential surfaces of lipases

In our laboratory we have developed a program, called TITRA, that allow us to compute charge distributions in a protein as a function of pH as well as of other physical chemical parameters. In Fig. 18 the computational strategies behind TITRA are given. Fig. 19 shows to what extent TITRA can predict the titration behavior of HEW Lysozyme. As is seen from the figure, the predictions are very closely matching the experimental values in the pH range from 4 to 11. In the range from pH 2–4 TITRA consistently predicts a higher charge than is actually observed. In this pH range, the carboxyl groups are titrating, and one can speculate whether small conformational changes in the protein are associated with this change. Since TITRA is computing the individual charges on each titratable group, a comparison is made between the TITRA data and NMR observations where the titration state of single residues can be followed in many cases [55]. In Table 6 such data are presented for 21 of the titratable groups in HEW lysozyme. In most cases the measured and calculated pK_a values are reasonably close with an average deviation of 0.56 pK_a unit, and only in 3 cases the deviation exceeds 1 pK_a unit.

In Fig. 20 TITRA has been used to generate the charge distribution as a function of pH for *Candida antarctica B*. In the figure are shown closeups of the active site environment at pH 4, 7 and 9. It is interesting to observe that the active site is predominantly negatively charged at all pH values, but that the active site His is associated with a pocket of positive potential at pH 4 and 7, but not at 9. In all electrostatic maps, the molecule is covered with a dot-surface, that highlights the surface. Crevases, such as the active site can be clearly seen at the top of the pictures, reaching from the surface down to the active site His. At pH 4 the active site is dominated by the positive potential surrounding the active site His, at pH 7 and 9 both the entrance as well as the crevase leading to the active site residues show a negative potential.

In Fig. 21 the electrostatic maps of the active site environment of human pancreatic lipase are shown at three pH values: 4, 7 and 9. The dot-surface (blue) is high-

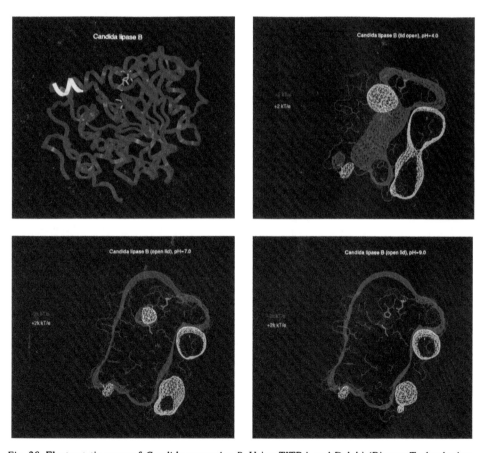

Fig. 20. Electrostatic maps of *Candida antarctica B*. Using TITRA and Delphi (Biosym Technologies, Inc) the electrostatic maps for *Candida antarctica B* have been calculated. In (A) is shown the fold of the lipase. The lid (which in this structure is open) is shown in yellow and the active site residues are displayed in green. In (B) the active site is shown at pH 4, in (C) at pH 7 and in (D) at pH 9. The red and blue isopotential surfaces represent the negative potential at $-2kT/e$ and $+2kT/e$, respectively.

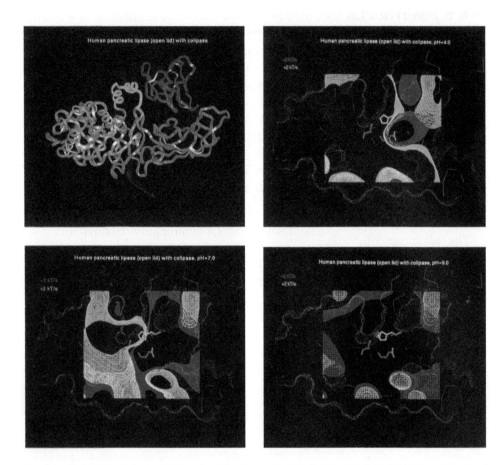

Fig. 21. pH dependent electrostatic maps of human pancreatic lipase with open lid in the presence of colipase. Using TITRA, InsightII and DelPhi (Biosym Technologies, San Diego) the electrostatic maps for human pancreatic lipase co-crystallized with porcine colipase have been calculated. In (A) is shown the fold of the lipase (beige) and the colipase (dark blue). The lid (which in this structure is open) is shown in light blue and the active site residues are displayed in green. In (B) the active site is shown at pH 4, in (C) at pH 7 and in (D) at pH 9. The red and blue isopotential surfaces represent the negative potential at $-2kT/e$ and $+2kT/e$, respectively.

lighting the active site. Similar to the observations made for the *Cand. B.* active site the human pancreatic lipase shows a positive potential for the whole of the active site at pH 4, whereas a clear dipolar character can be seen at pH 7, again with the active site His located in a high electric field gradient, as judged by the very closely positioned positive and negative isopotential surfaces. At pH 9 this feature is not visible, and the whole active site region appears predominantly negatively charged.

NMR Studies of Proteins

A prerequisite for rational Protein Engineering is 3D structural information about the protein. Besides X-ray crystallography, NMR is the most important method for protein structure determination for proteins with a molecular weight up to about 30 kDa [58].

X-Ray crystallography has several advantages when compared to NMR. Solving the crystal structure by X-ray crystallography is usually fast as soon as good crystals of the protein are obtained (even if it may not be easy to obtain crystals). It is also possible to determine the structure of very big proteins. The major disadvantage of X-ray crystallography is that it is the crystal structure that is determined. Molecular interactions between neighboring molecules in the crystal lattice (crystal contacts) may distort the structure [59,60]. Since active sites and other binding sites are usually located on the surface of the proteins, very important regions of the protein may be distorted. Some structures show large differences between NMR and X-ray structure [61,62].

The advantage of NMR is that it deals with protein molecules in solution, thus close to the natural environment can be achieved. It is possible to study the protein and the dynamic aspects of its interaction with other molecules like substrates, inhibitors, etc. It is also possible to obtain information about apparent pK_a values, hydrogen exchange rates, hydrogen bounds and conformational changes associated with change in the physical or chemical environment such as pH or temperature.

NMR: an established tool for biochemical investigations

All nuclei contain protons, and therefore they carry charge. In addition some nuclei also possesses a nuclear spin. This creates a magnetic dipole, and the nuclei will be oriented with respect to an external magnetic field. The most commonly studied nuclei in protein NMR (^1H, ^{13}C and ^{15}N) have two possible orientations, representing high and low energy states. The frequency of the transition between the two orientations is proportional to the magnetic field. At a magnetic field of 11.7 T the energy difference corresponds to about 500 MHz for protons. In an undisturbed system there will be an equilibrium population of the possible orientations, with a small difference in spin population between the high and low energy orientation.

The equilibrium population can be perturbed by a radio frequency pulse of a frequency at or close to the transition frequency. In addition, the spins will be brought into phase coherence (concerted motion) and a detectable magnetization will be created. The intensity of the NMR signal is proportional to the population difference between the levels the nuclei can possess.

Nuclei of the same type in different chemical and structural environments will experience different magnetic fields due to shielding from electrons. The shielding effect leads to different resonance frequencies for nuclei of the same type. The effect is measured as a difference in resonance frequency (in parts per million, ppm) between the nuclei of interest and a reference substance, and this is called the *chemical shift*. In molecules with low internal symmetry most atoms will experience different amounts of shielding, the resonance signals will be distributed over a well-defined range, and we obtain a typical NMR spectrum.

The process that brings the magnetization back to equilibrium may be divided into two parts, longitudinal and transverse relaxation. The longitudinal or T_1 relaxation describes the time it takes to reach the equilibrium population. The transverse or T_2 relaxation describes the time it takes before the induced phase coherence is lost. For macromolecules the T_2 relaxation is always shorter than the T_1 relaxation. Short T_2 relaxation leads to broad signals because of poor definition of the chemical shift. Most molecules have dipoles with magnetic moment, and the most important cause of relaxation is fluctuation of the magnetic field caused by the Brownian motion of molecular dipoles in the solution. How effective a dipole may relax the signal depends upon the size of the magnetic moments, the distance to the dipole, and the frequency distribution of the fluctuating dipoles.

A nucleus may also detect the presence of nearby nucleus through the chemical bounds connecting the nuclei (less than three bonds apart), and this will split the NMR signal from the nucleus into more components. Several nuclei in a coupling network is called a spin system.

By applying radio frequency pulses, it is possible to create and transfer magnetization to different nuclei. As an example, it is possible to create magnetization at one nucleus, and transfer the magnetization through bonds to other nuclei where it may be detected. The pulses are applied in a so-called pulse sequence [63,64].

Methods for structure determination by NMR

The methodology for determination of protein structure by two-dimensional NMR is described in several textbooks and review papers [65–67]. The standard method is based on two steps: sequential assignment, assignment of resonances from individual amino acids; and distance information, assignment of distance correlated peaks between different amino acids.

Assignment of resonances from individual amino acids

The first step involves acquiring coupling correlated spectra (COSY, TOCSY) in deuterium oxide to determine the spin system of correlated resonances. Some amino acids have spin systems that in most cases make them easy to identify (Gly, Ala, Thr, Ile, Val, Leu). The other amino acids have to be grouped into several classes, due to identical spin systems, even though they are chemically different. The spin systems can be correlated to the NH proton by acquiring COSY and TOCSY spectra in water.

The assigned NH resonance is then used in distance correlated spectra (NOESY) to assign correlations to protons (NH, H_a, H_b) at the previous amino acid residue (FIGA). By combining the knowledge of the primary sequence (which gives the spin system order) with the NMR data collected, it is possible to complete the sequential assignment.

Assignment of distance correlated peaks

When the sequential assignment is done, the assignment of short range nOe (up to four residues) will give information about secondary structure (a-helix, b-strand). Long-range correlations will serve as constraints (together with scalar couplings) to determine the tertiary structure of the protein. Excellent procedures describing these steps are available [66,68].

With large proteins there will be spectral overlap of resonance lines. The problem can be partially solved by uniformly labeling the protein with NMR active isotopes. The most common isotope exchange occurs with hydrogen ($^1H \rightarrow {}^2H$), carbon ($^{12}C \rightarrow {}^{13}C$) and nitrogen ($^{14}N \rightarrow {}^{15}N$) [68]. Isotope labeling can be used to simplify spectra. ^{15}N spectra of a protein will show only those nitrogens which have been isotope labeled. Isotope labeling may also be used for assignment of amino acid types (by selective labeling of one type amino acid) or sequence specific assignment (by labeling of a specific amino acid). It is also possible to monitor the hydrogen exchange rate of ^1H by measuring the decrease in signal with the protein dissolved in deuterium oxide.

Triple resonance multidimensional NMR methods may then be applied [68,69]. The resonances will then be spread out in two more dimensions (^{13}C and ^{15}N) and the problem with overlap is reduced. These methods depend upon the use of scalar couplings to perform the sequential assignment, the sequential assignment procedure will then be less prone to error. The NOESY spectra of such large proteins are often very crowded, but four-dimensional experiments like the ^{13}C–^{13}C edited NOESY spectrum [71] have been designed. Such experiments will spread the proton–proton distance correlated peaks by the chemical shift of its corresponding ^{13}C neighbor and reduce the spectral overlap. Secondary structure elements may also be predicted from the chemical shift of ^1H and ^{13}C [72–74].

Larger proteins

Obtaining NMR spectra of proteins has some aspects that should be considered.

Spectral overlap

As we move to larger proteins the probability of overlap of resonance lines increases. At some point it will become impossible to do sequential assignment due to this overlap. Development of three- and four-dimensional NMR methods (for a review see [75]) have made it possible to assign proteins in the 30 kDa range [17,58].

Fast relaxation

As the size of the protein is increased, the rate of tumbling in solution is reduced. This leads to a reduced transverse relaxation time (T_2), and broadening of the resonance lines in the NMR spectra. The intensities of the peaks are reduced and they may be difficult to detect. The short transverse relaxation time will also limit the length of the pulse sequences that it is possible to apply (because there will be no phase coherence left), and multidimensional methods become difficult.

Behavior of the protein

The proteins for which it is possible to determine a 3D structure by NMR or X-ray crystallography are probably a subset of all proteins [76]. Proteins may have regions with mobility and few cross peaks. The effective size of a protein is often increased by aggregation. The extent of aggregation can often be controlled by reducing the protein concentration. Thus very often the degree of aggregation will determine if it is possible to assign and solve a protein structure by NMR, by limiting the maximum concentration that may be used. The stability of the proteins is also a major issue. A sample may be left in solution for days, often at elevated temperatures, so denaturation may become a problem.

Other applications of NMR

Photochemically induced nuclear polarization (CIDNP) is an interesting technique for the study of surface positioned aromatic residues in proteins [77–80]. By introducing a dye and exciting it with a laser, it is possible to transfer magnetization to aromatic residues, where it can be observed.

Besides high-resolution NMR, solid-state NMR has also been applied to studies of proteins. Studies of active sites and conformation of bound inhibitors yields interesting information. The stability of proteins may be monitored under different conditions by detecting signals from transition intermediates bound to the active site [81,82]. Structural constraints on transition state conformation of bound inhibitors can be obtained [83,84]. Structural constraints of the fold and conformation of the amino sequence may be gathered by setting upper and lower distances for lengths between specific amino acids [85].

By solid-state NMR, it is possible to study membrane proteins and their orientation with respect to their membrane [86,87]. We expect such studies to give insight into ion channels in membranes [88].

Paramagnetic relaxation

An important mechanism for relaxation in high-resolution NMR is dipolar relaxation. Usually this is induced by the spin of nuclei in the immediate vicinity, and it is a function of the size of the dipole. The electron is also a magnetic dipole, and the magnitude of this dipole is about 700 times that of a proton. Paramagnetic compounds have an electron that will interact with nearby protons and increase the relaxation rate of these protons.

The widest use of paramagnetic compounds has been of Gd^{3+} bound to specific sites in a protein [89], but other compounds have also been used [90–92]. This will make it possible to identify resonance lines from residues in the vicinity of the binding site. It is also possible to calculate distances from the paramagnetic atom as the relaxation effect is distance dependent. An other interesting application of paramagnetic relaxation is to use it to estimate water exchange close to a paramagnetic center [93].

The paramagnetic broadening effect can also be used with a compound moving freely in solution [94–96]. In this way residues located on or close to the protein surface will give rise to broadened resonance lines compared to residues in the interior of the protein.

This method can be used to measure important nOe and chemical shifts inside the protein directly, or it can be used as a difference method to identify resonances at the surface by comparing spectra acquired with and without the paramagnetic relaxation agent (Fig. 22).

We have used the paramagnetic compound gadolinium diethylenetriamine pentaacetic acid (Gd-DTPA) as a relaxation agent. Gd-DTPA will increase both the longitudinal and the transverse relaxation rates of protons within the influence sphere. Suitable NMR experiments to highlight the relaxation effect may be NOESY, ROESY and TOCSY [97,98].

Gd-DTPA is widely used in magnetic resonance imaging (MRI) to enhance tissue contrast. It is assumed to be non-toxic and we do not expect it to bind to proteins.

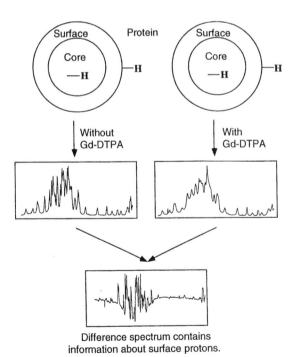

Difference spectrum contains
information about surface protons.

Fig. 22. Enhancement of signals from the protein surface using paramagnetic relaxation agents. The paramagnetic relaxation method. The protons located at the protein surface will be closer to the dissolved paramagnetic relaxation agent than the protons located in the protein interior, hence the resonance lines from protons at the surface will be broadened more than resonance lines stemming from protons located inside the protein (from [26]).

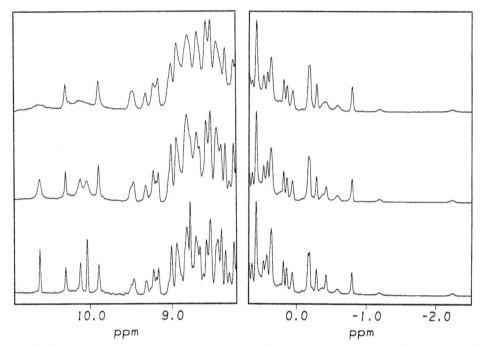

Fig. 23. 1D spectra of titration of lysozyme using GDTPA. Hen eggwhite lysozyme with Gd-DTPA. 1D proton spectra. High (right) and low field region (left) of 1D 500 Mhz ^{1}H-NMR spectra of HEW lysozyme with different concentrations of Gd-DTPA added. The lower trace shows the spectrum of pure lysozyme, the middle and upper trace show the spectra with Gd-DTPA added (middle trace, 5 mM lysozyme/0.25 mM Gd-DTPA; upper trace, 5 mM lysozyme/0.5 mM Gd-DTPA) (from [26]).

We used the well-studied protein hen egg-white lysozyme as a test protein. Both the structure and the NMR spectra of this protein are known [99,100], and the protein is extremely well suited for NMR experiments.

In Fig. 23 the 1D ^{1}H-NMR spectrum recorded in the presence and absence of Gd-DTPA is shown. Although it is evident that there is a selective broadening in the 1D spectrum, it is also clear that there are problems with overlapping spectral lines. We therefore applied two-dimensional NMR methods, and Fig. 24 shows the low field region of a NOESY spectrum of lysozyme. The region corresponds to the same region as shown in Fig. 23.

From Fig. 25 we see that the signals from W62, W 63 and W123 disappear with addition of Gd-DTPA, while the signals from W28, W108 and W111 are still observable. By examination of the solvent accessible surface of lysozyme, it is evident that the indole NH of W62, W63 and W123 is exposed to solvent, while the indole NH of W28, W108 and W111 is not exposed. This shows that the changes in the spectrum are as expected from the structure data.

The appearance of the NH–NH region of the spectrum (Fig. 24) also shows the reduction in the number of signals in the Gd-DTPA exposed spectrum.

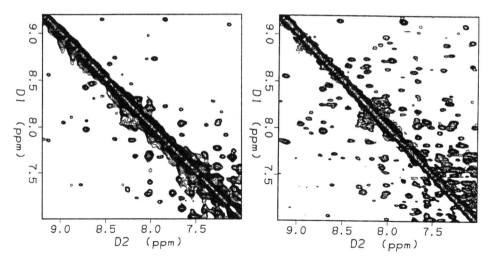

Fig. 24. 2D NOESY spectra of the amide proton region for HEW lysozyme. NH–NH region of NOESY spectra of lysozyme with and without added Gd-DTPA. In the left panel the NOESY spectrum without the addition of Gd-DTPA is shown. In the right panel the spectrum with Gd-DTPA is shown. The concentration of lysozyme was 5 mM, Gd-DTPA (when added) 0.25 mM. pH was 3.8 and the mixing time used 200 ms (from [26]).

Studies of enzyme kinetics

Studies of the reaction rate of enzymatic processes may gain useful information about enzyme structure and ionization states. For example, the pH dependency of the catalytic constant (k_{cat}) gives information about pK_a values of a particular enzyme–substrate complex while the specificity constant (k_{cat}/K_M) may provide information on the pK_a values for the free reactant [101,102]. This means that kinetic measurements may give information at atomic level regarding groups involved in substrate binding.

Different lipases may show different specificity for the three different ester bounds [103], even the active site of the different lipases is structurally very similar. The active site of the lipases is a serine proteinase-like catalytic triad consisting of a Ser, His and an Asp as the active site residue. Comparing the geometry of the region in the vicinity of the active site does not explain the difference in specificity of different lipases. Recent studies [104] also show that small changes in solvent conditions may create large differences in selectivity.

In order to monitor enzymatic reactions in vivo, conventional methods (e.g. UV, fluorescence) have been dependent of chromophore groups in the substrate. By monitoring the reaction with NMR no such constraints are laid upon the experimental procedure. Increased sensitivity may be gained by selective isotope labeling [105]. Several frequency regions may be followed simultaneously. Signals from both substrate and products may be fitted simultaneously.

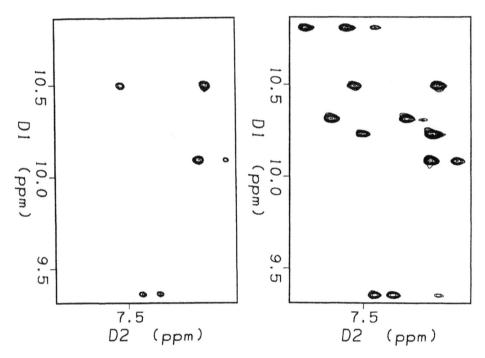

Fig. 25. The effect of GDTPA on the 2D COSY spectra of the tryptophan region. Low field region of the NOESY spectra of HEW lysozyme with and without added Gd-DTPA. Correlations between the indole NH of tryptophan residues with other protons from the same residue are shown. The left panel shows the NOESY spectrum in the absence of Gd-DTPA, the right panel the spectrum after the addition of Gd-DTPA (5 mM lysozyme/0.25 mM Gd-DTPA). The panels are plotted at identical contour levels. pH was 3.8 and the mixing time used 200 ms (from [26]).

Additional information may be obtained by studying substrate and/or products in the vicinity of the enzyme of interest. The conformation of the enzyme bound substrate and its interactions with residues in the enzyme may be determined.

Conclusion

The lipases and esterases compose a highly challenging group of enzymes. This is particularly so, since the physical state of their substrates are so poorly defined. The substrate evades crystallographical analysis, and the temporal dynamics of lipids is probably a contributing factor to this. Nevertheless nature has developed and optimized a range of proteins for degrading lipids and esters. Some of the lipases are clearly interfacially activated, and only achieve full catalytic potential in the presence of a lipophilic surface. The interesting fact, that this interface does not need to be a lipid, but could be, e.g. siliconized glass beads, indicates that the activation mechanism may be triggered by the low dielectric constant shared by many membrane-like environments. Lipases and esterases are inherent components of all living systems. They have very important medical roles, e.g. the presence of acetylcholine, acetyl-cholinesterase as well as the acetylcholine receptor is crucial for brain function. In order to provide an overview rather than to provide a comprehensive in-depth analysis of this large family of proteins, we have focused on finding common features, either at the sequence level or at the 3D structural level. Such features often have their roots in evolutionary relationships. Therefore we have covered in some detail the possible evolutionary links between the proteins, and we believe an interesting picture of these relations has been presented in this paper. The differences between closely related proteins often reside in the residue utilization on the surface of the protein. Here we have shown how the pH-dependent electrostatic features can be visualized using TITRA in conjunction with commercial molecular modeling software.

Acknowledgements

The authors acknowledge financial support from Junta Nacional de Investigação Científica, Portugal (AB, PM and MS), the Norwegian Research Council (BP 29345) as well as the French Norwegian Foundation (FNS). We want to express our gratitude to Christian Cambillau, CNRS Marseilles and Alwyn Jones, Biomedical Center, Uppsala for kindly sharing pre-release 3D structural data on cutinase, human pancreatic lipase as well as *Candida antarctica B* with us.

References

1. Patkar S, Björkling F. Lipase inhibitors. In Woolley P, Petersen SB (eds) Lipases: Their Structure Biochemistry and Application. Cambridge University Press, 1994;207.
2. Björkling F, Godtfredsen SE, Kirk O, Patkar SA, Andresen O. Lipase catalyzed organic synthesis. In Servi (ed) Microbial reagents in organic synthesis. The Netherlands: Kluwere Academic Publishers, 1992;249–260.
3. Laskowski M, Kato I. Protein inhibitors of proteinases. Annu Rev Biochem 1980;49:593–626.
4. Mori M, Yamaguchi K, Abe K. Purification of a lipoprotein lipase inhibiting protein produced by amelanoma cell line associated with cancer Cachexia. Biochem Biophys Res Commun 1989; 160:1085–1092.
5. Kgandopulo GB, Ruben EL. Effects of various activators and inhibitors on the lipase activity of fungus of the genus Geotrichum. Microbiologiya 1974;43:814–819.
6. Iwai M, Tsujisaka Y, Fukumoto J. Studies on lipase. V. Effect of iron ions on the *Aspergillus niger* lipase. J Gen Apl Microbiol 1970;16:81–90.
7. Entressangles B, Desnuelle P. Action of pancreatic lipase on monomeric tripropinin in the presence of water-miscible organic compounds. Biochim Biophys Acta 1974;341:437–446.
8. Brockman H, Law JH, Kezdy FJ. Catalysis by adsorbed enzymes. J Biol Chem 1973;248:4965–4970.
9. Verger R, de Haas GH. Enzyme reaction in a membrane model. 1: A new technique to study enzyme reactions in monolayers. Chem Phys Lipids 1973;10:127–136.
10. Verger R, Riviere C, Moreau H, Gargouri Y, Rogalska E, Nury S, Moulin A, Ferrato F, Ransac

S, Carriere F, Cudrey C, Tretout N. Enzyme kinetics of lipolysis. Lipase inhibition by proteins. In Alberghina L, Schmid RD, Verger R (eds) Lipases: Structure, Mechanism and Genetic Engineering, Vol 16. Gesellschaft für Biotechnologische Forschung mbH Monograph, 1990.

11. Martinelle M, Hult K. Kinetics of triglyceride lipases. In Woolley P, Petersen SB (eds) Lipases: Their Structure, Biochemistry And Applications. Cambridge University Press, 1994;158.

12. Aires-Barros MR, Taipa MA, Cabral JMS. Isolation and purification of lipases. In Woolley P, Petersen SB (eds) Lipases: Their Structure, Biochemistry and Application. Cambridge University Press, 1994;243.

13. Pires MJ, Martel P, Baptista A, Petersen SB, Wilson R, Cabral JMS. Improving protein extraction yield in reversed micellar systems through surface charge engineering. Biotech Bioeng 1994, in press.

14. Eigtved P. Enzymes and lipid modification. In Advances in Applied Lipid Research, vol 1, p. 1–64. JAI Press Ltd., 1992.

15. Hjort A, Carriere F, Cudrey C, Wöldike H, Boel E, Lawson DM, Ferrato F, Cambillau C, Dodson G, Thim L, Verger R. A structural domain (the lid) found in pancreatic lipases is absent in the guinea pig (phospho)lipase. Biochemistry 1993;32:4702–4707.

16. Sugihara A, Shimada Y, Nakamura M, Nagao T, Tominaga Y. Positional and fatty acid specificities of *Geotrichum candidum* lipases. Protein Eng 1994;7:585–588.

17. Stockman BJ, Nirmala NR, Wagner G, Delcamp TJ, DeYarman MT, Freisheim JH. Sequence-specific ^1H and ^{15}N resonance assignment for human dihydrofolate reductase in solution. Biochemistry 1992;31:218–229.

18. Sonnet PE, Baillargeon MW. Methyl-branched octanoic acids as substrates for lipase-catalyzed reactions. Lipids 1991;26:295–299.

19. Vainio P, Virtanen JA, Sparrow JT, Gotto AM, Kinnonen PK. Esterase type of activity possessed by human plasma apolipoprotein C-II and its synthetic fragments. Chem Phys Lipids 1983;33: 21–32.

20. Brady L, Brzozowksi AM, Derewenda ZS, Dodson E, Dodson G, Tolley S, Turkenburg JP, Christiansen L, Huge-Jensen B, Nørskov L, Thim L, Menge U. A serine protease triad forms the catalytic centre of a triacylglycerol lipase. Nature 1990;343:767–770.

21. Brzozowski AM, Derewenda U, Derewenda ZS, Dodson GG, Lawson DM, Turkenburg JP, Bjørkling F, Huge-Jensen B, Patkar SR, Thim L. A model for interfacial activation in lipases from the structure of a fungal lipase-inhibitor complex. Nature 1991;351:491–494.

22. Derewenda ZS, Derewenda U, Dodson GG. The crystal and molecular structure of the *Rhizomucor miehei* triacylglyceride lipase at 1.9 A resolution. J Mol Biol 1992;227:818–839.

23. Bourne Y, Martinez C, Kerfelec B, Lombardo D, Chapus C, Cambillau C. Horse pancreatic lipase. The crystal structure refined at 2.3 A resolution. J Mol Biol 1994;238:709–732.

24. van Tilbeurgh H, Egloff MP, Martinez C, Rugani N, Verger R, Cambillau C. Interfacial activation of the lipase-procolipase complex by mixed micelles revealed by X-ray crystallography. Nature 1993;362:814–820.

25. van Tilbeurgh H, Sarda L, Verger R, Cambillau C. Structure of the pancreatic lipase-colipase complex. Nature 1992;359:159–162.

26. Anthonsen HW, Baptista A, Drabløs F, Martel P, Petersen, SB. The blind watchmaker and rational protein engineering. J Biotech 1994;36:185–220.

27. Davies ME, McCammon JA. Electrostatics in biomolecular structure and dynamics. Chem Rev 1990;90:509–521.

28. Northrup SH. Hydrodynamic motions of large molecules. Curr Opinion Struct Biol 1984;4:269–274.

29. Warshel A, Naray-Szabo G, Sussman F, Hwang J-K. How do serine proteases really work? Biochemistry 1989;28:3629–3637.

30. Warshel A, Russel ST. Calculation of electrostatic interactions in biological systems and in solution. Q Rev Biophys 1984;17:283–422.

31. Matthew JB. Electrostatic effects in proteins. Annu Rev Biophys Biophys Chem 1985;14:387–417.

32. Rogers NK. The modelling of electrostatic interactions in the function of globular proteins. Prog Biophys Mol Biol 1986;48:37–66.

33. Harvey SC. Treatment of electrostatic effects in macromolecular modeling. Proteins 1989;5:78–92.

34. Sharp KA, Honig B. Electrostatic interactions in macromolecules: theory and applications. Annu Rev Biophys Biophys Chem 1990;19:301–332.

35. Bashford D. Electrostatic effects in biological molecules. Curr Opinion Struct Biol 1991;1:175–184.

36. van Gunsteren WF, Berendsen JC. Computer simulation of molecular dynamics: methodology, applications, and perspectives in chemistry. Angew Chem Int Ed Engl 1990;29:992–1023.

37. Brooks III CL, Pettitt BM, Karplus M. Structural and energetic effects of truncating long ranged interactions in ionic and polar fluids. J Chem Phys 1985;83:5897–5908.

38. Solmajer T, Mehler EL. Electrostatic screening in molecular dynamics simulations. Protein Eng 1991;4:911–917.

39. Northrup SH, Pear MR, Morgan JD, McCammon JA, Karplus M. Molecular dynamics of ferro-cytochrome c. Magnitude and anisotropy of atomic displacements. J Mol Biol 1981;153:1087–1109.

40. Still WC, Tempczyk A, Hawley RC, Hendrickson T. Semianalytical treatment of solvation for molecular mechanics and dynamics. J Am Chem Soc 1990;112:6127–6129.

41. Gilson MK, Honig B. The inclusion of electrostatic hydration energies in molecular mechanics calculations. J Computer-aided Mol Design 1991;5:5–20.

42. Gilson MK. Multiple-site titration and molecular modeling: two rapid methods for computing energies and forces for ionizable groups in proteins. Proteins 1993;15:266–282.

43. Gilson MK, Honig BH. On the calculation of electrostatic interactions in proteins. J Mol Biol 1985;183:503–516.

44. Jackson JD. Classical Electrodynamics. New York: Wiley, 1975.

45. Tanford C, Kirkwood JG. Theory of protein titration curves. I. General equations for impenetrable spheres. J Am Chem Soc 1957;79:5333–5339.

46. Shire SJ, Hanania GIH, Gurd FRN. Electrostatic effects in myoglobin. Hydrogen ion equilibria in sperm whale ferrimyoglobin. Biochemistry 1974;13:2967–2974.

47. States DJ, Karplus M. A model for electrostatic effects in proteins. J Mol Biol 1987;197, 122–130.

48. Warwicker J, Watson HC. Calculation of the electric potential in the active site cleft due to a-helix dipoles. J Mol Biol 1982;157:671–679.

49. Gilson MK, Sharp KA, Honig BH. Calculating the electrostatic potential of molecules in solution: Method and error assessment. J Comp Chem 1987;9:327–335.

50. Bashford D, Karplus M. pKa's of ionizable groups in proteins: atomic detail from a continuum electrostatic model. Biochemistry 1990;29:10219–10225.

51. Getzoff ED, Cabelli DE, Fisher CL, Parge HE, Viezzoli MS, Banci L, Hallewell RA. Faster superoxide dismutase mutants designed by electrostatic guidance. Nature 1992;358:347–351.

52. Gilson MK, Straatsma TP, McCammon JA, Ripoli DR, Faerman CH, Axelsen PH, Silman I, Sussman JL. Open "back door" in a molecular dynamics simulation of acetylcholinesterase. Science 1994;253:1276–1278.

53. Warshel A, Levitt M. Theoretical studies of enzymic reactions. J Mol Biol 1976;103:227–249.

54. Tanford C, Roxby R. Interpretation of protein titration curves. Application to lysozyme. Biochemistry 1972;11:2192–2198.

55. Bashford D, Karplus M. Multiple-site titration curves of proteins: an analysis of exact and approximate methods for their calculation. J Phys Chem 1991;95:9556–9561.

368

56. Beroza P, Fredkin DR, Okamura MY, Feher G. Protonation of interacting residues in a protein by a Monte Carlo method: application to lysozyme and the photosynthetic reaction center of *Rhodobacter sphaeroides*. Proc Natl Acad Sci USA 1991;88:5804–5808.

57. Yang A-S, Gunner MR, Sampogna R, Sharp K, Honig B. On the calculation of pK_as in proteins. Proteins 1993;15:252–265.

58. Foght RH, Schipper D, Boelens R, Kaptein R. ^{1}H, ^{13}C and ^{15}N NMR backbone assignments of the 269-residue serine protease PB92 from *Bacillus alcalophilus*. J Biomol NMR 1994;4:123–128.

59. Chazin WJ, Hugli TE, Wright PE. ^{1}H NMR studies of human C3a anaphylatoxin in solution: Sequential resonance assignments, secondary structure, and global fold. Biochemistry 1988;27:9139–9148.

60. Wagner G, Braun W, Havel T, Schaumann T, Go N, Wüthrich K. Protein structures in solution by nuclear magnetic resonance and distance geometry. J Mol Biol 1987;196:611–639.

61. Frey MH, Wagner G, Vasak M, Sørensen OW, Neuhaus D, Worgotter E, Kagi JHR, Ernst RR, Wüthrich K. Polypeptide – metal cluster connectivities in metallothionein 2 by novel [1]H-[113]Cd heteronuclear two dimensional NMR experiments. J Am Chem Soc 1985;107:6847–6851.

62. Klevit RE, Waygood EB. Two-dimensional ^{1}H NMR studies of histidine-containing protein from *Escherichia coli*. Secondary and tertiary structure as determined by NMR. Biochemistry 1986;25:7774–7781.

63. Ernst RR. Nuclear magnetic resonance fourier transform spectroscopy (Nobel lecture). Angew Chem 1992;31:805–930.

64. Kessler H, Gehrke M, Griesinger C. Two-dimensional spectroscopy: background and overview of the experiments. Angew Chem Int Ed Engl 1988;27:490–536.

65. Wagner G. NMR investigations of protein structure. Prog NMR Spectrosc 1990;22:101–139.

66. Wüthrich K. NMR of proteins and nucleic acids. New York: Wiley, 1986.

67. Wider G, Macura S, Kumar A, Ernst RR, Wüthrich K. Homonuclear two-dimensional ^{1}H NMR of proteins. Experimental procedures. J Magn Reson 1984;56:207–234.

68. Roberts GCK. NMR of macromolecules. A practical approach. In Rickwood D, Hames BD (eds) The Practical Approach Series, Vol 134. New York: Oxford University Press, 1993.

69. Griesinger C, Sørensen OW, Ernst RR. Three-dimensional fourier spectroscopy. Application to high-resolution NMR. J Magn Reson 1989;84:14–63.

70. Kay EL, Clore GM, Bax A, Gronenborg AM. Four-dimensional heteronuclear triple-resonance NMR spectroscopy of interleukin-1 in solution. Science 1990;249:411–414.

71. Clore GM, Wingfield PT, Gronenborn AM. High-resolution three-dimensional structure of inter-leukin 1 in solution by three- and four-dimensional nuclear magnetic resonance spectroscopy. Biochemistry 1991;30:2315–2323.

72. Spera S, Bax A. Empirical correlation between protein backbone conformation and C_a and C_b ^{13}C nuclear magnetic resonance chemical shifts. J Am Chem Soc 1991;113:5490–5492.

73. Williamson MP, Asakura T. Calculation of chemical shifts of protons on alpha carbons in proteins. J Magn Reson 1991;94:557–562.

74. Wishart DS, Sykes BD, Richards FM. The chemical shift index: a fast and simple method for the assignment of protein secondary structure through NMR spectroscopy. Biochemistry 1992;31:1647–1651.

75. Oschkinat H, Müller T, Dieckmann T. Protein structure determination with three- and four-dimensional nmr spectroscopy. Angew Chem Int Ed Engl 1994;33:277–293.

76. Wagner G. Prospects for NMR of large proteins. J Biomol NMR 1993;3:375–385.

77. Broadhurst RW, Dobson CM, Hore PJ, Radford SE, Rees ML. A photochemically induced dynamic nuclear polarization study of denatured states of lysozyme. Biochemistry 1991;30:405–412.

78. Cassels R, Dobson CM, Poulsen FM, Williams RJP. Study of the tryptophan residues of lysozyme using ^{1}H nuclear magnetic resonance. Eur J Biochem 1978;95:81–97.

79. Hore PJ, Kaptein R. Proton nuclear magnetic resonance assignment and surface accessibility of

tryptophan residues in lysozyme using photochemically induced dynamic nuclear polarization spectroscopy. Biochemistry 1983;22:1906–1911.

80. Scheffler JE, Cottrell CE, Berliner LJ. An inexpensive, versatile sample illuminator for photo–CIDNP on any NMR spectrometer. J Magn Reson 1985;63:199–201.

81. Burke PA, Griffin RG, Klibanov AM. Solid-state NMR assessment of enzyme active center structure under nonaqueous conditions. J Biol Chem 1992;267:20057–20064.

82. Gregory RB, Gangoda M, Gilpin RK, Su W. The influence of hydration on the conformation of lysozyme studied by solid-state ^{13}C-NMR spectroscopy. Biopolymers 1993;33:513–519.

83. Auger M, McDermott AE, Robinson V, Castelhano AL, Billedeau RJ, Pliura DH, Krantz A, Griffin RG. Solid-state ^{13}C NMR study of a transglutaminase – inhibitor adduct. Biochemistry 1993;32:3930–3934.

84. Christensen AM, Schaefer J. Solid-state NMR determination of intra- and intermolecular ^{31}P– ^{13}C distances for shikimate 3-phosphate and 1–^{13}C.glyphosate bound to enolpyruvylshikimate-3-phosphate synthase. Biochemistry 1993;32:2868–2873.

86. Killian JA, Taylor MJ, Koeppe RE. Orientation of the valine-1 side chain of the gramicidin transmembrane channel and implications for channel functioning. A ^{2}H NMR study. Biochemistry 1992;31:11283–11290.

87. Ulrich AS, Heyn MP, Watts A. Structure determination of the cyclohexene ring of retinal in bacteriorhodopsin by solid-state deuterium NMR. Biochemistry 1992;31:10390–10399.

88. Woolley GA, Wallace BA. Model ion channels: gramicidin and alamethicin. J Membr Biol 1992; 129:109–136.

89. Dobson CM, Ferguson SJ, Poulsen FM, Williams RJP. Complete assignment of aromatic ^{1}H nuclear magnetic resonances of the tyrosine residues of hen lysozyme. Eur J Biochem 1978;92: 99–103.

90. Chang CA, Brittain HG, Telser J, Tweedle MF. pH dependence of relaxivities and hydration numbers of Gadolinium(III) complexes of linear amino carboxylates. Inorg Chem 1990;29: 4468–4473.

91. Hernandez G, Brittain HG, Tweddle MF, Bryant RG. Nuclear magnetic relaxation in aqueous solutions of the Gd(HEDTA) complex. Inorg Chem 1990;29:985–988.

92. Hernandez G, Tweedle MF, Bryant RG. Proton magnetic relaxation dispersion in aqueous glycerol solutions of Gd(DTPA)$^{2-}$ and Gd(DOTA)$^-$. Inorg Chem 1990;29:5109–5113.

93. Martinez A, Olafsdottir S, Flatmark T. The cooperative binding of phenylalanine 4-monooxygenase studied by 1H-NMR paramagnetic relaxation. Eur J Biochem 1993;211:259–266.

94. Drayney D, Kingsbury CA. Free radical induced nuclear magnetic resonance shifts: comments on contact shift mechanism. J Am Chem Soc 1981;103:1041–1047.

95. Esposito G, Lesk AM, Molinari H, Motta A, Niccolai N, Pastore A. Probing protein structure by solvent perturbation of nuclear magnetic resonance spectra. J Mol Biol 1992;224:659–670.

96. Petros AM, Mueller L, Kopple KD. NMR identification of protein surfaces using paramagnetic probes. Biochemistry 1990;29:10041–10048.

97. Bax A, Davis DG. MLEV-17-based two-dimensional homonuclear magnetization transfer spectroscopy. J Magn Reson 1985;65:355–360.

98. Braunschweiler L, Ernst RR. Coherence transfer by isotropic mixing: Application to proton correlation spectroscopy. J Magn Reson 1983;53:521–528.

99. Diamond R. Real-space refinement of the structure of hen egg white lysozyme. J Mol Biol 1974; 82:371.

100. Redfield C, Dobson CM. Sequential ^{1}H-NMR assignments and secondary structure of hen egg white lysozyme in solution. Biochemistry 1988 ;27:122–136.

101. Brocklehurst K, Dixon HBF. pH-dependence of the steady-state of a two-step enzyme reaction. Biochem J 1976;155:61–70.

102. Brocklehurst K, Dixon HBF. The pH-dependence of second-order rate constants of enzyme modification may provide free-reactant pKa values. Biochem J 1977;167:859–862.

103. Rogalska E, Cudrey C, Ferrato F, Verger R. Stereoselective hydrolysis of triglycerides by animal and microbial lipases. Chirality 1993;5:24–30.

104. Hansen TV, Waagen V, Partali V, Anthonsen HW, Anthonsen T. Solvents effects in lipase catalysed racemate resolution. Abst Bridge Lipase T-Meeting, Bendor Island, September 1994.

105. Dimand RJ, Bradbury EM, Cox KL. Determination of triacylglycerol lipase activity using carbon-13-labeled triacylglycerols and nuclear magnetic resonance spectroscopy: evidence that hepatic lipase hydrolyzes medium-chain triacylglycerols. Magn Reson Med 1989;9:273–277.

106. Verheij HM, Dijkstra BK. Phospholipase A2 : mechanism and structure. In Woolley P, Petersen SB (eds) Lipases: Their Structure, Biochemistry and Application. Cambridge University Press, 1994;119.

107. Winkler FK, Gubernator K. Structure and mechanism of human pancreatic lipase. In Wooley P, Petersen SB (eds) Lipases: Their Structure, Biochemistry and Application. Cambridge University Press, 1994;139–157.

108. Higgins DG, Bleasby AJ, Fuchs R. CLUSTAL V: improved software for multiple sequence alignment. CABIOS 1992;8:189–191.

109. Barton GJ. ALSCRIPT a tool to format multiple sequence alignments. Prot Eng 1993;6:37–40.

110. Sander C, Schneider R. Database of homology-derived protein structures. Proteins 1991;9:56–68.

111. Pearson WR. Rapid and sensitive sequence comparison with FASTP and FASTA. Methods Enzymol 1990;183:63–98.

112. Pearson WR, Lipman DJ. Improved tools for biological sequence analysis. Proc Natl Acad Sci USA 1988;85:2444–2448.

113. Felsenstein J. PHYLIP - Phylogeny Inference Package (Version 3.2). Cladistics 1989;5:164–166.

114. Sussman JL, Harel M, Frolow F, Oefner C, Goldman A, Toker L, Silman I. Atomic Structure of Acetylcholinesterase from *Torpedo californica*: a prototypic Acetylcholine-Binding Protein. Science 1991;253:872–879.

115. Grochulski P, Li Y, Schrag JD, Bouthillier F, Smith P, Harrison D, Rubin B, Cygler M. Insights into interfacial activation from an open structure of *Candida rugosa* lipase. J Biol Chem 1993; 268:12843–12847.

116. Noble ME, Cleasby A, Johnson LN, Egmond MR, Frenken LG. The crystal structure of triacylglycerol lipase from *Psudomonas glumae* reveals a partially redundant catalytic aspartate. FEBS Lett 1993;331:123–128.

117. Uppenberg J, Hansen MT, Patkar S, Jones TA. The sequence, crystal structure determination and refinement of two crystal forms of Lipase B from *Candida antarctica*. Structure 1994;2:293–308.

118. Derewenda U, Brzozowski AM, Lawson DM, Derewenda ZS. Catalysis at the interface: the anatomy of a conformational change in a triglyceride lipase. Biochemistry 1992;31:1532–1541.

119. Schrag JD, Cygler M. 1.8 A refined structure of the lipase from *Geotrichum candidum*. J Mol Biol 1993;230:575–591.

120. Grochulski P, Li Y, Schrag JD, Cygler M. Two conformational states of *Candida rugosa* lipase. Protein Sci 1994;3:82–91.

121. Winkler FK, D'Arcy A, Hunziker W. Structure of human pancreatic lipase. Nature 1990;343:771–774.

122. Pathak D, Ollis D. Refined structure of denelactone hydrolase at 1.8 A. J Mol Biol 1990;214:497–525.

123. Martinez C, De Geus P, Lauwereys M, Matthyssens G, Cambillau C. Fusarium solani cutinase is a lipolytic enzyme with a catalytic serine accessible to solvent. Nature 1992;356:615–618.

124. Warden CH, Davis RC, Yoon MY, Hui DY, Svenson K, Xia YR, Diep A, Lusis AJ. Chromosomal location of lipolytic enzymes in the mouse: pancreatic lipase, colipase, hormone sensitive lipase, hepatic lipase and carboxyl ester lipase. J Lipid Res 1993;34:1451–1455.

125. Davis RC, Xia YR, Mohandas T, Schotz MC, Lusis AJ. "Assignment of the human pancreatic colipase gene to chromosome 6p21.1 to pter. Genomics 1991;10:262–265.

126. Sims HF, Lowe ME. The human colipase gene: isolation, chromosomal location and tissue-specific expression. Biochemistry 1992;31:7120–7125.
127. Davis RC, Diep A, Hunziker W, Klisak I, Mohandas T, Schotz MC, Sparkes RS, Lusis AJ. Assignment of human pancreatic lipase gene (PNLIP) to chromosome 10q24-q26. Genomics 1991;11:1164–1166.
128. Soreq H, Zakut H (eds). Human Cholinesterases and Anticholinesterase. Academic Press, 1993.
129. Holm C, Belfrage P, Østerlund T, Davis R, Schotz MC, Langin D. Hormone sensitive lipase: structure, function, evolution and overproduction in insect cells using the baculovirus expression system. Protein Eng 1994;7:537–541.
130. Sternby B, Engstrøm A, Hellman U. Purification and characterization of pancreatic colipase from the dogfish (Squalus acanthius). Biochim Biophys Acta 1984;789:159–163.
131. Pierrot M, Astier J-P, Astier M, Charles M, Drenth J. Pancreatic colipase: crystallographic and biochemical aspects. Eur J Biochem 1982;123:347–354.
132. Bonicel JJ, Couchoud PM, Foglizzo E, Desnuelle P, Chapus C. Amino acid sequence of horse colipase B. Biochem Biophys Acta 1981;669:39–45.
133. Fukuoka S, Zhang DE, Taniguchi Y, Scheele GA. Structure of the canine pancreatic colipase gene includes two protein-binding sites in the promoter region. J Biol Chem 1993;268:11312–11320.
134. Bosc-Bierne I, Rathelot J, Canioni P, Julien R, Bechis G, Gregoire J, Rochat H, Sarda L. Isolation and partial structural characterization of chicken pancreatic colipase. Biochem Biophys Acta 1981;667:225–232.
135. Sternby B, Engstrom A, Hellman U, Vihert AM, Sternby NH, Borgstrøm B. The primary sequence of human pancreatic colipase. Biochim Biophys Acta 1984;784:75–80.
136. Wicker C, Puigserver A. Rat Pancreatic colipase mRNA: nucleotide sequence of a cDNA clone and nutritional regulation by a lipidic diet. Biochem Biophys Res Commun 1990;167:130–136.

© 1995 Elsevier Science B.V. All rights reserved
Biotechnology Annual Review Volume 1
M.R. El-Gewely, editor

Solid-phase technology: magnetic beads to improve nucleic acid detection and analysis

Joakim Lundeberg[1,3] and Frank Larsen[2,3]
[1]*Department of Immunology, Institute for Cancer Research, The Norwegian Radium Hospital, Montebello, Oslo;* [2]*Biotechnology Centre of Oslo, Blindern, Oslo; and* [3]*Dynal AS, Skøyen, Oslo, Norway*

Introduction

The widespread use of solid-phase methods in molecular biology has had a major impact especially in the area of synthesis and analysis of peptides and nucleic acids. The solid-phase approaches have improved robustness by increased reproducibility with higher yields. Furthermore, automation is facilitated since reaction buffers and additional ingredients can be rapidly changed. Several alternative solid supports have been used including filters, the walls of microtiter wells and inorganic and organic polymer beads. The introduction of magnetic particles has proven to be an attractive solid support avoiding centrifugation and/or filtrations often encountered with the alternative solid phases.

The use of magnetic particles in various fields of biochemistry and medicine has been well documented and several magnetic particles are now commercially available for diagnostic, cell separation and drug delivery applications. Most of the these particles are paramagnetic, that is, the particles are only magnetic in a magnetic field and this characteristic is lost in a nonmagnetic environment. Different methods have been applied in the production of magnetic particles reviewed by Platsoucas [1]. Frequently the physical parameters such as size, shape and distribution of magnetic content varies greatly within each production of the commercial particles. Several particles are formed as treated flakes of magnetic oxides or as polydisperse beads with a content of magnetic oxide. There will be variation in size and form of these type of particles and thus an identical behaviour with respect to sedimentation and binding kinetics can not be achieved [1].

The preparation of the monosized paramagnetic polymer beads was solved by Ugelstad and co-workers [2,3] and differed from all other methods employed earlier. The resulting monodispersity ensures that each particle has the same size with identical surface activity. In addition, each bead contains the same amount of magnetic material and therefore each bead will be equally effected in a magnetic field. They prepared these defined magnetic beads from preformed macroporous polystyrene

---

Address for correspondence: J. Lundeberg, Dynal AS, P.O. Box 158, Skøyen, 0212 Oslo, Norway.

beads by in situ formation of magnetic iron oxides (magnetite Fe_3O_4 and/or magnemite γ-Fe_2O_3) evenly distributed inside the pores. The remaining space in the pores is filled with polymer and various methods can be applied for post-treatment of the magnetic beads to provide chemical groups on the surface, which allow direct coupling of the appropriate ligand by covalent bonds. Alternatively, the surface is prepared for activation processes using amino-, hydroxyl- or carboxylic spacers for subsequent coupling of various ligands. The beads have been commercialised as Dynabeads[R] (Dynal AS, Norway) and exist in two diameters 4.5 μm (M-450 Dynabeads) and 2.8 μm (M-280 Dynabeads).

Originally, Dynabeads were used for removal of tumour cells from bone marrow, and selective isolation of cells remains a very important area of application. Discussion of these applications falls outside the scope of the present review article. However, excellent reviews dealing with immunomagnetic cell separation with Dynabeads have appeared [4–7]. A rather new and rapidly increasing area of application of magnetic beads is in the field of molecular biology. Monodisperse paramagnetic beads have shown to be a valuable tool for the isolation, identification and genetic

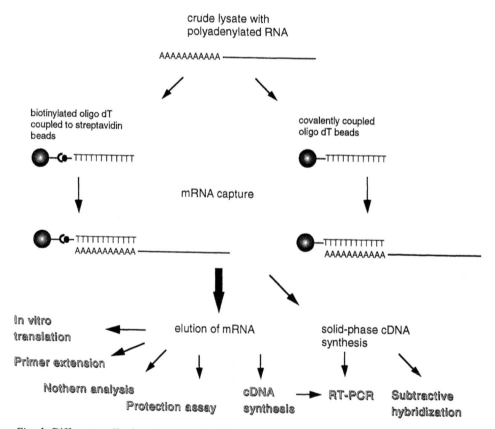

Fig. 1. Different applications in molecular biology using oligo(dT) sequences coupled to magnetic beads.

Biotinylated
oligonucleotide

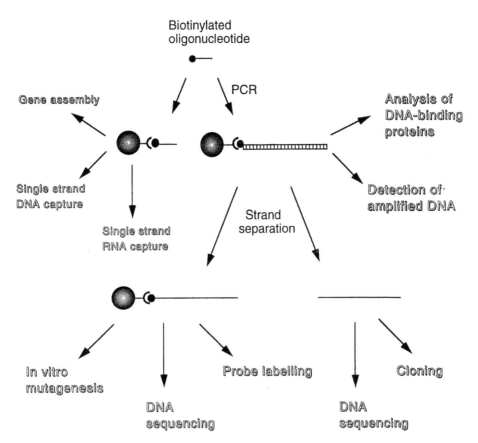

Fig. 2. Different applications in molecular biology using streptavidin covalently coupled to magnetic beads.

analysis of specific nucleic acid sequences including both RNA and DNA. Numerous applications exist for the use of DNA or RNA bound to the magnetic bead with either oligo(dT) sequences or streptavidin coupled to the surface (Figs. 1 and 2) [8,9]. Beads with covalently bound streptavidin can be used for directed immobilisation of both double-stranded and single-stranded biotinylated DNA. The principle of this approach is the use of the remarkable characteristics of the biotin-streptavidin binding: extremely strong ($K_d = 10^{-15}$ M), temperature stable (up to 80°C) and able to withstand alkali treatment (0.15 M NaOH). The most straightforward approach to introducing a biotin label into a double-stranded DNA fragment is to take advantage of the polymerase chain reaction (PCR) using a biotinylated primer.

The different solid-phase approaches are described starting with mRNA purification and solid-phase cDNA synthesis and other advanced RNA manipulations such as subtractive hybridisation and we continue with DNA related applications such as solid-phase DNA sequencing and DNA diagnostics and conclude with some special applications of magnetic beads in molecular biology.

Solid-Phase mRNA Isolation

Preparation of high-quality mRNA is an important step in the analysis of gene structure and gene regulation. It has been estimated that a mammalian cell typically expresses 10,000–50,000 different genes [10,11]. Many of these genes are common to most if not all cells of the body, the so-called "housekeeping" genes. However, a majority of the genes in the genome have tissue-restricted expression. Partly caused by this, the cell types of a multicellular organism differ in their mRNA populations. The average number of molecules of each mRNA per cell is called its abundance or representation. The mRNA population can be divided into general classes, according to their abundance. The abundant class typically consists of less than 100 different mRNAs present in 1,000–10,000 copies per cell and often corresponds to a major part of the total mRNA. About half of the mass of the mRNA consists of a large number of sequences, of the order of 10,000, each represented by only a small number of copies.

The underlying rationale for purifying mRNA is to increase the statistical representation of a particular mRNA in a sample, e.g. medium or low abundance mRNA. Any approach of this nature is an enrichment strategy to be able to observe specific transcripts by Northern analysis, S1-nuclease analysis, RNase-protection analysis or conversion of mRNA into complementary DNA (cDNA).

The existence of the poly(A) tail in most of the eukaryotic mRNAs has been known for many years. The length of the poly(A) tail is typically from 50 to 300 nucleotides long and all mRNA of this sort is said to be polyadenylated and referred to as poly(A)+ mRNA. In order to resolve the polyadenylated RNA from the nonadenylated RNA accounting for 95% or more of a cells total RNA, some type of affinity separation directed towards the poly(A) tail is performed. The conventional technology has involved purification of total RNA as a first step and selection of poly(A)+ RNA by affinity chromatography using oligo(dT)-cellulose as the second step [12]. This strategy is rather time-consuming and labour-intensive. Improvements to this traditional strategy have been to mix oligo(dT) cellulose and total RNA together in a microfuge tube without any prior formation of column or to use an oligo(dT) paper. An alternative strategy for mRNA purification is to use oligo(dT) linked to solid supports like microplate, latex, agarose or magnetic beads.

Over the past 4 years it has become increasingly popular to employ a magnetic bead assisted strategy for poly(A)+ RNA selection since such beads have proven to be favourable in mRNA manipulations (see, for example, Refs. [13–19]). In many approaches, the yield and the quality of the products depends on how rapidly the mRNA can be purified from nucleases and other contaminants. By using the magnetic bead separation technology, pure, intact poly(A)+ RNA can be obtained rapidly either from total RNA preparations or directly from lysates of solid tissues and cell lines [19–22]. The entire procedure can be carried out in a microfuge tube without phenol extractions or ethanol precipitations.

Two somewhat different strategies have been employed (Fig. 1); either hybridising poly(A)+ RNA to biotinylated oligo(dT) tracts linked by streptavidin to magnetic beads or hybridisation with oligo(dT) covalently linked to magnetic beads (Dyna-beads Oligo(dT)$_{25}$, Dynal, Oslo, Norway). Total RNA is mixed with the beads under ionic conditions that favour hybridisation between polyadenylated RNA and the oligo(dT) tracts linked to the beads. The hybridisation kinetics are similar to those found in free solution and complete hybridisation is observed within 5 min [20,21]. The magnetic beads are concentrated by using a strong magnet, whereas washing and elution are achieved by standard combination of temperature and lowering of salt concentrations of the buffers. The principal advantage of the technique is the rapid isolation of pure, intact RNA in a concentrated form. The average capacity of these paramagnetic beads is 2 μg poly(A)$^+$ RNA per mg of beads.

Direct mRNA Isolation from Lysates of Solid Tissues or Cells

Most mRNA purification strategies involve fractionation of total RNA, for example, by use of guanidinium salts and detergent or LiCl and detergent in combination with phenol extractions and ethanol precipitation. By using oligo(dT)-magnetic beads it is possible to purify poly(A)$^+$ RNA directly from crude lysates of cultured cells, animal and plant tissues [19–25]. These methods employ the lysis and hybridisation in LiCl and LiDS/SDS buffers and avoid extra steps such as phenol extraction or proteinase-K digestion. The whole direct mRNA isolation takes approximately 15 min and since the mRNA is stable for more than 30 min in the lysis buffer [21], this ensures the high quality of the mRNA purified. The yield from independent isolations from the same source give variations within ±10% which implies that mRNA isolated with this method can be used directly for quantitative expression studies. However, mRNA per weight unit of tissue is affected by the amount of tissue used. Above a critical threshold of lysed cells the yield of mRNA decreases probably due to the high viscosity of the lysate. The outcome of mRNA can be improved by shearing of DNA, but titration of the amount of tissue, the volume of lysis buffer and the amount of beads are important for optimal results. When working in Eppendorf tubes with volumes up to 1 ml, the rule of thumb is to use less than 100 mg plant tissue or less than 50 mg animal tissue and up to 1 mg of beads.

Most recently, direct mRNA purification was performed using guanidinium iso-thiocyanate (GTC) and sarkosyl [22]. Although, a GTC-buffer system is preferred by most researchers based on the ability of this chaetropic salt to inhibit RNases, the Li-salt buffer approach performs better [22]. The viscosity of the cell lysate in 4 M GTC is higher than in LiCl and the beads are not effectively attracted by the magnet. This may be the reason for lower yields than the LiCl-based approach. However, both approaches result in good quality mRNA as measured by Northern analysis. To test for biological activity, purified mRNA was used as template for in vitro translation [21]. The quality of the mRNA has also been tested by primer extension of mRNA from yeast and barley [22,26]. The quality of poly(A)$^+$ RNA is probably the most important factor in determining the quality of the final results in protocols utilising mRNA, especially for cDNA synthesis.

The combination of immunomagnetic separation of specific cells and subsequent oligo(dT)-magnetic purification of mRNA is a powerful technique for expression studies. Haire et al. [27] have used immunomagnetic beads to select CD19 positive cells from bone barrow to more than 95% purity. They isolated total RNA before using oligo(dT) magnetic beads to select for polyadenylated mRNA. The resulting mRNA was used in Northern blotting, RT-PCR and cDNA cloning. This approach can be shortened by lysing the immunocaptured cells while they are still attached to the beads, then to remove the IMS-beads and use oligo(dT) beads for direct isolation of poly(A)$^+$ mRNA.

Solid-Phase cDNA Synthesis

Detection of RNA transcripts using reverse transcriptase to synthesize complementary DNA (cDNA) and polymerase chain reaction to amplify cDNAs (RT-PCR) has proven to be a fast, sensitive and semi-quantitative assay for gene expression. RT-PCR of mRNA isolated with magnetic beads has been described in a number of reports [14,15,18,24–26,28–36].

An additional advantage of solid-phase mRNA isolation using magnetic beads, is that it is not necessary to elute off the captured mRNAs for subsequent construction of a cDNA library. It is in fact possible to produce solid-phase cDNA libraries specific for a particular cell type or tissue, directly on the bead surface [13,16,22,23,37–41]. The oligo(dT)-sequence bound to the bead surface that captures the mRNA is used as primer for the reverse transcriptase to synthesize the first strand cDNA (Fig. 3). This covalently linked or a streptavidin-biotin linked first-strand cDNA may be used for specific applications like cDNA amplification (RT-PCR) or subtractive hybridisation [13,39–41]. Solid-phase synthesis requires that the beads with the captured mRNA are washed properly before the enzymatic step, to remove detergent (e.g. LiDS or SDS) and salts. All reverse transcriptases tested were found to function in solid-phase cDNA synthesis, AMV (Promega, US), M-MLV (BRL, US), Superscript (BRL, US), rTth (Perkin Elmer, US), and Retrotherm (Epicentre, US) [F. Larsen, unpublished result; 13,23,37,40,41].

The construction of a reusable solid-phase cDNA library by single-sided PCR allows multiple copies of the cDNA of a specific mRNA (second-strand cDNA) to be generated using a single-sided PCR with a specific primer (Fig. 3). The second strand is melted off from the solid-phase template, the beads recovered by magnetic separation and the supernatant with the second strand cDNA is used for amplification. The beads can be used for several specific second-strand cDNA synthesis. Usually, only small amounts of cDNA beads are necessary for cDNA amplification and there may be no need for reuse of these beads. It is then possible to move directly from a cDNA synthesis to a PCR amplification with the beads present during the cycling reactions. The solid-phase approach simplifies the identification and amplification of specific cDNA molecules for downstream analysis and applications.

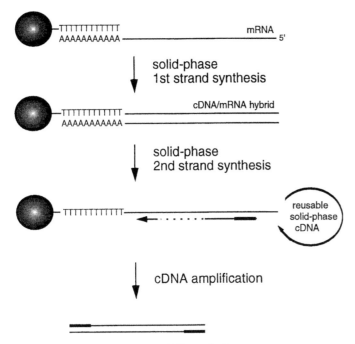

Fig. 3. Principle of solid-phase cDNA synthesis.

RT-PCR has proven sensitive enough to detect transcripts from a few or even a single cell [42–45]. However, these protocols typically require the lysis of cells in the presence of high concentrations of guanidinium thiocyanate and/or the purification of the RNA by caesium chloride ultracentrifugation and/or extraction and precipitation of the RNA. These steps are time-consuming and cumbersome and result in at least some loss of the RNA sample. By purifying mRNA with paramagnetic beads, RT-PCR has been used successfully to amplify transcripts from samples containing on average less than 2 cells (A. Deggerdal, unpublished) (Fig. 4). The abundant transcripts of β-2 microglobulin were isolated from the pre-B cell line Reh, and cDNA synthesis was done with the thermostable reverse transcriptase rTth for 15 min at 70°C (Perkin Elmer). RT-PCR detection was possible both by cDNA synthesis on purified and eluted mRNA and on solid-phase captured mRNA. The advantage of using magnetic beads on microscale mRNA isolations is that mRNA isolation and subsequent cDNA synthesis using a thermostable enzyme takes less than 1 h, probably without significant loss of RNA material.

Rapid amplification of cDNA ends (RACE) PCR has proven to be a valuable tool for obtaining the ends of cDNAs. Lee and Vacquier [37] made a solid-phase cDNA library for RACE by dG tailing of the cDNAs. 5′ RACE was done by amplification with a 3′ gene specific primer and a 5′ adaptor poly(dC) primer. For the 3′ end amplification, a 5′ end specific primer and a 3′ poly(dT) primer were used. The amplicons are separated from the cDNA magnetic beads using a magnetic stand, and the recovered dG-tailed cDNA beads can be reused for at least five rounds of PCR [37].

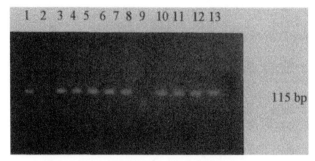

Fig. 4. Example of RT-PCR using minute amounts of sample. Solid-phase cDNA synthesis using mRNA isolated from Reh-cells (pre-B cell line) and PCR with β_2-microglobulin specific primers. Lane 1, 100 cells with RNase treatment prior to RT-PCR; lane 2, 100 cells; lane 3, negative control; lanes 4–13, mRNA from on the average 1.8 cells. Only lane 9 is negative. The probability of getting one negative reaction is less than 0.02 if single cells are not detectable. Courtesy of A. Deggerdal, The Norwegian Radium Hospital, Norway.

By using one biotinylated primer in the amplification, the amplicons can be sequenced by standard solid-phase protocol [46].

Magnetic Bead Assisted Cloning of cDNA Libraries

Construction of a cDNA library from tiny amounts of mRNA by PCR amplification of cDNA that has been tailed at the 3' end of the first strand cDNA has been a successful strategy [47]. By synthesizing a solid-phase cDNA library all manipulations can be carried out in one tube and drawbacks of other protocols like yield loss due to multiple transfers and precipitations can be overcome. Lambert and Williamson [38] were able to clone a representative library from about 5 ng of polyadenylated RNA from tomato root tips. After cDNA synthesis unprimed oligo(dT) on the beads were removed by T4 DNA polymerase before the cDNA strands were A-tailed by terminal transferase. The second cDNA strands were synthesised with an oligo(dT) primer with a tail sequence. The second strand cDNAs were released and amplified and subsequently cloned.

Solid-Phase Subtractive Hybridisation

Subtractive hybridisation approaches have been used to remove sequences common to two related tissues and to enrich for sequences or clones representing differentially expressed genes. Several different strategies have been developed to achieve this goal [48–53]. Traditionally, subtraction protocols have involved hybridisation of mRNA to first-strand cDNA and subsequent hydroxyapatite chromatography, screening of cDNA libraries from one cell type with cDNA probes from a different type of cell or subtractive hybridisations using biotinylated probes and avidin resins. Although successful to varying degrees, these systems often require large amounts of poly(A)$^+$ mRNA or highly purified single-stranded DNA.

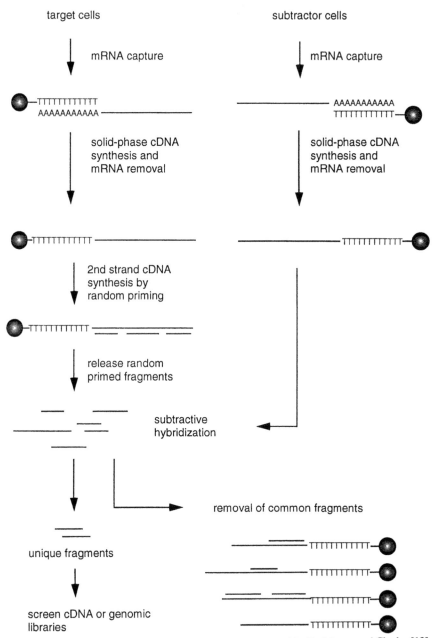

Fig. 5. Principle of subtractive hybridisation strategy suggested by Rodriguez and Chader [13].

A major improvement to a subtraction approach has been the use of magnetic bead assisted subtraction [13]. One principle is to make solid-phase cDNA libraries from both the target mRNA population and the subtractor mRNA (Fig. 5). The second strand cDNA is synthesised by random priming of the target cDNA. The second strand fragments are eluted and mixed with an excess of immobilised subtractor cDNA. The common fragments are allowed to anneal and are removed by magnetic attraction. The unique fragments left in the supernatant are used as a probe to screen a cDNA library [13,40].

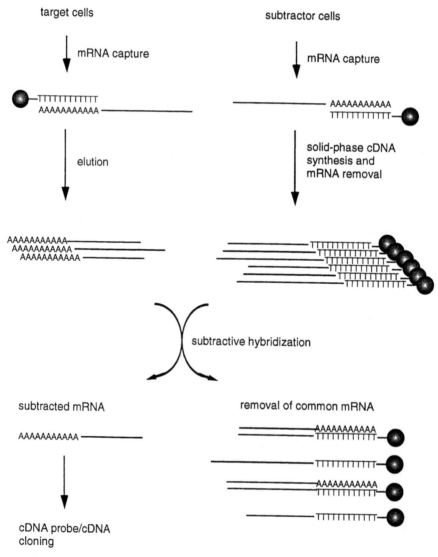

Fig. 6. Principle of subtractive hybridisation strategy suggested by Sharma et al. [39] and Aasheim et al. [41].

Another approach is to hybridise mRNA from a target cell-type with magnetic bead immobilised first-strand cDNA from a subtractor cell-type (Fig. 6) [39,41]. The subtracted mRNA is left in the supernatant after removal of the magnetic beads with the captured mRNA (common mRNA population). The subtracted mRNA is reverse transcribed to radiolabelled cDNA and used as a probe to screen existing cDNA libraries [41] or both first and second strand cDNA are synthesised for cDNA cloning [39]. The specificity of the subtraction procedure can be confirmed by Northern blot analysis. The successful subtraction strategy of Aasheim et al. [41] involved three cell types.

One of the problems with subtraction techniques is that specific mRNAs/cDNAs are subtracted away by common sequences (e.g. gene families) or repetitive sequences. By using short random primed fragments from target mRNAs, Rodriguez et al. [13] reduced this problem. The subtracted fragments used as a probe to screen a retina cDNA library, identified several retina-specific genes like rhodopsin, green visual pigment and interphotoreceptor retinoid-binding protein.

Solid-Phase DNA Sequencing

Since the original description of DNA sequencing using dideoxy nucleotide triphosphates as chain terminators by Sanger et al. [54], this method of sequencing has become well established and extremely popular and is now one of the fundamental techniques in molecular biology. The availability of synthetic oligonucleotides, reliable DNA polymerases and nonisotopic labels have led to automated systems for routine DNA sequencing.

In the past, the preparation of templates for DNA sequencing has relied on the cloning of the target sequence into phage or plasmid vectors. After cultivation of the host cell transformed with the vector (and in some cases helper phage), sufficient amounts of target DNA can be prepared to enable chain termination DNA sequencing. This indirect method, using cloning to prepare DNA for sequencing has become more efficient by the use of PCR [55,56]. The PCR primers can be designed to contain restriction handles enabling optimal, easy and rapid cloning into suitable sequencing vectors [57]. However, thermostable Taq DNA polymerase used in PCR lacks proofreading activity, and therefore multiple clones may have to be sequenced and analysed to find an unaltered sequence.

The advent of PCR also offers an alternative direct method for preparing template DNA which may not have already been cloned for sequencing. There are distinct advantages to producing a sequencing template by PCR; speed, the PCR is faster than bacterial or phage growth; purity, the DNA produced by PCR is free from contaminating cellular material; simplicity, the PCR is simple to set up and run; reproducibility, variation in DNA preparation is much reduced with PCR. Furthermore, the DNA sequence obtained will represent the sequence of the sample prior to amplification, as the errors produced by Taq polymerase will not significantly contribute to the resulting signal. Although these advantages are significant, there are several disadvantages. The first is the need to remove excess PCR primers and nucleotides. This can usually be overcome by a precipitation step or by centrifuge-driven spin dialysis. The second consideration is much more serious. The competition between sequence primer annealing and re-annealing of the complementary strand cause severe problems that leads to high background and other sequencing artefacts. Several solutions have been proposed and are discussed below.

There are two options to perform the direct approach for sequencing of PCR products depending on the nature of the generated template, being either double- or single-stranded DNA. The protocols for sequencing of double-stranded templates [58] are based on selective annealing and extension of the sequencing primer to a denatured target DNA. The essential problem is the reannealing of the complementary strand and the target. Therefore, the most optimal template for a DNA polymerase is a single-stranded DNA with no competition between the sequencing primer and the complementary strand. Several methods have been described for the generation of single-stranded templates from PCR products. Gyllensten et al. [59] described an asymmetric PCR protocol for the preferential production of single-stranded products. Phage lambda exonuclease III can also be used to generate single-stranded DNA [60], by selective digestion of one phosphorylated strand. In both cases the reaction conditions need to be carefully adjusted and the single strand needs to be purified before performing the subsequent sequencing reactions. Another approach uses the PCR primers to introduce RNA promoter sequences [61,62]. Thus following the amplification, the PCR product is used as template for RNA transcription using RNA polymerase. The single-strand RNA products generated can be used directly in a sequencing protocol. However, a RNA template is not as stable as a DNA template and problems can occur with RNase contamination.

The use of streptavidin-coated magnetic beads to capture and purify biotinylated PCR products circumvents many of the problems in the preparation of sequencing templates and enables a robust system for DNA sequencing (Fig. 7) [46,63]. The method allows for complete removal of one strand and all nucleotides and primers in a few simple steps, resulting in a pure single-strand template. Biotinylation of primers has been significantly simplified by the introduction of biotin phoshoramidites enabling a direct coupling onto to the 5' end during synthesis of the primer. For longer DNA fragments, when PCR cannot be employed, a restriction and fill-in reaction using biotin-dUTP can be performed [46].

The capture and immobilisation are accomplished by incubating the biotinylated DNA with the streptavidin-coated magnetic beads for 10 min. Once immobilised, the DNA is converted to a single-stranded template by addition of 0.15 M NaOH which denatures the double-stranded DNA structure. This results in the elution of the non-biotinylated strand into the supernatant while the biotinylated strand remains captured to the bead surface. After the magnetic separation, a pure single-stranded template is achieved. The benefit of the method is that all reaction components are removed, including the complementary strand, enabling optimal sequencing conditions with no reannealing problems. Note that after neutralisation, the eluted strand can also be used as a template. Furthermore, the solid-phase approach enables the devel-

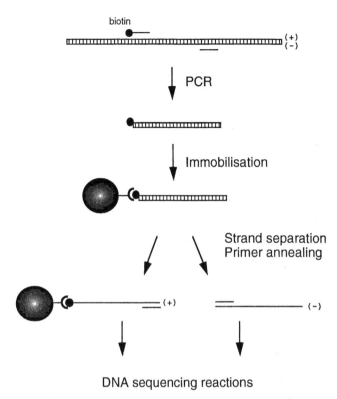

biotin

PCR

Immobilisation

Strand separation
Primer annealing

(+)   (−)

DNA sequencing reactions

Fig. 7. Principle of solid-phase DNA sequencing of both strands.

opment of integrated and automated methods for routine sequencing, facilitated by the defined and predictable behaviour of the monodisperse magnetic beads [46,63].

Solid-phase sequencing can be performed with every DNA polymerase suitable for sequencing, such as, T7, Klenow, Taq, TTh and Bst. The thermostable enzymes incorporate dideoxynucleotides poorly in an inconsistent pattern in comparison with native deoxynucleotides [64]. Despite the disadvantage of nonuniform signal intensities, the convenience of rapid sequence generation by just cycling the temperature has made Taq DNA polymerase useful for sequencing efforts involving cloned target DNA. However, in clinical applications, the T7 DNA polymerase is the enzyme of choice, because of better peak uniformity compared to the other polymerases. Peak uniformity is important for accurate base-calling, especially for the detection of heterozygosity in genomic material [64].

Through major advances in the chemistry and in the automation of DNA sequencing in recent years, the field of applications has broadened from large-scale sequencing projects to applications in clinical medicine. DNA sequencing is becoming one of the established techniques in disease diagnosis in a routine setting. The most important developments towards automated routine use is the introduction of PCR (as mentioned earlier) and the replacement of isotopic labelling by fluorescent dyes

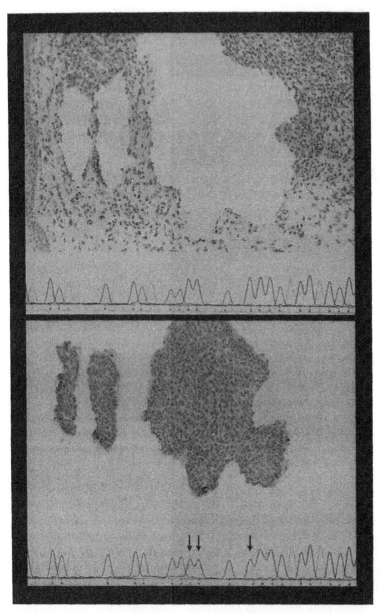

Fig. 8. Example of solid-phase sequencing of the p53 gene in a human basal cell carcinoma sample. Courtesy of A. Hedrum and M. Uhlén, Royal Institute of Technology, Stockholm, Sweden [78].

and on-line sequence reading [65,66]. The fluorescent sequencing bands are excited by a laser beam and detected in the gel during electrophoresis. There are two main methods to label the generated dideoxy DNA fragments, by fluorescent-labelled primer or dideoxy chain terminators, creating a flexible nonradioactive detection system. The commercially available instruments also enable reliable quantitation of polymorphic and heterozygous positions with the different software packages [67–69]. In addition, several systems have been described for automation of the sequencing reactions based on robotic workstations and magnetic beads [63,70]. These improvements have proven to be valuable in standardisation for clinical DNA sequencing.

There are numerous examples where solid-phase sequencing has proven to be a reliable tool for DNA analysis. Examples include: genomic analysis of viral variation and dynamics [28,68,71–74], drug resistance analysis [67,75], forensic investigations [76,77], point mutation analysis [64,78], typing using HLA loci or ribosomal genes [79–81] and many more. Noteworthy is the common feature among these approaches in their use of T7 DNA polymerase as sequencing enzyme to distinguish overlapping peaks originating from polymorphic sequences or heterogeneous complex of different species.

An example of direct solid-phase sequencing is shown in Fig. 8. It illustrates point mutations appearing in the p53 gene encoding a nuclear phosphoprotein which is among the most frequent genetic alterations in human solid tumours. These mutations are clustered in highly conserved domains in exons 4–9. Hedrum et al. [78] performed microdissection of a human basal cell carcinoma tissue stained with haematoxylin/eosin and analysed by sequence-based diagnostics. The DNA sequences obtained by microdissection of the normal cells (top) and cancer cells (below), followed by direct, solid-phase PCR sequencing are shown. The specific mutations of the cells of the tumour biopsy are indicated by arrows. These type of data can then be used to clarify risk factors as well as prognostic markers for tumour progression.

Ligation-Mediated PCR and Magnetic Separation

Ligation-mediated PCR (LM-PCR) provides a method for direct sequencing and cloning of fragments outside the boundaries of a known sequence [82]. The main application of such a genomic walking technique is the ability to clone the upstream flanking regions of genes. The approach is based on the ligation of an oligonucleotide cassette to specific restriction enzyme cleavage fragments [83], providing a generic sequence that can be used for amplification in combination with a specific primer for a known sequence. The specificity of the method is obtained by several rounds of primer extension using a biotinylated primer. The biotinylated target molecules are captured with streptavidin-coated beads and the nonspecific DNA removed prior to PCR amplification. The advantages of a solid-phase LM-PCR approach have been demonstrated by genomic walking, direct sequencing and in vivo footprinting in organisms with complex genomes [84–87]. In summary, the magnetic-bead capture of extension products improves the results and significantly reduces the labour and time requirements compared to other genomic walking methods.

Gene Hunting by Direct cDNA Selection

One of the major goals of the genome project is the identification and character-isation of novel genes. Rapid identification and isolation of transcribed sequences from defined but large genomic regions is an inherent part of positional cloning strategies of genes responsible for hereditary diseases. Conventional approaches for finding transcribed sequences, such as identifying conserved sequences, CpG islands or exons (exon trapping) are cumbersome when working with large genomic regions.

Recently, direct cDNA selection approaches ("hybrid selection") have been devised to circumvent the disadvantages of the traditional methods. These novel selection methods employ PCR combined with a selection of cDNAs by hybridisation to large genomic inserts. Target DNA (YAC, BAC or cosmids) is immobilised on solid support like nylon-filter discs [88] or magnetic beads [89–92]. These methods offer several advantages over library screening methods and are insensitive to the presence of introns or cryptic splice sites which can cause problems in exon-trapping approaches.

A direct cDNA selection approach was applied to a region of the human chromosome Xq28 [89,92]. Genomic DNA cloned in cosmids (around 40 kb) was fragmented by sonication and biotinylated. These fragments were preannealed with repetitive elements and hybridised with PCR-amplified cDNA inserts from a library. The captured cDNAs were eluted from the hybrids immobilised on streptavidin-coated magnetic beads, reamplified and subjected to two further rounds of hybridisation-amplification enrichment. The transcription map generated confirmed the high gene density, overall 11 genes in 300 kb. The direct cDNA selection method was shown to be a very simple, rapid and effective tool for the generation of a regional transcription map. It screens the genomic regions selected as a whole and does not require its precise analysis, selection of single probes, its subcloning into special vectors and the extensive screening of conventional cDNA libraries, if the full length cDNA is not required.

Magnetic bead capture is a powerful tool for the isolation of cDNAs encoded from large genomic regions. The efficiency and speed of the technique allows cDNAs or cDNA libraries from multiple tissues to be used in parallel. This will increase the probability of detecting tissue-specific transcripts. A possible disadvantage of the direct selection method is the isolation of transcripts from other regions of the genome captured by pseudogene fragments [89]. Furthermore, the method tends to normalise transcript levels, with a greater enrichment of rare messages than abundant ones [91].

To make a complete transcriptional map, a combination of approaches may be necessary. In addition to a direct cDNA selection and a CpG island screening [93], a magnetic bead selection of conserved sequences is possible [94]. PCR-amplified DNA fragments from the whole genome of one species is hybridised with biotinylated DNA fragments from a defined region of another species. The interspecies hybrids are subsequently immobilised on magnetic beads and PCR-amplifiable conserved sequences are eluted. By using this procedure, novel genes have been identified [94].

Detection of Amplified Material Using Magnetic Separation

Traditional techniques used for isolation and detection of pathogens have often been based on culturing with subsequent biochemical and immunological identification of the pathogen. These rather time-consuming procedures are now gradually being changed into nucleic acid-based amplification strategies. The introduction of polymerase chain reaction has dramatically changed and affected detection of nucleic acids. Numerous methods are available to analyse PCR-amplified material such as gel electrophoresis, hybridisation, restriction cleavage patterns and DNA sequencing. DNA sequencing is the most informative of all analysis techniques, however, it is not always needed when only a yes or no answer is required.

The analysis of specific nucleotide sequences using PCR has often overcome the limitations in sensitivity and specificity of conventional DNA detection methods such as probe hybridisation techniques. A recent development in PCR is the use of magnetic beads to capture the amplified material for subsequent detection. The detection of immobilised amplified material (DIANA) principle has been designed for colorimetric detection of amplified DNA [95,96]. It was first described using a lac operator handle in one of the PCR primers while the other primer contained a biotin molecule at the 5' end. Following amplification, the product is immobilised onto streptavidin-coated magnetic beads and the detection of DNA is achieved by using a reporter fusion protein consisting of the *Escherichia coli* lac repressor and β-galactosidase (Fig. 9). The analysis of the PCR product is thus, from a practical point

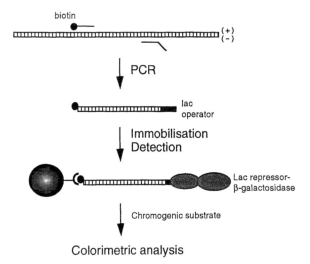

Fig. 9. Principle of the detection of immobilised amplified nucleic acids (DIANA) assay.

of view, identical to that in the enzyme linked immunosorbent assay (ELISA) with colorimetric response after substrate addition. The DIANA system has been used successfully to detect bacteria [95–98], virus [73] and parasites [75,95]. Alternative labelling methods of the primers include radioisotopic end labels [97] and digoxigenin (Boehringer Mannheim, Germany). The commercial system named QPCR from Perkin Elmer, is based on detection of PCR fragments by magnetic capture. The stability of the biotin-streptavidin system is used to separate the two DNA strands with alkali to enable the hybridisation with a labelled internal probe carrying a label for chemiluminescence detection.

Quantification of Pathogens using Magnetic Separation

The ability to determine the initial copy number of a specific RNA/DNA sequence in a sample is of great general interest. To allow quantitative determination of the initial number of target copies in a given sample the qualitative DIANA described earlier was modified [99]. The objective in the development of a quantitative assay was to achieve a system which enabled easy evaluation and to obviate the problems encountered for co-amplification systems based on the analysis in the narrow exponential "window" of PCR, but also to increase the dynamic range and reproducibility. A schematic illustration of the quantitative approach is shown in Fig. 10.

To obtain a colorimetric analysis a competitor DNA standard was used. The competitor construct was based on a cloned target region with a substitution of 21 bp into the middle of the target fragment sequence, corresponding to the *lac* operator sequence. The construct contained the same primer annealing sequences as the target, and resulted in an amplified product with the same size as the target product. The purpose of these manipulations was to minimise the differences during amplification of target and competitor. In addition, these types of constructs have two major advantages. First, the target regions are cloned into the multiple cloning sites of standard vectors enabling the production of RNA copies using the T7 promoter upstream of the insert. Thus RNA competitors containing a *lac* operator can be produced, quantified and used in a quantification scheme. Secondly, since the competitor and sample have identical sequences except for the *lac* operator sequence, a change in primer annealing sites can readily be performed without the need for extra cloning work.

The methodology using this type of competitor involves a titration of competitor and target prior to amplification. The nested primer approach is suggested to allow amplification until the plateau phase for all dilution steps of the competitor. One of the primers are biotinylated to allow the immobilisation of the amplified material onto streptavidin-coated magnetic beads. The incorporated *lac* operator sequence of the competitor is used as a recognition sequence for the subsequent binding of the reporter protein, LacI-β-galactosidase. A strong colour response reflects a high ratio of competitor to sample and a weak colour response a low ratio. At some point in the titration series, the starting concentrations of competitor and template is identical, and by assuming an identical amplification efficiency the ratio is kept constant

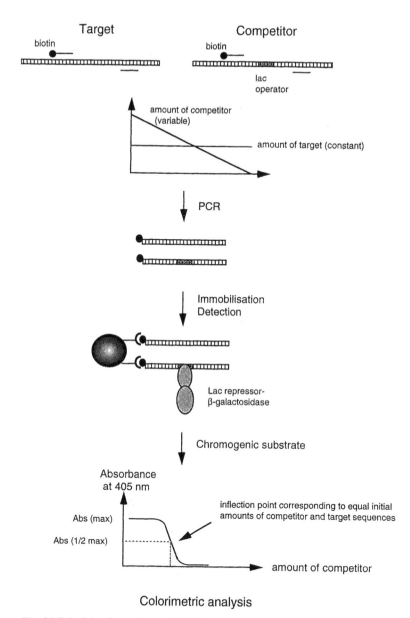

Fig. 10. Principle of quantitative DIANA.

throughout the amplification. In the subsequent colorimetric analysis this point is equivalent to the intermediate absorbance between the maximum and minimum values. Thus the initial target number can be easily determined by the point of inflection in the titration curve, where equal amounts of sample and competitor are present. The methodology has been used to quantitate the number of parasites in the blood of malaria patients [99], but also to analyse the viral load of HIV-1 and hepatitis C virus in blood [100,101].

Sample Preparation using Magnetic Separation

The application of PCR requires considerable attention in the preparation of samples. While PCR usually works well with highly purified DNA, it is more difficult to prepare quality DNA from blood, urine, stool and other samples which often contain inhibitors to the Taq polymerase. In addition, traditional sample preparation often includes steps such as phenol extractions, alcohol precipitations and centrifugations which are not simple to standardise and automate.

The paramagnetic beads coated with antibodies against surface antigens of pathogens have been shown to be efficient in sample preparation from different biological and environmental samples (reviewed in Refs. [6,102]). The procedure involves mixing beads with crude sample, and after incubation for 10–30 min, extracting the beads with the bound living bacteria/cells using a magnet. The immunomagnetic separation (IMS) technique results in a gentle capture of the pathogens which usually remains viable and can continue to grow if nutritional requirements are provided. Both polyclonal and monoclonal antibodies [103,104] can be used either directly coupled to the bead surface or indirectly using a pre-coated bead with anti-mouse or anti-rabbit antibodies on the bead surface (Fig. 11). Thus immunomagnetic separation has several advantages as the target pathogen can be separated from its environment (blood, urine, stool, etc.) and concentrated to a volume suitable for PCR. Potent PCR inhibitors present in the sample can be removed by this technique resulting in a more robust sample preparation. After IMS a simple boiling procedure (10–30 s) is generally sufficient to lyse most gram-negative bacteria [105,106]. Freeze-thawing and snap-cooling of the IMS samples can also be employed to open cells making the DNA available for PCR. The combined use of IMS and PCR has been

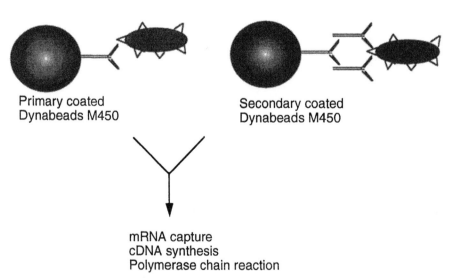

Primary coated
Dynabeads M450

Secondary coated
Dynabeads M450

mRNA capture
cDNA synthesis
Polymerase chain reaction

Fig. 11. Immunomagnetic separation of microorganisms. The specific antibody can be attached directly to the magnetic bead or through a secondary antibody.

shown to be a valuable help in the identification of food-borne pathogens such as *Escherichia coli*, *Salmonella*, *Listeria monocytogenes* with up to 100,000 times higher sensitivity compared with PCR alone (Ø. Olsvik, pers. commun.). The DIANA system has also been integrated with an initial immunomagnetic separation step for sample preparation both in blood samples [107] and urine samples [108] achieving a more robust system for diagnosis.

An alternative to antibody-coated beads for sample preparation is to use a capture probe coupled to the bead surface. Thereby solid-phase capturing of single-strand species can be achieved and nucleic acids are purified and concentrated by magnetic separation. Albert and co-workers [72] showed that this was a feasible and rapid approach for detection of polyadenylated HIV-1 RNA genomes in patient material using oligo(dT) beads, while others have used an approach with a specific probe attached by biotin to streptavidin-coated beads which they thereby hybridised specific with the viral RNA and used subsequently in an RT-PCR analysis [109,110].

Solid Phase Cloning

It is often desirable to clone PCR products into plasmid vectors. Unfortunately, such products have proven unusually resistant to standard cloning procedures, requiring significant vector and PCR product preparation. A solid phase approach has been described [111] which is based on sequence homology between the vector and the PCR product with no restriction enzymes or ligases involved. Briefly, a single-stranded vector is prepared by selective incorporation of a biotin into one of the strands of the vector DNA. The double-stranded DNA is bound to magnetic beads coated with streptavidin and the nonbiotinylated vector strand is eluted and neutralised. The DNA to be cloned is obtained by PCR, using primers with an additional handle sequence (complementary to the end of the single-strand vector). In addition one of the primers is biotinylated which enables the generation of single-strand PCR product in a similar procedure as with the vector. The two single-stranded fragments can then be mixed to form a gap-duplex molecule and is transformed directly into *E. coli* for in vivo gap repair. More than 90% correct recombinants were obtained when a human apolipoprotein E gene fragment was cloned from the human chromosome using this approach [111].

Solid Phase In Vitro Mutagenesis

A large number of approaches exist for in vitro mutagenesis based on PCR and mediating oligonucleotides to provide accurate and precise changes in a DNA sequence. Hultman et al. [112] described a solid phase approach with similarities to solid-phase cloning. A single-strand vector is prepared and mixed with a mutated single-strand fragment with complementary sequences at the ends and the gap-duplex is transformed into *E. coli* for propagation. The mutated single-strand fragment is prepared by amplification of the region with one of the primers biotinylated to enable capture to streptavidin coated magnetic beads and subsequent elution of the nonbiotinylated strand. The immobilised single strand is used as template for hybridisation of mutagenesis oligonucleotide with the desired change in nucleotide sequence. The mismatch primer is then extended by T4 DNA polymerase, creating a mutated copy of the target region. After elution of the mutated strand from the magnetic bead, this is directly mixed with the vector strand. After transformation, the clones are screened by conventional methods to find mutated plasmids. An important advantage with the approach is that the mutated strand region is not exposed to the mismatch repair system of *E. coli*.

Magnetic DNA-Affinity Purification of DNA-Binding Proteins

DNA-binding proteins are able to recognise and bind to a specific DNA sequence much more strongly than nonrelated sequences. This makes it possible to concentrate protein(s) of interest selectively by using a DNA fragment immobilised on solid support like magnetic beads [113 and references therein]. The principle of the method is that a specific DNA fragment is bound to magnetic beads through a streptavidin-biotin linkage. These beads are incubated with a crude extract containing a large excess of genomic DNA to avoid nonspecific DNA-binding proteins. After washing, the proteins are released by using a high-salt buffer. The specificity of the purified proteins must be confirmed by, for example, a band-shift assay. The complexity and time required for purification are reduced by using magnetic beads. The purification process may be carried out several times, as the beads are recyclable. Gabrielsen et al. [114] purified near to homogeneity a yeast transcription factor IIIC by using a magnetic bead approach. The complete purification including three cycles of binding was done within 1 h.

The confirmation of a pure protein can even be made directly by incorporating a restriction site in the specific DNA fragment and releasing the DNA-protein complex from the beads by restriction enzyme cleavage [115]. This complex can be checked by a direct band-shift assay.

Use of Magnetic Beads to Study DNA-Protein Complexes

The use of randomised oligonucleotides and PCR-based reiterative selection techniques have been used to define the consensus binding sites for transcription factors (reviewed in Ref. [116]). The principle of the method is to mix randomised DNA sequences which have PCR-amplifiable ends, with protein. DNA-protein complexes are isolated by a variety of methods such as immunoprecipitation, immobilisation on solid phases or isolation of the band from shift assays. The use of immunomagnetic separation is very convenient for a repeated enrichment. A monoclonal antibody specific for the DNA-binding protein can readily be immobilised on anti-Ig coated magnetic beads (Dynal, Norway). The resulting affinity matrix is then used to immunomagnetically isolate those sequences which have complexed with the DNA-binding protein [117]. The recovered DNA is PCR amplified and mixed with fresh protein and complexes are selected once more. After sufficient enrichment of specific sequences, the DNA fragments are cloned and sequenced to obtain the consensus sequence for a DNA-binding protein.

In most studies purified proteins have been used, but the method can be used with nuclear extracts. The use of such crude extracts rather than purified proteins permits multicomponent complexes to form, and allows information about protein interactions involved in gene regulation to be generated [116]. Wright et al. [117] described the use of this powerful technique to elucidate a 10-base pair consensus binding site sequence, for the muscle promoter factor myogenin, from a 14-base pair degenerate oligonucleotide to an enrichment factor of 450,000-fold.

Another interesting method for studying DNA-protein interaction is the reconstitution of chromatin on magnetic beads [118]. Nucleosomes was assembled on DNA fragments attached by biotin-streptavidin binding to the beads. The reconstituted chromatin must be purified away from the assembly reaction prior to in vitro transcription. In previously described systems, sucrose gradient sedimentations have generally been used, which is time-consuming and may change the state of the chromatin. The magnetic bead approach enabled an efficient purification of chromatin templates for transcription studies. This approach was used to study the contribution of histone H1 on transcriptional repression. An important step in the method is the immobilisation of very long DNA fragments of more than 6 kb by performing the coupling at room temperature overnight. The experiments also showed that RNA polymerase can use an magnetic bead immobilised DNA fragment as a template for transcription.

Concluding Remarks

Magnetic separation of DNA, RNA or proteins has proven to be very useful for many applications in molecular biology (Figs. 1 and 2). We have described procedures that are used for mRNA isolation, subtractive hybridisation, selective hybridisation, sequencing, cloning, in vitro mutagenesis, detection and quantification of amplified material, sample preparation and purification of DNA-binding proteins. Solid-phase DNA sequencing and mRNA purification have been used most frequently. However, novel solid-phase techniques are continually being published. Among these are gene assembly strategies [119]; new means to recover DNA binding fusion proteins [120] and several magnetic bead based mutation assays [121]. In addition there are procedures for recovery of single-stranded DNA (phage m13) [122], labelling of single-stranded DNA probes [123,124], direct purification of tRNAs [125,126] and reversible immobilisation of antibodies [127]. In many applications, magnetic bead technology provides simple, rapid and reliable alternatives to traditional techniques. Both manual and automated methods can be developed and some of these will probably have great impact on molecular biology in the near future.

References

1. Platsoucas CD. In: El Asser MS, Fitch RM (eds) Biomedical Applications of Polymer Beads with Emphasis on Cell Separation; Future Directions in Polymer Colloids, NATO ASI series E, No 138. Dordrecht, The Netherlands: M. Nijhoff, 1987;321–354.
2. Ugelstad J, Mørk PC, Kaggerud KH, Ellingsen T, Berge A. Swelling of oligomer-polymer particles. New methods of preparation of emulsions and polymer dispersions. Adv Colloid Interface Sci 1980;13:101–140.
3. Ugelstad J, Berge A, Ellingsen T, Aune O, Kilaas L, Nilsen TN, Schmid R, Stenstad P, Funderud S, Kvalheim G, Nustad K, Lea T, Vartdal F, Danielsen H. Monosized magnetic particles and their use in selective cell separation. Makromol Chem Macromol Symp 1988;17:177–211.
4. Lea T, Vartdal F, Nustad K, Funderud S, Berge A, Ellingsen T, Schmid R, Stenstad P, Ugelstad J. Monosized, Magnetic polymer particles: their use in separation of cells and subcellular components and the study of lymphocyte function *in vitro*. J Mol Recog 1988;1:9–18.
5. Padmanabhan R, Corsico C, Holter W, Howard T, Howard BH. Purification of transiently transfected cells by magnetic-affinity cell sorting. J Immunogenet 1989;16:91–102.
6. Ugelstad J, Berge A, Ellingsen T, Schmid R, Nilsen T-N, Mørk PC, Hornes E, Olsvik Ø. Preparation and Application of New Monosized Polymer Particles. Progress in Polymer Science. Oxford: Pergamon Press 1992;17:87–161.
7. Haukanes B-I, Kvam C. Application of magnetic beads in bioassays. Bio/Technology 1993;11:60–63.
8. Uhlén M. Magnetic separation of DNA. Nature 1989;340:733–734.
9. Uhlén M, Hultman T, Wahlberg J, Lundeberg J, Bergh S, Petersson B, Holmberg A, Ståhl S, Moks T. Semi-automated solid-phase DNA sequencing. Trends Biotechnol 1992;52–55.
10. Bishop JO. The gene numbers game. Cell 1974;2:81–86.
11. Hastie ND, Bishop JO. The expression of three abundant classes of messenger RNA in mouse tissues. Cell 1976;9:761–774.
12. Aviv H, Leder, P. Purification of biologically active globin messenger RNA on oligothymidylic acid-cellulose. Proc Natl Acad Sci USA 1972;69:1408.
13. Rodriguez IR, Chader GJ. A novel method for the isolation of tissue-specific genes. Nucleic Acids Res 1992;20:3528.
14. Cattanach BM, Barr JA, Evans EP, Burtenshaw M, Beechey CV, Leff SE, Brannan CI, Copeland NG, Jenkins NA, Jones J. A candidate mouse model for Prader-Willi syndrome which shows an absence of *Snrpn* expression. Nature Genet 1992;2:270–274.
15. Korpi ER, Kleingoor C, Kettenmann H, Seeburg PH. Benzodiazepine-induced motor impairment linked to point mutation in cerebellar GABA$_A$receptor. Nature 1993;361:356–359.
16. Raineri I, Senn HP. HIV-1 promoter insertion revealed by selective detection of chimeric provirus-host gene transcripts. Nucleic Acids Res 1992;20:6261–6266.
17. Meyer M, Schreck R, Baeuerle A. H_2O_2 and antioxidants have opposite effects on activation of NF-kB and AP-1 in intact cells, AP-1 as secondary antioxidant-responsive factor. EMBO J 1993; 12:2005–2015.
18. Orr HT, Chung M, Banfi S, Kwiatkowshi Jr TJ, Servadio A, Beaudet AL, McCall AE, Duvick LA, Ranum LPW, Zoghbi HY. Expansion of an unstable trinucleotide CAG repeat in spinocerebellar ataxia type 1. Nature Genet 1993;4:221–226.
19. Larsen F, Solheim J, Kristensen T, Kolstø A-B, Prydz H. A tight cluster of five unrelated human genes on chromosome 16q22.1. Hum Mol Genet 1993;2:1589–1595.

20. Hornes E, Korsnes L. Magnetic DNA hybridisation properties of oligonucleotide probes attached to superparamagnetic beads and their use in the isolation of poly(A) mRNA from eukaryotic cells. Genet Appl Tech Anal 1990;7:145–150.

21. Jacobsen KS, Breivold E, Hornes E. Purification of mRNA directly from crude plant tissues in 15 minutes using oligo dT microspheres. Nucleic Acids Res 1990;18:3669.

22. Jacobsen KS, Haugen M, Sæbøe-Larsen S, Hollung K, Espelund M, Hornes E. In: Uhlén M, Hornes E, Olsvik Ø (eds) Advances in Biomagnetic Separation. Direct mRNA Isolation Using Magnetic Oligo(dT) Beads: a Protocol for All Types of Cell Cultures, Animal and Plant Tissues. Eaton Publishing, 1994;61–71.

23. Raineri I, Moroni C, Senn HP. Improved efficiency for single-sided PCR by creating a reusable pool of first strand cDNA coupled to a solid phase. Nucleic Acids Res 1991;19:4010.

24. Specher E, Becker Y. Detection of IL-1, TNF- and IL-6 gene transcription by the polymerase chain reaction in keratinocytes:Langerhans cells and peritoneal exudate cells during infection with herpes simplex virus-1. Arch Virol 1992;126:253–269.

25. Spurkland A. Magnetic isolation of mRNA for in vitro amplification. Trends Genet 1992;8:225–226.

26. Faulkner JDB, Minton NP. Rapid small-scale isolation of mRNA from whole yeast cells. Biotechniques 1993;14:718–720.

27. Haire RN, Ohta Y, Lewis JE, Fu SM, Kroisel P, Litman GW. Txk, a novel human tyrosine kinase expressed in T cells shares sequence identity with Tec family kinases and maps to 4q12. Hum Mol Genet 1994;3:897–901.

28. Chiodi F, Keys B, Albert J, Lundeberg J, Uhlén M, Fenyö EM, Norkrans G. Human immunodeficiency virus type 1 is present in the cerebrospinal fluid of a majority of infected individuals. J Clin Microbiol 1992;30:1768–1771.

29. Ebling SB, Schutte MEM, Logtenberg T. Peripheral human CD5+ and CD5- B cells may express somatically mutated V_H5- and V6-encoded IgM receptors. J Immunol 1993;151:6891–6899.

30. Houssiau FA, Renauld JC, Stevens M, Lehmann F, Lethe B, Coulie PG, van Snick J. Human T cell lines and clones respond to IL-9. J Immunol 1993;150:2634–2640.

31. Kaldenhoff R, Kölling A, Richter G. A novel blue light- and abscisic acid-inducible gene of *Arabidopsis thaliana* encoding an intrinsic membrane protein. Plant Mol Biol 1993;23:1187–1198.

32. Moyal M, Berkowitz C, Rösen-Wolff A, Darai G, Becker Y. Mutations in the UL53 gene of HSV-1 abolish virus neurovirulence to mice by the intracerebral route of infection. Virus Res 1992;26:99–112.

33. Obata F, Tsunoda M, Ito K, Ito I, Kaneko T, Pawelec G, Kashiwagi N. A single universal primer for the T-cell receptor (TCR) variable genes enables enzymatic amplification and sequencing of TCR cDNA of various T-cell clones. Hum Immunol 1993;36:163–167.

34. Sprenger H, Bacher M, Rischkowsky E, Bender A, Nain M, Gemsa D. Characterization of a high molecular weight tumor necrosis factor-mRNA in influenza A virus-infected macrophages. J Immunol 1994;152:280.

35. Towner P, Gärtner W. cDNA cloning of 5′ terminal regions. Nucleic Acids Res 1992;20:4669–4670.

36. Villand P, Olsen OA, Kleczkowski LA. Molecular characterization of multiple cDNA clones for ADP-glucose pyrophosphorylase from Arabidopsis thaliana. Plant Mol Biol 1993;23:1279–1284.

37. Lee Y-H, Vacquier VD. Reusable cDNA libraries coupled to magnetic beads. Anal Biochem 1992; 206:206–207.

38. Lambert KN, Williamson VM. cDNA library construction from small amounts of RNA using paramagnetic beads and PCR. Nucleic Acids Res 1993;21:775–776.

39. Sharma P, Lönneborg A, Stougaard P. PCR-based construction of subtractive cDNA library using magnetic beads. BioTechniques 1993;15:610–611.

40. Schraml P, Shipman R, Stulz P, Ludwig CU. cDNA subtraction library construction using a magnet-assisted subtraction technique (MAST). Trends Genet 1993;3:70–71.

41. Aasheim H-C, Deggerdal A, Smeland EB, Hornes E. A simple subtraction method for the isolation of cell-specific genes using magnetic monodisperse polymer particles. Biotechniques 1994;16: 716–721.
42. Belyavsky A, Vinogradova T, Radjewsky K. PCR-based cDNA library construction: general cDNA libraries at the level of a few cells. Nucleic Acids Res 1989;17:919–932.
43. Lambolez BE, Audinat E, Bochet P, Crepel P, Rossier J. AMPA receptor subunits expressed by single purkinje cells. Neuron 1992;9:247–258.
44. Rappolee DA, Wang A, Mark D, Werb Z. Novel method for studying mRNA phenotypes in single or small number of cells. J Cell Biochem 1989;39:1–11.
45. Schriever F, Freeman G, Nadler L. Follicular dendritic cells contain a unique repertoire demonstrated by single-cell polymerase chain reaction. Blood 1991;77:787–791.
46. Hultman T, Ståhl S, Hornes E, Uhlén M. Direct solid phase sequencing of genomic and plasmid DNA using magnetic beads as solid support. Nucleic Acids Res 1989;17:4937–4946.
47. Gurr SJ, McPherson MJ, Scollan C, Atkinson HJ, Bowles DJ. Mol Gen Genet 1991;266:361–366.
48. Sargent TD, Dawid IB. Differential gene expression in the gastrula of Xenopus laevis. Science 1983;222:135–139.
49. Hedrick SM, Cohen DI, Nielsen EA, Davisw MM. Isolation of cDNA clones encoding T-cell specific membrane-associated proteins. Nature 1984;308:149–153.
50. Duguid JR, Rohwer RG, Seed B. Isolation of cDNAs of Scrapie-modulated RNAs by subtractive hybridisation of a cDNA library. Proc Natl Acad Sci USA 1988;85:5738–5742.
51. Rubenstein JL, Bruce AJ, Ciaranello D, Denney D, Porteus MH, Usdin TB. Subtractive hybridisation system using single-stranded phagemids with directional inserts. Nucleic Acids Res 1990;18:4833–4842.
52. Travis GH, Sutcliffe JG. Phenol emulsion-enhanced DNA-driven subtractive cDNA cloning: Isolation of low-abundance monkey cortex-specific mRNAs. Proc Natl Acad Sci USA 1988;85:1696–1700.
53. Timblin C, Battey J, Kuehl WM. Application for PCR technology to subtractive cDNA cloning: identification of genes expressed specifically in murine plasmacytoma cells. Nucleic Acids Res 1990;18:1578–1593.
54. Sanger F, Nicklen S, Coulson AR. DNA sequencing with chain-terminating inhibitors. Proc Natl Acad Sci USA 1977;74:5463–5467.
55. Saiki RK, Scharf S, Faloona F, Mullis KB, Horn GT, Erlich HA, Arnheim N. Enzymatic amplification of β-globin genomic sequences and restriction site analysis for diagnosis of sickle cell anemia. Science 1985;230:1350–1354.
56. Mullis K, Faloona F. Specific synthesis of DNA in vitro via a polymerase-catalysed chain reaction. Methods Enzymol 1987;155:335–350.
57. Scharf SJ, Horn GT, Erlich HA. Direct cloning and sequence analysis of enzymatically amplified genomic sequences. Science 1986;233:1076–1078.
58. Engelke DR, Hoener PA, Collins FS. Direct sequencing of enzymatically amplified human genomic DNA. Proc Natl Acad Sci USA 1988;85:544–548.
59. Gyllensten UB, Erlich HA. Generation of single-stranded DNA by the polymerase chain reaction and its application to direct sequencing of the HLA-DQA locus. Proc Natl Acad Sci USA 1988;85: 7652–7656.
60. Higuchi RG, Ochman H. Production of single-stranded DNA templates by exonuclease digestion following the polymerase chain reaction. Nucleic Acids Res 1989;17:5865.
61. Sarkar G, Sommer SS. RNA amplifications with transcript sequencing (RAWTS). Nucleic Acids Res 1988;16:5197.
62. Stoflet ES, Koeberl DD, Sarkar G, Sommer SS. Genomic amplification with transcript sequencing. Science 1988;239:491–494.
63. Hultman T, Bergh S, Moks T, Uhlén M. Bidirectional solid-phase sequencing of in vitro-amplified plasmid DNA. BioTechniques 1991;10:84–93.
64. Leren TP, Rødningen OK, Røsby O, Solberg K, Berg K. Screening for point mutations by semi-

automated DNA sequencing using Sequenase and magnetic beads. BioTechniques 1993;14:618–623.

65. Ansorge W, Sproat BS, Stegemann C, Schwager C. A nonradioactive automated method for DNA sequence determination. J Biochem Biophys Methods 1986;13:315–323.

66. Smith LM, Sanders JZ, Kaiser P, Hughes C, Dodd C, Conell CR, Heiner C, Kent SH, Hood LE. Fluorescent detection in automated DNA sequence analysis. Nature 1986;321:674–679.

67. Wahlberg J, Albert J, Lundeberg J, Cox S, Wahren B, Uhlén M. Dynamic changes in HIV-1 quasispecies from azidothymidine (AZT) treated patients. FASEB J 1992;6:2843–2847.

68. Leitner T, Halapi E, Scarletti G, Rossi P, Albert J, Fenyö E-M, Uhlén M. Analysis of hetereogeneous viral populations by direct DNA sequencing. BioTechniques 1993;15:120–127.

69. Larder BA, Kohli A, Kellam P, Kemp SD, Kronick M, Henfrey RD. Quantitative detection of HIV-1 drug resistance mutations by automated DNA sequencing. Nature 1993;365:671–673.

70. Holmberg A, Fry G, Uhlén M. In: Venter C (ed) Automated DNA Sequencing and Analysis Techniques. London: Academic Press, 1993;139–145.

71. Wahlberg J, Albert J, Lundeberg J, Fenyö E-M, Uhlén M. Analysis of the V3 loop in neutralization resistant human immunodeficiency virus type 2 variants by direct solid phase DNA sequencing. AIDS Res Hum Retrov 1991;7:983–990.

72. Albert J, Wahlberg J, Lundeberg J, Cox S, Sandström B, Wahren B, Uhlén M. Persistence of azidothymidine-resistant human immunodeficiency virus type 1 RNA genotypes in post-treatment sera. J Virol 1992;66:5627–5630.

73. Brytting M, Wahlberg J, Lundeberg J, Wahren B, Uhlén M, Sundquist V-A. Variation in the cytomegalovirus major immediate-early gene found by direct genomic sequencing. J Clin Microbiol 1992;25:955–960.

74. Scarletti G, Leitner T, Halapi E, Wahlberg J, Marchisio P, Clerici-Schoeller MA, Wigzell H, Fenyö E-M, Albert J. Comparison of variable region 3 sequences of human immunodeficiency virus type 1 from infected children with RNA and DNA sequences of the virus populations. Proc Natl Acad Sci USA 1993;90:1721–1725.

75. Holmberg M, Wahlberg J, Lundeberg J, Pettersson U, Uhlén M. Colorimetric detection of *Plasmodium falciparum* and direct sequencing of amplified gene fragments using a solid phase method. Mol Cell Probes 1992;6:201–208.

76. Albert J, Wahlberg J, Uhlén M. Forensic evidence by DNA sequencing. Nature 1993;361:595–596.

77. Hopgood R, Sullivan KM, Gill P. Strategies for automated sequencing of human mitochondrial DNA directly from PCR products. BioTechniques 1992;13:82–92.

78. Hedrum A, Pontén F, Ren Z, Lundeberg J, Pontén J, Uhlén M. Sequence-based analysis of the human p53 gene based on microdissection of tumor biopsy samples. BioTechniques 1994;17:1–9.

79. Kaneoka H, Lee DR, Hsu KC, Hoffman RW. Solid-phase direct DNA sequencing of allel-specific polymerase chain reaction-amplified HLA-DR genes. BioTechniques 1991;10:30–34.

80. Pettersson B, Johansson K-E, Uhlén M. Sequence analysis of 16S rRNA from Mycoplasmas by direct solid-phase DNA sequencing. Appl Environ Microbiol 1994 (in press).

81. Olsvik Ø, Wahlberg J, Petterson B, Uhlén M. Use of automated sequencing of polymerase chain reaction-generated amplicons to identify three types of cholera toxin subunit B in Vibrio cholerae o1 strains. J Clin Microbiol 1993;31:22–25.

82. Muller PR, Wold B. *In vivo* footprinting of a muscle specific enhancer by ligation mediated PCR. Science 1989;246:780–786.

83. Rosenthal A, Stephen D, Jones C. Genomic walking and sequencing by oligo-cassette mediated polymerase chain reaction. Nucleic Acids Res 1990;18:3095–3096.

84. Espelund M, Jakobsen K. Cloning and direct sequencing of plant promoters using primer-adapter mediated PCR on DNA coupled to a magnetic solid phase. BioTechniques 1992;13:74–81.

85. Warshawsky D, Miller L. A rapid genomic walking technique based on ligation-mediated PCR and magnetic separation technology. BioTechniques 1994;16:792–795.

86. Quivy J-P, Becker PB. Direct dideoxy sequencing of genomic DNA by ligation-mediated PCR. BioTechniques 1994;16:239–241.

87. Törmänen VT, Swiderski PM, Kaplan BE, Pfeifer GP, Riggs AD. Extension product capture improves genomic sequencing and DNase I footprinting by ligation mediated PCR. Nucleic Acids Res 1992;20:5487–5488.

88. Parimoo S, Patanjali SR, Shulka H, Chaplin DD, Weissman SM. Proc Natl Acad Sci USA 1991; 88:9623–9627.

89. Korn B, Sedlack Z, Manca A, Kioschis P, Konecki P. A strategy for the isolation of transcribed sequences in the Xq28 region. Hum Mol Genet 1992;1:235–242.

90. Morgan JG, Dolganov GM, Robbins SE, Hinton LM, Lovett M. The selective isolation of novel cDNAs encoded by the regions surrounding the human interleukin 4 and 5 genes. Nucleic Acids Res 1992;20:5173–5179.

91. Tagle DA, Swaroop M, Lovett M, Collins FS. Magnetic bead capture of expressed sequences encoded within large genomic segments. Nature 1993;361:751–753.

92. Sedlack Z, Korn B, Konecki DS, Siebenhaar R, Coy JF, Kioschis P, Poutska A. Construction of a transcription map of a 300 kb region around the human G6PD locus by direct cDNA selection. Hum Mol Genet 1993;2:1865–1869.

93. Larsen F, Gundersen G, Lopez R, Prydz H. CpG islands as gene markers in the human genome. Genomics 1992;13:1095–1107.

94. Sedlack Z, Konecki DS, Siebenhaar R, Kioschis P, Poutska A. Direct selection of DNA sequences conserved between species. Nucleic Acids Res 1993;21:3419–3425.

95. Lundeberg J, Wahlberg J, Holmberg M, Pettersson U, Uhlén M. Rapid colorimetric detection of in vitro amplified DNA sequences. DNA Cell Biol 1990;9:287–292.

96. Wahlberg J, Lundeberg J, Hultman T, Uhlén M. General colorimetric method for DNA diagnostics allowing direct solid phase genomic sequencing of the positive samples. Proc Natl Acad Sci USA 1990;87:6569–6573.

97. Wahlberg J, Lundeberg J, Hultman T, Holmberg M, Uhlén M. Rapid detection and sequencing of specific in vitro amplified DNA sequences using solid phase methods. Mol Cell Probes 1990;4: 285–297.

98. Lundeberg J, Bondesson L, Hedrum A, Grillner L, Stark M, von Krog G, Uhlén M. A colorimetric PCR method for large-scale screening of Chlamydia trachomatis. Scand J Infect Dis 1994 (in press).

99. Lundeberg J, Wahlberg J, Uhlén M. Rapid colorimetric quantification of PCR-amplified DNA. BioTechniques 1991;10:68–75.

100. Lundeberg J, Hedrum A, Seesod N, Uhlén M. An integrated colorimetric PCR-method for qualitative and quantitative diagnosis. Proc. 6th Eur Cong Biotechnol 1993;697–700.

101. Yun Z, Lundeberg J, Johansson B, Hedrum A, Weiland O, Uhlén M, Sönnerborg A. Colorimetric detection of competitive PCR products for quantification of hepatitis C viremia. J Virol Methods 1994;47:1–14.

102. Olsvik Ø, Popovic T, Skjerve E, Cudjoe KF, Hornes E, Ugelstad J, Uhlén M. Magnetic separation techniques in diagnostic microbiology. Clin Microbiol Rev 1994;43–54.

103. Lund A, Helleman AL, Vartdal F. Rapid isolation of K88+ Escherichia coli by using immunomagnetic particles. J Clin Microbiol 1988;26:2572–2575.

104. Skjerve E, Olsvik Ø. Immunomagnetic separation of Salmonella from foods. Int J Food Microbiol 1991;14:11–18.

105. Hornes E, Wasteson Y, Olsvik Ø. Detection of Escherichia coli heat-stable enterotoxin genes in pig stool specimens by an immobilised, colorimetric nested polymerase chain reaction. J Clin Microbiol 1991;29:201–208.

106. Islam D, Lindberg AA. Detection of Shigella dysenteriae type 1 and Shigella flexneri in feces by immunomagnetic isolation and polymerase chain reaction. J Clin Microbiol 1992;30:2801–2806.

107. Seesod N, Lundeberg J, Hedrum A, Åslund L, Holder A, Thaithong S, Uhlén, M. Immunomag-

netic purification to facilitate DNA diagnosis of *Plasmodium falciparum*. J Clin Microbiol 1993; 31:2715–2719.

108. Hedrum A, Lundeberg J, Påhlson C, Uhlén M. Immunomagnetic recovery of *Chlamydia trachomatis* from urine with subsequent colorimetric DMA detection. PCR Methods Appl 1992;2: 167–171.

109. Muir P, Nicholson M, Jhetam M, Neogi S, Banatval JE. Rapid diagnosis of enterovirus infection by magnetic bead extraction and polymerase chain reaction detection of enterovirus RNA in clinical specimens. J Clin Microbiol 1993;31:31–38.

110. van Doorn L-J, Kleter B, Voormans J, Maertens G, Brouwer H, Heijink R, Quint W. Rapid detection of hepatitis C virus RNA by direct capture from blood. J Med Virol 1994;42:22–28.

111. Hornes E, Hultman T, Moks T, Uhlén M. Direct cloning of the human genomic apolipoprotein E gene using magnetic separation of single-stranded DNA. BioTechniques 1990;9:730–737.

112. Hultman T, Murby M, Ståhl S, Hornes E, Uhlén M. Solid phase in vitro mutagenesis using plasmid DNA template. Nucleic Acids Res 1990;18:5107–5112.

113. Gabrielsen OS, Hornes E, Korsnes L, Ruet A, Øyen TB. Magnetic DNA affinity purification of a yeast transcription factor - a new purification principle for the ultrarapid isolation of near homogenous factor. Nucleic Acids Res 1989;17:6253–6267.

114. Gabrielsen OS, Huet J. Magnetic DNA affinity purification of a yeast transcription factor. Methods Enzymol 1993;218:508–525.

115. Ren L, Chen H, Sternberg EA. Tethered bandshift assay and affinity purification of a new DNA-binding protein. BioTechniques 1994;16:852–855.

116. Wright WE, Funk WD. CASTing for multicomponent DNA-binding complexes. Trends Biochem Sci 1993;16:7780.

117. Wright WE, Binder M, Funk W. Cyclic amplification and selection of targets (CASTing) for the myogenin consensus binding site. Mol Cell Biol 1991;11–4104–4110.

118. Sandaltzopoulos R, Blank T, Becker PB. Transcriptional repression by nucleosomes but not H1 in reconstituted preblastoderm *Drosophila* chromatin. EMBO J 1994;13:373–379.

119. Ståhl S, Hansson M, Ahlborg N, Ngoc Nguyen T, Liljekvist S, Lundeberg J, Uhlén M. Solid-phase gene assembly of constructs derived from the Plasmodium falciparum malaria blood-stage antigen Ag332. BioTechniques 1993;14:424–434.

120. Ljungquist C, Lundeberg J, Rasmussen A-M, Hornes E, Uhlén M. Immobilization and recovery of fusion proteins and B-lymphocyte cells using magnetic separation. DNA Cell Biol 1993;12:191–197.

121. Grompe M. The rapid detection of unknown mutations in nucleic acids. Nature Genet 1993;5:111–117.

122. Fry G, Lachmeier E, Mayrand E, Giusti B, Fisher J, Johnston-Dow L, Cathcart R, Finne E, Kilaas L. A new approach to template purification for sequencing applications using paramagnetic particles. BioTechniques 1992;13:124–131.

123. Espelund M, Stacy P, Jakobsen KS. A simple method for generating single-stranded DNA probes labeled to high activities. Nucleic Acids Res 1990;18:6157–6158.

124. Stacy JE, Ims RA, Stenseth NC, Jakobsen KS. Fingerprinting of diverse species with DNA probes generated from immobilized single-stranded DNA templates. Nucleic Acids Res 1991;21:4004.

125. Mörl M, Dorner M, Pääbo S. Direct purification of tRNAs using oligonucleotides coupled to magnetic beads. In: Advances in Biomagnetic Separation. Eaton Publishing, 1994;107–111.

126. Wakita K, Watanabe Y, Yokogawa T, Kumazawa Y, Nakamura S, Ueda T, Watanabe K, Nishikawa K. Higher-order structure of bovine mitochondrial tRNA^{Phe} lacking the 'conserved' GG and TΨCG sequences as inferred by enzymatic and chemical probing. Nucleic Acids Res 1994;22:347–353.

127. Scouten WH, Konecny P. Reversible immobilisation of antibodies on magnetic beads. Anal Biochem 1992;205:313–318.

© 1995 Elsevier Science B.V. All rights reserved
Biotechnology Annual Review Volume 1
M.R. El-Gewely, editor

Ultrasensitive enzyme immunoassay

Seiichi Hashida, Kazuya Hashinaka and Eiji Ishikawa
Department of Biochemistry, Medical College of Miyazaki, Kiyotake, Miyazaki, Japan

Abstract. Ultrasensitive enzyme immunoassay methods are reviewed not only for antigens but also for antibodies and haptens with emphasis on factors which limit the sensitivity. Ultrasensitive immunoassays can be developed by noncompetitive solid phase assay systems rather than competitive ones for antigens and antibodies. However, no noncompetitive immunoassays have been available for hapten molecules which cannot be bound simultaneously by two different antibody molecules. This has been overcome by developing methods to derivatize haptens with amino groups so that the derivatized haptens may be measured by two-site noncompetitive assays. For ultrasensitive noncompetitive solid phase immunoassays, the nonspecific binding of labeled reactants (background noise) should be minimized. This has achieved by developing methods to transfer the complex of analytes and labeled reactants from solid phase to solid phase with minimal dissociation of the complex. Thus, the sensitivity for antigens, haptens and antibodies has been markedly improved and some applications have been made.

Introduction

Radioimmunoassay, developed in 1959 [1], first made it possible to measure femtomole amounts of peptide hormones in the circulation and was almost generally accepted not to be exceeded in sensitivity by any other methods. However, in 1968, less than a decade later, the use of enzymes as labels in immunoassay was suggested [2], and the first attempts of enzyme immunoassay were reported in 1971 [3,4]. For many years following, the potential of enzyme immunoassay, particularly concerning sensitivity, was a focus of discussion by a number of investigators. In 1976, a typical view that enzyme immunoassays would replace radioimmunoassays in various fields was published [5]. One month later, arguments against this view appeared [6,7]. One argument stated that the suggestion that enzyme labeling would replace radioisotopic techniques (particularly in assays demanding highest sensitivity) was questionable [6]. The other stated that for the analysis of femtomole (1×10^{-15} mol) amounts of steroids, hormones, and so on to aid patient diagnosis and treatment there is only radioimmunoassay [7]. Less than 4 months after these arguments, detection by enzyme immunoassay of 1 attomole (1×10^{-18} mol; 600,000 molecules as calculated from Avogadro's number) of a macromolecular antigen, ornithine δ-amino-transferase from rat liver with a molecular weight of 170 kDa, was reported [8]. During the ensuing decade, enzyme immunoassay was successfully applied to the measurement of various antigens at attomole levels, which are below those

Address for correspondence: S. Hashida, Department of Biochemistry, Medical College of Miyazaki, Kiyotake, Miyazaki 889-16, Japan. Tel.: +81 985 85 0985; Fax: +81 985 85 2401.

detectable by radioimmunoassay [9–13]. Several years ago, the sensitivity was further improved to zeptomole (1×10^{-21} mol) levels [14,15]. For antibodies and haptens, both the assay system and the sensitivity of enzyme immunoassay were fundamentally unchanged for many years after the first reports. However, novel ultrasensitive enzyme immunoassay methods were also developed for antibodies [16–38] and haptens [39–46] and some applications have been made. This paper reviews ultrasensitive enzyme immunoassays for not only antigens but also antibodies and haptens with emphasis on factors which limit the sensitivity.

Ultrasensitive Enzyme Immunoassay for Antigens

Two-site enzyme immunoassay for attomole amounts of antigens

In 1976, one attomole of a macromolecular antigen, ornithine δ-aminotransferase from rat liver with a molecular weight of 170 kDa, was measured by enzyme immunoassay [8,9]. The reasons that this could be achieved are as follows.

The first reason is that a noncompetitive, two-site (or sandwich) enzyme immunoassay technique was used. An anti-ornithine δ-aminotransferase IgG-coated silicone rubber piece was incubated with the antigen and subsequently with anti-ornithine δ-

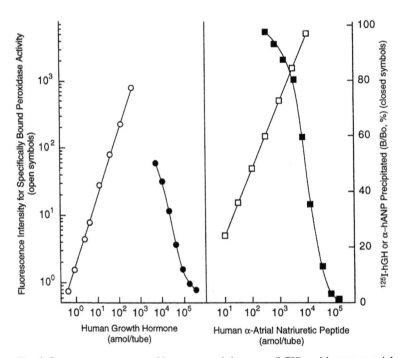

Fig. 1. Dose-response curves of human growth hormone (hGH) and human α-atrial natriuretic peptide (α-hANP) by competitive radioimmunoassay (closed symbols) and noncompetitive (two-site) enzyme immunoassay (open symbols).

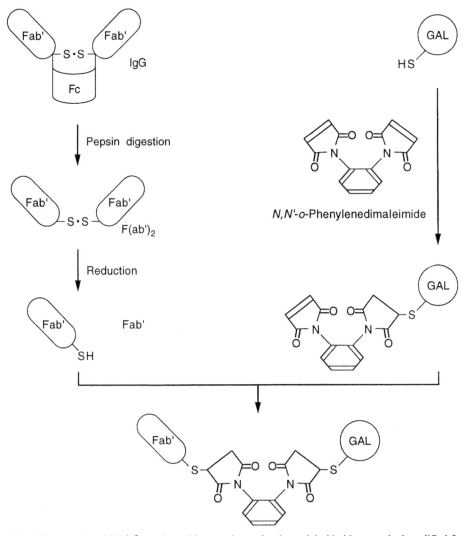

Fig. 2. Preparation of Fab'-β-D-galactosidase conjugate by the maleimide-hinge method modified from the original one. S–S, disulfide bond; GAL, β-D-galactosidase from *Escherichia coli*.

aminotransferase Fab' labeled with β-D-galactosidase from *Escherichia coli*. β-D-Galactosidase activity was correlated to the amount of the antigen to be measured.

In this noncompetitive immunoassay method, excess of enzyme-labeled antibody is efficiently eliminated by simple washing, and the amount of enzyme-labeled antibody nonspecifically bound to antibody-coated solid phase in the absence of antigen to be measured (background noise) can be reduced to a minimum. This makes it possible to achieve attomole sensitivities, provided that antibodies with sufficiently high affinity are used. By contrast, in the most widely used competitive immunoassay method, a certain amount of labeled antigen is reacted with the corresponding amount of antibody in the absence and presence of antigen to be measured. The amount of antigen to be measured is correlated to the amount of labeled antigen bound to the antibody, which is measured only with a certain range of error (approximately 5%). The lower the concentration of labeled antigen and antibody used, the higher the sensitivity. However, the concentration of labeled antigen and antibody should be sufficiently high so that more than 50% of labeled antigen and antibody used are in bound form in the absence of antigen to be measured. In other words, the minimal concentration of labeled antigen and antibody that can be used is limited by the affinity of the antibody used. In a modified competitive enzyme immunoassay using enzyme-labeled antibody, an analogous factor limits the sensitivity. Thus, noncompetitive immunoassay methods are theoretically more sensitive than

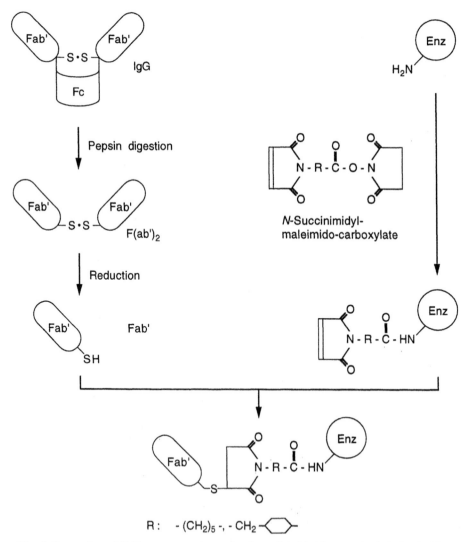

Fig. 3. Preparation of Fab'-enzyme conjugate by the maleimide-hinge method. S–S, disulfide bond; Enz, enzyme.

competitive ones, when identical antibody is used in both immunoassays, and the detection limit of antigens by competitive immunoassay methods is at femtomole or higher levels in most cases (Fig. 1). It makes no difference whether radioisotopes or enzymes are used as label.

The second reason is that Fab' was conjugated to β-D-galactosidase through thiol groups in the hinge of Fab' (the hinge method) (Fig. 2), while, in other reports, antibody IgG or its fragments were conjugated to enzymes through their amino groups (the nonhinge method) [10,11,47]. Fab'-β-D-galactosidase conjugate prepared by the hinge method retains original activities of Fab' and β-D-galactosidase and gives lower nonspecific binding and higher specific binding, resulting in higher sensitivity, than IgG, Fab' and Fab conjugated to β-D-galactosidase through their amino groups by the nonhinge method [47].

The third reason is that β-D-galactosidase activity bound to the solid phase was

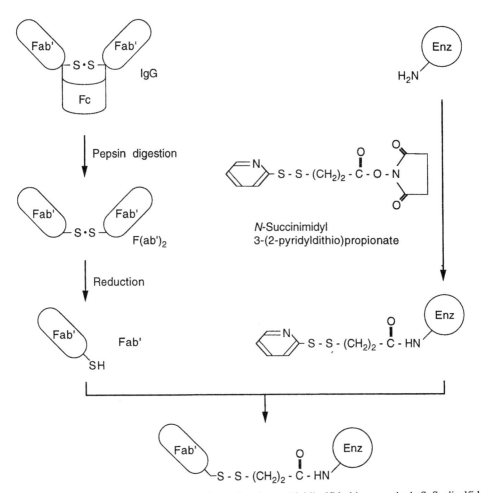

Fig. 4. Preparation of Fab'-enzyme conjugate by the pyridyldisulfide-hinge method. S–S, disulfide bond; Enz, enzyme.

assayed by fluorometry using 4-methylumbelliferyl-β-D-galactoside as substrate, while, in other reports, enzymes as label were assayed by colorimetry. β-D-Galactosidase from *Escherichia coli* can be assayed at least 1,000-fold more sensitively by fluorometry than by colorimetry, and attomole amounts of β-D-galactosidase cannot be measured by colorimetry [9–11].

Later, a more versatile enzyme-labeling method to conjugate Fab' to most enzymes by selective use of thiol groups in the hinge of Fab' was established (the hinge method) (Figs. 3 and 4) [47–51]. The hinge method made it possible to prepare monomeric Fab'-horseradish peroxidase conjugates [10–13,47–51]. The use of different cross-linking reagents provides equally useful Fab'-enzyme conjugates, when thiol groups in the hinge of Fab' are used for conjugation [47,51]. Fab'-enzyme conjugates prepared by the hinge method provide higher sensitivities of two-site enzyme immunoassays for antigens than IgG or its fragments labeled with enzymes through their amino groups (the nonhinge method) (Figs. 5–7). In addition, highly purified, usually affinity-purified, Fab'-enzyme conjugates were used. This lowered the nonspecific binding of Fab'-enzyme conjugates to antibody-coated solid phase

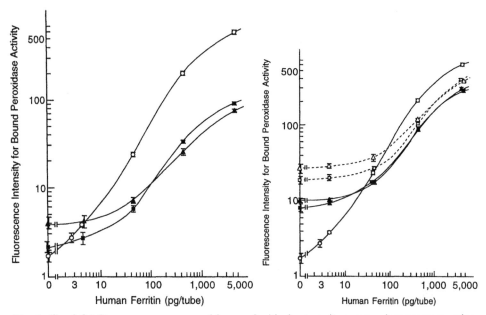

Fig. 5. (See left.) Dose-response curves of human ferritin by two-site enzyme immunoassay using horseradish peroxidase conjugates with Fab' and Fab. Open circles: Fab'-peroxidase conjugate prepared by the hinge method. Closed squares: Fab'-peroxidase conjugate prepared by the glutaraldehyde method (the nonhinge method). Closed triangles: Fab-peroxidase conjugate prepared by the glutaraldehyde method.

Fig. 6. (See right.) Dose-response curves of human ferritin by two-site enzyme immunoassay using horseradish peroxidase conjugates with Fab' and Fab. Open circles: Fab'-peroxidase conjugate prepared by the hinge method. Closed squares: monomeric Fab'-peroxidase conjugate prepared by the periodate method (the nonhinge method). Closed triangles: monomeric Fab-peroxidase conjugate by the periodate method. Open squares: polymeric Fab'-peroxidase conjugate prepared by the periodate method. Open triangles: polymeric Fab-peroxidase conjugate prepared by the periodate method.

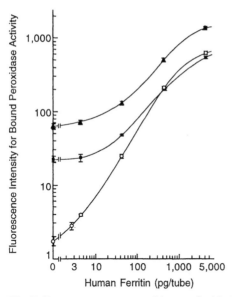

Fig. 7. Dose-response curves of human ferritin by two-site enzyme immunoassay using horseradish peroxidase conjugates with Fab' and IgG. Open circles: Fab'-peroxidase conjugate prepared by the hinge method. Closed squares: monomeric IgG-peroxidase conjugate prepared by the periodate method (the nonhinge method). Closed triangles: polymeric IgG-peroxidase conjugate by the periodate method.

(background noise) and improved the sensitivity of two-site enzyme immunoassay 10–100-fold, when polyclonal antibodies were used (Table 1) [10,11,13]. Several methods to prepare highly purified Fab'-enzyme conjugates from microgram quantities of Fab' were developed, since affinity-purified polyclonal Fab' was often available only in small quantities [13]. This facilitated the development of two-site enzyme immunoassays to measure attomole amounts of various antigens.

Besides β-D-galactosidase from *Escherichia coli*, horseradish peroxidase is also useful as label in two-site enzyme immunoassays with attomole sensitivity, since 0.5 amol of horseradish peroxidase can be detected by fluorometry using 3-(4-hydroxyphenyl)propionic acid as hydrogen donor [11–13]. It is also possible to measure attomole amounts of antigens using alkaline phosphatase assayed by either colorimetry coupled with enzymatic cycling [52] or luminometry [53–56]. By contrast, the sensitivity for the detection of radioisotopes used as label in radioimmunoassay is lower than those for the assay of enzymes as label described above [10]. The detection limit of radioisotopes depends upon the specific radioactivity, which can be calculated from the half-life. For example, the specific radioactivity of ^{125}I most widely used in radioimmunoassay is calculated to be 4.8 dpm/amol from its half-life of 60.2 days and, therefore, the detection limit of ^{125}I may be larger than 10 amol [10].

As a result, attomole amounts of clinically important antigens that cannot be detected by radioimmunoassay were measured by enzyme immunoassay [10–13]. These antigens include ferritin, IgE, α-fetoprotein, insulin, chorionic gonadotropin, thyroid-stimulating hormone, luteinizing hormone, α-atrial natriuretic peptide [11–

Table 1. Improved sensitivity of two-site enzyme immunoassay by using affinity-purified Fab'-enzyme conjugates

Antigen	Label enzyme	Affinity-purification of antibodies	Purify of Fab'-enzyme conjugate (%)	Detection limit of antigen (amol/assay)
IgE	GAL	–	13	100
		+	95	2
Ferritin	GAL	–	14	2
		+	93	0.05
TSH	GAL	–	7	250
		+	98	2
Growth hormone	HRP	–	35	200
		+	86	3
α-Fetoprotein	HRP	–	15	50
		+	84	2

Fab'-enzyme conjugates were prepared by the hinge method. Purify of Fab'-enzyme conjugates was expressed as percentages of the conjugates adsorbed to antigen-Sepharose 4B. GAL, β-D-galactosidase from *Escherichia coli*; HRP, horseradish peroxidase.

13]. Actually, growth hormone in urine (Fig. 8) [11,12], α-atrial natriuretic peptide in plasma of healthy subjects [13], thyroid-stimulating hormone in plasma of patients with Graves' disease [11], and luteinizing hormone in serum of children under the age of 9–10 years (Fig. 9) [13] could be first measured directly by enzyme

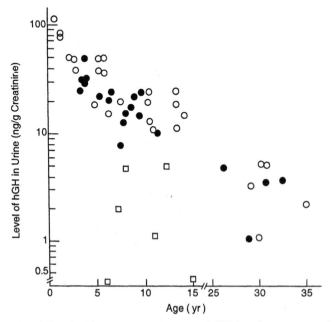

Fig. 8. Level of human growth hormone (hGH) in urine samples collected from healthy males (open circles), healthy females (closed circles), and dwarfism patients (open squares) before breakfast immediately after awakening early in the morning. hGH was measured by two-site enzyme immunoassay.

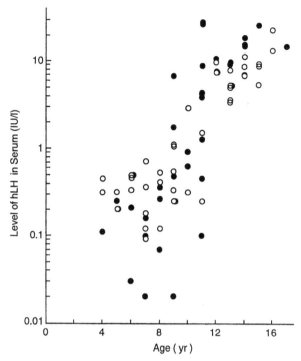

Fig. 9. Level of human luteinizing hormone (hLH) in serum samples collected from healthy children. Open and closed circles indicate males and females, respectively. hLH was measured by two-site enzyme immunoassay.

immunoassay without extraction and concentration. This is not possible using competitive radioimmunoassay as described above [11–13].

Two-site immune complex transfer enzyme immunoassay for zeptomole amounts of antigens

In the conventional two-site enzyme immunoassay for antigens described above, one of the greatest obstacles to hamper further improvement in sensitivity is the nonspecific binding of Fab'-enzyme conjugates to antibody-coated solid phase (background noise) (Fig. 10; left). An easy way to lower the background noise is to reduce the size of the solid phase surface. In fact, a small scale two-site enzyme immunoassay was developed using glass beads of 1 mm in diameter and a reaction mixture volume of 5μl for immunoreactions [57], while, in two-site enzyme immunoassays for antigens described above, polystyrene beads of 3.2 mm in diameter and a reaction mixture volume of 150μl were used for immunoreactions. By this small scale enzyme immunoassay, one zeptomole (zmol = 1×10^{-21} mol; 600 molecules as calculated from Avogadro's number) of human ferritin was detected. Smaller scale or microscale two-site enzyme immunoassays may detect even smaller amounts of antigens. However, the improvement in sensitivity thus achieved is only in terms of mol/assay or g/assay but not in terms of mol/l or g/l of samples such as serum, plasma or urine, since the volume of samples that can be used in smaller scale or microscale assays has to be reduced proportionally.

Several years ago, a novel method (the immune complex transfer method) was developed to lower the nonspecific binding of enzyme-labeled antibody without reducing the size of the solid phase surface and the reaction mixture volume for immunoreactions [14]. Antigen to be measured was reacted simultaneously with 2,4-dinitrophenylated antibody IgG and antibody Fab' labeled with β-D-galactosidase from *Escherichia coli*. The immune complex formed, comprising the three components, was trapped onto polystyrene beads coated with affinity-purified (anti-2,4-dinitrophenyl group) IgG. The polystyrene beads were washed to eliminate excess of the Fab'-β-D-galactosidase conjugate. The immune complex was eluted from the polystyrene beads with excess of εN-2,4-dinitrophenyl-L-lysine and was transferred to polystyrene beads coated with (anti-IgG Fc portion) IgG. By this transfer, the Fab'-β-D-galactosidase conjugate nonspecifically bound to polystyrene beads coated with affinity-purified (anti-dinitrophenyl group) IgG was eliminated more completely, reducing markedly the background noise of two-site enzyme immunoassay with less decrease in the specific binding and significantly improving the sensitivity.

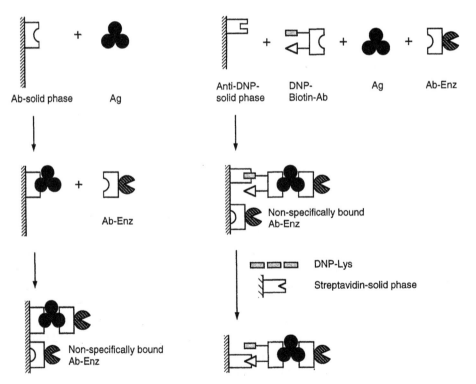

Fig. 10. Conventional two-site enzyme immunoassay (left) and two-site immune complex transfer enzyme immunoassay (right) for antigens. Ab, antibody; Ag, antigen; Enz, enzyme; DNP, 2,4-dinitrophenyl group; DNP-Lys, εN-2,4-dinitrophenyl-L-lysine.

By this two-site immune complex transfer enzyme immunoassay, the detection limit of human thyroid-stimulating hormone was 20 zmol/assay [14].

Later, 2,4-dinitrophenylated antibody IgG and (anti-IgG Fc portion) IgG-coated polystyrene beads were replaced by antibody IgG labeled with both 2,4-dinitrophenyl groups and biotin residues and streptavidin-coated polystyrene beads, respectively (Fig. 10; right) [15]. Ferritin was reacted simultaneously with affinity-purified 2,4-dinitrophenylated biotinylated anti-ferritin IgG and affinity-purified anti-ferritin Fab' labeled with β-D-galactosidase from *Escherichia coli*. The immune complex formed, comprising the three components, was trapped onto polystyrene beads coated with affinity-purified (anti-2,4-dinitrophenyl group) IgG, was eluted from the polystyrene beads with excess of εN-2,4-dinitrophenyl-L-lysine and was transferred to streptavidin-coated polystyrene beads. β-D-Galactosidase activity nonspecifically bound to streptavidin-coated polystyrene beads in the absence of ferritin (background noise) was 0.5–0.7% of that nonspecifically bound to polystyrene beads coated with affinity-purified (anti-2,4-dinitrophenyl group) IgG in the absence of ferritin before incubation with εN-2,4-dinitrophenyl-L-lysine. β-D-galactosidase activity specifically bound to streptavidin-coated polystyrene beads in the presence of ferritin was 43–66% of that specifically bound to polystyrene beads coated with affinity-purified (anti-2,4-dinitrophenyl group) IgG in the presence of ferritin before

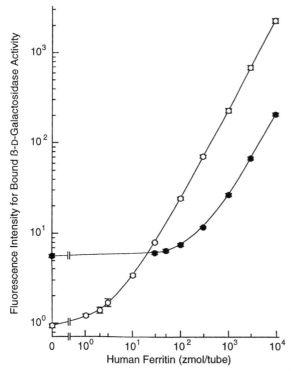

Fig. 11. Dose-response curves of human ferritin by two-site immune complex transfer enzyme immunoassay (open circles) and by the conventional two-site enzyme immunoassay (closed circles). Vertical bars indicate standard deviations of 3–5 determinations.

incubation with εN-2,4-dinitrophenyl-L-lysine. Namely, the nonspecific binding (background noise) was markedly lowered with much less decrease in the specific binding. As a result, the detection limit of ferritin was 1 zmol/assay (Fig. 11). This was 30-fold smaller than that by the conventional two-site enzyme immunoassay, in which an anti-ferritin IgG-coated polystyrene bead was incubated with ferritin and, after washing, with affinity-purified anti-ferritin Fab'-β-D-galactosidase conjugate (Fig. 10; left).

From β-D-galactosidase activity specifically bound to streptavidin-coated polystyrene beads in the presence of 1–10,000 zmol of ferritin, the average number of anti-ferritin Fab'-β-D-galactosidase conjugate molecules bound per ferritin molecule added was calculated to be 0.95–1.4. In addition, loss of specifically bound β-D-galactosidase activity during elution and transfer of the immune complex was only 34–57% as described above. Therefore, another transfer of the immune complex (double transfers) was strongly suggested to further improve the sensitivity, although the sensitivity of assay of label enzyme remains to be improved considerably, and an efficient method for another transfer of the immune complex also remains to be developed.

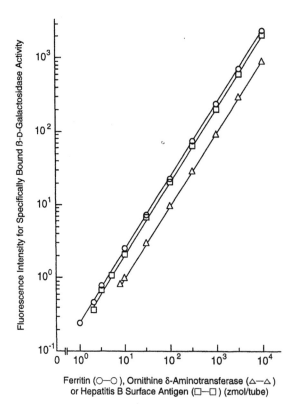

Ferritin (O—O), Ornithine δ-Aminotransferase (△—△)
or Hepatitis B Surface Antigen (□—□) (zmol/tube)

Fig. 12. Dose-response curves of human ferritin, rat ornithine δ-aminotransferase and hepatitis B surface antigen by two-site immune complex transfer enzyme immunoassay. Antibody IgG and Fab' were affinity-purified for ferritin but not for ornithine δ-aminotransferase. For hepatitis B surface antigen, mouse monoclonal IgG₃ was used.

As suggested by the above results for ferritin, the two-site immune complex transfer enzyme immunoassay technique may be capable of detecting zeptomole amounts of other antigens, which are measurable at attomole levels by the conventional two-site enzyme immunoassay. In a preliminary experiment, the detection limit of ornithine δ-aminotransferase from rat kidney was 10 zmol/assay by the two-site immune complex transfer enzyme immunoassay using 2,4-dinitrophenyl-biotinyl-bovine serum albumin-anti-ornithine δ-aminotransferase Fab' conjugate and anti-ornithine δ-aminotransferase Fab'-β-D-galactosidase conjugate without affinity-purification. On the other hand, the detection limit of this enzyme by the conventional two-site enzyme immunoassay was 300 zmol (Fig. 12). Affinity-purification may further lower the detection limit. In the same way, the detection limit of hepatitis B surface antigen (HBs Ag) is 2 zmol/assay using 2,4-dinitrophenyl-biotinyl-mouse monoclonal anti-HBs Ag IgG$_3$ and mouse monoclonal anti-HBs Ag Fab'-β-D-galactosidase conjugate, while its detection limit by the conventional two-site enzyme immunoassay using the same antibody is 100 zmol (Fig. 12).

Ultrasensitive Enzyme Immunoassay for Antibodies

Factors to limit the sensitivity of the conventional enzyme immunoassay for antibodies

As described for antigens and haptens, noncompetitive immunoassays potentially provide higher sensitivities for antibodies than competitive ones.

In the most widely used conventional enzyme immunoassay (so-called enzyme-linked immunosorbent assay, ELISA) for antibodies, antigen-coated solid phase is reacted with antibodies in samples such as serum and plasma and, after washing, with anti-immunoglobulin antibody-enzyme conjugate (Fig. 13). Enzyme activity bound to the solid phase is correlated to the amount or the concentration of antibodies to be measured. The sensitivity of this assay is seriously limited by the nonspecific binding of the conjugate to the solid phase, which is caused by the nonspecific binding of nonspecific immunoglobulins and probably other substance(s) in the samples.

In the second widely used conventional enzyme immunoassay for antibodies (so-called antibody capture enzyme immunoassay), anti-immunoglobulin IgG-coated solid phase is incubated with test samples to capture both specific and nonspecific immunoglobulins and, after washing, with enzyme-labeled antigen to measure specific immunoglobulin captured. In general, this assay is slightly more sensitive than the most widely used conventional enzyme immunoassay described above. However, the sensitivity of this assay is limited by the capacity of anti-immunoglobulin IgG-coated solid phase to capture immunoglobulins. The capacity becomes larger with increasing sizes of solid phase surfaces. However, larger sizes of solid phase surfaces suffer from higher nonspecific bindings of enzyme-labeled antigens, enhancing the background and limiting the sensitivity.

416

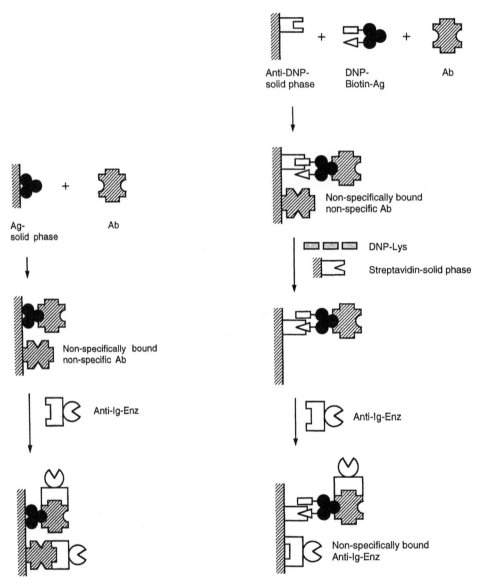

Fig. 13. (See left) Conventional enzyme-linked immunosorbent assay (ELISA) for antibodies. Ag, antigen; Ab, antibody; Ig, immunoglobulin; Enz, enzyme.

Fig. 14. (See right) Immune complex transfer enzyme immunoassay I (Method I). DNP, 2,4-dinitrophenyl group; Ag, antigen; Ab, antibody; DNP-Lys, εN-2,4-dinitrophenyl-L-lysine; Ig, immunoglobulin; Enz, enzyme.

In the third conventional enzyme immunoassay for antibodies (sandwich enzyme immunoassay), antigen-coated solid phase is reacted with antibodies to be measured and subsequently with enzyme-labeled antigen to measure antibodies trapped onto the solid phase. Enzyme activity bound to the solid phase is correlated to the amount or the concentration of antibodies to be measured. The sensitivity of this assay is limited by the nonspecific binding of enzyme-labeled antigen to the solid phase but not by the presence of nonspecific immunoglobulins which limit the sensitivity of the above two conventional methods. This assay is slightly more sensitive than the two conventional methods, although the classes of antibodies are not discriminated.

Various attempts were made to improve the sensitivity for antibody assay, and the most successful method is the immune complex transfer enzyme immunoassay described below, in which the immune complex of antibodies to be measured and labeled antigens is transferred from solid phase to solid phase to eliminate substances including nonspecific immunoglobulins, which limit the sensitivity [16,23–25].

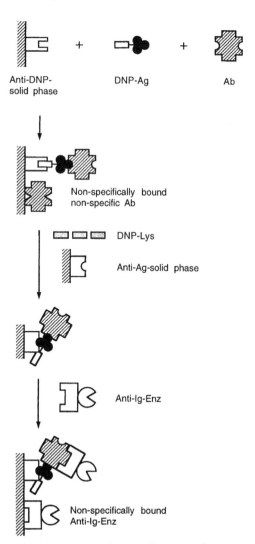

Fig. 15. Immune complex transfer enzyme immunoassay II (Method II). DNP, 2,4-dinitrophenyl group; Ag, antigen; Ab, antibody; DNP-Lys, εN-2,4-dinitrophenyl-L-lysine; Ig, immunoglobulin; Enz, enzyme.

Principle of the immune complex transfer enzyme immunoassay for antibodies

In the initially developed immune complex transfer enzyme immunoassay method (Method I, Fig. 14) [18], antibodies to be measured are reacted with 2,4-dinitro-phenyl-biotinyl-antigen, and the immune complex formed is trapped onto poly-styrene beads coated with affinity-purified (anti-2,4-dinitrophenyl group) IgG (the first solid phase). After washing, the immune complex is eluted from the first solid phase with excess of εN-2,4-dinitrophenyl-L-lysine and transferred onto polystyrene

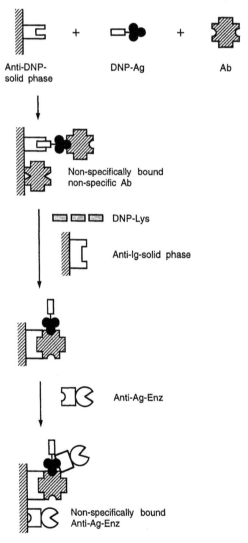

Fig. 16. Immune complex transfer enzyme immunoassay III (Method III). DNP, 2,4-dinitrophenyl group; Ag, antigen; Ab, antibody; DNP-Lys, εN-2,4-dinitrophenyl-L-lysine; Ig, immunoglobulin; Enz, enzyme.

beads coated with (strept)avidin (the second solid phase). Nonspecific immuno-globulins and other interfering substance(s) are minimized by transfer of the immune complex from the first solid phase to the second solid phase. The immune complex on the second solid phase is reacted with anti-immunoglobulin Fab' labeled with enzyme, and the enzyme activity bound to the second solid phase is correlated to the amount or the concentration of antibodies to be measured. This original method has been modified in various ways (Methods II–VIII).

In Method II (Fig. 15) [19], antibodies to be measured are reacted with 2,4-dinitrophenyl-antigen, and the immune complex formed is trapped onto polystyrene beads coated with affinity-purified (anti-2,4-dinitrophenyl group) IgG. After wash-

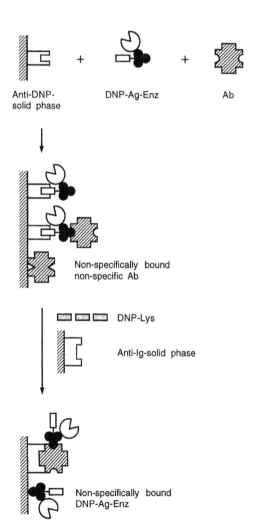

Fig. 17. Immune complex transfer enzyme immunoassay IV (Method IV). DNP, 2,4-dinitrophenyl group; Ag, antigen; Enz, enzyme; Ab, antibody; DNP-Lys, εN-2,4-dinitrophenyl-L-lysine; Ig, immuno-globulin.

420

ing, the immune complex is eluted from the polystyrene beads with excess of εN-2,4-dinitrophenyl-L-lysine and transferred to polystyrene beads coated with antibody IgG to the antigen. Antibodies in the immune complex on the last polystyrene beads are reacted with anti-immunoglobulin Fab' labeled with enzyme.

In Method III (Fig. 16) [20], antibodies to be measured are reacted with 2,4-dinitrophenyl-antigen, and the immune complex formed is trapped onto polystyrene beads coated with affinity-purified (anti-2,4-dinitrophenyl group) IgG. After washing, the immune complex is eluted from the polystyrene beads with excess of εN-2,4-dinitrophenyl-L-lysine and transferred to polystyrene beads coated with anti-immunoglobulin IgG. The antigen in the immune complex on the last solid phase is reacted with enzyme-labeled antibody Fab' to the antigen.

In Method IV (Fig. 17) [21], antibodies to be measured are reacted with 2,4-dinitrophenyl-antigen-enzyme conjugate, and the immune complex formed is trapped

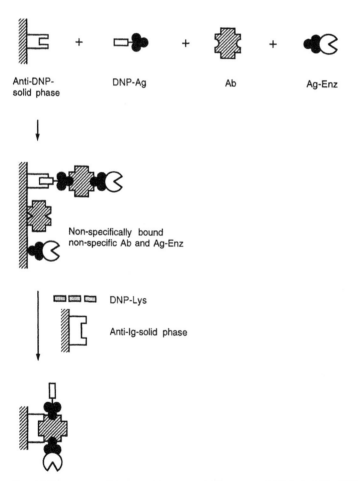

Fig. 18. Immune complex transfer enzyme immunoassay V (Method V). DNP, 2,4-dinitrophenyl group; Ag, antigen; Ab, antibody; Enz, enzyme; DNP-Lys, εN-2,4-dinitrophenyl-L-lysine; Ig, immunoglobulin.

onto polystyrene beads coated with affinity-purified (anti-2,4-dinitrophenyl group) IgG. After washing, the immune complex is eluted from the polystyrene beads with excess of εN-2,4-dinitrophenyl-L-lysine and transferred to polystyrene beads coated with anti-immunoglobulin IgG.

In Method V (Fig. 18) [22], antibodies to be measured are reacted simultaneously with 2,4-dinitrophenyl-antigen and antigen-enzyme conjugate. The immune complex formed, comprising the three components, is trapped onto polystyrene beads coated with affinity-purified (anti-2,4-dinitrophenyl group) IgG. After washing, the immune complex is eluted from the polystyrene beads with excess of εN-2,4-dinitrophenyl-L-lysine and transferred to polystyrene beads coated with affinity-purified anti-immunoglobulin IgG.

In Method VI (Fig. 19) [24,25,58], 2,4-dinitrophenyl-biotinyl-antigen and streptavidin-coated polystyrene beads are substituted for 2,4-dinitrophenyl-antigen and

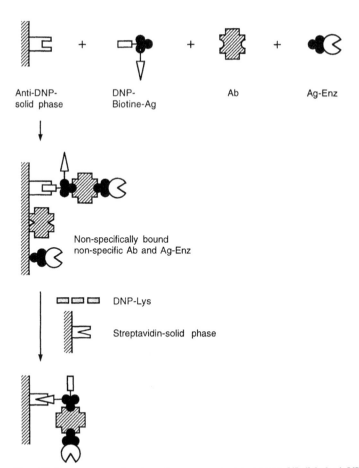

Fig. 19. Immune complex transfer enzyme immunoassay VI (Method VI). DNP, 2,4-dinitrophenyl group; Ag, antigen; Ab, antibody; Enz, enzyme; DNP-Lys, εN-2,4-dinitrophenyl-L-lysine.

affinity-purified anti-immunoglobulin IgG-coated polystyrene beads, respectively, in Method V.

In Method VII (Fig. 20) [25,59], antibodies to be measured are reacted with 2,4-

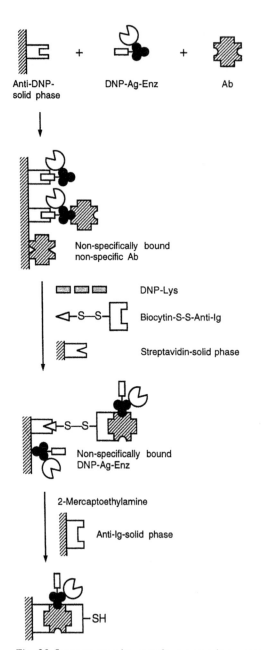

Fig. 20. Immune complex transfer enzyme immunoassay VII (Method VII). DNP, 2,4-dinitrophenyl group; Ag, antigen; Enz, enzyme; Ab, antibody; DNP-Lys, εN-2,4-dinitrophenyl-L-lysine; S–S, disulfide bond; Ig, immunoglobulin.

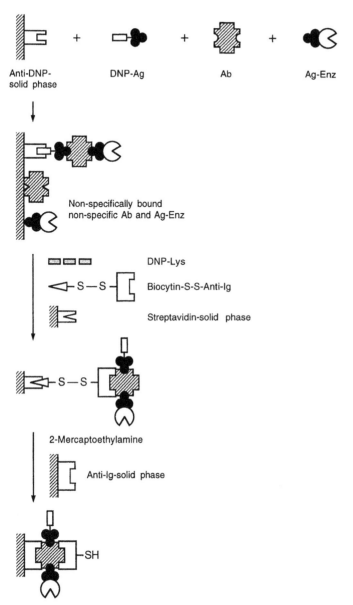

Fig. 21. Immune complex transfer enzyme immunoassay VIII (Method VIII). DNP, 2,4-dinitrophenyl group; Ag, antigen; Ab, antibody; Enz, enzyme; DNP-Lys, εN-2,4-dinitrophenyl-L-lysine; Ig, immunoglobulin; S–S, disulfide bond.

dinitrophenyl-antigen-enzyme conjugate, and the immune complex formed is trapped onto polystyrene beads coated with affinity-purified (anti-2,4-dinitrophenyl group) IgG. After washing, the immune complex is eluted from the polystyrene beads with excess of εN-2,4-dinitrophenyl-L-lysine and, after reaction with biocytin-S-S-anti-immunoglobulin IgG, transferred onto streptavidin-coated polystyrene beads. The immune complex is eluted from the streptavidin-coated polystyrene beads by reduction with 2-mercaptoethylamine and transferred onto polystyrene beads coated with anti-immunoglobulin IgG.

In Method VIII (Fig. 21) [25], antibodies to be measured are reacted simultaneously with 2,4-dinitrophenyl-antigen and antigen-enzyme conjugate. The immune complex formed, comprising the three components, is trapped onto polystyrene beads coated with affinity-purified (anti-2,4-dinitrophenyl group) IgG. After washing, the immune complex is eluted from the polystyrene beads with excess of εN-2,4-dinitrophenyl-L-lysine and, after reaction with biocytin-disulfide (S-S)-anti-immunoglobulin IgG, transferred onto streptavidin-coated polystyrene beads. The immune complex is eluted from the streptavidin-coated polystyrene beads by reduction with 2-mercaptoethylamine and transferred onto affinity-purified anti-immunoglobulin IgG-coated polystyrene beads.

Obviously from the principles described above, all the methods can discriminate

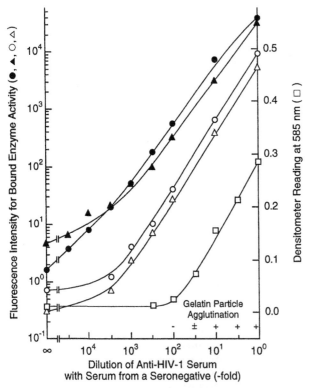

Fig. 22. Dilution curves of anti-HIV-1 serum by various methods. Anti-HIV-1 serum was serially diluted with serum from a seronegative subject and tested by various methods. Closed circles and triangles indicate dilution curves by Methods V and VI, respectively, using recombinant p24 of HIV-1 as antigen and β-D-galactosidase from *Escherichia coli* as label. Open circles and triangles indicate dilution curves by Methods V and VI, respectively, using recombinant p24 of HIV-1 as antigen and horseradish peroxidase as label. Open squares indicate a dilution curve by dot blotting with recombinant p24 of HIV-1 as antigen. Test results by gelatin particle agglutination using HIV-1 as antigen are also shown.

the immunoglobulin classes of antibodies to be measured except Method VI, which measures all antibodies regardless of the immunoglobulin classes.

Factors to limit the sensitivity of the immune complex transfer enzyme immunoassay for antibodies

Methods V–VIII are the most sensitive [23–25]. In these methods, the immune complexes containing enzyme-labeled antigens are transferred from one solid phase to another to minimize the nonspecific binding of enzyme-labeled antigens in the absence of antibodies to be measured (background noise). In Method I–III, however, enzyme-labeled reactants are reacted with the immune complexes on solid phase in the final step, and bound enzyme activities are assayed without transfer of the immune complexes. This is one of factors to limit the sensitivity of Method I–III.

The sensitivity of Methods V–VIII depends on the detection limit of enzymes as label. The use of β-D-galactosidase from *Escherichia coli*, which can be detected with higher sensitivity than other enzymes [11], improves the sensitivity of Methods V and VI approximately 10-fold compared with that of horseradish peroxidase (Fig. 22) [24,25]. Method V for anti-insulin IgG and anti-thyroglobulin IgG using β-D-galactosidase from *Escherichia coli* as label was shown to be 2,000 to 4,000-fold more sensitive than the corresponding conventional ELISA [24,25]. The high sensi-

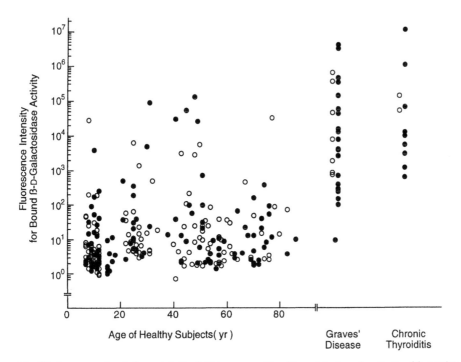

Fig. 23. Presence of anti-thyroglobulin IgG in serum of healthy subjects and patients with autoimmune thyroid diseases. Anti-thyroglobulin IgG was measured by Method V using β-D-galactosidase from *Escherichia coli* as label. Open and closed circles indicate males and females, respectively.

tivity of Method V using β-D-galactosidase from *Escherichia coli* as label was demonstrated in the measurement of anti-thyroglobulin IgG in serum from healthy subjects and patients with autoimmune thyroid diseases. Anti-thyroglobulin IgG was detected not only in all patients with Graves' disease and chronic thyroiditis but also in all healthy subjects (Fig. 23) [22,26], whereas anti-thyroglobulin antibodies were detected only in approximately 60% of the patients and in a few percent of healthy subjects by the conventional ELISA and hemagglutination test (Fig. 24) [22,26]. The difference between the highest level of anti-thyroglobulin IgG in serum of the patients and the lowest level in serum of healthy subjects shown by Method V was 10,000,000-fold (Fig. 23). The sensitivity of Methods VII and VIII remains to be evaluated.

The sensitivity of the immune complex transfer enzyme immunoassay, as in other immunoassays with solid phase, is limited by the nonspecific binding of enzyme-labeled reactants to the solid phase (background noise). Body fluids contain substances to enhance the nonspecific binding of enzyme-labeled reactants, and most of

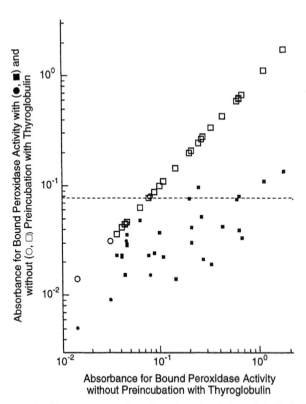

Fig. 24. Measurement and confirmation of anti-thyroglobulin IgG in serum of 31 patients with autoimmune thyroid diseases (16 patients with Graves' disease and 15 patients with chronic thyroiditis) by the conventional ELISA using horseradish peroxidase as label (MESACUP Anti-TG Test, Medical and Biological Laboratories Co., Ltd., Nagoya, Japan). Circles and squares indicate males and females, respectively. The dotted lines indicate a tentative cut-off value.

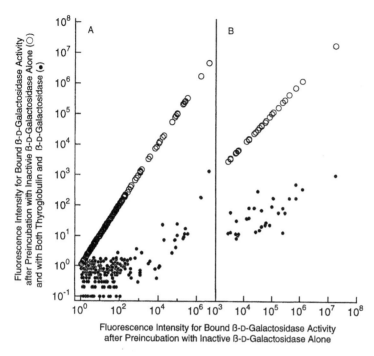

Fluorescence Intensity for Bound ß-D-Galactosidase Activity
after Preincubation with Inactive ß-D-Galactosidase Alone

Fig. 25. Confirmation of the presence of anti-thyroglobulin IgG in serum of healthy subjects (A) and patients with autoimmune thyroid diseases (B). Sera from 119 healthy males, 95 healthy females, 16 patients with Graves' disease and 15 patients with chronic thyroiditis were tested by Method V for anti-thyroglobulin IgG using β-D-galactosidase from *Escherichia coli* as label before and after preincubation with excess of thyroglobulin in the presence of excess of inactive β-D-galactosidase.

them are unknown in nature except nonspecific immunoglobulins and antibodies to enzymes as label. The sensitivity of Method I is obviously limited by the nonspecific binding of nonspecific immunoglobulins, although nonspecific immunoglobulins are eliminated more completely than in the conventional ELISA. In Methods V and VIII (Figs. 18 and 21), antibodies to enzymes as label in body fluids are bound specifically to antigen-enzyme conjugates. The complexes formed are bound nonspecifically to polystyrene beads coated with affinity-purified (anti-2,4-dinitrophenyl group) IgG, released in part and bound specifically to polystyrene beads coated with affinity-purified anti-immunoglobulin IgG in the final step. Antibody IgG to β-D-galactosidase from *Escherichia coli* is known to be present in body fluids of a significant part of healthy subjects and to limit the sensitivity of Method V [26,27,58]. The interference by anti-β-D-galactosidase IgG can be eliminated by using inactive β-D-galactosidase with an amino acid sequence different from that of the native enzyme, which is commercially available. When sera were tested by Method V using β-D-galactosidase from *Escherichia coli* as label for anti-thyroglobulin IgG [26] and anti-HTLV-I IgG [27] in the presence of excess of thyroglobulin and HTLV-I antigen, respectively, bound β-D-galactosidase activities varied to a relatively great extent but were uniformly lowered in the presence of excess of both thyroglobulin or HTLV-I antigen and inactive β-D-galactosidase.

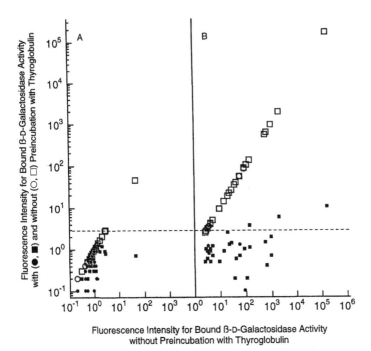

Fluorescence Intensity for Bound ß-D-Galactosidase Activity
without Preincubation with Thyroglobulin

Fig. 26. Measurement and confirmation of anti-thyroglobulin IgG in urine from 45 healthy subjects (A) and 31 patients with autoimmune thyroid diseases (16 patients with Graves' disease and 15 patients with chronic thyroiditis) (B) by Method V using β-D-galactosidase from *Escherichia coli* as label. Circles and squares indicate males and females, respectively. The dotted lines indicate tentative cut-off values.

Thus, the use of inactive β-D-galactosidase improves the sensitivity and specificity of Method V using β-D-galactosidase as label. It remains to be investigated whether antibodies to other enzymes limit the sensitivity of Method V in a similar manner.

Applications of the immune complex transfer enzyme immunoassay for antibodies

Method V has been applied to the measurement of anti-thyroglobulin IgG [22–26,32], anti-insulin IgG [23–25], anti-HTLV-I IgG [25,27–31,34] and anti-HIV-1 IgG [25,35–38] in serum and urine.

As described above, Method V using β-D-galactosidase from *Escherichia coli* as label demonstrated anti-thyroglobulin IgG in serum of not only all patients with Graves' disease and chronic thyroiditis but also all healthy subjects (Fig. 23) [22–25]. The presence of anti-thyroglobulin IgG at low concentrations in healthy subjects could be confirmed by preincubation of sera with excess of thyroglobulin in the presence of excess of inactive β-D-galactosidase (Fig. 25) [26,32]. By contrast, the conventional ELISA and hemagglutination test demonstrated anti-thyroglobulin antibodies only in 60% of the patients and a few per cent of healthy subjects (Fig. 24) [22–25].

Urine and saliva can be collected more easily with no invasive procedure, with less expense and less possibility of infection than serum, plasma or blood. However, the concentration of immunoglobulins is extremely low in urine and saliva [25]. The concentrations of IgA in saliva and urine of healthy subjects are approximately 20-fold and 2,000-fold, respectively, lower than that in serum. The concentrations of IgG in saliva and urine are approximately 400-fold and 4,000-fold, respectively, lower than that in serum. Therefore, a high sensitivity is required for detecting antibodies in saliva and urine samples, and Method V has been used for detecting anti-thyroglobulin IgG [32], anti-HTLV-1 IgG [34] and anti-HIV-1 IgG [35–38] in urine.

By Method V, using β-D-galactosidase from *Escherichia coli* as label, urinary anti-thyroglobulin IgG was measurable in 93.5% of patients with Graves' disease and chronic thyroiditis and 4.4% of healthy subjects (Fig. 26) [32]. The presence of urinary anti-thyroglobulin IgG could be confirmed by preincubation of urine with excess of thyroglobulin (Fig. 26) [32]. Those levels were well correlated to the corresponding serum levels, whether the values were corrected by the concentration of creatinine or IgG in urine or IgG in both urine and serum [32], indicating that anti-thyroglobulin IgG in urine derived from that in blood. By the conventional ELISA,

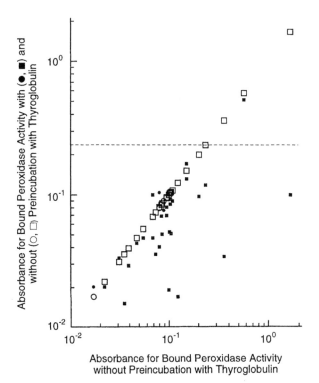

Fig. 27. Measurement and confirmation of anti-thyroglobulin IgG in urine from 31 patients with autoimmune thyroid diseases (16 patients with Graves' disease and 15 patients with chronic thyroiditis) by the conventional ELISA using horseradish peroxidase as label (MESACUP Anti-TG Test). Circles and squares indicate males and females, respectively. The dotted lines indicate a tentative cut-off value.

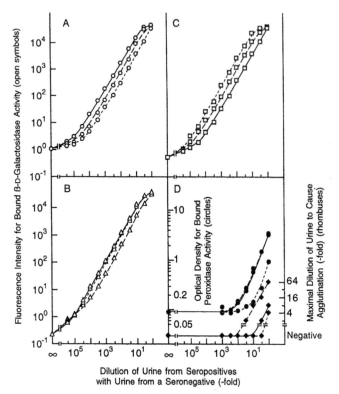

Fig. 28. Dilution curves of urine samples from seropositives by Method V using recombinant RT (A), p17 (B) and p24 (C) as antigens and β-D-galactosidase as label, the conventional ELISA using five recombinant proteins (p15, p17, p24, gp41 and gp120) as antigens and peroxidase as label (closed circles, D) and gelatin particle agglutination test using a lysate of HIV-1 as antigen (closed rhombuses, D). Urine samples from three seropositives were serially diluted with urine from a seronegative. Bound β-D-galactosidase activity was assayed for 25 h.

however, urinary anti-thyroglobulin IgG was measurable in only 10% of the patients and, therefore, only high levels of anti-thyroglobulin IgG in urine were correlated to those in serum (Fig. 27) [32].

Method V for anti-HIV-1 IgG using recombinant p17, p24 or reverse transcriptase (RT) as antigen and β-D-galactosidase from *Escherichia coli* as label was 3,000 to 100,000-fold more sensitive than conventional methods such as gelatin particle agglutination test using a lysate of HIV-1 as antigen and the conventional ELISA using five recombinant proteins (p15, p17, p24, gp41 and gp120) of HIV-1 as antigen and horseradish peroxidase as label (Fig. 28) [38].

When 83 urine samples from seropositives (60 asymptomatic carriers, 11 patients with AIDS-related complex (ARC) and 12 patients with AIDS) and 100 urine samples from seronegatives were tested by the immune complex transfer enzyme immunoassay using recombinant RT as antigen and β-D-galactosidase from *Escherichia coli* as label, all signals for seropositives were higher than those for seronegatives

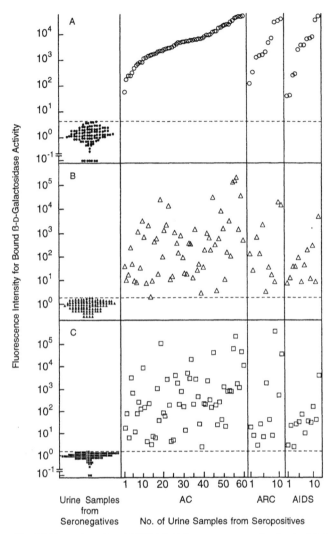

Fig. 29. Detection of anti-HIV-1 IgG in unconcentrated urine by Method V using recombinant RT (A), p17 (B) and p24 (C), as antigens and β-D-galactosidase as label. Eighty-three urine samples from seropositives and 100 urine samples from seronegatives were tested. Bound β-D-galactosidase activity was assayed for 25 h. Open and closed symbols indicate signals for seropositives and seronegatives, respectively. The broken lines indicate tentative cut-off values. AC, asymptomatic carriers; ARC, patients with AIDS-related complex; AIDS, patients with AIDS.

(Fig. 29) [37,38]. The lowest signals for the asymptomatic carriers and the patients with ARC and AIDS were 14-, 29- and 10-fold, respectively, higher than the highest signal for the seronegatives (Fig. 29).

The positivity with recombinant RT as antigen could be confirmed by demonstrating antibody IgG to p17 and p24 in most cases (Fig. 29) [38]. The signal was enhanced by using approximately 10-fold concentrated urine samples with little enhancement of signals for the seronegatives (Fig. 30) [35,38]. Low signals for some

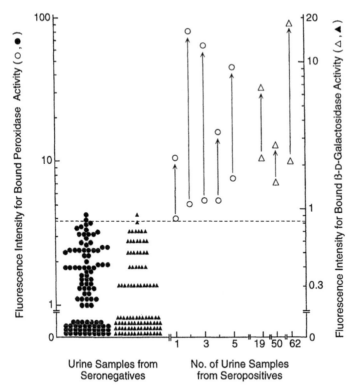

Fig. 30. Confirmation of the positivity and negativity by concentration of urine samples. Urine samples, which gave low signals by Method V using recombinant RT (circles) and p17 (triangles) as antigens and horseradish peroxidase and *Escherichia coli* β-D-galactosidase, respectively, as labels were concentrated approximately 10-fold and tested in the same way. Open and closed symbols indicate signals for seropositives and seronegatives, respectively. Arrows indicate changes by the concentration. The broken line indicates tentative cut-off values for unconcentrated urine samples.

seropositives with recombinant RT as antigen could be enhanced by the combined use of recombinant RT, p17 and p24 with no significant enhancement of signals for the seronegatives [38].

Thus, diagnosis of HIV-1 infection was possible by detecting anti-HIV-1 IgG in 100 μl of unconcentrated urine samples with the immune complex transfer enzyme immunoassay using recombinant RT as antigen or the three recombinant proteins (RT, p17 and p24) as antigens and β-D-galactosidase from *Escherichia coli* as label.

In contrast to the above results, the sensitivity and specificity for the seropositives and the seronegatives by the conventional ELISA and gelatin particle agglutination test were 77–89% and 97–99%, respectively (Fig. 31) [35,36,38]. With approximately 10-fold concentrated urine samples, the sensitivity was slightly improved, but the specificity was lowered to 57–76% due to enhanced nonspecific signals (Fig. 31) [35,38].

From the above results, the following possibilities were suggested. Diagnosis of HIV-1 infection may be possible with a smaller volume of oral fluids (whole saliva)

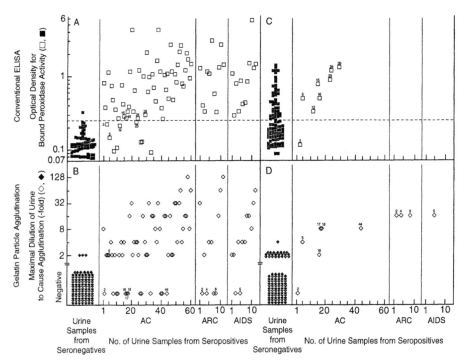

Fig. 31. Detection of anti-HIV-1 IgG or anti-HIV-1 antibodies in unconcentrated and concentrated urine by conventional methods. Eighty-three urine samples from seropositives and 100 urine samples from seronegatives were tested by the conventional ELISA using five recombinant proteins (p15, p17, p24, gp41 and gp120) of HIV-1 as antigens (A, C) and gelatin particle agglutination test using a lysate of HIV-1 as antigen (B, D). Urine samples, which gave low signals or were negative, were concentrated approximately 10-fold and tested in the same way (C, D). Open and closed symbols indicate results for urine samples from the seropositives and the seronegatives, respectively. The broken line indicates a tentative cut-off value.

and a much smaller volume of serum or blood than that of urine samples used above, since the concentrations of IgG in oral fluids and serum are approximately 10-fold and 4,000-fold, respectively, higher than that in urine [35,38]. In fact, diagnosis of HIV-1 infection has been possible with only 1 μl of oral fluids, collected by simple spitting without stimulation, using recombinant RT as antigen and *Escherichia coli* β-D-galactosidase as label. The volume of serum or blood required for diagnosis of HIV-1 infection may be 1–10 nl after seroconversion by conventional methods. By using 10–20 μl of serum samples, anti-HIV-1 IgG may be detected earlier than at the time of seroconversion by conventional methods and more rapidly after seroconversion than described above. Serum samples collected shortly after anti-HIV-1 antibody seroconversion by conventional methods show no positive band or only one positive band by Western blotting and are taken to be indeterminate according to the criteria of WHO [60]. However, these samples are being shown to be definitely positive by demonstrating antibody IgGs to three different antigens

(RT, p17 and p24) with the immune complex transfer enzyme immunoassay described above, which is much more sensitive than Western blotting.

For improving the specificity by detecting antibodies to increasing numbers of antigens or epitopes of HIV-1, antigens other than RT, p17 and p24 (e.g. *env* proteins) remain to be prepared. By this, the sensitivity may also be improved.

Difficulties to overcome

The sensitivity of Methods V and VI depends upon the nonspecific binding of antigen-enzyme conjugates to solid phase (background noise). In order to minimize the nonspecific binding of the conjugates, the molar ratio of antigen to enzyme in the conjugate molecules should be within a narrow range. This is possible only after many trials under different conditions, since an optimal condition of conjugation is different for each antigen. The yield of the conjugates is often low. Thus, a significant amount of purified antigen is required to develop an ultrasensitive immune complex transfer enzyme immunoassays for antibodies. Antigen conjugates are prepared usually by random use of amino groups of antigen molecules. This may limit the reactivity of some antigens with antibodies.

β-D-Galactosidase from *Escherichia coli* used as label has been assayed by fluorometry using 4-methylumbelliferyl-β-D-galactoside as substrate, and the sensitivity is the highest among those of enzymes used as label in enzyme immunoassay [11]. However, this fluorometric assay is no more sufficiently sensitive in Methods V–VIII, in which the nonspecific binding of antigen-enzyme conjugates is markedly lowered. A more sensitive assay method for label enzymes or an alternative label which can be detected more sensitively and more rapidly remains to be developed.

In the immune complex transfer enzyme immunoassay for antibodies described above, polystyrene beads as solid phase are transferred from test tubes to test tubes with tweezers, and bound enzyme activities are measured one by one with a spectrofluorophotometer. This is tedious and time-consuming. Therefore, testing many samples is difficult. In addition, the use of tweezers is causative of false positivity by carryover. These drawbacks are being minimized by substituting microplates and a fluororeader for test tubes and a spectrofluorophotometer, respectively.

Ultrasensitive Enzyme Immunoassay for Haptens

Sensitivity of competitive immunoassay for haptens

Currently available immunoassay methods for haptens may be divided into two groups: competitive and noncompetitive immunoassay methods. In a typical competitive assay system, a certain amount of labeled hapten is reacted with the corresponding amount of antibody in the absence and presence of hapten to be measured. The amount of hapten to be measured is correlated to the amount of labeled hapten bound to the antibody, which is measured only with a certain range of error (approximately 5%). The lower the concentration of labeled hapten and antibody used is, the higher the sensitivity is. However, the concentration of labeled hapten and antibody should be sufficiently high so that more than 50% of labeled hapten and antibody used are in bound form in the absence of hapten to be measured. In other words, the minimal concentration of labeled hapten that can be used is limited by the affinity of the antibody. Thus, the detection limit of haptens by competitive immunoassay is at femtomole or higher levels in most cases [61–68]. By contrast, in noncompetitive two-site immunoassay, antigen to be measured is trapped onto antibody-coated solid phase and reacted with labeled antibody. The amount of antigen to be measured is correlated to the amount of labeled antibody bound to the solid phase. The detection limit of antigens by noncompetitive two-site enzyme immunoassay with appropriate techniques is at attomole levels [10–13].

Molecular size of haptens measured by noncompetitive (two-site) enzyme immunoassay

As evident from the principle of two-site enzyme immunoassay, hapten molecules to be measured have to have two or more epitopes, which are sufficiently separated from each other to allow simultaneous binding of two antibody molecules. The smallest peptide that has been measured with attomole sensitivity by two-site enzyme immunoassay is human α-atrial natriuretic peptide (α-hANP), a single chain polypeptide consisting of 28 amino acids with a ring structure formed by an intramolecular disulfide bond [39]. Ten attomoles of this peptide were measured by two-site enzyme immunoassay using peroxidase-labeled Fab' directed to the C-terminus of the peptide and polystyrene beads coated with IgG against the N-terminal half of the ring structure. This was 100-fold more sensitive than competitive radioimmunoassay [61]. The distance between the two epitopes recognized by the two antibody molecules appeared to correspond to 12–15 amino acid single chain peptides. From this, peptides consisting of more than 12–15 amino acids are strongly suggested to be measurable at attomole levels using antibodies with sufficiently high affinity. However, smaller peptides may not be measured with attomole sensitivity by two-site enzyme immunoassay. Namely, there is only competitive assay for measuring smaller peptides and haptens with similar or smaller sizes that cannot be bound simultaneously by two antibody molecules. The sensitivity of competitive immunoassay is at femtomole or higher levels for most haptens. In addition, many peptides consisting of more than 12–15 amino acids may not have two epitopes which are sufficiently separated from each other and can simultaneously bind two antibody molecules with sufficiently high affinity.

Principle of the hetero-two-site (immune complex transfer) enzyme immunoassay for haptens

A novel noncompetitive two-site enzyme immunoassay (hetero-two-site enzyme immunoassay) was developed for smaller peptides or haptens with amino groups [40–46]. The principle of this method is as follows.

First, the hapten to be measured is labeled with an appropriate substance, so that one antibody molecule directed to the hapten to be measured and another binding molecule for the label may be simultaneously bound to the labeled hapten molecule, thus allowing a two-site assay. An appropriate label and its binding substance may be chosen from a variety of combinations such as biotin-avidin, hapten-anti-hapten antibody (distinct from those to be measured), antigen-antibody, hormone-receptor and nucleotide hybrids. Labeling of haptens to be measured may be performed chemically using excess of labels activated with appropriate functional groups such as N-hydroxysuccinimide esters, anhydride groups and aldehyde groups which are reactive with haptens to be measured. Alternatively, haptens may be labeled by direct conjugation to enzymes. Subsequently, excess of the labels partly unreacted and partly bound to substances other than haptens to be measured should be eliminated prior to two-site assay. This may be performed using a solid phase coated with antibodies directed to the structure of haptens to be measured. Finally, a two-site assay can be performed using anti-hapten antibodies and binding substances for the labels. Anti-hapten antibodies are used for trapping labeled haptens onto the solid phase, and binding substances for the labels are conjugated with enzymes. Alternatively, anti-hapten antibodies are conjugated with enzymes, and binding substances for the labels are used for trapping labeled haptens onto the solid phase. Enzymes may be replaced by fluorescent or luminescent substances. Alternatively, haptens to be measured may be labeled directly with fluorescent and luminescent substances such as europium and acridinium which can be measured with high sensitivity. Various versions of the hetero-two-site (immune complex transfer) enzyme immunoassay have been tested.

In hetero-two-site enzyme immunoassay I (Fig. 32) [40–45], peptides with amino groups to be measured are biotinylated by using N-hydroxysuccinimide ester of biotin and trapped onto polystyrene beads coated with anti-peptide antibody. The polystyrene beads are washed to eliminate unreacted biotin and other biotinylated substances. Biotinylated peptides are eluted from the polystyrene beads with acid (pH 1.0), and the eluate is neutralized. Biotinylated peptides in the neutralized eluate are reacted simultaneously or sequentially with anti-peptide antibody-enzyme conjugate and (strept)avidin-coated polystyrene beads. Finally, the enzyme activity bound to the polystyrene beads is correlated to the amount of peptides to be measured.

In hetero-two-site enzyme immunoassay II (Fig. 33) [43,44], peptides to be measured are biotinylated and eluted from anti-peptide antibody-coated polystyrene beads as in hetero-two-site enzyme immunoassay I. Subsequently, biotinylated peptides in the neutralized eluate are reacted simultaneously or sequentially with anti-peptide antibody-coated polystyrene beads and (strept)avidin-enzyme conjugate.

In hetero-two-site (immune complex transfer) enzyme immunoassay III (Fig. 34) [43], peptides to be measured are biotinylated and eluted from anti-peptide antibody-coated polystyrene beads as in hetero-two-site enzyme immunoassay I. Biotinylated peptides in the neutralized eluate are reacted with 2,4-dinitrophenylated anti-peptide antibody and trapped onto polystyrene beads coated with affinity-purified (anti-2,4-dinitrophenyl group) IgG. The polystyrene beads are washed to more completely

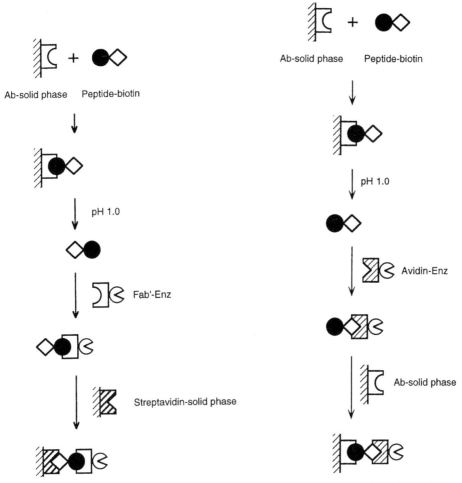

Fig. 32. (See left) Hetero-two-site enzyme immunoassay I of biotinylated peptides. Ab, anti-peptide IgG; Fab′, anti-peptide Fab′; Enz, enzyme.

Fig. 33. (See right) Hetero-two-site enzyme immunoassay II of biotinylated peptides. Ab, anti-peptide IgG; Enz, enzyme.

eliminate unreacted biotin and other biotinylated substances and subsequently re-acted with (strept)avidin-enzyme conjugate. After washing to eliminate excess of the conjugate, the complex of the three components (2,4-dinitrophenylated anti-peptide antibody, biotinylated peptides and (strept)avidin-enzyme conjugate) was eluted from the solid phase with excess of εN-2,4-dinitrophenyl-L-lysine and transferred onto polystyrene beads coated with affinity-purified (anti-rabbit IgG) IgG. The en-zyme activity bound to the last polystyrene beads is correlated to the amount of pep-tides to be measured.

438

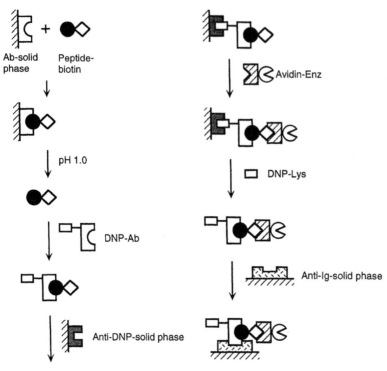

Fig. 34. Hetero-two-site (immune complex transfer) enzyme immunoassay III of biotinylated peptides. Ab, anti-peptide IgG; DNP, 2,4-dinitrophenyl group; Enz, enzyme; DNP-Lys, εN-2,4-dinitrophenyl-L-lysine; Ig, immunoglobulin.

In hetero-two-site (immune complex transfer) enzyme immunoassay IV (Fig. 35) [44], peptides to be measured are biotinylated and eluted from anti-peptide antibody-coated polystyrene beads as in hetero-two-site enzyme immunoassay I. Biotinylated peptides in the neutralized eluate are reacted with 2,4-dinitrophenylated-fluorescein disulfide-bovine serum albumin (DNP-FL-SS-BSA)-anti-peptide antibody and trapped onto polystyrene beads coated with affinity-purified (anti-dinitrophenyl group) IgG. The polystyrene beads are washed to more completely eliminate unre-acted biotin and other biotinylated substances and reacted with (strept)avidin-enzyme conjugate. After washing to eliminate excess of the conjugate, the complex of the three components (DNP-FL-SS-BSA-anti-peptide antibody, biotinylated peptides and (strept)avidin-enzyme conjugate) was eluted from the polystyrene beads with excess of εN-2,4-dinitrophenyl-L-lysine and transferred to anti-fluorescein IgG-coated polystyrene beads. After washing, the complex was released from the polystyrene beads by reduction with 2-mercaptoethylamine and transferred to poly-styrene beads coated with affinity-purified (anti-rabbit IgG) IgG. The enzyme activ-ity bound to the last polystyrene beads is correlated to the amount of peptides to be measured.

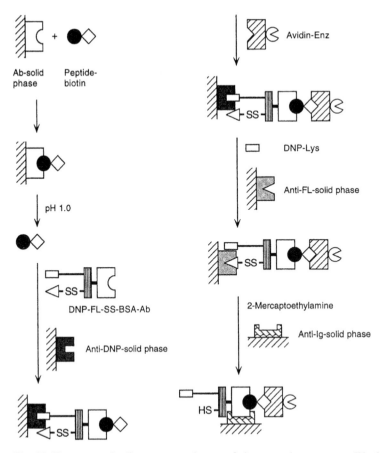

Fig. 35. Hetero-two-site (immune complex transfer) enzyme immunoassay IV of biotinylated peptides. Ab, anti-peptide IgG; DNP, 2,4-dinitrophenyl group; FL, fluorescein; SS, disulfide bond; BSA, bovine serum albumin; Enz, enzyme; DNP-Lys, εN-2,4-dinitrophenyl-L-lysine; Ig, immunoglobulin.

Factors to limit the sensitivity of hetero-two-site (immune complex transfer) enzyme immunoassay for haptens

Hetero-two-site enzyme immunoassay I (Fig. 32) has been applied to the measurement of angiotensin I, a 10 amino acid single chain peptide with no lysine residue [40–42] and arginine vasopressin (AVP), a nine amino acid single chain peptide with an intramolecular disulfide bridge (Fig. 36) [43]. Anti-angiotensin I IgG used was specific for the C-terminus of the peptide, which was confirmed by the finding that there was no significant cross-reaction with angiotensin II which is produced from angiotensin I by deletion of two C-terminal amino acids. Both peptides were biotinylated directly with *N*-hydroxysuccinimidobiotin and trapped onto anti-angiotensin I or anti-AVP IgG-coated polystyrene beads. The polystyrene beads were washed to eliminate unreacted biotin and other biotinylated substances and were subsequently treated at pH 1.0 to elute biotinylated angiotensin I or AVP.

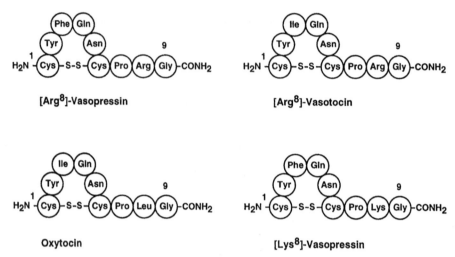

Fig. 36. Primary structure of arginine vasopressin and related peptides.

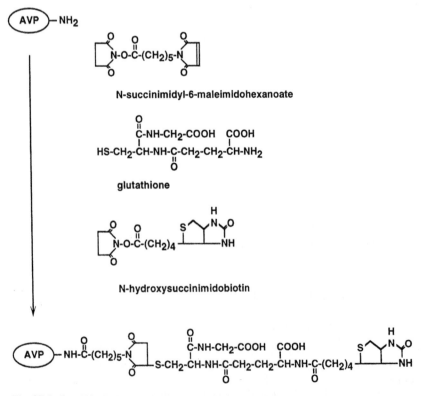

Fig. 37. Indirect biotinylation of arginine vasopressin (AVP).

The eluate was neutralized, reacted with anti-angiotensin I or anti-AVP Fab'-horseradish peroxidase conjugate and subsequently with streptavidin-coated polystyrene beads. The detection limits of angiotensin I and AVP were 13 fg (10 amol)/tube and 54 fg (50 amol)/tube, respectively.

As a reason for the difference between the detection limits of the two peptides, the following possibility was suggested. Biotin residues bound to AVP molecules through N-terminal amino groups might have been fairly close to the epitopic sites recognized by anti-AVP Fab'-peroxidase conjugate used. Therefore, the biotinylated AVP molecules, which had been bound to the conjugate molecules, might not have efficiently reacted with streptavidin-coated polystyrene beads due to steric hindrance. This possibility was tested by indirect biotinylation as illustrated in Fig. 37. Maleimide groups were introduced into the peptide molecules using N-succinimidyl-6-maleimidohexanoate and subsequently reacted with thiol groups of glutathione molecules. Finally, amino groups of glutathione residues bound to AVP molecules were reacted with N-hydroxysuccinimidobiotin (indirect biotinylation). As a result, the detection limit of AVP was lowered 5-fold to 11 fg (10 amol)/tube (Fig. 38). This supported the above possibility and suggested that haptens with amino groups could also be measured with high sensitivity in a similar manner. (Maleimidohexanoate and glutathione, which were used as spacers between biotin residues and the binding site of anti-AVP antibody to ensure the simultaneous binding of avidin molecules

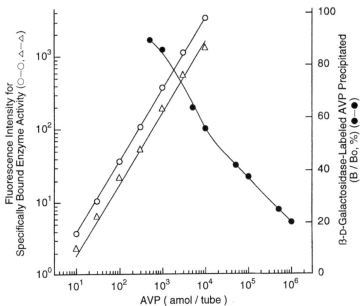

Fig. 38. Dose-response curves of AVP by competitive enzyme immunoassay using β-D-galactosidase-arginine vasopressin conjugate (closed circles) and by hetero-two-site enzyme immunoassays I (open circles) and hetero-two-site (immune complex transfer) enzyme immunoassay III (open triangles) after indirect biotinylation.

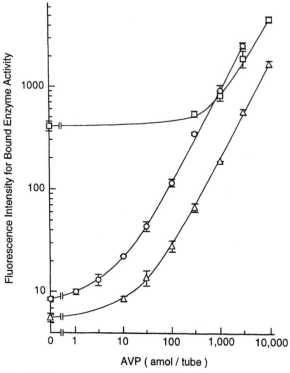

Fig. 39. Dose-response curves of AVP by hetero-two-site (immune complex transfer) enzyme immuno-assay I, II, and IV. Triangles, squares, and circles indicate dose-response curves by hetero-two-site (immune complex transfer) enzyme immunoassay I, II, and IV, respectively.

and anti-AVP antibody molecules to biotinylated AVP, may be replaced by biotinyl peptide with cysteine at the C-terminus.)

For the measurement of biotinylated AVP after elimination of unreacted *N*-hydroxysuccinimidobiotin and other biotinylated substances, various methods are possible in theory as described above, and some of them have been tested (Fig. 39).

In hetero-two-site enzyme immunoassay II, the sensitivity for AVP was extremely low (approximately 2 fmol) even by indirect biotinylation (Fig. 39). One of the reasons for this was that the amount of biotin residues remaining in the eluate from an anti-AVP IgG-coated polystyrene bead was fairly large (approximately 1 pmol) and, therefore, a large amount of (strept)avidin-enzyme conjugate is required, causing high background and low sensitivity. This difficulty was partly overcome in the following way. Biotinylated AVP eluted from an anti-AVP IgG-coated polystyrene bead was trapped on the second anti-AVP IgG-coated polystyrene bead, eluted, trapped on the third anti-AVP IgG-coated polystyrene bead and reacted with avidin-β-D-galactosidase conjugate. However, the detection limit of AVP was still high (approximately 0.3 fmol) largely due to a high nonspecific binding of avidin-β-D-galactosidase conjugate. By contrast, in hetero-two-site enzyme immunoassay I, the detection limit of AVP was 10 amol (Fig. 39). This high sensitivity was possible because streptavidin-coated polystyrene beads had a large capacity to bind biotin resi-dues and the nonspecific binding of anti-AVP Fab'-peroxidase conjugate is low. However, no further improvement in the sensitivity was expected.

In hetero-two-site (immune complex transfer) enzyme immunoassay III and IV, biotinylated AVP eluted from an anti-AVP IgG-coated polystyrene bead was reacted with DNP-(FL-SS-) BSA-Anti-AVP IgG conjugate, and the complex formed was trapped on polystyrene beads coated with affinity-purified (anti-2,4-dinitrophenyl group) IgG. The polystyrene beads were washed to eliminate unreacted N-hydroxysuccinimidobiotin and other biotinylated substances more completely. Sub-sequently, the complex on the polystyrene beads was reacted with avidin-β-D-galactosidase conjugate. When the amount of added AVP was 1 fmol, β-D-galactosidase activity specifically bound to polystyrene beads coated with affinity-purified (anti-2,4-dinitrophenyl group) IgG, depending on the efficiency of the complex formation, was equivalent to 0.053 fmol (5.3% of added AVP on a molar basis), and that nonspecifically bound was 0.02% of avidin-β-D-galactosidase conju-gate added. The detection limit of AVP was approximately 0.3 fmol. In the next step, the complex was transferred to the second solid phase (anti-rabbit IgG) IgG or anti-fluorescein IgG-coated polystyrene beads). When the amount of added AVP was 10–

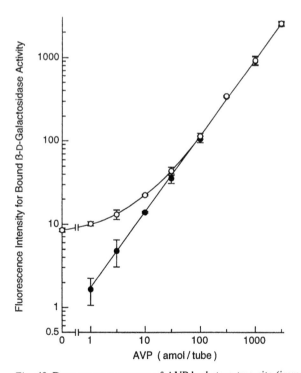

Fig. 40. Dose-response curves of AVP by hetero-two-site (immune complex transfer) enzyme immuno-assay IV. Open and closed circles indicate bound β-D-galactosidase activity (specifically bound, β-D-galactosidase activity plus nonspecifically bound β-D-galactosidase activity) and specifically bound β-D-galactosidase activity, respectively. Vertical bars indicate standard deviations of 3–5 determinations.

30 amol, β-D-galactosidase activity specifically bound to the second solid phase was equivalent to 0.14–0.42 amol (1.4% of added AVP on a molar basis), and that non-specifically bound was 0.0001% of avidin-β-D-galactosidase conjugate added. The detection limit of AVP was improved 53-fold to approximately 5 amol. A relatively small decrease in the specific binding with a larger decrease in the nonspecific binding might have been due to the following facts. Not only the complex of the three components but also free DNP-(FL-SS-) BSA-Anti-AVP IgG conjugate was transferred, and the affinity between biotinylated AVP and avidin-β-D-galactosidase conjugate was extremely high [69]. Both factors might have been effective in minimizing the dissociation of the complex comprising the three components. In the final step (hetero-two-site (immune complex transfer) enzyme immunoassay IV), the complex was transferred to (anti-rabbit IgG) IgG-coated polystyrene beads. When the amount of AVP added was 3–10 amol, β-D-galactosidase activity specifically bound to anti-rabbit IgG IgG-coated polystyrene beads was equivalent to 0.02–0.065 amol (0.65% of added AVP on a molar basis), and that nonspecifically bound was 0.00001% of avidin-β-D-galactosidase conjugate added. The detection limit was improved 4.5-fold to 1 amol (Figs. 39 and 40).

Thus, the decrease in the nonspecific binding of avidin-β-D-galactosidase conjugate by the second complex transfer was not satisfactory. This might have been due to nonspecific association between avidin-β-D-galactosidase conjugate and DNP-FL-SS-BSA-Anti-AVP IgG conjugate. This possibility was supported by the finding that nonspecifically bound β-D-galactosidase activity in the final step decreased 13–26-fold, when nonspecific rabbit IgG-coated polystyrene beads were substituted for

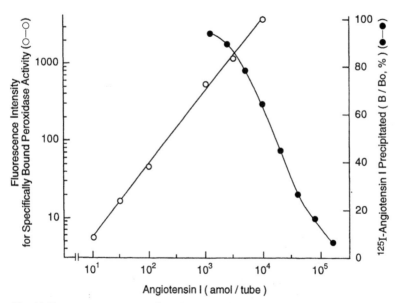

Fig. 41. Dose-response curves of angiotensin I by hetero-two-site enzyme immunoassay I after direct biotinylation and by competitive radioimmunoassay using ^{125}I-angiotensin I.

anti-2,4-dinitrophenyl group IgG-, anti-fluorescein IgG-, or anti-rabbit IgG IgG-coated polystyrene beads or when DNP-FL-SS-BSA-Anti-AVP IgG conjugate was omitted. The nonspecific association between DNP-FL-SS BSA-Anti-AVP IgG conjugate and avidin-β-D-galactosidase conjugate may be reduced as follows. One way is to affinity-purify DNP-FL-SS-BSA-Anti-AVP IgG conjugate, which may reduce the amount to be added and consequently the amount nonspecifically associated with avidin-β-D-galactosidase conjugate. Another method is to prepare DNP-FL-SS-BSA-Anti-AVP IgG conjugate in such a manner that the nonspecific association may be minimized.

Application of the hetero-two-site (immune complex transfer) enzyme immunoassay for haptens

Hetero-two-site enzyme immunoassay I has been applied to the measurement of angiotensin I [40–42], α-hANP [45] and AVP [42–44]. The detection limit of angiotensin I was 13 fg (10 amol)/tube, which was 100-fold lower than that by competitive radioimmunoassay using the same antiserum and ^{125}I-angiotensin I (Fig. 41) and 80–480-fold lower than those previously reported by competitive radioimmunoassay [63] and competitive enzyme immunoassay [64,65]. The detection limit of α-hANP was 60 fg (20 amol)/tube, which was 50-fold lower than that by a competitive enzyme immunoassay using same antiserum (Fig. 42) [45]. The detection limit of AVP by indirect biotinylation was 11 fg (10 amol)/tube. This was

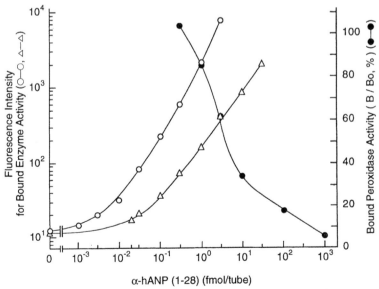

Fig. 42. Dose-response curves of α-hANP by hetero-two-site enzyme immunoassay I (open triangles) and IV (open circles) after direct biotinylation and by competitive enzyme immunoassay (closed circles).

45-fold lower than that by competitive enzyme immunoassay using the same antiserum (Fig. 38) [63] and 22–360-fold lower than those previously reported by competitive radioimmunoassays [67,68]. Hetero-two-site (immune complex transfer) enzyme immunoassay IV has also been applied to the measurement of α-hANP [46] and AVP [44]. The detection limits of these peptides were both 1 amol/tube, which was 450–500-fold lower than those by competitive enzyme immunoassays using the same antisera (Figs. 40 and 42) [45,66].

The maximal volume of plasma that could be used with satisfactory recoveries of peptides added to plasma in hetero-two-site (immune complex transfer) enzyme immunoassays was 5–10 μl, when plasma was directly biotinylated. This is because the concentration of amino groups in plasma is much higher (approximately 60 mmol/l) than the concentration of N-hydroxysuccinimidobiotin added to the reaction mixture for biotinylation (3–6 mmol/l) (Fig. 37) [42–46]. The detection limits of angiotensin I, α-hANP and AVP using 5 μl of plasma were 2.6 ng/l [41,42], 12 ng/l [45,46] and 2.2 ng/l [43,44], respectively.

Peptides to be measured can be extracted from larger volumes of plasma using anti-peptide IgG-coated polystyrene beads. When angiotensin I was extracted from 50 μl of plasma using angiotensin I IgG-coated polystyrene beads, the detection limit of angiotensin I in plasma was improved to 0.8 ng/l [41,42] which was 40–300-fold lower than those previously reported [63–65]. The concentration of angiotensin I in the plasma of seven healthy subjects was 154 ± 38 (SD) ng/l (range 97–206 ng/l) [41].

Alternatively, peptides to be measured can be separated from proteins in plasma by filtration through a molecular sieve, which is usually done by brief centrifugation in a commercially available microconcentrator with a polysaccharide membrane. The concentration of amino groups in plasma filtrates of healthy subjects was 1.7–2.1 mmol/l [42–46]. When 100 μl of plasma filtrates corresponding to 75 μl of plasma were used for biotinylation, the detection limit of AVP in plasma was

Table 2. Detection limit of antigens and peptides by various enzyme immunoassays

Analyte	Enzyme immuno-assay (EIA)	Approximate level of detection limit
Protein		
Antigens	Competitive EIA	Femtomole (1×10^{-15} mol)
	Two-site EIA	Attomole (1×10^{-18} mol)
	Two-site immune complex transfer EIA	Zeptomole (1×10^{-21} mol)
Peptides		
Peptides consisting of more than 12–15 amino acids	Competitive EIA	Femtomole
	Two-site EIA	10 amol
Peptides consisting of less than 12–15 amino acids	Competitive EIA	Femtomole
	Hetero-two-site EIA	10 amol
	Hetero-two-site immune complex transfer EIA	1 amol

Table 3. Sensitivity of various enzyme immunoassays for anti-insulin antibody and anti-thyroglobulin antibodies in serum

Enzyme immunoassay (EIA)		Solid phase (SP)	Conjugate	Improvement of sensitivity (-fold)	
				Anti-insulin antibodies	Anti-thyroglobulin antibodies
Conventional EIA (ELISA)		SP-Ag	Anti-Ig-Enz	1	1
Immune complex transfer EIA	Method I	SP-anti-DNP SP-avidin	DNP⟩Biotin Ag Anti-Ig-Enz	150	–
	Method II	SP-anti-DNP SP-anti-Ag	DNP-Ag Anti-Ig-Enz	–	40
	Method III	SP-anti-DNP SP-anti-Ig	DNP-Ag Anti-Ag-Enz	–	120
	Method IV	SP-anti-DNP SP-anti-Ig	DNP-Ag-Enz	150	20
	Method V	SP-anti-DNP SP-anti-Ig	DNP-Ag Ag-Enz	4,000	2,000
	Method VI	SP-anti-DNP SP-streptavidin	DNP⟩Biotin Ag Ag-Enz	4,000	–

Ag, antigen; Ig, immunoglobulin; Enz, enzyme; DNP, 2,4-dinitrophenyl group; S–S, disulfide bond.

0.14 ng/l [43,44] and 0.014 ng/l [44] by hetero-two-site (immune complex transfer) enzyme immunoassay I and IV, respectively. These were 2–14-fold and 20–140-fold lower than those previously reported only after extraction from as much as 1–5 ml of plasma using acetone and octadecyl silica followed by concentration [66–68]. The detection limit of α-hANP was 0.8 ng/l [45] and 0.04 ng/l [46] by hetero-two-site (immune complex transfer) enzyme immunoassay I and IV, respectively. These were 13-fold and 250-fold lower than that previously reported by competitive radioimmunoassay using 0.1 ml of plasma [62]. The concentrations of AVP and α-hANP in the plasma of healthy subjects aged 24–42 years with *ad libitum* water intake and normal activity approximately 4 h after breakfast were 0.72 ± 0.22 (SD) ng/l (range 0.42–1.04 ng/l) [43] and 30 ± 11 (SD) ng/l (range 11–56 ng/l) [46], respectively.

Thus, some peptides such as AVP in plasma can be measured without extraction by hetero-two-site (immune complex transfer) enzyme immunoassays, whereas tedious and time-consuming extraction and concentration processes are inevitable in the measurement by competitive radioimmunoassay and enzyme immunoassay.

The detection limits of antigens and haptens and the sensitivity for antibodies by various methods described above are summarized in Tables 2 and 3.

References

1. Berson SA, Yalow RS. Species-specificity of human anti-beef, pork insulin serum. J Clin Invest 1959;38:2017–2025.
2. Miles LEM, Hales CN. Labelled antibodies and immunological assay systems. Nature 1968;219: 186–189.
3. Engvall E, Perlmann P. Enzyme-linked immunosorbent assay (ELISA). Quantitative assay of immunoglobulin G. Immunochemistry 1971;8:871–874.
4. van Weemen BK, Schuurs AHWM. Immunoassay using antigen-enzyme conjugates. FEBS Lett 1971;15:232–236.
5. Editorial. ELISA: a replacement for radioimmunoassay? Lancet 1976;ii:406–407.
6. Ekins R. ELISA: a replacement for radioimmunoassays? Lancet 1976;ii:569–570.
7. Watson D. ELISA: a replacement for radioimmunoassays? Lancet 1976;ii:570.
8. Kato K, Hamaguchi Y, Okawa S, Ishikawa E, Kobayashi K, Katunuma N. Enzyme immunoassay in rapid progress. Lancet 1977;i:40.
9. Ishikawa E, Kato K. Ultrasensitive enzyme immunoassay. Scand J Immunol 1978;8(Suppl. 7):43–55.
10. Ishikawa E, Imagawa M, Hashida S. Ultrasensitive enzyme immunoassay using fluorogenic, luminogenic, radioactive and related substrates and factors to limit the sensitivity. Dev Immunol 1983; 18:219–232.
11. Ishikawa E. Development and clinical application of sensitive enzyme immunoassay for macromolecular antigens - a review. Clin Biochem 1987;20:375–385.
12. Ishikawa E, Hashida S, Kato Y, Imura H. Sensitive enzyme immunoassay of human growth hormone for clinical application: a review. J Clin Lab Anal 1987;1:238–242.
13. Ishikawa E, Hashida S, Tanaka K, Kohno T. Ultrasensitive enzyme immunoassay for antigens: technology and applications - a review. Clin Chem Enzym Comms 1989;1:199–215.
14. Hashida S, Tanaka K, Kohno T, Ishikawa E. Novel and ultrasensitive sandwich enzyme immunoassay (sandwich transfer enzyme immunoassay) for antigens. Anal Lett 1988;21:1141–1154.
15. Hashida S, Ishikawa E. Detection of one milliattomole of ferritin by novel and ultrasensitive enzyme immunoassay. J Biochem 1990;108:960–964.

16. Ishikawa E, Yoshitake S, Endo Y, Ohtaki S. Highly sensitive enzyme immunoassay of rabbit (anti-human IgG) IgG using human IgG-β-D-galactosidase conjugate. FEBS Lett 1980;111:353–355.

17. Kohno T, Ruan K-h, Ishikawa E. A highly sensitive enzyme immunoassay of anti-insulin antibodies in guinea pig serum. Anal Lett 1986;19:1083–1095.

18. Kohno T, Ishikawa E. A novel enzyme immunoassay of anti-insulin IgG in guinea pig serum. Biochem Biophys Res Commun 1987;147:644–649.

19. Kohno T, Mitsukawa T, Matsukura S, Ishikawa E. Novel enzyme immunoassay (immune complex transfer enzyme immunoassay) for anti-thyroglobulin IgG in human serum. J Clin Lab Anal 1988; 2:209–214.

20. Kohno T, Mitsukawa T, Matsukura S, Ishikawa E. Novel and sensitive enzyme immunoassay (immune complex transfer enzyme immunoassay) for anti-thyroglobulin IgG in human serum. Clin Chem Enzym Comms 1988;1:89–96.

21. Kohno T, Ishikawa E. Novel enzyme immunoassay (immune complex transfer enzyme immunoassay) for anti-insulin IgG in guinea pig serum. Anal Lett 1988;21:1019–1031.

22. Kohno T, Mitsukawa T, Matsukura S, Tsunetoshi Y, Ishikawa E. More sensitive and simpler immune complex transfer enzyme immunoassay for antithyroglobulin IgG in serum. J Clin Lab Anal 1989;3:163–168.

23. Ishikawa E, Kohno T. Development and applications of sensitive enzyme immunoassay for antibodies: a review. J Clin Lab Anal 1989;3:252–265.

24. Ishikawa E, Hashida S, Kohno T. Development of ultrasensitive enzyme immunoassay reviewed with emphasis on factors which limit the sensitivity. Mol Cell Prob 1991;5:81–95.

25. Ishikawa E, Hashida S, Kohno T, Hirota K, Hashinaka K, Ishikawa S. Principle and applications of ultrasensitive enzyme immunoassay (immune complex transfer enzyme immunoassay) for antibodies in body fluids. J Clin Lab Anal 1993;7:376–393.

26. Kohno T, Katsumaru H, Nakamoto H, Yasuda T, Mitsukawa T, Matsukura S, Tsunetoshi Y, Ishikawa E. Use of inactive β-D-galactosidase for elimination of interference by anti-β-D-galactosidase antibodies in immune complex transfer enzyme immunoassay for anti-thyroglobulin IgG in serum using β-D-galactosidase from *Escherichia coli* as label. J Clin Lab Anal 1991;5:197–205.

27. Kohno T, Sakoda I, Ishikawa E. Immune complex transfer enzyme immunoassay for (anti-human T-cell leukemia virus type I) IgG in serum using a synthetic peptide, *env* gp46(188–209), as antigen. J Clin Lab Anal 1991;5:25–37.

28. Kohno T, Sakoda I, Suzuki M, Izumi A, Ishikawa, E. Immune complex transfer enzyme immunoassay for (anti-human T-cell leukemia virus type I) IgG in serum using a synthetic peptide, cys-*gag* p19(100–130), as antigen. J Clin Lab Anal 1991;5:307–316.

29. Kohno T, Sakoda I, Ishikawa E. Sensitive enzyme immunoassay (immune complex transfer enzyme immunoassay) for (anti-human T-cell leukemia virus type I) IgG in serum using a synthetic peptide, cys-*env* gp46(188–224), as antigen. J Clin Lab Anal 1992;6:105–112.

30. Kohno T, Sakoda I, Ishikawa E. Sensitive enzyme immunoassay (immune complex transfer enzyme immunoassay) for (antihuman T-cell leukemia virus type I) immunoglobulin G in serum using a synthetic peptide, ala-cys-*env* gp46(237–262), as antigen. J Clin Lab Anal 1992;6:162–169.

31. Kohno T, Hirota K, Sakoda I, Yamasaki M, Yokoo Y, Ishikawa E. Sensitive enzyme immunoassay (immune complex transfer enzyme immunoassay) for anti-anti-human T-cell leukemia virus type I) IgG in serum using recombinant *gag* p24(14–214) as antigen. J Clin Lab Anal 1992;6:302–310.

32. Yogi Y, Hirota K, Kohno T, Toshimori H, Matsukura S, Setoguchi T, Ishikawa E. Measurement of anti-thyroglobulin IgG in urine of patients with autoimmune thyroid diseases by sensitive enzyme immunoassay (immune complex transfer enzyme immunoassay). J Clin Lab Anal 1993;7:70–79.

33. Hirota K, Kohno T, Ishikawa E, Sugiyama S. Simpler and more sensitive immune complex transfer enzyme immunoassay for anti-insulin IgG in serum using β-D-galactosidase as label. Clin Chem Enzym Comms 1991;4: 27–38.

34. Hashida S, Hirota K, Kohno T, Ishikawa E. Anti-HTLV-I IgG in urine detected by sensitive enzyme immunoassay (immune complex transfer enzyme immunoassay) using a synthetic peptide, cys-*env* gp46(188–224), as antigen. J Clin Lab Anal 1994;8:149–156.

35. Hashida S, Hirota K, Hashinaka K, Saitoh A, Nakata A, Shinagawa H, Oka S, Shimada K, Mimaya J, Matsushita S, Ishikawa E. Detection of antibody IgG to HIV-1 in urine by sensitive enzyme immunoassay (immune complex transfer enzyme immunoassay) using recombinant proteins as antigens for diagnosis of HIV-1 infection. J Clin Lab Anal 1993;7:353–364.

36. Hashida S, Hashinaka K, Hirota K, Saitoh A, Nakata A, Shinagawa H, Oka S, Shimada K, Mimaya J, Matsushita S, Ishikawa E. Detection of antibody IgG to HIV-1 in urine by ultrasensitive enzyme immunoassay (immune complex transfer enzyme immunoassay) using recombinant p24 as antigen for diagnosis of HIV-1 infection. J Clin Lab Anal 1994;8:86–95.

37. Hashinaka K, Hashida S, Hirota K, Saitoh A, Nakata A, Shinagawa H, Oka S, Shimada K, Ishikawa E. Detection of anti-human immunodeficiency virus type 1 (HIV-1) immunoglobulin G in urine by an ultrasensitive enzyme immunoassay (immune complex transfer enzyme immunoassay) with recombinant reverse transcriptase as an antigen. J Clin Microbiol 1994;32:819–822.

38. Hashida S, Hashinaka K, Saitoh A, Takamizawa A, Shinagawa H, Oka S, Shimada K, Hirota K, Kohno T, Ishikawa S, Ishikawa E. Diagnosis of HIV-1 infection by detection of antibody IgG to HIV-1 in urine with ultrasensitive enzyme immunoassay (immune complex transfer enzyme immunoassay) using recombinant proteins as antigens. J Clin Lab Anal 1994;8, (in press).

39. Hashida S, Ishikawa E, Nakao K, Mukoyama M, Imura H. Enzyme immunoassay for α-human atrial natriuretic polypeptide - direct measurement of plasma level. Clin Chim Acta 1988;175:11–18.

40. Tanaka K, Hashida S, Kohno T, Yamaguchi K, Ishikawa E. Novel and sensitive noncompetitive enzyme immunoassay to measure peptides by biotinylation. Anal Lett 1989;22:353–363.

41. Tanaka K, Hashida S, Kohno T, Yamaguchi K, Ishikawa E. Novel and sensitive noncompetitive enzyme immunoassay for peptides. Biochem Biophys Res Commun 1989;160:40–45.

42. Ishikawa E, Tanaka K, Hashida S. Novel and sensitive noncompetitive (two-site) immunoassay for haptens with emphasis on peptides. Clin Biochem 1990;23:445–453.

43. Hashida S, Tanaka K, Yamamoto N, Ishikawa E. Ultrasensitive enzyme immunoassay for peptides. In: Parvez SH, Naoi M, Nagatsu T, Parvez S (eds) Methods in Neurotransmitter and Neuropeptide Research. New York: Elsevier 1993;279–304.

44. Hashida S, Tanaka K, Yamamoto N, Uno T, Yamaguchi K, Ishikawa E. Detection of one attomole of [Arg^8]-vasopressin by novel noncompetitive enzyme immunoassay (hetero-two-site complex transfer enzyme immunoassay). J Biochem 1991;110:486–492.

45. Hashida S, Yamamoto N, Ishikawa E. Novel and sensitive noncompetitive enzyme immunoassay (hetero-two-site enzyme immunoassay) for α-human atrial natriuretic peptide in plasma. J Clin Lab Anal 1991;5:324–330.

46. Hashida S, Ishikawa E. Novel and ultrasensitive noncompetitive enzyme immunoassay (hetero-two-site complex transfer enzyme immunoassay) for α-human atrial natriuretic peptide (α-hANP). J Clin Lab Anal 1992;6:201–208.

47. Ishikawa E, Imagawa M, Hashida S, Yoshitake S, Hamaguchi Y, Ueno T. Enzyme-labeling of antibodies and their fragments for enzyme immunoassay and immunohistochemical staining. J Immunoassay 1983;4:209–327.

48. Hashida S, Imagawa M, Inoue S, Ruan K-h, Ishikawa E. More useful maleimide compounds for the conjugation of Fab' to horseradish peroxidase through thiol groups in the hinge. J Appl Biochem 1984;6:56–63.

49. Ishikawa E, Hashida S, Kohno T, Tanaka K. Methods for enzyme-labeling of antigens, antibodies and their fragments. In: Ngo TT (ed) Nonisotopic Immunoassay. New York: Plenum, 1988;27–55.

50. Ishikawa E, Hashida S, Kohno T, Kotani T, Ohtaki S. Modification of monoclonal antibodies with enzymes, biotin, and fluorochromes and their applications. In: Schook LB (ed) Monoclonal Antibody Production Techniques and Applications. New York: Marcel Dekker, 1987;113–137.

51. Ishikawa E, Yoshitake S, Imagawa M, Sumiyoshi A. Preparation of monomeric Fab'-horseradish peroxidase conjugate using thiol groups in the hinge and its evaluation in enzyme immunoassay and immunohistochemical staining. Ann NY Acad Sci 1983;420:74–89.

52. Johannsson A, Ellis DH, Bates DL, Plumb AM, Stanley CJ. Enzyme amplification for immunoassays - detection limit of one hundredth of an attomole. J Immunol Methods 1986;87:7–11.

53. Geiger R, Hauber R, Miska W. New bioluminescence-enhanced detection systems for use in enzyme activity tests, enzyme immunoassays, protein blotting and nucleic acid hybridization. Mol Cell Prob 1989;3:309–328.

54. Miska W, Geiger R. Luciferin derivatives in bioluminescence-enhanced enzyme immunoassays. J Biolumin Chemilumin 1989;4:119–128.

55. Bronstein I, Edwards B, Voyta JC. 1,2-Dioxetanes: novel chemiluminescent enzyme substrates. Applications to immunoassays. J Biolumin Chemilumin 1989;4:99–111.

56. Bronstein I, Voyta JC, Thorpe GHG, Kricka LJ, Armstrong G. Chemiluminescent assay of alkaline phosphatase applied in an ultrasensitive enzyme immunoassay of thyrotropin. Clin Chem 1989;35:1441–1446.

57. Ruan K-h, Hashida S, Tanaka K, Ishikawa E, Niitsu Y, Urushizaki I, Ogawa H. A small scale sandwich enzyme immunoassay for macromolecular antigens using β-D-galactosidase from *Escherichia coli* and horseradish peroxidase as labels. Anal Lett 1987;20:587–601.

58. Kohno T, Hirota K, Ishikawa E. Presence of antibodies against β-D-galactosidase from *Escherichia coli* in serum of healthy subjects examined by immune complex transfer enzyme immunoassay. Clin Chem Enzym Comms 1991;4:39–49.

59. Ishikawa S, Ishikawa E. Novel and sensitive enzyme immunoassay for urinary antibody IgG to β-D-galactosidase from *Escherichia coli*. Anal Lett 1994;27:337–350.

60. WHO News and Activities. AIDS: proposed WHO criteria for interpreting Western blot assays for HIV-1, HIV-2, and HTLV-I/HTLV-II. Bull WHO 1991;69:127–130.

61. Mukoyama M, Nakao K, Yamada T, Itoh H, Sugawara A, Saito Y, Arai H, Hosoda K, Shirakami G, Morii N, Shiono S, Imura H. A monoclonal antibody against N-terminus of α-atrial natriuretic polypeptide (α-ANP): a useful tool for preferential detection of naturally circulating ANP. Biochem Biophys Res Commun 1988;151:1277–1284.

62. Tikkanen I, Fyhrquist F, Metsärinne K, Leidenius R. Plasma atrial natriuretic peptide in cardiac disease and during infusion in healthy volunteers. Lancet 1985;2:66–69.

63. Fyhrquist F, Soveri P, Puutula L, Stenman U-H. Radioimmunoassay of plasma renin activity. Clin Chem 1976;22:250–256.

64. Scharpé S, Verkerk R, Sasmito E, Theeuws M. Enzyme immunoassay of angiotensin I and renin. Clin Chem 1987;33:1774–1777.

65. Aikawa T, Suzuki S, Murayama M, Hashiba K, Kitagawa T, Ishikawa E. Enzyme immunoassay of angiotensin I. Endocrinology 1979;105:1–6.

66. Uno T, Uehara K, Motomatsu K, Ishikawa E, Kato K. Enzyme immunoassay for arginine vasopressin. Experientia 1982;38:786–787.

67. Morton JJ, Padfield PL, Forsling ML. A radioimmunoassay for plasma arginine-vasopressin in man and dog: application to physiological and pathological states. J Endocrinol 1975;65:411–424.

68. Glänzer K, Appenheimer M, Krück F, Vetter W, Vetter H. Measurement of 8-arginine-vasopressin by radioimmunoassay: development and application to urine and plasma samples using one extraction method. Acta Endocrinol 1984;106:317–329.

69. Wilchek M, Bayer EA. The avidin-biotin complex in bioanalytical applications. Anal Biochem 1988;171:1–32.

© 1995 Elsevier Science B.V. All rights reserved
Biotechnology Annual Review Volume 1
M.R. El-Gewely, editor

The politics of patent legislation in biotechnology: an international view

G. Kristin Rosendal

The Fridtjof Nansen Institute, Lysaker, Norway

Abstract. The realization of the economic value of the genetic resources has prompted an international debate about property rights to genetic resources. The international debate pertaining to patenting of genetic material is the main theme of this chapter. As a backdrop for the international debate, the chapter starts out with a summary of the main events and arguments in the expanding scope of patent legislation in biotechnological inventions. Summing up, the new biotechnologies represent a tool which meets the legal requirements for patenting biological material. From the industry's point of view, biotechnology also necessitates patenting. On the negative side, defending a patent is often a long and costly business, and the trend is that patenting will mainly benefit the bigger and stronger companies and thus weaken public control over the rapid developments in biotechnology. A central argument in the chapter is that without sophisticated biotechnological tools, trained scientists, and adequate infrastructure, patenting is, as yet, hardly a viable solution for the majority of developing countries. Gene-rich developing countries fear that developments in patent legislation will pave the way for increased Northern control over Third World natural resources. The International Convention on Biological Diversity goes some way in making amends to this situation, but the gene-poor, least developed countries may still have reason to fear that they will lose access to breeding material. In a long-term perspective, the implications may be detrimental for resource conservation in developing countries. In conclusion, the patent question seems to remain unresolved and may still be one of the most likely stumbling blocks for future ratifications and implementation of the Biodiversity Convention.

Introduction

The conservation of biological diversity constitutes one of today's greatest challenges, as environmental degradation worldwide has led to species extinction at a hitherto unprecedented rate. There are an estimated 5–30 million biological organisms in the world, of which only some 1.4 million have been described scientifically [1]. Biodiversity encompasses species diversity, ecosystem diversity and diversity within species – genetic diversity. Realizing that this genetic material may have economic value, it is also commonplace to speak about genetic resources.

As the new biotechnologies make it possible to utilize the full potential of the world's genetic resources, the economic incentive to conserve biological diversity increases [2]. Hence, the interest in genetic material is arising from environmental concerns, as well as being based on developments in the technological field. By the year 2000, farm-level sales of products of agricultural biotechnology are expected to

---

Address for correspondence: G.K. Rosendal, The Fridtjof Nansen Institute, P.O. Box 326, Fridtjof Nansens vei 17, N-1324 Lysaker, Norway.

grow to some 100 billion US$. The value of global trade in plant-based pharmaceuticals was estimated to 20 billion US$ for the year 1986 alone [3]. Besides the ethical and aesthetical value of species diversity, people depend on genetic resources for food, medicines and for raw materials in the chemical industries. Genetic diversity or variability is a necessary condition to sustain vitality in both wild and domesticated plants and animals and also for the development of new and improved products.

The realization of the economic value of genetic resources has also prompted an international debate about property rights to genetic resources. Genetic resources may be subject to different types of property rights, one of which is intellectual property rights, including patents. The international debate pertaining to patenting of genetic material is the main theme of this chapter.

As a backdrop for the international debate, I will start with a summary of the main events in the expanding scope of patent legislation in biotechnological inventions. These milestones, include inter alia, the Budapest Convention of 1977 on depositions of micro-organisms, the US "Chakrabaty-case" of 1980 granting a patent on a naturally occurring microorganism, and the 1987 case of the "Onco-mouse", the first patent given on a higher category of animals. In 1992 the US Government's National Institutes of Health applied for patents on a large number of human gene sequences, and in 1993 Agracetus Company claimed patent protection over all genetically engineered cotton varieties, regardless of production method.

The main arguments pro et contra patent protection in biotechnology are also presented. These include socio-economic, environmental and ethical concerns, as well as economic, technical and scientific considerations. A conflict can be identified along a "profit versus preservation" line: on the one side there are private and public interests linked to the biotechnology industry advocating stronger patent protection in order to boost innovation as well as getting compensation for research expenditures in biotechnology. On the other side environmental interest groups and smaller farmers' organisations are citing environmental and socio-economic concerns to stop patents on organic and breeding material. To some extent, this conflict coincides with the classical North/South controversy with regard to control over and access to genetic resources. Neither of these conflict lines are clear-cut, of course.

While patent legislation has traditionally belonged to the national sphere, the last decade has seen an increased preoccupation with the subject also in the international arena. The type of forum is also changing; from those occupied with legal, technical and industrial issues, to those that focus on issues such as North/South relations, environment and development and food production. Most significantly, the issue was brought up in the Uruguay round of the General Agreement on Tariffs and Trade negotiations (GATT). In a special negotiation group on Trade Related Aspects of Intellectual Property Rights (TRIPs) a main objective of some GATT members has been to make patent legislation a compulsory part of the international trade system. Failure to comply with these regulations will be regarded as an infringement of GATT rules, and may thus be subject to economic sanctions. In the mid-1980s the issue of intellectual property rights gave rise to a heated debate in the UN Food and Agricultural Organization (FAO), as the member states were negotiating the FAO

Undertaking on Plant Genetic Resources. Subsequently, the topic has recurred in the UN Environment Programme (UNEP), in the international negotiations for the Convention on Biological Diversity, which was signed in Rio at the UN Conference on Environment and Development (UNCED) in June, 1992. In this section, I will show how the questions related to intellectual property rights in biotechnology were handled in these fora. Consequently, some examples are given regarding proposals to resolve the controversy.

The Expanding Scope of Patent Protection in Biotechnology

The employment of intellectual property rights, and more specifically industrial patent legislation is socially justified by the deal it establishes between society and the patentee: The scientist is making her or his innovation known to the public, while society allows the patentee to monopolize the invention and claim royalties for its use during the patent period, usually about 20 years. Thus, scientific research is encouraged for the benefit of society as a whole.

When industrial patenting was first introduced with the industrial revolution in the middle of the 19th century, it was not, however, without a heated public debate. The opponents had misgivings that it would involve too strong elements of control and power, and also that it constituted an unacceptable breach with the basic principle of free access to utilize, exchange and enhance inventions and solutions among scientists. In this spirit, basic research has been exempted from patenting. With the introduction of compulsory licensing, public opinion finally went in favour of industrial patents [4].

The inventor may choose to keep the invention secret, rather than file for a patent. This has provided another argument in favour of patenting, as secrecy obviously does not enhance scientific exchange. Another alternative to an often costly patent application process, is to publish the invention in some "obscure" journal. This prevents others from patenting the invention, and it may (hopefully) keep them ignorant of its design as well. Besides industrial patents, intellectual property rights include, inter alia, trade marks, copyrights or breeders' rights.

An invention may be patented either as a product, a process or an application. When patent protection is granted for a product it gives the patentee a monopoly to produce, market and utilize the product. A patent on a process may also provide an indirect product protection, as it stops others from marketing and utilizing products which are produced by means of the same patented process. A patent on an application confers a monopoly only for one particular use of a product.

In order that an invention may be patented, it must meet four fundamental criteria. Firstly, the invention must be novel, meaning basically that the invention has not been published anywhere before. Secondly, there is the criterion of nonobviousness, in other words, the invention must display an inventive step. The third criterion says that the invention must have an industrial application, i.e. it must have a practical utility. In this respect, the US patent legislation is less rigid, merely requiring that the invention may be useful. One function of the utility requirement is to distinguish between basic research, which is believed to belong in the public domain, and applied technology, which is eligible for patenting. Finally, the patent application must fulfil the criterion of reproduceability, in the sense that it must describe the invention in such detail that other experts may repeat the experiment. In addition to these criteria, patent legislation commonly excludes from patentability inventions whose utilization would run counter to "public order or morality".

Most patent legislation is of a national character, and patent protection is thus applicable only in the country where it has been granted. One exception is the European Patent Convention (EPC) of 1973, which applies primarily in the European Communities, and which is administered by the European Patent Office (EPO). National patent legislation in the various countries tend to have great similarities, as it is largely drawn from international conventions, administered by the World Intellectual Patent Organization (WIPO).

National and international patent legislation draws no a priori distinction between various sectors of technology. Traditionally, however, the patent system has been limited to technologies dealing with nonorganic material. Biological materials have been regarded as natural products rather than industrial products, and thus, discoveries rather than inventions [5]. Food and medicinals were traditionally excluded from patentability due to these products' fundamental importance for basic human needs. Also, biological products or processes were originally excluded from patentability on the basis that such inventions could not meet all the patent criteria.

These barriers have now been largely overcome by developments in the new biotechnologies. These developments have not only made patenting a practical possibility, they have also created a need for it, as seen from the perspective of many scientists and corporations in the US, Japan and the EC. Research in biotechnology often involves high costs; compared to traditional breeding methods, competition is fierce, and research is increasingly being carried out by the private sector. The biotechnology sector has been arguing strongly for compensation in terms of royalties, along the lines of other fields of technology.

The principal ruling on the patentability of biological material appeared in the German Federal Supreme Court in 1969 (the Red Dove Case), determining that a breeding process for animals was patentable [6]. Plant varieties could be protected by "plant breeders' right", as under the US Plant Variety Protection Act of 1970. In the Chakrabaty Case of 1980 the US Supreme Court of Justice decided, five against four, to allow industrial patents for naturally occurring living matter, including both asexually and sexually reproduced plants [7]. A judge from this case was later hired by the EC Commission in drawing up their formulation of a directive on industrial patents in biotechnology.

The following sections deal with how the patent barriers have been overcome.

Reproduceability

A fundamental restraint on patenting biological material would be the requirement of reproduceability. New breeders' lines in agriculture or horticulture could meet the softer criteria of plant breeders' rights (novel, uniform and stable), but a true replication could not be guaranteed. The very nature of experiments on living material were too arbitrary to meet this criterion of reproduceability. Following the developments in the new biotechnologies it has become possible to give such a detailed description of the invention that an exact repetition of the experiment may be guaranteed [8]. Some uncertainty remains as there is little control over exactly where the gene is being transferred, nor is it possible to guarantee against mutations over time. Still, genetic engineering has paved the way for the application of the patent system also for biotechnological inventions. Thus, the barrier has been overcome.

Preceding this scientific breakthrough, the criterion of reproduceability had been circumvented in the case of inventions involving microorganisms. The Budapest Convention of 1967 solved the problem by allowing the scientist to deposit cultures of the patented microorganisms, rather than file a detailed description.

The transformations in patent legislation are to a large extent based on reinterpretations of existing regulations. Thus, both the European Patent Office (EPO) and the patent offices in the Nordic countries interpret their patent legislation to apply to microorganisms. The scope for patent protection may still vary, however, according to the actual definition of microorganisms applied. Thus, the Norwegian Panel on Biotechnology [9] was sceptical to the inclusion in the EC Commission's patent directive [10] of, inter alia, algae and cells in the definition. As no "upper limit" is given in the definition of cells, a US court recently concluded that a patent could be granted for a human somatic cell line (*Moore vs University of California*, 1990).

It is the reproduceability clause which primarily distinguishes patenting from alternative protection by means of secrecy. A recent criticism, however, claims that patent applications within sectors like biotechnology and computers are no longer a true alternative to secrecy. According to these critics, the publicity clause is being undermined as the technology is becoming increasingly complicated and hard to describe accurately, and accordingly hard to control. It has become easier to "deceive" the patent offices by submitting inadequate patent applications. Thus, it has become harder to replicate an invention on the background of the patent description (Per T. Lossius, the Norwegian Patent Office, personal communication, 25 July, 1992).

The exemption for plant and animal varieties

Most patent systems explicitly exclude plant and animal varieties from patenting. The original interpretation of this ban is to regard it as an exception for plants and animals in general and products derived thereof. This stems, inter alia, from the concern for basic human needs, which was similarly inherent in the common exception for food and medicinal products from patentability. A general trend internationally has been to reinterpret this exclusion narrowly, so that it includes varieties only [11]. In practice, this may signify the end of the exclusion altogether, as the biologists have no exact definition of a variety [12]. Thus in 1988 the EPO granted their first patent on a genetically modified plant, arguing that the plant in question was not to be regarded as a variety. The ban against patenting plant and animal varieties continues to apply in the EPC, but this exception is greatly diluted in the EC Commission's proposal for a directive on patents in biotechnology.

For plant varieties, "plant breeders' rights" has been the alternative to industrial patents. The 1967 Union for the Protection of New Varieties of Plants (UPOV Convention) has been the major international instrument in this regard. UPOV had 20 member states in 1988, mainly from the developed world [13]. Protection under the UPOV Convention of 1978 only pertains to the final product, the variety, and not to subsequent varieties bred on the basis of the protected one. Farmers are thus free to use such seeds for next year's sowing, and breeders and scientists may use UPOV-protected material for developing new products, without paying royalties. These provisions are known as the "farmers' privilege" and the "breeders' exemption". Industrial patents differ from UPOV's "plant breeders' rights" in both respects [14]. In the 1991 revision of UPOV the farmers' privilege and breeders' exemption has been relaxed in order to adjust the UPOV Convention to the patent system. The 1978 version still holds, however, as not enough parties have ratified the 1991 version. Many plant breeders fear that the strengthened UPOV protection, like that of patenting, may reduce scientific exchange, and thus hamper breeding of new varieties. Moreover, the original UPOV Convention barred "double protection"; i.e. prohibiting the protection of "one and the same botanical genus of species" through both regular patenting and plant breeders' rights protection. This provision was also dropped in the 1991 revision of the UPOV Convention.

As "farmers' privilege" is losing its long standing position in agriculture, we see the emergence of the problem of enforcing and administering patent protection on subsequent generations of plants and animals in farming [15]. The practical barriers to such a system seem insurmountable, and they remain to be worked out by the patent lawyers. The situation is not, however, a completely novel one. Another method by which the seeds industry may secure a steady return from their products, is by the introduction of hybrid plants, which are either sterile or produce inferior seeds.

The strong criticism from the agricultural sector now seems to have made its impact on the EC Commission patent directive. In an amended proposal for the directive, the principle of "farmers' privilege" is redeemed [16]. The amendment was adopted by a large majority by the European Parliament.

US practice has long been more lenient in this respect. Already in 1985 the US Board of Patent Appeals resolved to permit plant varieties to be protected under regular patent systems [17]. Three years later, in 1988, the American patent office granted the first patent on a transgenic vertebrate. This was the Harvard "onco-mouse", which is carrying an onco-sequence disposing it for breast cancer [18]. After an initial rejection of the mouse, the European Patent Office (EPO) reluctantly followed suit, arguing that the medical advantages presented by the onco-mouse outweigh ethical considerations for suffering animals. As Vines [19] has argued, this decision may, however, still be contested. In the EC, the thrust for biotechnology patenting has been met with strong opposition from the environmental movement, as well as farmers' organisations. This is best illustrated by the fact that after 5 years, the EC Commission's proposal for a patent directive is still held at bay by these interest groups.

Small-scale farmers worry that they will lose access to breeding material, if patenting becomes the norm in agriculture. Defending a patent is often a long and costly business, and there is a tendency to concentrate knowledge and resources in fewer and bigger seeds- and pharmaceutical companies [20]. The companies' opportunity to patent their inventions is believed to strengthen their position (op. cit). The small farmers fear the potential monopolies of the bigger and stronger units. This concern is common for small-scale farmers in both the developed and the developing world. There is also international consensus that intellectual property rights systems in relation to plant genetic resources are ineffective in providing incentives to informal innovation, even though they may increase general research in breeding and biotechnology [21].

Environmental interest groups have similar fears that patenting will add to the weakening public control over the rapid developments in biotechnology [22]. The same environmentalist groups go on to argue that patenting may act indirectly in reducing genetic diversity, as patent protection is mainly an asset of the developed world [23]. As only a handful of multinational corporations dominate the seeds industry, the result, as illustrated by the "Green Revolution", tends to be fewer varietites and a reduced diversity in food crops. Moreover, the developing countries, where most of the biological diversity is found, are left with no incentives to conserve their genetic heritage (Fig. 1). Thus, they hardly regard as legitimate the industrialized world's request for conservation of biological diversity in the South.

While the patent system is concerned with private rights, environmental agreements deal essentially with common rights, relating to the long-term survival of humankind. The question of access to environmentally sound technologies has been raised in the recent international negotiations, preceding the UN Conference on Environment and Development (UNCED) in Rio, 1992. This question of course far exceeds that of patenting in biotechnology.

Japanese and American patent legislation are currently at the forefront with regard to granting patent protection for biotechnological products. Both the EPC and the World Organization for Intellectual Property Rights (WIPO) have had a more restrictive orientation towards the issue. The WIPO-secretariat seems, however, to be moving towards a wider scope in biotechnology patenting [24]. The Paris Convention of 1883, on which WIPO rests, is currently in the process of revision. It is still not decided whether the exclusions for plant and animal varieties should be upheld, and neither is it clear whether countries adhering to WIPO will be permitted to exclude certain types of technologies from patentability [25].

The Nordic countries have been unwilling to grant patents for plant and animal varieties. Thus, the Danish *Folketinget* in their negotiation mandate on the proposed patent directive of the EC Commission, voted that the directive must (i) include an explicit ban on patenting human beings, (ii) ban patenting of animals, (iii) restore "farmers' privilege" to reuse seeds for free, (iv) not run counter to efforts to conserve biodiversity, and (v) include in it consideration for the developing countries. Two

Norwegian white papers have made similar recommendations regarding Norwegian legislation, adding, inter alia, the exclusion from patentability of naturally occurring biological material [26]. This issue is discussed further in the next section.

Invention or discovery? The criteria of novelty and inventive step

Living organisms were traditionally assumed to be excluded from patentability as being products of nature rather than products of manufacture, and thus not constituting an invention, nor having the proscribed novelty [27]. Hence "naturally occurring biological material" has traditionally been excluded from patent protection as it does not comply with the criteria of inventive step, and hardly with that of novelty either. Aside from the jurisdictional hesitations, this issue has not least raised concerns of an ethical nature. Opponents claim that patenting of living material displays a mechanical view of life in general.

These views no longer persist with industrial and business authorities in most industrialized countries. The very process of identifying and isolating a gene or a microorganism is now deemed to involve the necessary level of scientific challenges to be regarded as an invention. In line with these developments, patent legislation is presently being redefined so that it may be applied to genetic material [28]. Thus the "inventive step" barrier seems to have been overcome, though the question remains a controversial one. Many biologists and geneticians regard the isolation of a gene and the insertion of it into another organism to be a routine process, however, which no longer involves great scientific challenges [29].

An emerging problem has also been identified with regard to competition between patent offices to attrach patent application. The patentees are free to chose between the national patent authorities and that of the EPO as a recipient of their patent applications, and thus there is a tendency for the national offices to soften the standards in order to retain customers and income [30]. Obviously, this tendency is putting additional strain on the employment of the inventive step criterion.

As most of the world's genetic resources have not been described scientifically (and may thus be eligible for patenting under the emerging reinterpretation of the patent criteria), the question of invention or discovery has not least activated the North-South conflict with regard to property rights to genetic resources. The tendency is enhanced by the US patent criteria, which merely demand that the invention "may be useful", as opposed to European legislation which asks for an "industrial application". Gene-rich developing countries fear escalated gene-hunting by Northern companies. This conflict is elaborated upon below.

A related area which may be subject to new interpretations is that of patents on processes. In general, "primarily biological processes" for producing or modifying plants or animals are excluded from patentability. Hence, traditional breeding methods are not eligible for patent protection. The question in point is to what extent the process must be of a technical, as opposed to a biological, nature. In the EC Commission's proposal for a patent directive, the whole process may be patented if one step is of a microbiological nature.

In spite of their oft declared intention of excluding naturally occurring biological material from patent protection, Norway recently granted a patent on a naturally occurring microorganism (*Lactobacillus plantarum* DSM 3676 and DSM 3677) to a German corporation (Patent no. 169573, 15.7.92).

Industrial application

The most far-reaching patent application to date has come from the US National Institutes of Health (NIH). The NIH has been filing patent applications for close to 3,000 human genes from the Human Genome Project [31]. This action has caused turmoil in the biotechnology industry, causing the UK Medical Research Council to retaliate by filing patent applications for more than 1,000 genes. This in its turn caused France, Germany and Italy to put a stop to scientific information exchange within the international DNA database, recently set up in London with EC funding. French scientists, Daniel Cohen and his colleagues have gone one step further, promising to donate their gene map of the human genome to the United Nations as a gift to the world, thereby ensuring all scientists unrestricted access to the data [32].

In the first round, the NIH application was turned down, on account that their application did not specify how the genes were supposed to be useful, as the functions of the DNA sequences in question are not known [33]. As the American Patent Office did not argue along the lines of inventive step or "naturally occurring organisms", this may not however be the final verdict in the case. The NIH will appeal the decision, and if at a later point they succeed in demonstrating the potential usefulness of the genes, there is still a possibility that patents will be granted.

The NIH move can be regarded as pre-emptive in order to prevent the private sector from "snatching" their discoveries and patenting them. A similar strategy can be discerned in a recent proposal by the Consultative Group for International Agricultural Research (CGIAR), the main vehicle behind the "Green Revolution". CGIAR decided at their mid-term meeting in Istanbul in May 22, 1992, to embark on "defensive patenting", in order to forestall pre-emptive protection by others of CGIAR-generated technology. The group insisted, however, that germplasm derived from the genebanks of the CGIAR's International Agricultural Research Centres (IARCs) should still be subject to the principle of free availability, following the policy of the CGIAR [34]. In a recent report, the Stripe Study of Genetic Resources in the CGIAR, it is recommended that the IARC germplasm collections should be held in trust and in accordance with the provisions of the Convention on Biological Diversity, and that the IARCs should not seek to benefit financially from the commercialization of germplasm, but, in order to maintain trust with germplasm donors, should work with national agricultural research systems (NARS) at their request, should opportunities for commercialization occur [35]. The possible effects of patent protection on the deposition of and access to germplasm from genebanks, remains however an unresolved question.

The motivation behind the CGIAR Istanbul decision can be found in the developments in biotechnology, in the "privatization" of agricultural research and in the strengthening of national and international patent legislation [36]. Agricultural research is tending to move from the public sphere to that of private enterprise. This also makes it increasingly necessary for developing countries to introduce IPR-systems in order to attract funding of agricultural research. In addition, patent legislation is generally expanding not only geographically, but even more so functionally, thus providing ever broader possibilities to issue patent protection for plants and seeds.

The most far-reaching patent granted to date seems to be the one claimed by the American biotechnology company, Agracetus Company. Their patent pertains to all genetically engineered cotton varieties and, being the first patent to claim protection of an entire species, is likely to be challenged. The patent was granted in 1991 for a method to produce transgenic cotton, and in the USA Agracetus has already been able to expand the patent to cover all genetically engineered cotton (in 1992). This means that they have been given a product patent on transgenic cotton, requiring anybody else genetically modifying cotton to get a licence from Agracetus. This is different from having a patent on the production method, in which case others might produce the cotton by a different method, without paying royalties to Agracetus. Fearing a similar development in India, and arguing that inventive processes related to agriculture cannot be patented according to Indian patent law, the Ministry of Agriculture and the Prime Minister's Office in India is currently trying to stop Agracetus' Indian patent [37]. Meanwhile, Agracetus has secured patent rights in Europe to all transgenic soya beans.

The broad scope of the Agracetus patents is reported to cause surprise and distress also among patent experts in Europe and the USA [38]. The questions are raised how one company can own the rights to all genetically altered varieties of two entire species, and what may be the effects for small companies and poorer countries who want to improve these crops [39]. One way of looking at the situation, is that common sense must eventually generate a resumed balance in biotechnology patenting, discouraging freak incidents like the Agracetus case.

The new biotechnologies, both in pharmacy and agriculture, are dominated by the private sector, and they are clearly the technologies of highly industrialized countries [40]. The larger firms possess the economic strength to defend intellectual property rights over their end-products, and the scope for patent protection is widening fast. These circumstances, among others, are aggravating the already brewing North/South conflict over genetic resources, a situation into which we will look further in the next two sections.

The North/South Controversy: Property Rights to Genetic Resources

The raw material for biotechnology is the world's genetic resources. The major bulk of the world's biological diversity is to be found in tropical countries [41]. Biodiversity includes diversity at genetic, species and ecosystem level, and genetic resources are generally understood as genetic material of actual or potential value [42].

Genetic resources may be subject to private or public control, or be regarded as part of the Common Heritage of Mankind. As a general rule, the Common Heritage principle has been applied to the wild plant material and land races of the South, while elaborated material of the North is often subject to some kind of intellectual property rights (patents or plant breeders' rights). In the following sections, four broad categories of property rights will be briefly presented and discussed with regard to their application to genetic resources.

Common heritage of mankind

Common property resources are usually defined by their character of nonrivalry and nonexclusiveness. Nonrivalry implies that it is possible for more than one person to use or consume the good without diminishing the amount of good available to others. The second characteristic concerns how easy it is to exclude others from using or consuming the good. The air we breathe is generally regarded as an example of a nonrival and nonexclusive good. This was traditionally the case with clean water too, but its character of nonrivalry is rapidly declining.

In the idea of common heritage, there is always an element of open access: the absence of well-defined property rights. This was also the case with ocean fisheries in the past century, in the Grotian doctrine of the freedom of the high seas [43]. Common heritage, however, is not necessarily identical with the idea of open access as practised under the high seas doctrine [44]. Open access merely implies that nobody can be excluded from using the resources, save by lack of economic and technological capacity. The common heritage principle conversely may imply that everybody (all mankind) has a right to benefit from the exploitation of the resources.

The common heritage principle was first introduced at the UN Conference on the Law of the Sea in 1967 by the Portuguese Ambassador, Arvid Pardo, as a guiding principle in governing the exploitation of minerals in the deep seabed. The idea was to secure more equity between developed and developing countries in the exploitation of a "common" resource. The majority of industrialized nations contested the principle as being legally diffuse and practically impossible.

During the debate on plant genetic resources in FAO, two main arguments were voiced in favour of placing genetic resources in the common heritage category. First, the resource is of basic significance to all mankind in terms of food and medicines. Second, international use of plant genetic resources has traditionally had the character of a common heritage regime.

Compared to the traditional use of the concept of common heritage in international negotiations, there are two discrepancies: Firstly, it is the industrialized countries who have been most eager to advocate its application, this time. There is a parallel in the debate in UNESCO on the Convention on the Means of Prohibiting and Preventing the Illicit Import, Export and Transfer of Ownership of Cultural Property. The developed countries maintained that the Third World's artifacts formed part of a common heritage for all humanity, and that they could be better and more widely enjoyed if they were located in the major centres of the world [45].

Secondly, a closer examination of its practical contents reveals that it is more concerned with open access to genetic resources, than actually regarding them as a common heritage. This implies that nobody can be excluded from utilizing the resource, except by lack of economic and technological means.

Private property

Private property issues from a situation in which an individual or corporation has the right to exclude others from using the resources and to regulate their use. Such rights may be embedded in various kinds of intellectual property rights. A patent is a property right which excludes others from the use or benefit of the patented invention without the consent of the patentee [46].

In the first place, such intellectual property rights may be granted through the "plant breeders' rights" of the 1961 UPOV Convention (The Union for the Protection of New Varieties of Plants). In order to be subject to UPOV protection, a plant variety must be "uniform, stable and distinct from existing varieties". In order to attain protection by breeders' rights or patents, some kind of systematic breeding is required. This is seldom the case with Third World breeders' lines.

A much cited case from medicine is the Rosy Periwinkle, a native plant of Jamaica and Madagascar (Fig. 2). The plant has been turned into a medicine for treatment of Hodgkin's disease and certain types of leukaemia by the American pharmaceutical company, Eli Lilly. The company's annual return from the invention is about £60 million, none of which is returned to the country of origin [47].

Put briefly, it has been felt that the developments in patent legislation will pave the way for increased Northern control over Third World natural resources, owing to these resources being regarded as a "common heritage". Patenting genetic material is not itself compatible with the common heritage principle, as it infers exclusive rights over the material.

State property

Thirdly, a resource may be held in public trust as state property. This implies that rights to control access and levels of exploitation are vested exclusively in government. Fish stocks inside the 200-mile exclusive economic zones may be an example of such resources. Many developing countries have recently been claiming national sovereignty over their genetic heritage, regarding it as a national asset along the lines of other natural resources, such as oil and minerals. According to them, to regard genetic resources as part of the common heritage of mankind, is problematic for two reasons. First, the resources are for a large part situated within national borders. Second, extracted natural resources within national borders are generally treated as commodities (for instance oil and minerals). Genetic resources differ, however, from oil and minerals in being nonrival goods. Furthermore, species distribution is not necessarily confined to national borders.

Nevertheless, developing countries started to question the practice of applying the

Fig. 1.The main bulk of the world's biological diversity is found in the developing countries. Tropical rain forests, which occupy only 6% of the earth's land surface, are thought to contain more than half the biodiversity on earth.

Fig. 2. The Rosy Periwinkle, a native plant of i.a. Madagascar, is used for a high profit medicine by an American pharmaceutical company, but the country of origin obtains no benefits from its use.

common heritage principle solely to the resources from the South. Their first line of reasoning was that the elaborated material of the industrialized countries is to a great extent based on material from the South, and should thus also belong to the common heritage. This view met with strong resistance from the industrialized countries, on the basis that such an arrangement would not be compatible with Northern "breeders' rights" and patent legislation, which recognize restrictions on, and payment for, patented plants and animals.

Finally, the Third World abandoned the claim for an all-embracing common heritage regime and turned the argument around. As a counterpart to northern patenting they insisted on national sovereignty over genetic resources. The principle of sovereignty is now asserted in the International Convention on biological diversity.

Communal property

A less noted, but widely applied category is communal property. Here the resources are held by an identifiable community of users who can regulate use. Pastoral and hunting communities traditionally abide by customary laws that regulate grazing and hunting in communal areas. The concept of "farmers' rights", pertaining to local varieties of crops, comes close to this ideal type. It is meant to cover valuable genetic material that has not been subject to systematic breeding, but rather been subject to breeding by farmers through generations. This concept will be further explored in the next section on the FAO debate.

The North/South Controversy: International Negotiations

FAO and the undertaking for plant genetic resources

A central forum for the discussion of this controversy has been the UN Food and Agricultural Organization, FAO. The issue of control and access to genetic material was put on the FAO agenda in the early 1980s, and has so far resulted in three concrete outputs; an International Undertaking, a Commission and a Fund for Plant Genetic Resources.

The Undertaking (1983) lays down the principle that "all" categories of plant genetic resources (both elaborated and wild) shall be subject to free exchange for exploration, preservation, evaluation, plant breeding and scientific research. With regard to elaborated material subject to legal protection under national legislation, this material may be made available "on mutually agreed terms", according to the Undertaking. The "Agreed Interpretation" of the Undertaking (1989) states that the mutually agreed terms may allow for national "breeders' rights", and consequently involve some payment for legally protected varieties. Since the 1989 "Agreed Interpretation" regards intellectual property rights as compatible with the Undertaking, the developing countries have abandoned the "common heritage of mankind" strategy.

The Commission constitutes the administrative body and serves as a forum for continuous debate among the delegates. The main idea of the Fund is crystallized in the concept of "farmers' rights", to represent a counterpart to "plant breeders' rights". The Fund aims to provide compensation for "the enormous contributions generations of farmers have made to the conservation, selection, domestication and development of plant genetic resources. If breeders, who provide the finishing touches to this process, can secure a title and handsome profits for their efforts, then the farmers too should receive compensation..." [48]. The Fund is meant to assist training in plant breeding in the developing countries, and to work as an incentive to maintain utilization of a diversity of local varieties of plant crops. It will thus directly promote a main objective of the Undertaking: the conservation of genetic resources.

To date, the concept of "farmers' rights" has not had much practical effect. This may primarily be put down to the inability of the FAO Fund to attract funding. A related explanation may be found in the difficulties of applying the concept to practical politics. First, there is the problem of tracing the "contributor" to which compensation should be made. Seeds have crossed borders and been developed in so diverse parts of the world over the years, that such a system would be hard to design, let alone to administer.

Trade-Related Aspects of Intellectual Property Rights (TRIPs) in GATT

While the developing world seemed to find approval for some of their argumentation within the FAO arena, other international arenas present a different and more difficult situation. Most significantly, questions concerned with widening the scope of industrial patents have been subject to debate in the General Agreement on Tariffs and Trade – the GATT negotiations.

In no other sector in the GATT negotiations has the north/south controversy been more outspoken than with regard to patents. The US, Japan and the EC advocate the principle that all countries should provide and respect intellectual property protection in all technical fields, including biotechnology. They go on to say that disregarding this principle should constitute a contravention of GATT regulations, and thus be liable to economic sanctions.

The developing world has been strongly opposed to these proposals, maintaining that patents benefit technologically and economically strong states only [49]. This point is hard to refute, as Third World nations hold no more than 1–3% of all patents worldwide [50]. India has argued against patenting of plant and animal varieties as well as food and pharmaceutical products, citing concern for basic human needs [51].

For a large number of developing countries, this situation is not only a matter of contesting theoretical principles, however. With the introduction of industrial patents, access to improved breeding material may be hampered as prices for seed increase [52]. On the same note the FAO Commission on Plant Genetic Resources has warned, "if the patent system is applied universally to living matter, including plant and animals, and their genetic resources, then the principle of unrestricted access will be severely eroded" [53]. A major complaint of the South is that the Western patent system fails to recognize the innovations of Third World farmers in selecting, improving and breeding a diversity of crop varieties [54]. Some fear that patents will place financial constraints on technology transfers, in general. With these possible results in mind, the Nordic countries have argued that plant and animal varieties should be exempted from patenting, and they also emphasise the need to strengthen provisions for compulsory licensing.

The opposition has had some success in TRIPs. Section 5, article 27 in TRIPs, Annex III, grants the parties the right to "exclude from patentability (a) diagnostic, therapeutic and surgical methods for the treatment of humans and animals, (b) plants and animals other than microorganisms, and essentially biological processes..." [55].

The parties are, however, bound to introduce some kind of intellectual property rights for plants, as the article continues; "However, Members shall provide for the protection of plant varieties either by patents or by an effective *sui generis* system or by any combination thereof."

The latter part of this article has prompted, inter alia, the farmers' movement in India to propose that such a *sui generis* system should focus on the rights of farmers in protecting and improving plant genetic resources: "Common Intellectual Property Rights" [56]. India is now on the way to developing two *sui generis* systems for protecting plant varieties. One legislation is for the protection of breeders' rights in line with the TRIPs intentions. The other law is for the recognition of the rights of local communities to their landraces and wild biodiversity [57].

Anyhow, TRIPs is unlikely to have harmful effects for Third World farmers in a short term perspective. The inherent threat in the expanding patent legislation that farmers must pay royalties for reusing seeds, is still a long way from being enforcable. A much more harmful effect of TRIPs is its bolstering of a North/South conflict line in an area in which common solutions and cooperation are paramount.

The Biodiversity Convention

The Biodiversity Convention was signed by more than 150 states during the UNCED Conference in Rio, 1992. The Convention will be managed by an interim secretariat under UNEP, and it came into effect 29 December, 1993, 90 days after ratification by the first 30 nations.

The International Convention on Conservation of Biological Diversity sets out obligations and objectives for nations to combat the destruction of plant and animal species and ecosystems [58]. It also deals with a number of underlying controversial items, although for perhaps the most controversial one – on regulating "modified organisms" – it was left to the parties to consider the need for such a protocol in the future.

The question of access and property rights in relation to various types of genetic resources has been a crucial element throughout the negotiations for the Biodiversity

Convention [59]. Parelleling the FAO debate, the developing countries were still concerned that genetic resources from the South were labelled "a common heritage of mankind", while elaborated genetic material in the North, albeit often originating in the South, was seen as commercial products. The developing countries insisted on technology transfers as well as financial mechanisms as compensation for conserving biodiversity and as payment for the North's use of their genetic material.

In its final version, the Biodiversity Convention does obligate industrialized countries to share the financial burden of funding national conservation programmes in developing nations. It also calls on such donor nations to share relevant technology and research (with gene-donor nations). Moreover, an important outcome of the negotiations was the consensus to abandon the common heritage principle and accept intellectual property rights as well as national sovereignty over genetic resources. The questions related to intellectual property rights remain, however, the most controversial item in the Convention.

The US delegation refused to sign the Biodiversity Convention in Rio, mainly because of its formulations on intellectual property rights [60]. The US argument is primarily tied to articles 15, 16, 18 and 19 of the Biodiversity Convention, dealing with the controversial issues of access to genetic resources, technology transfer, and sharing of benefits from utilization of genetic resources. When the US finally did sign, in June 1993, this was coupled with the promotion of an alternative, reinterpretative statement on these articles in the Convention. I will expand somewhat on the main elements in these articles.

Article 15 states that each country has the sovereign authority to determine access to its genetic resources (paragraph 1), that access to genetic resources requires prior informed consent and must be on mutually agreed terms (paras 3 and 4), and that a country providing genetic resources is entitled to benefit from the commercial use of its resources (paras 6 and 7). Three basic mechanisms are envisaged by the Convention by which a country may benefit from use of its genetic resources: participation in the research using the resources (15.6), receiving technology which embodies or utilizes the resources (16, 18 and 19), and sharing the financial benefits realized from commercial exploitation of the genetic material or resource (15.7 and 19.2).

As far as patenting is concerned, paragraphs 2 and 5 of article 16 are the most important. In article 16.2, which deals with technology transfer, it is stated that: "In the case of technology subject to patents and other intellectual property rights, such access and transfer shall be provided on terms which recognize, and are consistent with, the adequate and effective protection of intellectual property rights". This is all in line with the interests of the US, until 16.2 goes on to declare that: "The application of this paragraph shall be consistent with paragraphs 3, 4, and 5". It is the content of paragraph 5 that so alienated the Bush administration, which states: "The Contracting Parties, recognizing that patents and other intellectual property rights may have an influence on the implementation of this Convention, shall cooperate in this regard subject to national legislation and international law in order to ensure that *such rights do not run counter to its objectives*" (author's italics). Moreover, this is still a major concern with the Clinton administration, as the decision to sign the convention was followed by an interpretative statement addressing article 16.5 as well as the provisions for financial mechanisms.

There are several reasons, however, why the Biodiversity Convention will probably not impose severe limitations on biotechnology companies. First of all, there is a certain limitation inherent in the Convention which restricts its application to future dealings with genetic resources. Hence, it does not induce any kind of payment for germplasm already existing in national and international genebanks. To some developing countries and NGOs this is regarded as a weakness of the Biodiversity Convention [61], but in fact the legal basis of an international convention requires that it cannot have a retroactive impact. This question remains one of the most controversial items in the Convention, however, and it must certainly be addressed in the future negotiations on protocols. A possible way to go might be to look to the FAO Undertaking and seek for solutions along the lines of "farmers' rights".

Secondly, patenting is evidently catching on in ever wider circles, and there are few indications that the Convention has had any prior diminishing effects on this practice. This phenomenon may be observed in a number of developing countries, even some of those most zealously fighting the international patent quest in the TRIPs negotiations in the GATT Uruguay round. Most of those developing countries that can possibly sustain a patent system, like India, Brazil and Mexico, are well on their way to accepting parts of it [62]. A major explanation lies in the different levels of technological development. Unlike Brazil, India and newly industrialized countries in general, however, the majority of developing countries still lack the infrastructure to introduce effective patent systems. An indication that the BioConvention may have had an impact, however, may be found in the decision by the Consultative Group for International Agricultural Research (CGIAR) to embark on "defensive patenting", a decision that has met with strong opposition.

A third and important reason why industry should have little to fear is the provision in article 15, which states that access to genetic resources should be according to "mutually agreed terms". This provision can at best pave the way for more orderly transactions of genetic resources – a situation which may in the end prove beneficial to both actors. At "worst" its implementation will rest on the mutual strengths of the transacting parties, which is hardly likely to be harmful for the multinational seeds and pharmaceutical corporations. This provision also works to provide the Biodiversity Treaty with a bias in favour of bilateral arrangements, and again it is hard to see why the US' industry or any other industry should object to this.

Moreover, as John Duesing from CIBA-GEIGY Seeds points out, the expression "mutually agreed terms" also means that "intellectual property rights issues and the commercial use of derived technology will have been negotiated before that genetic resource is transferred and exploited by the requesting party" [63]. Duesing further argues that the formulation in paragraph 16.5 ("*patents and IPR ... not run counter to its* (the Conventions) *objectives*") which so alienated the Americans, really depends on whether one believes IPRs to be supportive of the objectives in the Biodiversity Convention or not. Duesing believes patenting to be promoting these objectives, as it provides an incentive to evaluate and develop new genetic resources.

Still, the Biodiversity Convention would hardly have achieved the consent of the majority of the developing countries if it had been void of realistic distributional provisions. To the extent that the treaty makes arrangements for a more equal distribution of benefits from the utilization of genetic resources, it will, of course, detach an equivalent revenue from the transacting parties. Hence, some developing countries reacted strongly to the US rejection of the Treaty. As a direct response Venezuela threatened to stop signing new agreements with the US scientific institutions who were collecting and screening biodiversity in the country [64]. Similar policies are emerging in several Central American countries, and Ethiopia has for some time been had a policy of strict control over exports of genetic resources.

Finally, when it comes to the actual interpretation of the Biodiversity Convention, it is interesting to note that the British biotechnology industry displayed a much more relaxed reaction to the patent-formulations than did their counterparts in the US. The BioIndustry Association (BIA), representing 160 British biotechnology companies, made an immediate appeal to President Bush to back the Convention. Said Louis Da Gama, executive of the BIA: "Having reviewed the convention, we do not share the concerns expressed by the US government" [65]. Moreover, the reaction from the European Parliament was actually to block once again the EC Commission's proposed directive on patents in biotechnology. The EP's environment committee at first expressed concern that the text of the proposed directive would not be compatible with the Biodiversity Convention [66,67]. Hence, the EP gave a similar interpretation as the US, and the BIA a slightly different interpretation from the US, but in both cases they insisted on backing the Biodiversity Convention.

Scope for Reconciliation?

In light of the wide variety of interests clustering the biotechnology and genetic resources issue, as well as the admitted global interdependence on these resources, what strategies can be envisaged to overcome some of the main conflicts? The countries providing genetic resources can no longer be expected to give them away for free, any more than the biotechnology industry can be expected to turn into primarily philanthropic enterprises. This is not to say that industry is devoid of philanthropic activities: a relevant example could be Merch, Sharp and Dohme, a US pharmaceutical company, which in 1989 decided to donate Mechtizan, a medicinal remedy against river blindness, to the World Health Organization for free distribution [68]. On the other hand, one could also question the ability of the Third World to control access to their genetic material. Genes are obviously easier to smuggle than a lot of other commodities. An important factor facilitating control of genetic resources, however, is the need for knowledge about their qualities.

Essentially, this concerns the development of good working relations between providers of genetic resources and utilizers of genetic resources. Examples of how such working relations may be achieved are already appearing. I will first cite three examples from the private sector and then describe a proposal for national legislation.

The first is the well-known deal between the American multinational corporation, Merck & Co. Pharmaceuticals, and the Costa Rican National Institute of Biodiversity (InBio). InBio provides plant and animal species for screening and drug research which Merck gets exclusive rights to develop. In return Merck gives an initial payment of US$1 million as well as royalties for any drug developed. Training of local "parataxonomists" and institutional capacity-building in Costa Rica are also part of the deal. Similar deals could be developed, adapted to the specific needs of provider and user. In a situation which is clearly characterized by unequal bargaining positions between a poor country and a multinational corporation, the Biodiversity Convention may become an instrumental framework for guiding the formulation of the deals.

Two examples show how collecting agencies are presently changing their attitude towards the countries providing genetic material: The UK Royal Botanic Gardens Kew (RBG) now states that "any net profits derived by RBG from ... collaboration will be shared equally between RBG and the Supplier". There is also the case of Biotics, a private British for-profit company, that acts as a broker between companies and in-country collectors, granting the latter 50% of Biotics' royalties [69]. For a more extensive analysis of the contents and variety of such biodiversity deals, the interested reader is recommended to consult the World Resources Institute book, "Biodiversity Prospecting" [70].

Finally, an improved relationship between provider and utilizer of genetic resources may also come about through national legislation. According to Hendrickx, Koester and Prip such legislation may be nested in paragraph 5 of article 15 in the BioConvention [71]. The article maintains that access to genetic resources shall be subject to the prior informed consent (PIC) of the Contracting Party. Thus, the country providing genetic resources must provide national legislation regulating the appropriation of genetic material. There will be a need for specially designated bodies to conduct the deals, and to establish information databases on the genetic material. Moreover, the providing country will have to make priorities with regard to compensation mechanisms for its genetic material, and also make clear the relationship to gene resources held by local communities and genebanks. Obviously, this may represent an impediment to countries, especially the least developed ones, which lack administrative capacity both to enact and enforce a legal framework. Hendrickx, Koester and Prip also make suggestions for what user countries may do to improve the effectiveness of PIC on the import side: National legislation could be tailored to prohibit illegal import of genetic resources, i.e. collections conflicting with PIC export rules in the providing country. On the same note, companies and other importers could be obliged to keep records of imported genetic material, in order to facilitate monitoring by government authorities. Another interesting suggestion is to make it a requirement of patent applications to give information about how genetic material was obtained.

Concluding Remarks

Summing up, the new biotechnologies represent a tool which meets the legal requirements for patenting biological material. From the industry's point of view, biotechnology also necessitates patenting. As research becomes more costly, requiring significant financial investment, the patent is seen as a reward for undertaking the risk of such research which may not always achieve useful results. The primary function of a patent lies in the "deal" between the inventor and society, the inventor being reimbursed for his or her investment and, in return, making the invention public. On the negative side, defending a patent is often a long and costly business, and the trend is that patenting will mainly benefit the bigger and stronger companies and thus weaken public control over the rapid developments in biotechnology.

Without sophisticated biotechnological tools, trained scientists, and adequate infrastructure, patenting is, as yet, hardly a viable solution for the majority of developing countries. On account of the patent criteria, only the systematically bred material may thus be protected, and this is rarely the case with germplasm in the South. Generich developing countries fear that developments in patent legislation will pave the way for increased Northern control over Third World natural resources. Some countries, like Ethiopia, have embarked on a policy of denying exports of their genetic resources, seeing how developing countries at present are not getting any benefits from the technological utilization of their genetic material. The Biodiversity Convention goes some way in making amends to this situation, but the gene-poor, least developed countries may still have reason to fear that they will lose access to breeding material.

In a long-term perspective the implications may be detrimental for resource conservation in developing countries. At present, it is doubtful whether the South will benefit from an extended patent system. Many developing countries fear that patent legislation will work as a constraint on technology transfers from the North to the South, and also reduce the Third World's chances of reaping the benefits from their own resources. This may put an increased strain on the Third World's capacity to preserve its own genetic heritage, and it may also jeopardize the legitimacy of Northern requests for it to do so.

The provisions in the Biodiversity Convention go some way in establishing common ground between the North and the South in the patent question. There are several trends that may work to subdue the North/South conflict in this issue area. Primarily, the international patent regime is evidently expanding rapidly, and there are no indications that the Convention has had any prior diminishing effects on this process. As the patent system reaches worldwide acceptance, it will be increasingly hard to argue with the American biotechnology industry, that only US biotechnology industry will be harmed by restrictions on it. This assumption seems to be supported by the European interpretations of the formulations in the Biodiversity Convention. Secondly, there are certain limitations inherent in the Convention itself, which restrict its application to *future* dealings with genetic resources. Hence, germplasm in international genebanks (deposited prior to the first 30 ratifications of the Convention) may still be utilized without compensation.

This last item remains an unresolved question, which may still create tension in international exchange of and access to genetic resources. Moreover, as the examples from the US National Institutes of Health and the International Agricultural Research Centres indicated, much uncertainty is still tied to the introduction of patents on naturally occurring genetic material. A crucial factor is the point at which the pendulum will turn with regard to the ever-broadening scope of patent protection in biotechnology. Perhaps the natural diversity and genetic variation which is continuously subject to change and evolution, in itself will make the practical application of patents on living organisms an impossible task. In conclusion, the patent question seems to remain unresolved and may still be one of the most likely stumbling blocks for future ratifications and implementation of the Biodiversity Convention. As the last section showed, however, there are a number of ways in which the controversies may be overcome. In light of the rapidly dwindling biodiversity, the winners in this game would seem to be those who succeed in inventing arrangements which include both access to and use of valuable genetic material today as well as conservation for the future. This kind of invention could turn out to become even more profitable than the biotechnology inventions themselves.

References and notes

1. Wilson EO. (ed) Biodiversity, Washington DC: National Academy Press, 1988.
2. While the "old" biotechnology includes traditional activities like brewing beer and baking bread, the concept of "new biotechnologies" refers to activities like tissue culture and recombinant-DNA (r-DNA) techniques.
3. Report of Panel II, UNEP/Bio.Div/Panels/Inf.2, Nairobi, 28 April, 1993.
4. Compulsory licensing allows the government to force the patentee either to sell the patent or start to utilize his or her invention, if important social concerns should require it.
5. Crespi RS. Patents: a Basic Guide to Patenting in Biotechnology. UK: Cambridge University Press, 1988.
6. Mooney PR et al. The Laws of Life. Another Development and the New Biotechnologies", Development Dialogue, Dag Hammarskjøld Foundation, Uppsala, 1988;1–2.
7. Bent SA, Schwaab RL, Conlin DG, Jeffery DD. Intellectual Property Rights in Biotechnology Worldwide. US and Canada: Stockton Press, 1987.
8. Crespi RS. 1988 (see no. 3, above).
9. NOU. Bioteknologi og patentering, Delinnstilling fra Bioteknologiutvalget, Oslo: Norges Offentlige Utredninger 1989:8.
10. Commission of the European Communities, Proposal for a Council Directive on the Legal Protection of Biotechnological Inventions (17 October, 1988). The directive will be integrated in the EPC.
11. NOU. 1989 (see no. 7, above)
12. NORD. Bioteknologiska uppfinningar och immaterialrätten i Norden - del II ("Biotechnology inventions and intellectual property rights in the Nordic countries - part II"),København: Nordisk Ministerråd, 1992:8.
13. Barton JH,Siebeck WE. Intellectual Property Issues for the International Agricultural Research

Centres. What are the Options?, Issues in Agriculture, no. 4, Consultative Group on International Agricultural Research, 1992.

14. Hindar K, Rosendal GK, Trønnes HNO. Bioteknologi og norsk tilpassning til EFs indre marked, Norsk Institutt for Naturforskning NINA), Trondheim, Norway 1990.
15. NORD. 1992 (see no. 10, above)
16. COM(92) 589 final - SYN 159. Brussels, 16 December, 1992.
17. Barton JH, Siebeck WE. 1992 (see no. 11, above)
18. NOU. 1989 (see no. 7, above)
19. Vines G. Guess what's coming to dinner?, New Scientist, 14 December:1992;14.
20. NORD. 1992 (see no. 10, above)
21. Keystone International Dialogue on Plant Genetic Resources.1991, Final Consensus Report, Oslo Plenary Session, 31 May–4 June, Keystone, CO.
22. Mooney PR et al. 1988 (see no. 4, above)
23. NORD. 1992 (see no. 10, above)
24. NOU. 1989 (see no. 7, above)
25. Barton JH, Siebeck WE. , 1992 (see no. 11, above)
26. NORD. 1992 (see no. 10, above)
27. Crespi RS. 1988 (see no. 3, above)
28. NOU. 1989 (see no. 7, above)
29. Bent SA,Schwaab RL, Conlin DG, Jeffery DD. 1987 (see no. 5, above)
30. Statement by Professor Mogens Kogtvedtgaard, University of Copenhagen, at the meeting: 'Patents and Biotechnology', organized by the Norwegian Biotechnology Advisory Board, 3. November:1993.
31. Financial Times, "Don't patent human genes", April 24:1992;16.
32. Time. February 8:1993;43.
33. Coghlan A. Genome patents 'a waste of time'. New Scientist 1992;December 5:6.
34. CGIAR. Working Document on Genetic Resources and Intellectual Property, Draft, Istanbul, May 22, 1992.
35. Consultative Group on International Agricultural Research, 1994, Stripe Study of Genetic Resources in the CGIAR, AGR/TAC:IAR/94/2.1
36. Barton JH, Siebeck WE. 1992 (see no. 11, above)
37. Down to Earth, March 31, 1994;5–7.
38. Rosie Mestel. Bean patent sweeps the field. New Scientist 1994;142(1923):7.
39. Comment. New Scientist 1994;142(1923):3.
40. See for instance, OECD,1988, Economic and wider impacts of biotechnology, SPT (88)18, Paris, 23 July.
41. Kloppenburg JR. First seed. Cambridge, UK: Cambridge University Press, 1988.
42. This definition was used in the International Convention on Biological Diversity, which was signed in Rio de Janeiro, 5 June, 1992.
43. Grotius Hugo. Mare Liberum sive de iure quod Batavis competit ad Indicana commercia dissertatio. Leiden: Elzevier, 1609.
44. Bilder RB. International law and natural resources policies, Nat Res J 1980;20(July):451–486.
45. Mooney PR. The Law of the Seed. Another Development and Plant Genetic Resources. Development Dialogue. Uppsala: Dag Hammarskjøld Foundation, 1983:1–2.
46. Crespi RS. 1988 (see no. 3, above)
47. Pierce F. Brazil, where the ice cream comes from. New Scientist 1992;17 July:47.
48. FAO. Working Group of the Commission for PGR, CPGR/89/9, 1989.
49. GATT. Communication from India: Standards and principles concerning the availability, scope and use of Trade-Related Intellectual Property Rights. GATT Secretariat, MTN GNG/NG11/W/37, July 1989.

50. WCED. World Commission for Environment and Development, Our Common Future. Tiden Norsk Forlag, Oslo, 1987.

51. GATT. 1989 (see no. 43, above).

52. Keystone International Dialogue on Plant Genetic Resources, Final Consensus Report, Madras Plenary Session, 14 February,1990, Keystone, CO.

53. FAO. Harvesting nature's diversity. The Information Division, 1993.

54. Shiva, Vandana. The need for sui generis rights, Seedling,1994;March:11–15.

55. GATT Document. Multilateral Trade Negotiations, the Uruguay Round. MNT.TNC/W/124, 13 December, 1993; MTN/FA II-A1C, p. 13.

56. Shiva, Vandana. 1994 (see no. 47, above)

57. Down to Earth April 15:1994;14–16.

58. The Convention defines biological diversity as the variability among all living organisms, including diversity within species, between species and of ecosystems.

59. Rosendal GK. International conservation of biological diversity: the quest for effective solutions. FNI – Report 012–1991.

60. Rosendal GK. The Biodiversity Convention: Analysing the footwork of Bush in Rio. International Challenges. FNIs Newslett 1992;12(3).

61. GRAIN. Genetic Resources Action International. Briefing on Biodiversity. No. 3, August, 1992, Barcelona.

62. van Wijk J. Diminishing National Sovereignty in Intellectual Property Protection: Enforced Global Recognition of Patents in Biotechnology. Paper presented at the International Symposium and Experts Workshop: "Property Rights, Biotechnology and Genetic Resources", African Centre for Technology Studies and World Resources Institute, 10–14 June, 1991, Nairobi, Kenya.

63. Duesing J. The Convention on Biological Diversity: Its impact on biotechnology research. AGRO-food-Industry Hi-Tech, Ciba-Geigy. no. 4, July/August:1992;19–23.

64. Reid W. Bush Biodiversity Policy Risks Dangerous Side Effects, The Wall Street Journal October 8:1992. See also U.S. Access Could Be Blocked. International Environment Reporter, November 4:1992;705.

65. Coghlan A. Biodiversity Convention a 'lousy deal' says US. New Scientist 1992;July 4;9.

66. Vote on proposal to protect biotech patents delayed to ensure no conflict with Rio Treaty, June 17. International Environment Reporter 1992;398.

67. The Parliament later concluded that the amended proposed directive of 16 December, 1992, would be compatible with the Biodiversity Convention's objectives (COM(92)589 final - SYN 159, Brussels, 16 December, 1992).

68. Illustrert Vitenskap, Første medisin mot uhelbredelig blindhet ("First medicinal against incurable blindness") 1989;9:44-47.

69. Laird Sarah A. Contracts for biodiversity prospecting. In: Biodiversity Prospecting. A World Resources Institute Book, 1993.

70. World Resources Institute. 1993.

71. This article has been studied in detail by Frederic Hendrickx, Veit Koester and Christian Prip: "Convention on Biological Diversity. Access to genetic resources: A Legal Analysis", Environmental Policy and Law, 23/6, 1993. Veit Koester played a central role throughout the biodiversity negotiation process, as Chairman and Vice Chairman, i.a. leading Working Group II.

Index of authors

Keyword index

484